METHODS IN MOLECULAR BIOLOGY

Series Editor
John M. Walker
School of Life and Medical Sciences
University of Hertfordshire
Hatfield, Hertfordshire, AL10 9AB, UK

For further volumes:
http://www.springer.com/series/7651

Adeno-Associated Virus Vectors

Design and Delivery

Edited by

Michael J. Castle

Department of Neurosciences, University of California, San Diego, La Jolla, CA, USA

Humana Press

Editor
Michael J. Castle
Department of Neurosciences
University of California, San Diego
La Jolla, CA, USA

ISSN 1064-3745 ISSN 1940-6029 (electronic)
Methods in Molecular Biology
ISBN 978-1-4939-9138-9 ISBN 978-1-4939-9139-6 (eBook)
https://doi.org/10.1007/978-1-4939-9139-6

Library of Congress Control Number: 2019931903

This Humana Press imprint is published by the registered company Springer Science+Business Media, LLC, part of Springer Nature.
The registered company address is: 233 Spring Street, New York, NY 10013, U.S.A.

Preface

Adeno-associated virus (AAV) vectors are essential tools for experimental gene transfer to living animals and for the clinical delivery of gene therapies. The FDA recently approved the first AAV-based gene therapy, Luxturna, for treatment of inherited retinal disease, and many ongoing clinical trials show great promise. The widespread use of AAV vectors in the laboratory and clinic over the past two decades has thoroughly confirmed their safety and efficacy. The usage of AAV vectors will continue to expand as new AAV technologies emerge and more AAV-based therapies are approved for the treatment of human disease.

This volume provides a complete and timely guide to the use of AAV vectors for genetic manipulation of mammalian tissues. The opening chapters describe the design of AAV vectors for RNAi, for large gene delivery, and for CRISPR-mediated genome editing, as well as droplet digital PCR titration and single-strand sequencing of AAV preparations, ligand coupling to AAV capsids, and in situ detection of AAV mRNA expression. These methods for the design and characterization of AAV vectors are followed by protocols for AAV delivery to various components of the central nervous system, to a number of sensory systems, and to a broad range of other tissues. Novel techniques such as ultrasound-targeted delivery to the brain, subpial delivery to the spinal cord, and subILM delivery to the retina are accompanied by chapters that provide an overview and comparison of current methods for AAV delivery to tissues such as brain, heart, liver, and lung.

In addition, several different protocols for the production of AAV vectors are provided throughout this volume, including iodixanol gradient purification (Chapters 3, 19, and 23), CsCl gradient purification (Chapter 7), a simplified approach that does not require gradient ultracentrifugation (Chapter 22), and heparin column purification of heparin-binding serotypes such as AAV2, AAV6, and AAV-DJ (Chapter 21).

Collectively, the protocols in this volume allow the reader to produce, characterize, and deliver AAV vectors to any tissue of interest using both well-established and innovative methods. These vital techniques will enhance the utility of AAV vectors for targeted gene transfer to living animals and support the ongoing development of novel AAV-based gene therapies for human disease.

La Jolla, CA, USA *Michael J. Castle*

Contents

Contributors

OMAR AKIL • *Department of Otolaryngology-Head and Neck Surgery, University of California San Francisco, San Francisco, CA, USA*
JOHN J. ALEXANDER • *Department of Human Genetics, Emory University, Atlanta, GA, USA*
ISABELLE AUBERT • *Brain Sciences, Biological Sciences, Sunnybrook Research Institute, Toronto, ON, Canada; Department of Laboratory Medicine and Pathobiology, University of Toronto, Toronto, ON, Canada*
EDUARD AYUSO • *INSERM UMR1089, University of Nantes, Centre Hospitalier Universitaire de Nantes, Nantes, France*
GREG L. BAIN • *Department of Biology, Texas A&M University, College Station, TX, USA*
DAVID C. BLOOM • *Department of Molecular Genetics and Microbiology, University of Florida College of Medicine, Gainesville, FL, USA*
FLORIE BOREL • *Horae Gene Therapy Center, University of Massachusetts Medical School, Worcester, MA, USA*
SANFORD L. BOYE • *Department of Pediatrics and the Powell Gene Therapy Center, University of Florida College of Medicine, Gainesville, FL, USA*
SHANNON E. BOYE • *Department of Ophthalmology, University of Florida College of Medicine, Gainesville, FL, USA*
MARIANA BRAVO-HERNÁNDEZ • *Neuroregeneration Laboratory, Department of Anesthesiology, University of California, San Diego, La Jolla, CA, USA*
CHRISTIAN J. BUCHHOLZ • *Paul-Ehrlich-Institut, Molecular Biotechnology and Gene Therapy, Langen, Germany*
MATTHEW CAMPBELL • *Neurovascular Genetics Laboratory, Smurfit Institute of Genetics, University of Dublin, Trinity College, Dublin, Ireland*
LEI CAO • *Department of Cancer Biology and Genetics, The Ohio State University, Columbus, OH, USA; The Comprehensive Cancer Center, The Ohio State University, Columbus, OH, USA*
MICHAEL J. CASTLE • *Department of Neurosciences, University of California, San Diego, La Jolla, CA, USA*
JAMES M. CONNER • *Molecular Neurobiology Laboratory, Salk Institute for Biological Studies, La Jolla, CA, USA*
JAKOB M. DOMM • *Department of Pathobiology, University of Guelph, Guelph, ON, Canada*
DONGSHENG DUAN • *Department of Molecular Microbiology and Immunology, School of Medicine, University of Missouri, Columbia, MO, USA; Department of Biomedical Sciences, College of Veterinary Medicine, University of Missouri, Columbia, MO, USA; Department of Neurology, School of Medicine, University of Missouri, Columbia, MO, USA; Department of Bioengineering, University of Missouri, Columbia, MO, USA*
JENNIFER N. DULIN • *Department of Biology, Texas A&M University, College Station, TX, USA*
CAROLYN A. FAIRBANKS • *Department of Neuroscience, University of Minnesota, Minneapolis, MN, USA; Department of Pharmacology, University of Minnesota, Minneapolis, MN, USA; Department of Pharmaceutics, University of Minnesota, Minneapolis, MN, USA*

PAUL D. GAMLIN • *Department of Ophthalmology and Visual Sciences, University of Alabama at Birmingham, Birmingham, AL, USA*

GUANGPING GAO • *Horae Gene Therapy Center, University of Massachusetts Medical School, Worcester, MA, USA; Li Weibo Institute for Rare Diseases Research, University of Massachusetts Medical School, Worcester, MA, USA; Department of Microbiology and Physiological Systems, University of Massachusetts Medical School, Worcester, MA, USA*

DOMINIC J. GESSLER • *Horae Gene Therapy Center, University of Massachusetts Medical School, Worcester, MA, USA; Li Weibo Institute for Rare Diseases Research, University of Massachusetts Medical School, Worcester, MA, USA; Department of Microbiology and Physiological Systems, University of Massachusetts Medical School, Worcester, MA, USA*

GEORGE GITTES • *University of Pittsburgh/Children's Hospital of Pittsburgh of UPMC, Pittsburgh, PA, USA*

RESHMA GORE • *Graduate Program in Neuroscience, University of Minnesota, Minneapolis, MN, USA*

PING GUO • *University of Texas Health Science Center at Houston, Houston, TX, USA*

BRITTNEY L. GURDA • *Research Institute of the Children's Hospital of Philadelphia, Philadelphia, PA, USA*

ROLAND W. HERZOG • *Department of Pediatrics, University of Florida, Gainesville, FL, USA; Department of Pediatrics, Indiana University, Indianapolis, IN, USA*

WEI HUANG • *Department of Cancer Biology and Genetics, The Ohio State University, Columbus, OH, USA; The Comprehensive Cancer Center, The Ohio State University, Columbus, OH, USA*

YUEXI HUANG • *Physical Sciences, Sunnybrook Research Institute, Toronto, ON, Canada*

PETER HUMPHRIES • *Ocular Genetics Unit, Smurfit Institute of Genetics, University of Dublin, Trinity College, Dublin, Ireland*

JACQUELINE E. HUNTER • *Research Institute of the Children's Hospital of Philadelphia, Philadelphia, PA, USA*

KULLERVO HYNYNEN • *Physical Sciences, Sunnybrook Research Institute, Toronto, ON, Canada; Department of Medical Biophysics, University of Toronto, Toronto, ON, Canada*

MAKIKO KASHIO • *Department of Physiology, Aichi Medical University, Nagakute, Japan*

BRIAN K. KASPAR • *AveXis, San Diego, CA, USA*

MICHAEL G. KATZ • *Cardiovascular Institute, Icahn School of Medicine at Mount Sinai, New York City, NY, USA*

KELLEY F. KITTO • *Department of Neuroscience, University of Minnesota, Minneapolis, MN, USA*

ERIK KOHLBRENNER • *Cardiovascular Institute, Icahn School of Medicine at Mount Sinai, New York City, NY, USA*

SEBASTIAN KÜGLER • *Department of Neurology, University Medical Center Göttingen, Göttingen, Germany*

YI LAI • *Department of Molecular Microbiology and Immunology, School of Medicine, University of Missouri, Columbia, MO, USA*

EMILIE LECOMTE • *INSERM UMR1089, University of Nantes, Centre Hospitalier Universitaire de Nantes, Nantes, France*

ADRIEN LEGER • *INSERM UMR1089, University of Nantes, Centre Hospitalier Universitaire de Nantes, Nantes, France; European Molecular Biology Laboratory-European Bioinformatics Institute (EMBL-EBI), Cambridge, UK*

JIA LI • *Horae Gene Therapy Center, University of Massachusetts Medical School, Worcester, MA, USA; Li Weibo Institute for Rare Diseases Research, University of Massachusetts*

Medical School, Worcester, MA, USA; Department of Microbiology and Physiological Systems, University of Massachusetts Medical School, Worcester, MA, USA

ESTRELLA LOPEZ-GORDO • *Cardiovascular Institute, Icahn School of Medicine at Mount Sinai, New York City, NY, USA*

LAWRENCE LUSTIG • *Otolaryngology-Head and Neck Surgery, Columbia University Medical Center, New York, NY, USA*

DAVID M. MARKUSIC • *Department of Pediatrics, Indiana University, Indianapolis, IN, USA*

MARTIN MARSALA • *Neuroregeneration Laboratory, Department of Anesthesiology, University of California, San Diego, La Jolla, CA, USA*

JEFFREY R. MARTENS • *Department of Pharmacology and Therapeutics, University of Florida, Gainesville, FL, USA; Center for Smell and Taste, University of Florida College of Medicine, Gainesville, FL, USA*

CHRISTIAN MUELLER • *Horae Gene Therapy Center, University of Massachusetts Medical School, Worcester, MA, USA; Department of Pediatrics, University of Massachusetts Medical School, Worcester, MA, USA*

ALEXANDER MUIK • *Paul-Ehrlich-Institut, Molecular Biotechnology and Gene Therapy, Langen, Germany*

DONNA M. NEUMANN • *Department of Pharmacology and Experimental Therapeutics, Louisiana State University Health Sciences Center, New Orleans, LA, USA*

ZEINAB NOROOZIAN • *Brain Sciences, Biological Sciences, Sunnybrook Research Institute, Toronto, ON, Canada; Department of Laboratory Medicine and Pathobiology, University of Toronto, Toronto, ON, Canada*

JEFFREY O'CALLAGHAN • *Ocular Genetics Unit, Smurfit Institute of Genetics, University of Dublin, Trinity College, Dublin, Ireland*

BRETT PALASCHAK • *Department of Pediatrics, University of Florida, Gainesville, FL, USA*

AMAN PATEL • *Department of Molecular Microbiology and Immunology, School of Medicine, University of Missouri, Columbia, MO, USA*

MAGALIE PENAUD-BUDLOO • *INSERM UMR1089, University of Nantes, Centre Hospitalier Universitaire de Nantes, Nantes, France*

CRISTINA D. PETERSON • *Department of Neuroscience, University of Minnesota, Minneapolis, MN, USA*

NICHOLAS J. QUEEN • *Department of Cancer Biology and Genetics, The Ohio State University, Columbus, OH, USA; The Comprehensive Cancer Center, The Ohio State University, Columbus, OH, USA*

JOHANNA REUL • *Paul-Ehrlich-Institut, Molecular Biotechnology and Gene Therapy, Langen, Germany*

MAUREEN S. RIEDL • *Department of Neuroscience, University of Minnesota, Minneapolis, MN, USA*

JULIO SANMIGUEL • *Grousbeck Gene Therapy Center, Schepens Eye Research Institute and Massachusetts Eye and Ear Infirmary, Boston, MA, USA; Harvard Stem Cell Institute, Harvard University, Cambridge, MA, USA*

ALEXANDER G. J. SKORPUT • *Department of Neuroscience, University of Minnesota, Minneapolis, MN, USA*

TAKAHIRO TADOKORO • *Neuroregeneration Laboratory, Department of Anesthesiology, University of California, San Diego, La Jolla, CA, USA*

PHILLIP W. L. TAI • *Horae Gene Therapy Center, University of Massachusetts Medical School, Worcester, MA, USA; Li Weibo Institute for Rare Diseases Research, University of*

Massachusetts Medical School, Worcester, MA, USA; Department of Microbiology and Physiological Systems, University of Massachusetts Medical School, Worcester, MA, USA

AKIYUKI TARUNO • *Department of Molecular Cell Physiology, Kyoto Prefectural University of Medicine, Kyoto, Japan*

CEDRIC R. UYTINGCO • *Department of Pharmacology and Therapeutics, University of Florida, Gainesville, FL, USA; Center for Smell and Taste, University of Florida College of Medicine, Gainesville, FL, USA*

LAURA P. VAN LIESHOUT • *Department of Pathobiology, University of Guelph, Guelph, ON, Canada*

LUK H. VANDENBERGHE • *Grousbeck Gene Therapy Center, Schepens Eye Research Institute and Massachusetts Eye and Ear Infirmary, Boston, MA, USA; Department of Ophthalmology, Harvard Medical School, Boston, MA, USA; Harvard Stem Cell Institute, Harvard University, Cambridge, MA, USA; The Broad Institute of Harvard and MIT, Cambridge, MA, USA*

LUCY VULCHANOVA • *Department of Neuroscience, University of Minnesota, Minneapolis, MN, USA*

ZACHARY L. WATSON • *Department of Obstetrics and Gynecology, University of Colorado School of Medicine, Aurora, CO, USA*

THOMAS WEBER • *Cardiovascular Institute, Icahn School of Medicine at Mount Sinai, New York City, NY, USA; Graduate School of Biomedical Sciences, Icahn School of Medicine at Mount Sinai, New York City, NY, USA*

JOHN WIERSCH • *University of Texas Health Science Center at San Antonio, San Antonio, TX, USA*

GEORGE L. WILCOX • *Department of Neuroscience, University of Minnesota, Minneapolis, MN, USA; Department of Pharmacology, University of Minnesota, Minneapolis, MN, USA; Department of Dermatology, University of Minnesota, Minneapolis, MN, USA*

C. DOUGLAS WITHERSPOON • *Department of Ophthalmology and Visual Sciences, University of Alabama at Birmingham, Birmingham, AL, USA*

JOHN H. WOLFE • *Research Institute of the Children's Hospital of Philadelphia, Philadelphia, PA, USA; W.F. Goodman Center for Comparative Medical Genetics, School of Veterinary Medicine, University of Pennsylvania, Philadelphia, PA, USA; Department of Pediatrics, Perelman School of Medicine, University of Pennsylvania, Philadelphia, PA, USA*

SARAH K. WOOTTON • *Department of Pathobiology, University of Guelph, Guelph, ON, Canada*

ZHIJIAN WU • *Ocular Gene Therapy Core, National Eye Institute, NIH, Bethesda, MD, USA*

KRISTIANA XHIMA • *Brain Sciences, Biological Sciences, Sunnybrook Research Institute, Toronto, ON, Canada; Department of Laboratory Medicine and Pathobiology, University of Toronto, Toronto, ON, Canada*

XIANGWEI XIAO • *University of Pittsburgh/Children's Hospital of Pittsburgh of UPMC, Pittsburgh, PA, USA*

SEA YOUNG YOON • *Research Institute of the Children's Hospital of Philadelphia, Philadelphia, PA, USA; W.F. Goodman Center for Comparative Medical Genetics, School of Veterinary Medicine, University of Pennsylvania, Philadelphia, PA, USA*

WENHAN YU • *Ocular Gene Therapy Core, National Eye Institute, NIH, Bethesda, MD, USA*

JUNLING ZHAO • *Department of Molecular Microbiology and Immunology, School of Medicine, University of Missouri, Columbia, MO, USA*

Part I

AAV Vector Design

Chapter 1

Design of AAV Vectors for Delivery of RNAi

Florie Borel and Christian Mueller

Abstract

Adeno-associated viral vectors have emerged as an important tool for human gene therapy, having demonstrated high transduction efficiency in a broad range of target tissues, a good safety profile in animal models and human clinical trials, and prospective long-lasting gene expression. First discovered 20 years ago, RNA interference (RNAi) has become another important tool for human gene therapy, enabling scientists to move on from classical gene transfer to gene silencing approaches, or combinations thereof. In this chapter, we describe a simple step-by-step method that will allow gene silencing novices to design their own artificial miRNAs against a target of their choice, clone these miRNAs into an AAV-based vector, and rapidly screen for highly efficient artificial miRNAs. The described method takes into consideration recent advances in the field including miRNA processing from various cellular miRNA backbones, choice between polymerase II and III promoters, and the potential impact of these factors on toxicity as it relates to off-targeting and to saturation of the endogenous RNAi machinery.

Key words AAV, Adeno-associated virus, RNA interference, RNAi, miRNA, shRNA, Silencing, Artificial miRNA, ddRNAi

1 Introduction

RNA interference (RNAi) was described for the first time in 1998 by Andrew Fire and Craig Mello [1], a discovery for which they were later awarded the Nobel Prize in Physiology or Medicine in 2006. RNAi is defined as sequence-specific, posttranscriptional gene silencing. Four years after Fire and Mello's publication in *C. elegans*, Mark Kay demonstrated DNA-derived gene silencing in vivo in mice [2]. Since then, the gene therapy field has realized the potential of this technology for the treatment of formerly undruggable disorders and embraced this new strategy as a tool that complements traditional gene transfer approaches.

Adeno-associated viral (AAV) vector-mediated gene therapy is currently thriving, and 2017 was a landmark year for the field. In November, results from a Phase 1 clinical trial of AVXS-101 from AveXis were published. The investigators documented that treating spinal muscular atrophy type 1 patients with a single dose

Michael J. Castle (ed.), *Adeno-Associated Virus Vectors: Design and Delivery*, Methods in Molecular Biology, vol. 1950, https://doi.org/10.1007/978-1-4939-9139-6_1,

of AAV9-based AVXS-101 improved motor and respiratory functions as well as dramatically extended survival, with 100% survival at 20 months instead of 8% in a historical cohort [3]. Furthermore, the year auspiciously ended with the first FDA approval of AAV-mediated gene therapy for a genetic disease, AAV2-based Luxturna from Spark Therapeutics [4].

These two products are based on the classic gene transfer approach. However, gene silencing is another approach that has proven valuable, even though no such therapy has gained FDA approval yet. Nevertheless, countless proof-of-concepts have demonstrated the potential of gene silencing for the treatment of gain-of-function disorders. One case that exemplifies the field is Huntington's disease (HD), a devastating neurodegenerative disorder that affects up to 12 in 100,000 people worldwide. The condition is caused by an expansion of a CAG repeat region in exon 1 of the Huntingtin gene. The resulting protein is toxic to striatal medium spiny neurons, and leads to clinical symptoms such as depression and chorea. It is therefore anticipated that silencing the mutant transcript would reduce the levels of mutant protein and alleviate some of the symptoms. Early work by the group of Beverley Davidson laid the foundations of RNA interference-based silencing for HD, initially using shRNAs [5]. With growing toxicity-related concerns, the same group then demonstrated the advantage of using artificial miRNAs to achieve gene silencing [6, 7]. This approach is now used in several biotech/pharma preclinical programs for HD, superoxide dismutase 1-linked amyotrophic lateral sclerosis (SOD1-linked ALS), and alpha-1 antitrypsin deficiency, including those of Spark Therapeutics, Voyager Therapeutics, AVeXis, Apic Bio, and uniQure [8–10], as well as in academic programs including our own at UMass Medical [11–15].

The ground for AAV-based gene silencing, and more broadly for any DNA-directed gene silencing, is the cellular RNAi pathway. As reviewed in [15], the only difference is the source of gene expression, which in the case of endogenous miRNAs will be the genome of the cell itself, and in the case of artificial miRNAs will be the AAV genome. From this point on, the fate of all miRNAs will be identical, starting with nuclear transcription and 5′ capping and 3′ polyadenylation of the stem-loop structure. The resulting ~1000–4000 nucleotides (nt) transcript is termed primary miRNA or "pri-miRNA" [16]. The next step involves the so-called Microprocessor complex, which is composed of RNAse III Drosha and RNA-binding protein DiGeorge syndrome critical region gene 8 (DGCR8), and cleaves the pri-miRNA into its pre-miRNA form, or miRNA precursor form, which is a shorter, ~60–80 nt-long stem-loop structure. It is at this step that the pre-miRNA is recognized by Exportin-5, which guides it through the nuclear pore complex into the cytoplasm. Once in the cytoplasm, the pre-miRNA is recognized by the Dicer/PACT/TBRP complex, and the loop is cleaved by RNAse III Dicer, resulting in its final form, the miRNA-miRNA* duplex. As the name indicates, this duplex is composed of two strands, one designated as the guide strand (miRNA) and one designated as the passenger strand (miRNA*). They are ~22 nt-long with 2 nt overhangs. One strand is degraded, which by definition is the passenger strand. The thermodynamically favored strand, which by definition is the guide strand, is loaded into the RNA-induced silencing complex (RISC), and mediates target recognition. Partial complementarity between the target and the seed sequence, which is composed of 6–8 nt at the 5′ end of the mature miRNA, causes translational repression. Complete complementarity between the target and the mature miRNA leads to site-specific mRNA cleavage and degradation. The use of artificial miRNAs is therefore a simple and elegant manner to achieve gene silencing.

In this chapter, we will describe a step-by-step, easy-to-implement protocol that will allow the beginner to effortlessly design, clone, and test AAV vectors to achieve gene silencing. Of note, while this work is directed toward the use of AAV vectors, the design and cloning strategy we describe here could be adapted to achieve silencing using a variety of other vectors including plasmids, lentiviral vectors, and more. We urge readers to read Subheading 3.1 of this chapter closely prior to starting, and to carefully consider all conditions in which the vectors generated herein will be tested. In our experience, it is often more efficient to start by designing and cloning 10 or more artificial miRNA candidates in expression plasmids, screen them, and subsequently subclone only the lead candidate into an AAV plasmid. The different iterations one may want to generate to allow for maximal versatility may include constructs with and without a reporter gene or tag of choice for early testing versus more advanced development phases, respectively, with various ubiquitous promoters including pol II and pol III as well as possible enhancer elements and other cis-regulatory modules for the largest range of expression, and finally with tissue-specific promoters as well as possible de-targeting elements for fine-tuning of expression.

2 Materials

1. Expression plasmid and/or AAV plasmid of choice.
2. Cloning software (optional).
3. Annealing buffer such as nuclease-free duplex buffer (Integrated DNA Technologies).
4. Plasmid purification kits such as Qiagen mini kit and Qiagen plasmid plus midi kit (Qiagen).

5. Shaking incubator.
6. Restriction enzymes.
7. Electrophoresis-grade agarose.
8. Gel electrophoresis equipment.
9. Razor blades.
10. Gel extraction kit such as QIAquick gel extraction kit (Qiagen).
11. Ligation kit such as Quick Ligation kit (New England Biolabs).
12. Competent cells of choice such as SURE-2 supercompetent cells (Agilent Technologies).
13. NZY^+ broth (Teknova).
14. Selection plates with antibiotic of choice.
15. Culture broth with antibiotic of choice.
16. Mammalian cell line of choice.
17. Transfection method or reagent of choice.
18. RNA isolation kit, such as Trizol reagent (Life Technologies).
19. Nuclease-free water (Life Technologies).
20. RNA quantification method, such as Nanodrop spectrophotometer.
21. Retrotranscription kit, such as High capacity RNA-to-cDNA (Life Technologies).
22. PCR tubes or strips.
23. PCR machine.
24. Taqman primers and probe for target and housekeeping gene.
25. Taqman mastermix, such as Fast Advanced Mastermix (Life Technologies).
26. RT-qPCR plates and optical adhesive film.
27. RT-qPCR machine.
28. Spreadsheet software for data analysis.

3 Methods

3.1 General Considerations for Vector Design

1. *Transgene size.* Artificial miRNAs are much shorter than regular cDNA-based transgenes. It is therefore important to consider early on the totality of the cassette that will be used: the promoter, the termination signal, as well as the reporter gene or cDNA cassette for concomitant gene transfer (a dual-function vector). If the entire AAV genome is very small, such as 2 kb or below, stuffer sequences should be added to allow for efficient AAV packaging.

2. *Promoter.* Artificial miRNAs may be expressed from polymerase II (pol II) or polymerase III (pol III) promoters. We recommend selecting a pol III promoter for the initial screening of artificial miRNA candidates in vitro, as these will lead to high expression levels of the artificial miRNA and will allow for a clear readout. However, a pol II promoter may be preferred for further testing in vivo, for reasons of tissue-specific expression as well as safety. The issue of RNAi-related toxicity has been extensively reviewed in [17]. For convenience, common ubiquitous promoter sequences are shown in Table 1 (including pol III H1 and U6 [18], as well as variants of the pol II chicken beta-actin promoter [19]), and common tissue-specific promoter sequences are shown in Table 2 (including liver-specific human TBG promoter [20], neuron-specific SYN promoter [21], astrocyte-specific GFAP promoter, oligodendrocyte-specific MBP promoter, and photoreceptor-specific GRK1 promoter [22]).
3. *Cellular miRNA backbone.* The cellular miRNA backbone will dictate processing of the artificial miRNA and the mature miRNA sequence. This is therefore an element of crucial importance in order to avoid unintended off-targeting. We recommend using the backbone of miR-155. In our hands, this miRNA has demonstrated efficient and accurate processing in vivo [11] and is therefore highly suitable for this application. Another commonly used backbone is that of miR-30a. However, our experience with this miRNA backbone tends to indicate less efficient processing in vivo [11]. Both sequences are provided in Table 3.
4. *Target sequence.* If more than one species is likely to be used during preclinical development, this should be taken into consideration early on in the design. For instance, the AAV vector may be initially tested in a mouse model of a given disease; however, further development may warrant a study in macaques, and ultimately it may be tested in humans. In this situation, designing artificial miRNAs that target all three species will streamline preclinical development by using a single candidate for all studies. In this case, the mouse, macaque, and human sequences can be aligned, and conserved stretches of sequences identified. These are ideal targets for artificial miRNA candidates, provided further considerations detailed in Subheading 3.2 are followed. Conversely, the AAV vector may be tested in a transgenic mouse model in order to achieve specific silencing of the human transgene but not of the wild-type murine gene. In this situation, designing artificial miRNAs that only target the human sequence is critical. In this case, artificial miRNAs can be designed to target the human sequence, and a Blast search against murine transcripts can then indicate which candidates are suitable.

Table 1
Ubiquitous promoters

Name	Type	Sequence 5′-3′
H1	Pol III	GGAATTCGAACGCTGACGTCATCAACCCGCTCCAAGGAATCGCGGGCCCAGTGTCACTAGGCGGGAACA CCCAGCGCGCGTGCGCCCTGGCAGGAAGATGGCTGTGAGGGACAGGGGAGTGGCGCCCTGCAATATTT GCATGTCGCTATGTGTTCTGGGAAATCACCATAAACGTGAAATGTCTTTGGATTTGGGAATCTTATAAGTTCT GTATGAGACCACTCTTTCCC *Source: Gao et al. 2017 [19]*
U6	Pol III	GAGGGCCTATTTCCCATGATTCCTTCATATTTGCATATACGATACAAGGCTGTTAGAGAGATAATTGGAATTAA TTTGACTGTAAACACAAAGATATTAGTACAAAATACGTGACGTAGAAAGTAATAATTTCTTGGGTAGTTTGCA GTTTTAAAATTATGTTTTAAAATGGACTATCATATGCTTACCGTAACTTGAAAGTATTTCGATTTCTTGGC TTTATATATCTTGTGGAAAGGACGAAACACC *Source: Gao et al. 2017 [19]*
CB full length (CB)	Pol II	GATCTTCAATATTGGCCATTAGCCATATTATTCATTGGTTATATAGCATAAATCAATATTGGCTATTGGCCATT GCATACGTTGTATCTATATCATAATATGTACATTTATATTGGCTCATGTCCAATATGACCGCCATGTTGGC ATTGATTATTGACTAGTTATTAATAGTAATCAATTACGGGGTCATTAGTTCATAGCCCATATATGGAGTTCCGC GTTACATAACTTACGGTAAATGGCCCGCCTGGCTGACCGCCCAACGACCCCCGCCCATTGACGTCAATAAT GACGTATGTTCCCATAGTAACGCCAATAGGGACTTTCCATTGACGTCAATGGGTGGAGTATTTACGGT AAACTGCCCACTTGGCAGTACATCAAGTGTATCATATGCCAAGTCCGCCCCCTATTGACGTCAATGACGGT AAATGGCCCGCCTGGCATTATGCCCAGTACATGACCTTACGGGACTTTCCTACTTGGCAGTACATCTACGT ATTAGTCATCGCTATTACCATGGTCGAGGTGAGCCCCACGTTCTGCTTCACTCTCCCCATCTCCCCCCCCT CCCCACCCCCAATTTTGTATTTATTTATTTTTTAATTATTTTGTGCAGCGATGGGGGCGGGGGGGGGGG GGGGGCGCGCGCCAGGCGGGGCGGGGCGGGGCGAGGGGCGGGGCGGGGCGAGGCGGAGAGGTGCGGC GGCAGCCAATCAGAGCGGCGCGCTCCGAAAGTTTCCTTTTATGGCGAGGCGGCGGCGGCGGCGGCCC TATAAAAAGCGAAGCGCGCGGCGGGCGGGAGTCGCTGCGACGCTGCCTTCGCCCCGTGCCCCGCTCCGCC GCCGCCTCGCGCCGCCCGCCCCGGCTCTGACTGACCGCGTTACTCCCACAGGTGAGCGGGCGGGAC GGCCCTTCTCCTCCGGGCTGTAATTAGCGCTTGGTTTAATGACGGCTTGTTTCTTTTCTGTGGCTGCGTG AAAGCCTTGAGGGGCTCCGGGAGGGCCCTTTGTGCGGGGGGGAGCGGCTCGGGGGGTGCGTGCGTGTG TGTGTGCGTGGGGAGCGCCGCGTGCGGCCCGCGCTGCCCGGCGGCTGTGAGCGCTGCGGGCGCGGCG CGGGGCTTTGTGCGCTCCGCAGTGTGCGCGAGGGGAGCGCGGCCGGGGGCGGTGCCCCGCGGTG CGGGGGGGGCTGCGAGGGGAACAAAGGCTGCGTGCGGGGTGTGTGCGTGGGGGGGTGAGCAGGGGGTG TGGGCGCGGCGGTCGGGCTGTAACCCCCCCCTGCACCCCCCTCCCCGAGTTGCTGAGCACGGCCCGGCTT CGGGTGCGGGGCTCCGTACGGGGCGTGGCGCGGGGCTCGCCGTGCCGGGCGGGGGGTGGCGGCAGGT GGGGGTGCCGGGCGGGGCGGGGCCGCCTCGGGCCGGGGAGGGCTCGGGGGAGGGGCGCGGCGG CCCCCGGAGCGCCGGCGGCTGTCGAGGCGCGGCGAGCCGCAGCCATTGCCTTTTATGGTAATCGTGCGAG

		AGGGCGCAGGGACTTCCTTTGTCCCAAATCTGTGCGGAGCCGAAATCTGGGAGGCGCCGCCGCACCCCC TCTAGCGGGCGCGGGGCGAAGCGGTGCGGCGCCGGCAGGAAGGAAATGGGCGGGGAGGGCCTTCGTGC GTCGCCGCGCCGCCGTCCCCTTCTCCCTCTCCAGCCTCGGGGCTGTCCGCGGGGGGACGGCTGCCTTC GGGGGGGACGGGGCAGGGCGGGGTTCGGCTTCTGGCGTGTGACCGGCGGCTCTAGAGCCTCTGCTAACC ATGTTCATGCCTTCTTCTTTTTCCTACAGCTCCTGGGCAACGTGCTGGTTATTGTGCTGTCTCATCATTTT GGCAAAGAATTCGATATCA *Source: Patent US20090186002*
CB hybrid (CBh)	Pol II	AGATGTACTGCCAAGTAGGAAAGTCCCGTAAGGTCATGTACTGGGCATAATGCCAGGCGGGCCATTTACCG TCATTGACGTCAATAGGGGGCGTACTTGGCATATGATACACTTGATGTACTGCCAAGTGGGCAGTTTACCG TAAATACTCCACCCATTGACGTCAATGGAAAGTCCCTATTGGCGTTACTATGGGAACATACGTCATTA TTGACGTCAATGGGCGGGGGTCGTTGGGCGGTCAGCCAGGCGGGCCATTTACCGTAAGTTATG TAACGCGGAACTCCATATATGGGCTATGAACTAATGACCCCGTAATTGATTACTATTAACCACGTTCTGC TTCACTCTCCCCATCTCCCCCCCCTCCCCACCCCCAATTTTGTATTTATTTATTTTTTAATTATTTTG TGCAGCGATGGGGGCGGGGGGGGGGGGGCGCGCGCCAGGCGGGGCGGGGCGGGGCGAGGGGCG GGGCGGGGCGAGGCGGAGAGGTGCGGCGGCAGCCAATCAGAGCGGCGCGCTCCGAAAGTTTCCTTTTA TGGCGAGGCGGCGGCGGCGGCGGCCCTATAAAAAGCGAAGCGCGCGGCGGGGAGTCGCTGCGTTGCC TTCGCCCCGTGCCCCGCTCCGCGCCGCCTCGCGCCGCCCGCCCCGGCTCTGACTGACCGCGTTAC TCCCACAGGTGAGCGGGCGGGACGGCCCTTCTCCTCCGGGCTGTAATTAGCGCTTAAGAGGTAAGGG TTTAAGGGATGGTTGGTTGGTGGGGTATTAATGTTTAATTACCTGTTTTACAGGCCTGAAATCACTTGG TTTTAGGTTGG *Source: Gray et al. 2011 [20]*
small CBA (smCBA)	Pol II	AATTCGGTACCCTAGTTATTAATAGTAATCAATTACGGGGTCATTAGTTCATAGCCCATATATGGAGTTCCGCG TTACATAACTTACGGTAAATGGCCCGCCTGGCTGACCGCCCAACGACCCCCGCCCATTGACGTCAATAA TGACGTATGTTCCCATAGTAACGCCAATAGGGACTTTCCATTGACGTCAATGGGTGGACTATTTACGGTAAAC TGCCCACTTGGCAGTACATCAAGTGTATCATATGCCAAGTACGCCCCCTATTGACGTCAATGACGGTAAA TGGCCCGCCTGGCATTATGCCCAGTACATGACCTTATGGGACTTTCCTACTTGGCAGTACATCTACGTA TTAGTCATCGCTATTACCATGGTCGAGGTGAGCCCCACGTTCTGCTTCACTCTCCCCATCTCCCCCCCC TCCCCACCCCCAATTTTGTATTTATTTATTTTTTAATTATTTTGTGCAGCGATGGGGGCGGGGGG GGGGGGGGGCGCGCGCCAGGCGGGGCGGGGCGGGGCGA GGGGCGGGGCGGGGCGAGGCGGAGAGG TGCGGCGGCAGCCAATCAGAGCGGCGCGCTCCGAAAGTTTCCTTTTATGGCGAGGCGGCGGCGGC GGCGGCCCTATAAAAAGCGAAGCGCGCGGCGGGCGGGAGTCGCTGCGACGCTGCCTTCGCCCCG TGCCCCGCTCCGCCGCCGCCTCGCGCCGCCCGCCCCGGCTCTGACTGACCGCGTTACTCCCACAGG TGAGCGGGCGGGACGGCCCTTCTCCTCCGGGCTGTAATTAGCGCTTGGTTTAATGACGGCTTGTTTC TTTTCTGTGGCTGCGTGAAAGCCTTGAGGGGCTCCGGGAGCTAGAGCCTCTGCTAACCATGTTCATGCC TTCTTCTTTTTCCTACAGCTCCTGGGCAACGTGCTGGTTATTGTGCTGTCTCATCATTTTGGCAAAG *Source: Patent US8298818*

The sequence of commonly used ubiquitous promoters is provided here. These may be used to express artificial miRNAs

Table 2
Tissue-specific promoters

Name	Sequence 5′-3′
Thyroxine binding globulin (TBG) promoter(liver-specific)	TGCATGTATAATTTCTACAGAACCTATTAGAAAGGATCACCCAGCCTC TGCTTTTGTACAACTTTCCCTTAAAAAACTGCCAATTCCACTGCTG TTTGGCCCAATAGTGAGAACTTTTTCCTGCTGCCTCTTGGTGC TTTTGCCTATGGCCCCTATTCTGCCTGCTGAAGACACTC TTGCCAGCATGGACTTAAACCCCTCCAGCTCTGACAATCCTCTTTC TCTTTTGTTTTACATGAAGGGTCTGGCAGCCAAAGCAATCAC TCAAAGTTCAAACCTTATCATTTTTTGCTTTGTTCCTCTTGGCC TTGGTTTTGTACATCAGCTTTGAAAATACCATCCCAGGGTTAATGC TGGGGTTAATTTATAACTAAGAGTGCTCTAGTTTTGCAA TACAGGACATGCTATAA *Source: Yan et al. 2012 [21]*
Human Synapsin I (SYN) promoter (neuron-specific)	AGTGCAAGTGGGTTTTAGGACCAGGATGAGGCGGGGTGGGGGTGCC TACCTGACGACCGACCCCGACCCACTGGACAAGCACCCAACCCCC ATTCCCCAAATTGCGCATCCCCTATCAGAGAGGGGGAGGGGAAACA GGATGCGGCGAGGCGCGTGCGCACTGCCAGCTTCAGCACCGCG GACAGTGCCTTCGCCCCCGCCTGGCGGCGCGCGCCACCGCCGCC TCAGCACTGAAGGCGCGCTGACGTCACTCGCCGGTCCCCCGCA AACTCCCCTTCCCGGCCACCTTGGTCGCGTCCGCGCCGCCGCCG GCCCAGCCGGACCGCACCACGCGAGGCGCGAGATAGGGGGGCAC GGGCGCGACCATCTGCGCTGCGGCGCCGGCGACTCAGCGCTGCC TCAGTCTGCGGTGGGCAGCGGAGGAGTCGTGTCGTGCC TGAGAGCGCAG *Source: Kugler et al. 2003 [22]*
Human Glial Fibrillary Acidic Protein (GFAP) promoter (astrocyte-specific)	GAGCTCCCACCTCCCTCTCTGTGCTGGGACTCACAGAGGGAGACC TCAGGAGGCAGTCTGTCCATCACATGTCCAAATGCAGAGCATACCC TGGGCTGGGCGCAGTGGCGCACAACTGTAATTCCAGCACTTTGG GAGGCTGATGTGGAAGGATCACTTGAGCCCAGAAGTTCTAGACCA GCCTGGGCAACATGGCAAGACCCTATCTCTACAAAAAAAGTTAAAA AATCAGCCACGTGTGGTGACACACACCTGTAGTCCCAGCTATTCAG GAGGCTGAGGTGAGGGGATCACTTAAGGCTGGGAGGTTGAGGCTG CAGTGAGTCGTGGTTGCGCCACTGCACTCCAGCCTGGGCAACAG TGAGACCCTGTCTCAAAAGACAAAAAAAAAAAAAAAAAAAAAAG AACATATCCTGGTGTGGAGTAGGGGACGCTGCTCTGACAGAGGC TCGGGGGCCTGAGCTGGCTCTGTGAGCTGGGGAGGAGGCAGACA GCCAGGCCTTGTCTGCAAGCAGACCTGGCAGCATTGGGCTGGCC GCCCCCCAGGGCCTCCTCTTCATGCCCAGTGAATGACTCACCTTG GCACAGACACAATGTTCGGGGTGGGCACAGTGCCTGCTTCCCGCC GCACCCCAGCCCCCCTCAAATGCCTTCCGAGAAGCCCATTGAGCA GGGGGCTTGCATTGCACCCCAGCCTGACAGCCTGGCATCTTGGGA TAAAAGCAGCACAGCCCCCTAGGGGCTGCCCTTGCTGTGTGGCGC CACCGGCGGTGGAGAACAAGGCTCTATTCAGCCTGTGCCCAGG AAAGGGGATCAGGGGATGCCCAGGCATGGACAGTGGGTGGCAG GGGGGGAGAGGAGGGCTGTCTGCTTCCCAGAAGTCCAAGGACA CAAATGGGTGAGGGGACTGGGCAGGGTTCTGACCCTGTGGGACCA GAGTGGAGGGCGTAGATGGACCTGAAGTCTCCAGGGACAACA GGGCCCAGGTCTCAGGCTCCTAGTTGGGCCCAGTGGCTCCAGCG TTTCCAAACCCATCCATCCCCAGAGGTTCTTCCCATCTCTCCAGGC

(continued)

Table 2
(continued)

Name	Sequence 5′-3′
	TGATGTGTGGGAACTCGAGGAAATAAATCTCCAGTGGGAGACGG AGGGGTGGCCAGGGAAACGGGGCGCTGCAGGAATAAAGACGAGC CAGCACAGCCAGCTCATGTGTAACGGCTTTGTGGAGCTG TCAAGGCCTGGTCTCTGGGAGAGAGGCACAGGGAGGCCAGACAAG GAAGGGGTGACCTGGAGGGACAGATCCAGGGGCTAAAGTCCTGA TAAGGCAAGAGAGTGCCGGCCCCCTCTTGCCCTATCAGGACC TCCACTGCCACATAGAGGCCATGATTGACCCTTAGACAAAGGGC TGGTGTCCAATCCCAGCCCCCAGCCCCAGAACTCCAGGGAATGAA TGGGCAGAGAGCAGGAATGTGGGACATCTGTGTTCAAGGGAAGGA CTCCAGGAGTCTGCTGGGAATGAGGCCTAGTAGGAAATGAGG TGGCCCTTGAGGGTACAGAACAGGTTCATTCTTCGCCAAA TTCCCAGCACCTTGCAGGCACTTACAGCTGAGTGAGATAATGCC TGGGTTATGAAATCAAAAAGTTGGAAAGCAGGTCAGAGGTCATC TGGTACAGCCCTTCCTTCCCTTTTTTTTTTTTTTTTTTGTGAGAC AAGGTCTCTCTCTGTTGCCCAGGCTGGAGTGGCGCAAACACAGC TCACTGCAGCCTCAACCTACTGGGCTCAAGCAATCCTCCAGCC TCAGCCTCCCAAAGTGCTGGGATTACAAGCATGAGCCACCCCAC TCAGCCCTTTCCTTCCTTTTTAATTGATGCATAATAATTGTAAGTA TTCATCATGGTCCAACCAACCCTTTCTTGACCCACCTTCCTAGAG AGAGGGTCCTCTTGCTTCAGCGGTCAGGGCCCCAGACCCATGGTC TGGCTCCAGGTACCACCTGCCTCATGCAGGAGTTGGCGTGCCCAG GAAGCTCTGCCTCTGGGCACAGTGACCTCAGTGGGGTGAGGGGA GCTCTCCCCATAGCTGGGCTGCGGCCCAACCCCACCCCCTCAGGC TATGCCAGGGGGTGTTGCCAGGGGCACCCGGGCATCGCCAGTC TAGCCCACTCCTTCATAAAGCCCTCGCATCCCAGGAGCGAGC *Source: GenBank M67446.1*
Mouse Myelin Basic Protein (MBP) promoter (oligodendrocyte-specific)	AAGCTTTGAGAGAAAAGGGACCAGATCTTATTCCTCACCGTGGC TTTAACACTTAGAGAAAATGCATCCCCTCTAATCAATAAGTCA TCGACAGTGGGTAGATGGAGGAACGGCAGTGCGTAGTAGGATGCG TGCTAAGCATAGTCTCGTGCATGGGTGCATAGATCGCTGGGCAGG TGGACAAGGTGGGGGTGGATAAAGAAGTGGGTAGATGATTGATG TTAGGTAAATATCACTGGGTGGACAGATGGGTGGTAGGTGGATGGA TGGTTAGAATAGTCAGAAGAGGGATGGATTGATAAGGTGAACAGA TGATAAATGGGTGATAGACTGGAAGGGTTGTCAAAAGAGGA TAAGGGAAGTGTGAGCTAGCCGTATTTCTAAGGTCAGTAATAGAG TTGGGAGAAGAGGTTAAGTTAC ATCCATTTAAACCTCACACGAAGCTGAGTGGGAATGGACTTGC TGCCGTTGGTGAGGAAAGCGTTGCATTTCCCGTGTGCTTGGTTG TGGAAGTGCTCAGGTCCCACATGAAGCAGTCAGGTTACTGCGGC TTACAGAGGAGCCAGATCCAAATGCCCCGAGTAAGCACG TCCCCGAGCCAGAGGCCTCCAGCGGAATCCGGGAGAGGGATTGC TCAGTGCCCTGCTTCCCTGGACTGTAAGCTGCAGAAAGATG TGGGAAGTCCTGTTCTCCACTGAGAACACTAAAAGCACCTTTTG TCAAACGACCGCTTCACATCTGGGGCTTGTGCACTGGTGGCC TTTTAAACCAGAGACAACCCACAAGATACCTAACCTGCGGGGCTC TCTGGTACAGTGAGCAACTCAGGAAATGCTTTGGCTTGATTGCTG TGGGCTCTCAGGCCATCGCCCTCTGGAGTGGTTCTTTTAA TGAGAACCTGAAGATTGGCCCCTGAGCCATGTATACCAAGCAAGC

(continued)

Table 2
(continued)

Name	Sequence 5′-3′
	TCAATCCAGGTTAGCTCCCTCTGGTTGGGGCAAGCTAACGTGCTC CTTGGGCCCCGCGCGTAACTGTGCGTTTTATAGGAGACAGCTAG TTCAAGACCCCAGGAAGAAAGCGGCTTTGTCCCCCTCTAGGCC TCGTACAGGCCCACATTCATATCTCATTGTTGTTGCAGGGGAGG CAGATGCGATCCAGAACAATGGGACCTCGGCTGAGGACACGGCGG TGACAGACTCCAAGCACACAGCAGACCCAAAGAATAACTGGCAA GGCGCCCACCCAGCTGACCCAGGGAACCGCCCCCACTTGA TCCGCCTCTTTTCCCGAGATGCCCCGGGAAGGGAGGACAACACC TTCAAAGACAGGCCCTCAGAGTCCGACGAGCTTCAGACCA TCCAAGAAGACCCCACAGCAGCTTCCGGAGGCCTGGATGTG *Source: GenBank M24410.1*
Human Rhodopsin kinase (GRK1) promoter (photoreceptor-specific)	GGGCCCCAGAAGCCTGGTGGTTGTTTGTCCTTCTCAGGGGAAAAG TGAGGCGGCCCCTTGGAGGAAGGGGCCGGGCAGAATGATCTAA TCGGATTCCAAGCAGCTCAGGGGATTGTCTTTTTCTAGCACCTTC TTGCCACTCCTAAGCGTCCTCCGTGACCCCGGCTGGGATTTAGCC TGGTGCTGTGTCAGCCCCGGGCTCCCAGGGGCTTCCCAGTGG TCCCCAGGGAACCCTCGACAGGGCCAGGGCGTCTCTCTCG TCCAGCAAGGGCAGGGACGGGCCACAGGCAAGGGC *Source: Beltran et al. 2010 [23]*

The sequence of commonly used tissue-specific promoters is provided here. These may be used to express artificial miRNAs

Table 3
Artificial miRNA backbones

Name	Sequence 5′-3′
miR-30a 5′ arm	GCTCGAGTGAGCGAG
miR-30a loop	GTAAAGCCACAGATGGG
miR-30a 3′ arm	TCGCCTACTAGT
miR-155 5′ arm	CCTGGAGGCTTGCTGAAGGCTGTA
miR-155 loop	GTTTTGGCCACTGACTGAC
miR-155 3′ arm	CAGGACACAAGGCCTGTTACTAGCACTCACATGGAACAAATGGCC

The sequences of commonly used miRNA "backbones" miR-30a and miR-155 are indicated in this table. These may be used as a basic scaffold with which to generate artificial miRNAs

5. *Location of the artificial miRNA in the AAV vector.* The location will be closely linked to the choice of promoter and to the final composition of the transgene cassette. Generally speaking, it is better to avoid placing the artificial miRNA in close proximity to the ITRs, as this may interfere with vector production

[23]. In the context of a pol II promoter, previous work indicates that placing the miRNA either in the intron or in the 3′UTR is comparable [24].

6. *In vitro screening*. An ideal cell line for in vitro screening will have the following characteristics: (1) the target transcript is expressed and can be reproducibly detected (RT-qPCR, ddPCR or other method of choice) and (2) the cell line can be transfected at high efficiency. If such a cell line is not available, alternative strategies include: (1) use a cell line lacking the target gene but with high transfection efficiency, and co-transfect a transgene-expressing plasmid along with the miRNA plasmid or (2) use a cell line that expresses the transgene but with low transfection efficiency, and include a reporter gene (such as a GFP) in your miRNA expression plasmid which will allow sorting of transfected cells and gene expression analysis in this subset only. While screening at the mRNA level allows for rapid, high-throughput analysis, it may be desirable to further confirm silencing efficiency by determining protein levels by western blot, ELISA, or another method of choice.

3.2 Artificial miRNA Design

1. Retrieve the target mRNA sequence(s) from NCBI at http://www.ncbi.nlm.nih.gov.
2. Fold the sequence in a RNA folding program (such as RNA-fold, currently hosted at http://rna.tbi.univie.ac.at/cgi-bin/RNAWebSuite/RNAfold.cgi): determine stretches of sequence that are predicted to be inaccessible and eliminate them, and determine stretches of sequence that are predicted to be accessible and select them (*see* **Note 1**). To ensure robustness of the predictions, the folding of a given sequence can be repeated using several algorithms, and the results compiled.
3. Within the accessible sequences, select several (*see* **Note 2**) 22-nt long miRNA guide sequences with U in position 1, and either A or U in position 2–3 (*see* **Note 3**), with a GC content of 30–60%. Sequences of 4+ U should strictly be avoided, in order for the design to be compatible with pol III promoters.
4. Further refine the selection by analyzing the predicted off-target profile of the sequence by performing a nucleotide blast (blastn suite, http://blast.ncbi.nlm.nih.gov) search against the transcripts of all species in which the construct will be used. Candidates which present full-length exact matches to additional targets should be excluded, as well as candidates which present an exact match to additional targets in the seed region.
5. Incorporate the miRNA guide sequence in the miRNA backbone of choice in the following order:
   (a) 5′ arm of miRNA backbone of choice

 (b) artificial miRNA guide strand
 (c) loop
 (d) artificial miRNA passenger strand (*see* **Note 4**)
 (e) 3′ arm of miRNA backbone of choice

6. In order to obtain cohesive ends when annealing the oligos, add restriction sites of choice at 5′ and 3′ ends, based on the selected plasmid that will be used for cloning.

3.3 Cloning

1. Model cloning strategy using cloning software (optional).
2. Synthesize oligos corresponding to the forward and reverse strands of the sequence designed in **steps 5** and **6** of Subheading 3.2.
3. Anneal the oligos by mixing 1 μL of each oligo (forward and reverse) at 100 μM in 8 μL of annealing buffer. Heat to 95 °C for 5 min, then cool down slowly (*see* **Note 5**).
4. Prepare the expression plasmid DNA. Digest at least 1 μg of the expression plasmid with the selected restriction enzyme(s) for 2 h at the manufacturer's recommended temperature.
5. Run 2 μL of the annealed oligos and 2 μL of the digested expression plasmid on a 1% agarose gel with molecular ladders of the appropriate size (1 kb and 100 bp size are suggested for the plasmid and oligos, respectively) to verify proper annealing of the oligos and proper digestion of the expression plasmid (*see* **Note 6**).
6. Cut the band of interest from the expression plasmid digestion using a razor blade. Proceed to gel extraction according to the manufacturer's instructions (*see* **Note 7**).
7. Ligate the annealed oligos into the expression plasmid using a ligation kit according to the manufacturer's instructions. Briefly, combine:
   (a) 10 μL of 2× quick ligase buffer
   (b) 1 μL of quick ligase
   (c) 50 ng of digested, gel-purified expression plasmid (estimate concentration based on gel)
   (d) x μL of annealed oligos (estimate concentration based on gel; using a 3:1 oligos:plasmid molar ratio)
   (e) bring to 20 μL with dH_2O

   Mix, spin down, and incubate for 5 min at RT.
8. Transform in SURE-2 supercompetent cells according to the manufacturer's instructions (*see* **Note 8**) and using the following:
   (a) 100 μL of SURE-2 supercompetent cells
   (b) 2 μL of 1.22 M beta-mercaptoethanol

   (c) 1 μL of ligation
   (d) 0.9 mL of preheated NZY$^+$ broth
9. Inoculate minipreps and grow overnight at 30 °C, with 250 rpm shaking (*see* **Note 9**).
10. Isolate plasmid DNA. Check for presence of the miRNA insert in the plasmid using the restriction enzymes used for cloning and running the digestion on a 1% agarose gel.
11. Send the positive clone(s) for sequencing to confirm the absence of mutation in the miRNA sequence (*see* **Note 10**).
12. Prepare transfection-grade plasmid DNA using a plasmid purification kit.

3.4 In Vitro Screening of Artificial miRNA Constructs

1. Seed cell line of choice in 6-well plates so that they will reach 70–90% confluence at the time of transfection (*see* **Note 11**).
2. Transfect 2.5 μg of plasmid DNA per well using the transfection method of choice according to the manufacturer's instructions. Include a mock transfection control (no DNA) and a miRNA control (an artificial miRNA designed to target another mRNA).
3. At 48 h post-transfection, collect cells and isolate RNA according to your method of choice. Resuspend RNA in 200 μL of nuclease-free water.
4. Quantify RNA using your method of choice and homogenize all concentrations to 200 ng/μL in order to normalize for retrotranscription efficiency.
5. Retrotranscribe RNA samples using the High Capacity RNA-to-cDNA kit (Life Technologies), according to the manufacturer's instructions. Briefly, combine:
   (a) 10 μL of 2× RT buffer mix
   (b) 1 μL of 20× RT enzyme mix
   (c) 9 μL of RNA, 200 ng/μL

   Incubate for 60 min at 37 °C, for 5 min at 95 °C, then store at 4 °C.
6. Quantify transcripts of the target as well as of a housekeeping mRNA of choice by RT-qPCR, according to the manufacturer's instructions. For example, combine:
   (a) 5 μL of 2× Taqman fast advanced mastermix (Life Technologies)
   (b) 0.5 μL of 20× Taqman probe of choice
   (c) 4.5 μL of cDNA, 30 ng/μL (3× dilution)
7. Select the artificial miRNA(s) that leads to the best silencing of the target mRNA, as determined in **step 6**, for delivery by AAV vectors as described in the following chapters of this book.

4 Notes

1. If multiple sequences are to be used, it is recommended to fold each complete sequence individually prior to mapping the conserved sequences onto the folded sequence.
2. We recommend selecting around 10 sequences in order to obtain at least one highly potent artificial miRNA.
3. This is based on recent work by Fang and Bartel [25].
4. The sequence of the passenger strand is designed by reverse complementing the guide strand sequence, and removing nucleotides 10–11 to create a bulge.
5. This can be done either in a PCR machine (−1 °C/min steps) or manually, by heating up a water-filled beaker to 95 °C, and subsequently letting it cool down slowly at room temperature. We recommend using at least 200 mL of water to ensure a slow cool down.
6. This confirmation step is critical and should be carefully performed before proceeding with the cloning. The annealed oligos should appear as one band at the expected size. If this is not the case, anneal the oligos again.
7. For a higher purity and higher subsequent ligation efficiency, we recommend following the gel extraction with PCR purification such as QIAquick PCR Purification Kit. To increase DNA recovery and concentration, pre-warm 30 μL of elution buffer to 50 °C prior to elution.
8. The use of SURE-2 supercompetent cells is recommended to minimize rearrangement and deletion of secondary structures.
9. A lower culture temperature of 30 °C (instead of the usual 37 °C) is recommended in order to minimize potential rearrangement of the plasmid caused by the artificial miRNA secondary structure.
10. Due to the presence of secondary structure, proper reads may be difficult to obtain. We recommend ordering "power reads" if available or consulting with your sequencing service provider for optimal results.
11. Determining silencing may be challenging in difficult-to-transfect cells lines. For these, we recommend including a GFP in the vector, sorting cells, and comparing silencing in GFP-positive cells against GFP-negative cells.

References

1. Fire A, Xu S, Montgomery MK, Kostas SA, Driver SE, Mello CC (1998) Potent and specific genetic interference by double-stranded RNA in Caenorhabditis elegans. Nature 391 (6669):806–811. https://doi.org/10.1038/35888
2. McCaffrey AP, Meuse L, Pham TT, Conklin DS, Hannon GJ, Kay MA (2002) RNA interference in adult mice. Nature 418 (6893):38–39. https://doi.org/10.1038/418038a
3. Mendell JR, Al-Zaidy S, Shell R, Arnold WD, Rodino-Klapac LR, Prior TW, Lowes L, Alfano L, Berry K, Church K, Kissel JT, Nagendran S, L'Italien J, Sproule DM, Wells C, Cardenas JA, Heitzer MD, Kaspar A, Corcoran S, Braun L, Likhite S, Miranda C, Meyer K, Foust KD, Burghes AHM, Kaspar BK (2017) Single-dose gene-replacement therapy for spinal muscular atrophy. N Engl J Med 377(18):1713–1722. https://doi.org/10.1056/NEJMoa1706198
4. Russell S, Bennett J, Wellman JA, Chung DC, Yu ZF, Tillman A, Wittes J, Pappas J, Elci O, McCague S, Cross D, Marshall KA, Walshire J, Kehoe TL, Reichert H, Davis M, Raffini L, George LA, Hudson FP, Dingfield L, Zhu X, Haller JA, Sohn EH, Mahajan VB, Pfeifer W, Weckmann M, Johnson C, Gewaily D, Drack A, Stone E, Wachtel K, Simonelli F, Leroy BP, Wright JF, High KA, Maguire AM (2017) Efficacy and safety of voretigene neparvovec (AAV2-hRPE65v2) in patients with RPE65-mediated inherited retinal dystrophy: a randomised, controlled, open-label, phase 3 trial. Lancet 390(10097):849–860. https://doi.org/10.1016/S0140-6736(17)31868-8
5. Harper SQ, Staber PD, He X, Eliason SL, Martins IH, Mao Q, Yang L, Kotin RM, Paulson HL, Davidson BL (2005) RNA interference improves motor and neuropathological abnormalities in a Huntington's disease mouse model. Proc Natl Acad Sci U S A 102 (16):5820–5825. https://doi.org/10.1073/pnas.0501507102
6. Boudreau RL, Martins I, Davidson BL (2009) Artificial microRNAs as siRNA shuttles: improved safety as compared to shRNAs in vitro and in vivo. Mol Ther 17 (1):169–175. https://doi.org/10.1038/mt.2008.231
7. McBride JL, Boudreau RL, Harper SQ, Staber PD, Monteys AM, Martins I, Gilmore BL, Burstein H, Peluso RW, Polisky B, Carter BJ, Davidson BL (2008) Artificial miRNAs mitigate shRNA-mediated toxicity in the brain: implications for the therapeutic development of RNAi. Proc Natl Acad Sci U S A 105 (15):5868–5873. https://doi.org/10.1073/pnas.0801775105
8. Miniarikova J, Zanella I, Huseinovic A, van der Zon T, Hanemaaijer E, Martier R, Koornneef A, Southwell AL, Hayden MR, van Deventer SJ, Petry H, Konstantinova P (2016) Design, characterization, and lead selection of therapeutic miRNAs targeting Huntingtin for development of gene therapy for Huntington's disease. Mol Ther Nucleic Acids 5:e297. https://doi.org/10.1038/mtna.2016.7
9. Miniarikova J, Zimmer V, Martier R, Brouwers CC, Pythoud C, Richetin K, Rey M, Lubelski J, Evers MM, van Deventer SJ, Petry H, Deglon N, Konstantinova P (2017) AAV5-miHTT gene therapy demonstrates suppression of mutant huntingtin aggregation and neuronal dysfunction in a rat model of Huntington's disease. Gene Ther 24(10):630–639. https://doi.org/10.1038/gt.2017.71
10. Foust KD, Salazar DL, Likhite S, Ferraiuolo L, Ditsworth D, Ilieva H, Meyer K, Schmelzer L, Braun L, Cleveland DW, Kaspar BK (2013) Therapeutic AAV9-mediated suppression of mutant SOD1 slows disease progression and extends survival in models of inherited ALS. Mol Ther 21(12):2148–2159. https://doi.org/10.1038/mt.2013.211
11. Pfister EL, Chase KO, Sun H, Kennington LA, Conroy F, Johnson E, Miller R, Borel F, Aronin N, Mueller C (2017) Safe and efficient silencing with a Pol II, but Not a Pol lII, promoter expressing an artificial miRNA targeting human Huntingtin. Mol Ther Nucleic Acids 7:324–334. https://doi.org/10.1016/j.omtn.2017.04.011
12. Pfister EL, DiNardo N, Mondo E, Borel F, Conroy F, Fraser C, Gernoux G, Han X, Hu D, Johnson E, Kennington L, Liu P, Reid SJ, Sapp E, Vodicka P, Kuchel T, Morton AJ, Howland D, Moser R, Sena-Esteves M, Gao G, Mueller C, DiFiglia M, Aronin N (2018) Artificial miRNAs reduce human mutant Huntingtin throughout the striatum in a transgenic sheep model of Huntington's disease. Hum Gene Ther 29(6):663–673. https://doi.org/10.1089/hum.2017.199
13. Borel F, Gernoux G, Cardozo B, Metterville JP, Toro Cabrera GC, Song L, Su Q, Gao GP, Elmallah MK, Brown RH Jr, Mueller C (2016)

Therapeutic rAAVrh10 mediated SOD1 silencing in adult SOD1(G93A) mice and nonhuman primates. Hum Gene Ther 27(1):19–31. https://doi.org/10.1089/hum.2015.122

14. Borel F, Tang Q, Gernoux G, Greer C, Wang Z, Barzel A, Kay MA, Shultz LD, Greiner DL, Flotte TR, Brehm MA, Mueller C (2017) Survival advantage of both human hepatocyte xenografts and genome-edited hepatocytes for treatment of alpha-1 antitrypsin deficiency. Mol Ther 25(11):2477–2489. https://doi.org/10.1016/j.ymthe.2017.09.020
15. Borel F, Gernoux G, Sun H, Stock R, Blackwood M, Brown RH Jr, Mueller C. (2018). Safe and effective superoxide dismutase 1 silencing using artificial microRNA in macaques. Sci Transl Med. 10(465). pii: eaau6414. https://doi.org/10.1126/scitranslmed.aau6414. PMID: 30381409
16. Borel F, Kay MA, Mueller C (2014) Recombinant AAV as a platform for translating the therapeutic potential of RNA interference. Mol Ther 22(4):692–701. https://doi.org/10.1038/mt.2013.285
17. Saini HK, Griffiths-Jones S, Enright AJ (2007) Genomic analysis of human microRNA transcripts. Proc Natl Acad Sci U S A 104 (45):17719–17724. https://doi.org/10.1073/pnas.0703890104
18. Grimm D (2011) The dose can make the poison: lessons learned from adverse in vivo toxicities caused by RNAi overexpression. Silence 2:8. https://doi.org/10.1186/1758-907X-2-8
19. Gao Z, Harwig A, Berkhout B, Herrera-Carrillo E (2017) Mutation of nucleotides around the +1 position of type 3 polymerase III promoters: the effect on transcriptional activity and start site usage. Transcription 8 (5):275–287. https://doi.org/10.1080/21541264.2017.1322170
20. Gray SJ, Foti SB, Schwartz JW, Bachaboina L, Taylor-Blake B, Coleman J, Ehlers MD, Zylka MJ, McCown TJ, Samulski RJ (2011) Optimizing promoters for recombinant adeno-associated virus-mediated gene expression in the peripheral and central nervous system using self-complementary vectors. Hum Gene Ther 22(9):1143–1153. https://doi.org/10.1089/hum.2010.245
21. Yan Z, Yan H, Ou H (2012) Human thyroxine binding globulin (TBG) promoter directs efficient and sustaining transgene expression in liver-specific pattern. Gene 506(2):289–294. https://doi.org/10.1016/j.gene.2012.07.009
22. Kugler S, Kilic E, Bahr M (2003) Human synapsin 1 gene promoter confers highly neuron-specific long-term transgene expression from an adenoviral vector in the adult rat brain depending on the transduced area. Gene Ther 10(4):337–347. https://doi.org/10.1038/sj.gt.3301905
23. Beltran WA, Boye SL, Boye SE, Chiodo VA, Lewin AS, Hauswirth WW, Aguirre GD (2010) rAAV2/5 gene-targeting to rods: dose-dependent efficiency and complications associated with different promoters. Gene Ther 17(9):1162–1174. https://doi.org/10.1038/gt.2010.56
24. Xie J, Mao Q, Tai PWL, He R, Ai J, Su Q, Zhu Y, Ma H, Li J, Gong S, Wang D, Gao Z, Li M, Zhong L, Zhou H, Gao G (2017) Short DNA hairpins compromise recombinant adeno-associated virus genome homogeneity. Mol Ther 25(6):1363–1374. https://doi.org/10.1016/j.ymthe.2017.03.028
25. Mueller C, Tang Q, Gruntman A, Blomenkamp K, Teckman J, Song L, Zamore PD, Flotte TR (2012) Sustained miRNA-mediated knockdown of mutant AAT with simultaneous augmentation of wild-type AAT has minimal effect on global liver miRNA profiles. Mol Ther 20(3):590–600. https://doi.org/10.1038/mt.2011.292
26. Fang W, Bartel DP (2015) The menu of features that define primary MicroRNAs and enable de novo design of MicroRNA genes. Mol Cell 60(1):131–145. https://doi.org/10.1016/j.molcel.2015.08.015

Chapter 2

Design of AAV Vectors for Delivery of Large or Multiple Transgenes

Aman Patel, Junling Zhao, Dongsheng Duan, and Yi Lai

Abstract

Adeno-associated virus (AAV)-mediated gene therapy has evolved from bench to bedside, and now is the therapy of choice for certain inherited diseases. However, the small packaging capacity of AAV vectors prevents this technique from treating genetic diseases with mutations of large genes. Multiple strategies, including split AAV gene delivery and oversized AAV gene delivery, have been explored to deliver large gene expression cassettes. These strategies have gained some success in animal experiments. In this chapter, we review the progress of AAV-mediated delivery of large expression cassettes. We also review using AAV to deliver multiple transgenes.

Key words AAV, Dual vectors, Triple vectors, Large gene expression cassette, Multiple gene expression cassette, Oversized AAV, Trans-splicing AAV vectors, Overlapping AAV vectors, Fragmented AAV gene delivery, Split AAV vectors

1 Introduction

Over half a century ago, adeno-associated virus (AAV) was identified as a naturally replication-defective parvovirus [1]. Due to its non-pathogenicity, low immunogenicity, and high efficiency in gene transfer, recombinant AAV vectors were engineered by replacing all viral coding sequences with a gene expression cassette of interest. With more than 30 years of development, AAV has become one of the most promising and widely used gene transfer tools [2]. Originally, AAV was implemented in gene replacement therapy to treat monogenic diseases. More recently, AAV has been expanded to new applications, including RNAi and gene editing, further extending the applicability of AAV vectors (*see* Chapter 1). As a tool for gene replacement, AAV has gained remarkable success in early clinical trials. In the European Union, the first AAV-based medicine (Glybera) was approved for clinical use to treat lipoprotein lipase deficiency in 2012. In the United States, the first AAV drug was approved by the Food and Drug Administration in 2018

Michael J. Castle (ed.), *Adeno-Associated Virus Vectors: Design and Delivery*, Methods in Molecular Biology, vol. 1950, https://doi.org/10.1007/978-1-4939-9139-6_2,

for the treatment of congenital blindness (inherited retinal dystrophy) [3].

Despite its increasing popularity, one limitation of AAV vectors impedes broad application of AAV gene therapy. AAV usually measures 20–25 nm in diameter. It can readily enter target cells, and efficiently transfers the gene expression cassette into the nuclei of cells to induce transgene expression. However, AAV has a relatively small packaging capacity, and cannot effectively package more than 5 kb of DNA. Obviously, the small packaging capacity of AAV poses a significant challenge to treating genetic diseases with mutations of large genes.

Multiple strategies have been explored to overcome the size limit of AAV vectors. One strategy is to split large transgenes into two or three parts, generating dual or triple AAV vectors that can deliver large gene expression cassettes. Another strategy is to harness factors that promote the expression of large transgenes from a single AAV vector. In this chapter, we will review the development of AAV vectors to deliver large genes. We also discuss co-delivery of two or three AAV vectors to transfer multiple gene expression cassettes.

2 Split AAV Vectors

Typically, a standard-sized AAV vector contains a complete gene expression cassette, including the promoter, the transgene coding sequence, polyadenylation (polyA) signal, and other cis-transcriptional signals. The gene expression cassette is flanked by an inverted terminal repeat (ITR) at both ends. To deliver a large gene expression cassette using split AAV vectors, including dual and triple AAV vectors, the large expression cassette is split into two or three parts, which are carried by two or three individual AAV vectors. After simultaneous delivery of dual or triple AAV vectors to target cells, the whole large gene expression cassette is reconstituted within the cells, and the complete large protein is expressed. Three different split AAV vectors have been explored with different reconstitution mechanisms: trans-splicing, overlapping, and hybrid AAV vectors [4].

2.1 Trans-splicing AAV Vectors

Trans-splicing AAV vectors take advantage of inherent AAV concatamerization by ITR-mediated intermolecular recombination. One AAV vector carries the promoter, the 5′ portion of the coding sequence, and a splicing donor signal, while the other AAV vector carries a splicing acceptor signal, the remaining coding sequence, and the polyA signal. If ITR-mediated concatamerization connects the two individual AAV vector genomes in head-to-tail orientation, the intact gene expression cassette will be restored when the splicing signals trigger removal of the intervening recombinant ITR sequence [5–10]. Earlier studies showed that the transduction efficiency of trans-splicing AAV vectors is much lower than that of a single AAV vector [11]. Since appropriate concatamerization orientation is critical for reconstitution of the expression cassette, strategies that manipulate the directionality of AAV concatamerization were tested to improve the efficiency of trans-splicing AAV vectors. Strategies such as the use of nonhomologous AAV2 and AAV5 ITRs, as well as engineering single-stranded DNA oligonucleotides, were used to direct concatamerization in the appropriate head-to-tail orientation. These strategies indeed significantly enhanced the efficiency of trans-splicing AAV vectors [12, 13]. Another study analyzed the limiting factors in trans-splicing AAV vectors and found that splicing efficiency and stability of pre-mRNA and mRNA transcripts are critical [14].

Building on these observations, a pair of trans-splicing AAV vectors has been generated to successfully deliver a 6 kb mini-dystrophin gene to the muscle of Duchenne muscular dystrophy (DMD) mice. Robust expression of mini-dystrophin improved muscle phenotype and muscle function in the mouse model [15]. This success has stimulated other investigators to explore trans-splicing AAV vectors for delivering other large therapeutic genes. To date, trans-splicing AAV vectors have been successfully used to deliver 8.4 kb von Willebrand factor for von Willebrand disease [16], 6.8 kb ATP-binding cassette 4 (ABCA4) for Stargardt disease [17, 18], 6.7 kb myosin VIIa (MYO7A) for Usher syndrome type 1B [18, 19], 6.2 kb dysferlin [20], and 4.5 kb B-domain-deleted human factor VIII for hemophilia A [21, 22] (Table 1). Before translating these successes to human patients, there is a need to test trans-splicing AAV vectors in large animal models. Trans-splicing AAV vectors have successfully expressed the large genes ABCA4 and MYO7A in the retinas of pigs [23]. Some genetic diseases, such as DMD, require systemic delivery to all affected tissues in the body. Two studies have addressed this issue and successfully expressed alkaline phosphatase (AP) and mini-dystrophin in mice by intravenous administration of trans-splicing AAV vectors [24, 25]. Nevertheless, it remains unclear whether systemic trans-splicing AAV vector delivery can work in large mammals.

2.2 Overlapping AAV Vectors

Overlapping AAV vectors exploit homologous recombination within the coding sequence to reconstitute a large gene expression cassette. One AAV vector contains the promoter and 5′ portion of the coding sequence, while the other AAV vector carries the 3′ portion of the coding sequence and polyA signal. The coding sequence in these two AAV vectors shares a homologous region. Homologous recombination of this overlapping region restores the large gene expression cassette and thereby induces the expression of large proteins in target cells. This overlapping approach successfully

Table 1
Summary of split AAV vectors for delivering large therapeutic genes

Dual/ Triple	Reconstitution mechanism	Transgene	Disease	References
Dual	Trans-splicing	Factor VIII	Hemophilia A	[21, 22]
Dual	Trans-splicing	EPO and EGFP	Renal failure-induced anemia	[7]
Dual	Trans-splicing	Dystrophin	DMD	[15]
Dual	Trans-splicing	Dysferlin	Dysferlinopathy	[20]
Dual	Trans-splicing	AP and dystrophin	DMD	[24, 25]
Dual	Trans-splicing	von Willebrand factor	von Willebrand disease	[16]
Dual	Overlapping	Dysferlin	Dysferlinopathy	[30–32]
Dual	Overlapping	GDE	Glycogen storage disease type III	[34]
Dual	Overlapping	Dystrophin	DMD	[28, 29, 35]
Dual	Overlapping and fragmented	MYO7A	Usher syndrome type 1B	[33]
Dual	Overlapping and hybrid	Dystrophin	DMD	[38]
Dual	Hybrid	CFTR	Cystic fibrosis	[39]
Dual	Hybrid	ABCA4	Stargardt's disease	[45]
Dual	Hybrid	LacZ and dystrophin	DMD	[37]
Dual	Trans-splicing, overlapping, and hybrid	ABCA4 and MYO7A	Stargardt's disease and Usher syndrome type 1B	[17, 18, 23, 40]
Dual	Trans-splicing, overlapping, hybrid, and fragmented	Dysferlin	Dysferlinopathy	[30]
Dual	Trans-splicing, overlapping, hybrid, and fragmented	MYO7A	Usher syndrome type 1B	[19]
Triple	Hybrid ITRs and Trans-splicing	Dystrophin	DMD	[41]
Triple	Trans-splicing and hybrid	CDH23 and ALMS1	Inherited retinal degenerations	[43]
Triple	Overlapping and hybrid	Dystrophin	DMD	[42]

ABCA4 ATP-binding cassette 4, *MYO7A* myosin VIIa, *CFTR* cystic fibrosis transmembrane conductance regulator, *GDE* glycogen-debranching enzyme, *AP* alkaline phosphatase, *LacZ* β-galactosidase, *CDH23* cadherin-related family member 23, *ALMS1* Alstrom syndrome 1, *EPO* erythropoietin

induces the expression of reporter genes such as the AP gene [26] and the β-galactosidase (LacZ) gene [11] in multiple cell types. The efficiency of overlapping AAV vectors heavily relies on homologous recombination of the overlapping region of the coding sequence. Hence, AP overlapping AAV vectors are more efficient than LacZ overlapping AAV vectors [27] because the AP gene contains a highly recombinogenic region. Overlapping AAV vectors have been convincingly used to deliver large therapeutic genes, including 6 kb mini-dystrophin [28, 29], 6.2 kb dysferlin [30–32], 6.8 kb ABCA4 [17, 18], 6.7 kb MYO7A [18, 19, 33], and 4.6 kb glycogen-debranching enzyme (GDE) for glycogen storage disease type III (GSDIII) [34] (Table 1). Recently, overlapping AAV vectors have been used to deliver the 7 kb mini-dystrophin gene to the muscle of a canine DMD model [35] (Table 1). Efficient expression of mini-dystrophin protein in canine dystrophic muscle paves the way for translating overlapping AAV vectors for clinical applications.

2.3 Hybrid AAV Vectors

Hybrid AAV vectors combine the mechanisms of both trans-splicing and overlapping AAV vectors, and are expected to have higher efficiency in delivering large gene expression cassettes. To drive homologous recombination, a highly recombinogenic region, such as a fragment from the AP gene or from the phage F1 genome (AK), serves as the homologous overlapping sequence. To incorporate the trans-splicing mechanism, a splicing donor signal is engineered at the 5′ end of the overlapping region in the 5′ AAV vector, while a splicing acceptor signal is included at the 3′ end of the overlapping region in the 3′ AAV vector. Homologous recombination either at the overlapping region or at the ITR can rejoin the two AAV vectors, and the pre-engineered splicing signal then removes the overlapping region and/or recombinant ITR structure to restore the large gene expression cassette. Hybrid AAV vectors can exploit highly recombinogenic sequences, such as AP [36] or AK [17], for homologous recombination, while overlapping vectors must rely on a homologous region in the coding sequence. This feature enables broad application of hybrid vectors to the delivery of large genes. Compared with trans-splicing vectors, homologous recombination of the overlapping region in hybrid vectors is more efficient and dramatically increases the correct directionality of dual vectors. Thus, hybrid AAV vectors are more efficient than trans-splicing AAV vectors [37]. Hybrid AAV vectors have been successfully used to transfer large therapeutic genes, including dystrophin [37, 38], CFTR [39], ABCA4 and MYO7A [17–19, 23, 40], and dysferlin [30] (Table 1).

2.4 Triple AAV Vectors

Dual AAV can expand the packaging capacity of AAV vectors up to 10 kb, but this is still insufficient for gene transfer of some giant genes, such as the 11 kb dystrophin coding sequence, the 10.1 kb

CDH23 gene and the 12.5 kb ALMS1 gene. Triple AAV vectors have been developed to transfer such giant expression cassettes. These vectors are designed to transfer up to 15 kb of DNA. To date, three proof-of-principle studies have shown the feasibility of delivering full-length giant genes with triple AAV vectors. Triple AAV vectors utilizing hybrid ITRs and trans-splicing successfully induced expression of the full-length 11-kb dystrophin gene in dystrophic muscle [41]. Another study showed that overlapping and hybrid triple AAV vectors can express full-length dystrophin in muscle [42]. In addition, using trans-splicing and hybrid triple AAV vectors, full-length CDH23 and ALMS1 have been expressed in the retina [43] (Table 1). Although the transduction efficiency of current triple AAV vectors may be too low to be therapeutic in many diseases, the ability of triple AAV vectors to deliver full-length giant genes still renders this strategy an attractive approach for large gene transfer.

2.5 Summary

Delivering large genes with split AAV vectors holds great promise for eventually treating many diseases. Some dual vectors have already shown immense translational potential, and can lead to efficient transgene expression via systemic delivery [24, 25] or in large animal models [23, 35]. To achieve safe and efficient gene expression in a clinical setting, several factors need to be considered. First, the efficiency of triple AAV vectors needs to be improved to meet therapeutic needs. Second, additional studies are needed to understand potential toxicity from expression of undesirable gene products such as those produced from the cryptic polyA signal in the 5′ AAV vector or from the cryptic transcription initiation site in the 3′ AAV vector [44]. One study has explored the inclusion of a CL1 degradation sequence after the splicing donor signal to abrogate the buildup of unwanted proteins from the 5′ AAV vector [45]. This strategy, together with prescreening of cryptic polyA and transcription start sites, should help prevent the accumulation of undesirable truncated proteins. Third, split AAV gene transfer requires co-delivery of dual or triple AAV vectors, which inevitably doubles or triples the vector dose. A recent study has shown that high dose AAV delivery can induce severe toxicity in large animal models, including nonhuman primates and piglets [46]. Hence, it is critical to further improve the transduction efficiency of split AAV vectors to achieve efficient transgene expression at a low dose.

3 Oversized AAV Gene Transfer

In addition to split AAV vectors, other strategies have been explored to deliver a large expression cassette using a single AAV vector. These oversized AAV vectors have been shown to result in detectable expression of large transgenes.

3.1 Fragmented Oversized AAV Gene Delivery

Initially, one study reported that an 8.9 kb oversized AAV genome was effectively packaged by the AAV5 capsid [47]. However, further studies from several groups demonstrated that an oversized AAV genome cannot be completely packaged [48–50]. During the process of packaging an oversized genome, packaging starts at the 3′ end of the AAV genome and stops when the maximal packaging capacity is reached or a termination signal is encountered. Thus, heterogeneous fragmented genomes with different sizes are produced. When AAVs containing mixed DNA fragments of oversized AAV genomes are delivered into target cells, homologous recombination of the fragmented genomes restores the whole expression cassette and can mediate the expression of large proteins [4, 51].

Despite the ability to induce large gene expression, the transduction efficiency of fragmented oversized AAV gene delivery is lower than intact AAV gene delivery. However, fragmented AAV gene delivery is more efficient than split AAV vectors in some cases. For example, using luciferase as the reporter, fragmented AAV gene delivery induced more robust expression in the muscle and the retina than split AAV gene delivery [52]. Another study comprehensively compared the transduction efficiency of split and fragmented AAV gene transfer in delivering the dysferlin gene, and found that overlapping vectors were the most efficient, followed by fragmented, hybrid, and trans-splicing [30]. However, in delivering another large gene, MYO7A, expression from overlapping AAV vectors was significantly lower than that of fragmented AAV gene delivery [33]. The transduction efficiency of fragmented AAV vectors largely depends on the homologous recombination efficiency of the coding sequence of the large gene. Hence, it is not surprising to see discrepancy in the transduction efficiency of overlapping and fragmented AAV vectors when different transgenes are delivered. Fragmented AAV gene delivery can mediate detectable large gene expression and thus provides an alternative and viable option to deliver large genes (Tables 1 and 2).

3.2 Other Strategies

Several additional strategies have been investigated to deliver an oversized AAV genome. These include capsid manipulation, improvement of virus production, administration of proteasome inhibitors, and optimization of transcriptional elements.

1. Manipulating capsids. One study has shown that deletion of the VP2 capsid protein can increase packaging capacity of AAV vectors to 6 kb [53]. VP2-null AAV capsids have been tested for packaging of CFTR and mini-dystrophin. Full-length CFTR and mini-dystrophin were completely packaged by VP2-null capsids, and the expression of these two genes was successfully induced, albeit at reduced efficiency [5, 53]. In another study, by adopting capsid proteins from human bocavirus virus-1 (HBoV1), which is an autonomous parvovirus

Table 2
Summary of oversized AAV gene delivery

Strategies	Transgene	Diseases	References
Capsid	CFTR	Cystic fibrosis	[53, 54]
Capsid	Dystrophin	DMD	[5]
Producer cell line	Factor VIII	Hemophilia A	[55]
Intron and promoter	GluN2A and GluN2B	Psychiatric disorders	[57]
Promoter and enhancers	CFTR	Cystic fibrosis	[58]
Proteasome inhibitors	Factor VIII	Hemophilia A	[56]
Fragmented	MYO7A	Usher syndrome 1B	[63]
Fragmented	Factor VIII	Hemophilia A	[64]
Fragmented	ABCA4, MYO7A, and CEP290	Stargardt's disease, Usher syndrome type 1B, and Leber congenital amaurosis	[47]
Fragmented	Dysferlin	Dysferlinopathy	[52, 65]

with a 5.5 kb genome, an oversized AAV.CFTR genome was successfully packaged, and the transduction efficiency of oversized AAV.CFTR in human-polarized airway epithelia was significantly improved [54].

2. Virus-producing cell lines. To improve the expression of oversized AAV.VIII, a producer cell line containing either a 5.1 or 5.4 kb AAV.VIII was established. Compared with traditional transient triple-plasmid transfection, oversized AAV.VIII vectors generated by the producer cell line resulted in twofold higher plasma factor VIII activity [55].
3. Administration of proteasome inhibitors. One study suggests that administration of a proteasome inhibitor at the time of oversized AAV.VIII delivery enhanced factor VIII expression by three- to sixfold [56].
4. Optimizing transcriptional elements. Inclusion of an intron in the transgene, optimizing promoters, and addition of an enhancer improved the expression of oversized AAV.GluN2 and AAV.CFTR [57, 58] (Table 2).

3.3 Summary

Oversized AAV gene delivery, assisted by fragmented AAV gene delivery or other strategies, has gained some success in delivering large transgenes. These methods offer feasible ways to transfer oversized AAV vectors.

4 AAV-Mediated Delivery of Multiple Transgenes

Although AAV has a small packaging capacity, multiple transgene delivery can still be achieved with AAV vectors. Depending on the size of the expression cassette, multiple transgenes can be delivered by one, two, or three AAV vectors.

4.1 Single AAV Vectors

If the size of two transgenes is within the 5 kb packaging limit, the coding sequence of these two transgenes can be engineered into a single AAV vector. The expression of these two genes can be mediated by two independent promoters in two different expression cassettes, or by a single promoter in one expression cassette, with two transgenes separated by an internal ribosome entry site (IRES) or a P2A sequence [59].

4.2 Co-delivery of Two AAV Vectors

If the size of two transgenes is greater than the 5 kb AAV packaging capacity, two AAV vectors carrying two different intact gene expression cassettes can be used. This requires that multiple AAV vectors carrying different gene expression cassettes enter the same cell and mediate the expression of two or three transgenes. Independent studies have confirmed that two vectors carrying two different reporter genes can mediate the expression of these two proteins in the same cells of muscle [14], retina, and hippocampus [60]. Building on this finding, co-delivery of two AAV vectors has been used to express two different therapeutic genes in various disease models (Table 3).

4.3 Co-delivery of Three AAV Vectors

Patients with Parkinson's disease may benefit from co-expression of three therapeutic proteins: tyrosine hydroxylase (TH), aromatic-L-amino-acid decarboxylase (AADC), and GTP cyclohydrolase I (GCH). Simultaneous delivery of three AAV vectors containing TH, AADC, and GCH gene expression cassettes led to the expression of these three different proteins in neurons and improved the disease phenotype in mouse models [61, 62].

Recently, clustered regularly interspaced short palindromic repeat (CRISPR)-Cas9 has gained popularity in gene editing due to its simplicity and versatility. The CRISPR-Cas9 system is composed of endonuclease Cas9 and single guide RNAs (sgRNAs). Because AAV is one of the most widely used gene transfer tools, AAV-mediated gene transfer has been extensively used in CRISPR-Cas9 gene editing. For most Cas9 genes (such as the SpCas9 gene), the expression of both Cas9 and sgRNAs is beyond the packaging capacity of a single AAV vector. Hence, two AAV vectors are commonly used: one vector is responsible for the expression of Cas9 and the other contains sgRNA expression cassettes and other DNA sequences for reporter genes or donor templates for homology-directed repair. To date, the CRISPR-Cas9 system

Table 3
Summary of multiple gene delivery by two or three AAV vectors

Vectors	Transgenes	Diseases	References
Two	GFP and FGF14	Cerebellar Purkinje neurons	[59]
Two	IL-10 and STAT3	Atherosclerosis	[66]
Two	Liver-specific and ubiquitous expression of GAA	Pompe disease	[67]
Two	BMP2 and rtTA	Bone regeneration and repair	[68]
Two	SOX9 and OP1	Intervertebral disc degeneration	[69]
Two	Rhodopsin suppressor and replacement genes	Dominant retinitis pigmentosa	[70]
Two	Liver and muscle-specific expression of GAA	Pompe disease	[71]
Two	SaCas9, sgRNA, and OTC donor template	OTC deficiency	[72]
Two	SpCas9, sgRNAs, and EGFP	Pathological angiogenesis	[73]
Two	SpCas9, sgRNAs, and mCherry	DMD	[74]
Two	SpCas9 and sgRNAs	Leber congenital amaurosis 10	[75]
Two	SpCas9, sgRNAs, and TdTomato	Retinitis pigmentosa	[76]
Two	SpCas9, sgRNAs, and EGFP	Huntington disease	[77]
Two	SaCas9 and sgRNAs	DMD	[78, 79, 81]
Two	SpCas9, sgRNAs, and GFP	DMD	[80]
Three	TH, AADC, and GCH	Parkinson's disease	[61]
Three	TH, AADC, and GCH	Parkinson's disease	[62]

FGF fibroblast growth factor, *IL-10* interleukin 10, *STAT3* signal transducer and activator of transcription 3, *GAA* acid alpha-glucosidase, *rtTA* Tet-responsive transactivator, *BMP2* bone morphogenetic protein-2, *OP1* osteogenic protein-1, *OTC* ornithine transcarbamylase, *DsR* DsRed-Express2, *SaCas9* Staphylococcus aureus Cas9, *SpCas9* Streptococcus pyogenes Cas9, *sgRNA* single guide RNA, *TH* tyrosine hydroxylase, *AADC* aromatic-L-amino-acid decarboxylase, *GCH* GTP cyclohydrolase I

delivered with two AAV vectors has been successfully used to correct gene mutations in multiple diseases (Table 3).

5 Conclusion

AAV gene therapy has become the leading clinical candidate for several diseases. With increasing success in clinical trials, AAV gene therapy will be expanded to more diseases. Although the small packaging capacity of AAV vectors limits their applicability, exploration of multiple strategies, including split AAV gene delivery and oversized AAV gene delivery, may provide a solution to this issue. Some strategies to deliver large genes have gained remarkable progress, such as split AAV vectors in large animal models and systemic delivery. With our continued effort in developing and fine-tuning AAV vectors, there is no doubt that AAV gene therapy will be useful for treating many more diseases.

Acknowledgments

This work was supported by grants from Jackson Freel DMD Research Fund (D.D.), National Institutes of Health (AR-69085) (D.D.), Department of Defense (MD150133) (D.D.), Hope for Javier (D.D.), and Duchenne Parent Project (the Netherlands) (Y.L.).

Disclosure: D.D. is a member of the scientific advisory board for Solid Biosciences and an equity holder of Solid Biosciences. D.D. has received research supports from Solid Biosciences, which are unrelated to this book chapter.

References

1. Berns KI (1974) Molecular biology of the adeno-associated viruses. Curr Top Microbiol Immunol 65:1–20
2. Berns KI, Muzyczka N (2017) AAV: an overview of unanswered questions. Hum Gene Ther 28(4):308–313
3. Dunbar CE, High KA, Joung JK, Kohn DB, Ozawa K, Sadelain M (2018) Gene therapy comes of age. Science 359(6372). https://doi.org/10.1126/science.aan4672
4. Hirsch ML, Wolf SJ, Samulski RJ (2016) Delivering transgenic DNA exceeding the carrying capacity of AAV vectors. Methods Mol Biol 1382:21–39
5. Lai Y, Yue Y, Bostick B, Duan D (2010) Delivering Large Therapeutic Genes for Muscle Gene Therapy. In: Duan D. (eds) Muscle Gene Therapy. Springer, New York, NY, pp. 205–218
6. Duan D, Yue Y, Yan Z, Engelhardt JF (2000) A new dual-vector approach to enhance recombinant adeno-associated virus-mediated gene expression through intermolecular cis activation. Nat Med 6(5):595–598
7. Yan Z, Zhang Y, Duan D, Engelhardt JF (2000) Trans-splicing vectors expand the utility of adeno-associated virus for gene therapy. Proc Natl Acad Sci U S A 97(12):6716–6721
8. Sun L, Li J, Xiao X (2000) Overcoming adeno-associated virus vector size limitation through viral DNA heterodimerization. Nat Med 6 (5):599–602
9. Nakai H, Storm TA, Kay MA (2000) Increasing the size of rAAV-mediated expression cassettes in vivo by intermolecular joining of two complementary vectors. Nat Biotechnol 18 (5):527–532
10. Reich SJ, Auricchio A, Hildinger M, Glover E, Maguire AM, Wilson JM, Bennett J (2003) Efficient trans-splicing in the retina expands the utility of adeno-associated virus as a vector for gene therapy. Hum Gene Ther 14 (1):37–44
11. Duan D, Yue Y, Engelhardt JF (2001) Expanding AAV packaging capacity with trans-splicing or overlapping vectors: a quantitative comparison. Mol Ther 4(4):383–391
12. Yan Z, Lei-Butters DC, Zhang Y, Zak R, Engelhardt JF (2007) Hybrid adeno-associated virus bearing nonhomologous inverted terminal repeats enhances dual-vector reconstruction of minigenes in vivo. Hum Gene Ther 18 (1):81–87
13. Hirsch ML, Storici F, Li C, Choi VW, Samulski RJ (2009) AAV recombineering with single strand oligonucleotides. PLoS One 4(11): e7705
14. Xu Z, Yue Y, Lai Y, Ye C, Qiu J, Pintel DJ, Duan D (2004) Trans-splicing adeno-associated viral vector-mediated gene therapy is limited by the accumulation of spliced mRNA but not by dual vector coinfection efficiency. Hum Gene Ther 15(9):896–905

15. Lai Y, Yue Y, Liu M, Ghosh A, Engelhardt JF, Chamberlain JS, Duan D (2005) Efficient in vivo gene expression by trans-splicing adeno-associated viral vectors. Nat Biotechnol 23(11):1435–1439
16. Wang L, Rosenberg JB, De BP, Ferris B, Wang R, Rivella S, Kaminsky SM, Crystal RG (2012) In vivo gene transfer strategies to achieve partial correction of von Willebrand disease. Hum Gene Ther 23(6):576–588
17. Trapani I (2018) Dual AAV vectors for Stargardt disease. Methods Mol Biol 1715:153–175
18. Trapani I, Colella P, Sommella A, Iodice C, Cesi G, De Simone S, Marrocco E, Rossi S, Giunti M, Palfi A, Jane Farrar G, Polishchuk R, Auricchio A (2014) Effective delivery of large genes to the retina by dual AAV vectors. EMBO Mol Med 6:194–211
19. Dyka FM, Boye SL, Chiodo V, Hauswirth W, Boye SE (2014) Dual AAV vectors result in efficient in vitro and in vivo expression of an oversized gene, MYO7A. Hum Gene Ther Methods 25:166–177
20. Lostal W, Bartoli M, Bourg N, Roudaut C, Bentaib A, Miyake K, Guerchet N, Fougerousse F, McNeil P, Richard I (2010) Efficient recovery of dysferlin deficiency by dual adeno-associated vector-mediated gene transfer. Hum Mol Genet 19(10):1897–1907
21. Chao H, Sun L, Bruce A, Xiao X, Walsh CE (2002) Expression of human factor VIII by splicing between dimerized AAV vectors. Mol Ther 5(6):716–722
22. Zhu F, Liu Z, Miao J, Qu H, Chi X (2012) Enhanced plasma factor VIII activity in mice via cysteine mutation using dual vectors. Sci China Life Sci 55(6):521–526
23. Colella P, Trapani I, Cesi G, Sommella A, Manfredi A, Puppo A, Iodice C, Rossi S, Simonelli F, Giunti M, Bacci ML, Auricchio A (2014) Efficient gene delivery to the cone-enriched pig retina by dual AAV vectors. Gene Ther 21(4):450–456
24. Ghosh A, Yue Y, Long C, Bostick B, Duan D (2007) Efficient whole-body transduction with trans-splicing adeno-associated viral vectors. Mol Ther 15(4):750–755
25. Ghosh A, Yue Y, Shin JH, Duan D (2009) Systemic Trans-splicing adeno-associated viral delivery efficiently transduces the heart of adult mdx mouse, a model for duchenne muscular dystrophy. Hum Gene Ther 20 (11):1319–1328
26. Halbert CL, Allen JM, Miller AD (2002) Efficient mouse airway transduction following recombination between AAV vectors carrying parts of a larger gene. Nat Biotechnol 20 (7):697–701
27. Ghosh A, Yue Y, Duan D (2006) Viral serotype and the transgene sequence influence overlapping adeno-associated viral (AAV) vector-mediated gene transfer in skeletal muscle. J Gene Med 8(3):298–305
28. Odom GL, Gregorevic P, Allen JM, Chamberlain JS (2011) Gene therapy of mdx mice with large truncated dystrophins generated by recombination using rAAV6. Mol Ther 19 (1):36–45
29. Zhang Y, Yue Y, Li L, Hakim CH, Zhang K, Thomas GD, Duan D (2013) Dual AAV therapy ameliorates exercise-induced muscle injury and functional ischemia in murine models of Duchenne muscular dystrophy. Hum Mol Genet 22:3720–3729
30. Pryadkina M, Lostal W, Bourg N, Charton K, Roudaut C, Hirsch ML, Richard I (2015) A comparison of AAV strategies distinguishes overlapping vectors for efficient systemic delivery of the 6.2 kb Dysferlin coding sequence. Mol Ther Methods Clin Dev 2:15009
31. Potter RA, Griffin DA, Sondergaard PC, Johnson RW, Pozsgai ER, Heller KN, Peterson EL, Lehtimäki KK, Windish HP, Mittal PJ, Albrecht DE, Mendell JR, Rodino-Klapac LR (2017) Systemic delivery of dysferlin overlap vectors provides long-term gene expression and functional improvement for dysferlinopathy. Hum Gene Ther. https://doi.org/10.1089/hum.2017.062
32. Sondergaard PC, Griffin DA, Pozsgai ER, Johnson RW, Grose WE, Heller KN, Shontz KM, Montgomery CL, Liu J, Clark KR, Sahenk Z, Mendell JR, Rodino-Klapac LR (2015) AAV.Dysferlin overlap vectors restore function in dysferlinopathy animal models. Ann Clin Transl Neurol 2(3):256–270
33. Lopes VS, Boye SE, Louie CM, Boye S, Dyka F, Chiodo V, Fofo H, Hauswirth WW, Williams DS (2013) Retinal gene therapy with a large MYO7A cDNA using adeno-associated virus. Gene Ther 20(8):824–833
34. Vidal P, Pagliarani S, Colella P, Costa Verdera H, Jauze L, Gjorgjieva M, Puzzo F, Marmier S, Collaud F, Simon Sola M, Charles S, Lucchiari S, van Wittenberghe L, Vignaud A, Gjata B, Richard I, Laforet P, Malfatti E, Mithieux G, Rajas F, Comi GP, Ronzitti G, Mingozzi F (2018) Rescue of GSDIII phenotype with gene transfer requires liver- and muscle-targeted GDE expression. Mol Ther 26(3):890–901

35. Kodippili K, Hakim C, Pan X, Yang HT, Yue Y, Zhang Y, Shin JH, Yang NN, Duan D (2018) Dual AAV gene therapy for Duchenne muscular dystrophy with a 7-kb mini-dystrophin gene in the canine model. Hum Gene Ther 29 (3):299–311
36. Ghosh A, Yue Y, Duan D (2011) Efficient transgene reconstitution with hybrid dual AAV vectors carrying the minimized bridging sequences. Hum Gene Ther 22(1):77–83
37. Ghosh A, Yue Y, Lai Y, Duan D (2008) A hybrid vector system expands adeno-associated viral vector packaging capacity in a transgene-independent manner. Mol Ther 16 (1):124–130
38. Zhang Y, Dongsheng D (2012) Novel mini-dystrophin gene dual AAV vectors restore nNOS expression at the sarcolemma. Hum Gene Ther 23(1):98–103
39. Song Y, Lou H, Boyer J, Limberis M, Vandenberghe L, Hackett N, Leopold P, Wilson J, Crystal R (2009) Functional CFTR expression in cystic fibrosis airway epithelial cells by AAV6.2-mediated segmental trans-splicing. Hum Gene Ther 20(3):267–281
40. Carvalho LS, Turunen HT, Wassmer SJ, Luna-Velez MV, Xiao R, Bennett J, Vandenberghe LH (2017) Evaluating efficiencies of dual AAV approaches for retinal targeting. Front Neurosci 11:503
41. Koo T, Popplewell LJ, Athanasopoulos T, Dickson G (2014) Triple trans-splicing AAV vectors capable of transferring the coding sequence for full-length dystrophin protein into dystrophic mice. Hum Gene Ther 25 (2):98–108
42. Lostal W, Kodippili K, Yue Y, Duan D (2014) Full-length dystrophin reconstitution with adeno-associated viral vectors. Hum Gene Ther 25(6):552–562
43. Maddalena A, Tornabene P, Tiberi P, Minopoli R, Manfredi A, Mutarelli M, Rossi S, Simonelli F, Naggert JK, Cacchiarelli D, Auricchio A (2018) Triple vectors expand AAV transfer capacity in the retina. Mol Ther 26:524–541
44. McClements ME, MacLaren RE (2017) Adeno-associated virus (AAV) dual vector strategies for gene therapy encoding large transgenes. Yale J Biol Med 90(4):611–623
45. Trapani I, Toriello E, de Simone S, Colella P, Iodice C, Polishchuk EV, Sommella A, Colecchi L, Rossi S, Simonelli F, Giunti M, Bacci ML, Polishchuk RS, Auricchio A (2015) Improved dual AAV vectors with reduced expression of truncated proteins are safe and effective in the retina of a mouse model of Stargardt disease. Hum Mol Genet 24 (23):6811–6825
46. Hinderer C, Katz N, Buza EL, Dyer C, Goode T, Bell P, Richman L, Wilson JM (2018) Severe toxicity in nonhuman primates and piglets following high-dose intravenous administration of an AAV vector expressing human SMN. Hum Gene Ther 29 (3):285–298. https://doi.org/10.1089/hum.2018.015
47. Allocca M, Doria M, Petrillo M, Colella P, Garcia-Hoyos M, Gibbs D, Kim SR, Maguire A, Rex TS, Di Vicino U, Cutillo L, Sparrow JR, Williams DS, Bennett J, Auricchio A (2008) Serotype-dependent packaging of large genes in adeno-associated viral vectors results in effective gene delivery in mice. J Clin Invest 118(5):1955–1964
48. Wu Z, Yang H, Colosi P (2010) Effect of genome size on AAV vector packaging. Mol Ther 18(1):80–86
49. Dong B, Nakai H, Xiao W (2010) Characterization of genome integrity for oversized recombinant AAV vector. Mol Ther 18 (1):87–92
50. Lai Y, Yue Y, Duan D (2010) Evidence for the failure of adeno-associated virus serotype 5 to package a viral genome > or = 8.2 kb. Mol Ther 18(1):75–79
51. Hirsch ML, Agbandje-McKenna M, Samulski RJ (2010) Little vector, big gene transduction: fragmented genome reassembly of adeno-associated virus. Mol Ther 18(1):6–8
52. Hirsch ML, Li C, Bellon I, Yin C, Chavala S, Pryadkina M, Richard I, Samulski RJ (2013) Oversized AAV transduction is mediated via a DNA-PKcs independent, Rad51C-dependent repair pathway. Mol Ther 21(12):2205–2216
53. Grieger JC, Samulski RJ (2005) Packaging capacity of adeno-associated virus serotypes: impact of larger genomes on infectivity and postentry steps. J Virol 79(15):9933–9944
54. Yan Z, Keiser NW, Song Y, Deng X, Cheng F, Qiu J, Engelhardt JF (2013) A novel chimeric adenoassociated virus 2/human bocavirus 1 parvovirus vector efficiently transduces human airway epithelia. Mol Ther 21 (12):2181–2194
55. Nambiar B, Cornell Sookdeo C, Berthelette P, Jackson R, Piraino S, Burnham B, Nass S, Souza D, O'Riordan CR, Vincent KA, Cheng SH, Armentano D, Kyostio-Moore S (2017) Characteristics of minimally oversized adeno-associated virus vectors encoding human factor VIII generated using producer cell lines and

triple transfection. Hum Gene Ther Methods 28(1):23–38

56. Monahan PE, Lothrop CD, Sun J, Hirsch ML, Kafri T, Kantor B, Sarkar R, Tillson DM, Elia JR, Samulski RJ (2010) Proteasome inhibitors enhance gene delivery by AAV virus vectors expressing large genomes in hemophilia mouse and dog models: a strategy for broad clinical application. Mol Ther 18 (11):1907–1916
57. Holehonnur R, Lella SK, Ho A, Luong JA, Ploski JE (2015) The production of viral vectors designed to express large and difficult to express transgenes within neurons. Mol Brain 8:12
58. Yan Z, Sun X, Feng Z, Li G, Fisher JT, Stewart ZA, Engelhardt JF (2015) Optimization of rAAV-mediated expression for large transgenes using a synthetic promoter and tandem array enhancers. Hum Gene Ther 26(6):334–346
59. Bosch MK, Nerbonne JM, Ornitz DM (2014) Dual transgene expression in murine cerebellar Purkinje neurons by viral transduction in vivo. PLoS One 9(8):e104062
60. Palfi A, Chadderton N, McKee AG, Blanco-Fernandez A, Humphries P, Kenna PF, Farrar GJ (2012) Efficacy of co-delivery of dual AAV2/5 vectors in the murine retina and hippocampus. Hum Gene Ther 23(8):847–858
61. Shen Y, Muramatsu SI, Ikeguchi K, Fujimoto KI, Fan DS, Ogawa M, Mizukami H, Urabe M, Kume A, Nagatsu I, Urano F, Suzuki T, Ichinose H, Nagatsu T, Monahan J, Nakano I, Ozawa K (2000) Triple transduction with adeno-associated virus vectors expressing tyrosine hydroxylase, aromatic-L-amino-acid decarboxylase, and GTP cyclohydrolase I for gene therapy of Parkinson's disease. Hum Gene Ther 11:1509–1519
62. Muramatsu S, Fujimoto K, Ikeguchi K, Shizuma N, Kawasaki K, Ono F, Shen Y, Wang L, Mizukami H, Kume A, Matsumura M, Nagatsu I, Urano F, Ichinose H, Nagatsu T, Terao K, Nakano I, Ozawa K (2002) Behavioral recovery in a primate model of Parkinson's disease by triple transduction of striatal cells with adeno-associated viral vectors expressing dopamine-synthesizing enzymes. Hum Gene Ther 13 (3):345–354
63. Colella P, Sommella A, Marrocco E, Di Vicino U, Polishchuk E, Garcia Garrido M, Seeliger MW, Polishchuk R, Auricchio A (2013) Myosin7a deficiency results in reduced retinal activity which is improved by gene therapy. PLoS One 8:e72027
64. Lu H, Chen L, Wang J, Huack B, Sarkar R, Zhou S, Xu R, Ding Q, Wang X, Wang H, Xiao W (2008) Complete correction of hemophilia A with adeno-associated viral vectors containing a full-size expression cassette. Hum Gene Ther 19(6):648–654
65. Grose WE, Clark KR, Griffin D, Malik V, Shontz KM, Montgomery CL, Lewis S, Brown RHJ, Janssen PM, Mendell JR, Rodino-Klapac LR (2012) Homologous recombination mediates functional recovery of dysferlin deficiency following AAV5 gene transfer. PLoS One 7(6):e39233
66. Cao M, Khan JA, Kang BY, Mehta JL, Hermonat PL (2012) Dual AAV/IL-10 plus STAT3 anti-inflammatory gene delivery lowers atherosclerosis in LDLR KO mice, but without increased benefit. Int J Vasc Med 2012:524235
67. Zhang P, Sun B, Osada T, Rodriguiz RM, Yang XY, Luo X, Kemper AR, Clay TM, Koeberl D (2012) Immunodominant, liver-specific expression suppresses transgene-directed immune responses in murine Pompe disease. Hum Gene Ther 23(5):460–472
68. Gafni Y, Pelled G, Zilberman Y, Turgeman G, Apparailly F, Yotvat H, Galun E, Gazit Z, Jorgensen C, Gazit D (2004) Gene therapy platform for bone regeneration using an exogenously regulated, AAV-2-based gene expression system. Mol Ther 9(4):587–595
69. Ren S, Liu Y, Ma J, Liu Y, Diao Z, Yang D, Zhang X, Xi Y, Hu Y (2013) Treatment of rabbit intervertebral disc degeneration with co-transfection by adeno-associated virus-mediated SOX9 and osteogenic protein-1 double genes in vivo. Int J Mol Med 32 (5):1063–1068
70. Millington-Ward S, Chadderton N, O'Reilly M, Palfi A, Goldmann T, Kilty C, Humphries M, Wolfrum U, Bennett J, Humphries P, Kenna PF, Farrar GJ (2011) Suppression and replacement gene therapy for autosomal dominant disease in a murine model of dominant retinitis pigmentosa. Mol Ther 19 (4):642–649
71. Doerfler PA, Todd AG, Clément N, Falk DJ, Nayak S, Herzog RW, Byrne BJ (2016) Copackaged AAV9 vectors promote simultaneous immune tolerance and phenotypic correction of Pompe disease. Hum Gene Ther 27 (1):43–59
72. Yang Y, Wang L, Bell P, McMenamin D, He Z, White J, Yu H, Xu C, Morizono H, Musunuru K, Batshaw ML, Wilson JM (2016) A dual AAV system enables the Cas9-mediated correction of a metabolic liver disease

in newborn mice. Nat Biotechnol 34 (3):334–338

73. Wu W, Duan Y, Ma G, Zhou G, Park-Windhol-C, D'Amore PA, Lei H (2017) AAV-CRISPR/Cas9-mediated depletion of VEGFR2 blocks angiogenesis in vitro. Invest Ophthalmol Vis Sci 58(14):6082–6090
74. Bengtsson NE, Hall JK, Odom GL, Phelps MP, Andrus CR, Hawkins RD, Hauschka SD, Chamberlain JR, Chamberlain JS (2017) Muscle-specific CRISPR/Cas9 dystrophin gene editing ameliorates pathophysiology in a mouse model for Duchenne muscular dystrophy. Nat Commun 8:14454
75. Ruan GX, Barry E, Yu D, Lukason M, Cheng SH, Scaria A (2017) CRISPR/Cas9-mediated genome editing as a therapeutic approach for Leber congenital Amaurosis 10. Mol Ther 25 (2):331–341
76. Yu W, Mookherjee S, Chaitankar V, Hiriyanna S, Kim JW, Brooks M, Ataeijannati Y, Sun X, Dong L, Li T, Swaroop A, Wu Z (2017) Nrl knockdown by AAV-delivered CRISPR/Cas9 prevents retinal degeneration in mice. Nat Commun 8:14716
77. Monteys AM, Ebanks SA, Keiser MS, Davidson BL (2017) CRISPR/Cas9 editing of the mutant Huntingtin allele in vitro and in vivo. Mol Ther 25(1):12–23
78. Tabebordbar M, Zhu K, Cheng JK, Chew WL, Widrick JJ, Yan WX, Maesner C, Wu EY, Xiao R, Ran FA, Cong L, Zhang F, Vandenberghe LH, Church GM, Wagers AJ (2016) In vivo gene editing in dystrophic mouse muscle and muscle stem cells. Science 351 (6271):407–411
79. Nelson CE, Hakim CH, Ousterout DG, Thakore PI, Moreb EA, Rivera RM, Madhavan S, Pan X, Ran FA, Yan WX, Asokan A, Zhang F, Duan D, Gersbach CA (2016) In vivo genome editing improves muscle function in a mouse model of Duchenne muscular dystrophy. Science 351(6271):403–407
80. Long C, Amoasii L, Mireault AA, McAnally JR, Li H, Sanchez-Ortiz E, Bhattacharyya S, Shelton JM, Bassel-Duby R, Olson EN (2016) Postnatal genome editing partially restores dystrophin expression in a mouse model of muscular dystrophy. Science 351(6271):400–403
81. Hakim CH, Wasala NB, Nelson CE, Wasala LP, Yue Y, Louderman JA, Lessa TB, Dai A, Zhang K, Jenkins GJ, Nance ME, Pan X, Kodippili K, Yang NN, Chen SJ, Gersbach CA, Duan D (2018) AAV CRISPR editing rescues cardiac and muscle function for 18 months in dystrophic mice. JCI Insight 3 (23) http://doi.org/10.1172/jci.insight.124297

Chapter 3

Ligand Coupling to the AAV Capsid for Cell-Specific Gene Transfer

Johanna Reul, Alexander Muik, and Christian J. Buchholz

Abstract

Cell entry of AAV vectors is initiated by contacting the cell surface attachment receptor. This process can be rationally engineered through mutating the contact residues on the AAV capsid and covalently coupling targeting ligands to the capsid surface that exhibit high affinity for a cell surface protein of choice. This way, selective gene delivery to target-receptor positive cell types has been achieved. Two methods for coupling targeting ligands to the AAV capsid can be distinguished. Genetic coupling is achieved through expressing fusion proteins composed of the capsid protein VP2 and the targeting ligand in packaging cells. Biochemical coupling involves split-intein-mediated protein *trans*-splicing between the mutated AAV capsid and the targeting ligand. While genetic coupling is restricted to designed ankyrin repeat proteins as targeting ligand, biochemical coupling tolerates single-chain antibody fragments as well.

Key words Receptor targeting, Split intein, DARPin, Protein *trans*-splicing

1 Introduction

Restricting gene delivery to the cell type relevant for a particular application in therapy or basic research is among the prime goals in vector development. While vector-encoded genetic elements can restrict transgene expression to particular cell types, only modifications on the capsid surface can influence the biodistribution of AAV vector particles. Altering biodistribution and receptor usage can be achieved by directed evolution approaches with libraries covering large repertoires of AAV variants [1, 2]. Alternatively, a rational approach can be taken. Here, a targeting ligand exhibiting high affinity for a cell surface protein of choice is displayed on the vector particle surface to enhance attachment of the particle to the cell type of interest. Further specificity in gene delivery can be achieved by mutating the contact residues of the natural attachment receptor, as has been successfully demonstrated for the heparan-sulfate proteoglycan (HSPG) binding sites in AAV2 [3].

Michael J. Castle (ed.), *Adeno-Associated Virus Vectors: Design and Delivery*, Methods in Molecular Biology, vol. 1950, https://doi.org/10.1007/978-1-4939-9139-6_3, © Springer Science+Business Media, LLC, part of Springer Nature 2019

In order to couple the targeting ligand with the particle surface different strategies can be distinguished. Non-covalent coupling can be achieved through adaptors such as bispecific antibodies bridging the capsid surface with the target receptor [4]. A more stable coupling, resistant against competition with serum components and thus better compatible with in vivo gene delivery, is mediated by covalent coupling of the targeting ligand. This is usually achieved by the genetic approach, i.e., expressing a fusion protein composed of a capsid subunit, such as VP2, and the targeting ligand in the AAV packaging cells. Initially, small peptides have been incorporated into the capsid this way [5]. More recently, we demonstrated that designed ankyrin repeat proteins (DARPins) also become incorporated into the AAV2 capsid when expressed as an N-terminal fusion with VP2 in packaging cells [3]. Acting as alternatives to antibodies, DARPins can be selected from libraries as high-affinity ligands for basically any protein of interest [6]. A detailed protocol for the ribosomal display-based, completely cell-free, selection process has been published [7]. Combining the insertion of a receptor-specific DARPin together with mutations abolishing natural receptor binding resulted in AAV particles that enabled efficient receptor-specific transduction in vitro and in vivo. This system is flexible since the incorporation of different DARPins, specific for the cell surface receptors Her2/*neu*, CD4 or EpCAM, allowed selective gene delivery into the respective target receptor-positive cell types [8]. Selective gene delivery was also achieved when the target cell type was highly underrepresented or the vector particles were intravenously administered in different mouse models.

Since AAV particles are assembled in the nucleus, targeting ligands requiring oxidizing conditions to form disulfide bonds, e.g., single-chain variable fragments (scFv) cannot be used in the genetic coupling approach. Therefore, covalent coupling of targeting ligands to fully assembled AAV capsids via split-intein-mediated protein *trans*-splicing (PTS) was established [9]. This strategy enables the coupling of not only DARPins but also scFvs, thus broadening the repertoire of targeting ligands for the generation of receptor-targeted AAV vectors. The split-intein system is based on the DNA polymerase III (DnaE) of *Nostoc punctiforme* (Npu) [10]. For the covalent coupling of the targeting ligand to the AAV capsid, the C- and the N-terminal fragments of the Npu intein are fused to the N-terminus of VP2 and the desired targeting ligand, respectively. When both intein domains are brought into close contact, an autocatalytic protein *trans*-splice reaction takes place, in which the two split-intein fused polypeptides (i.e., the targeting ligand and VP2) are covalently linked by the formation of a peptide bond under physiological conditions, while the split-intein sequences excise themselves from the protein precursors.

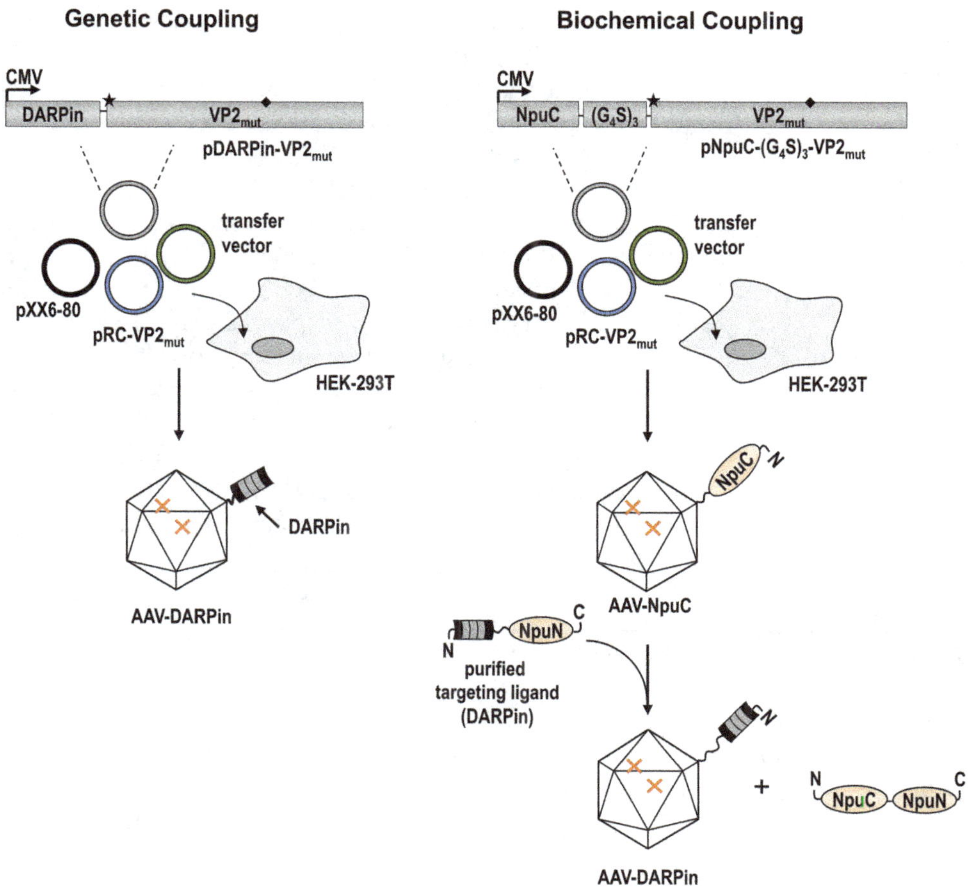

Fig. 1 Genetic and biochemical coupling of targeting ligands to the AAV capsid. For the production of cell-type-specific AAV vectors, an adenoviral helper plasmid (pXX6-80), an AAV helper plasmid (pRC-$VP2_{mut}$), a transfer vector plasmid, and a plasmid encoding VP2 either fused to a DARPin (genetic coupling) or to the NpuC intein (biochemical coupling) are co-transfected into HEK-293T producer cell lines. The black star and diamond shown in the pDARPin-$VP2_{mut}$ and pNpuC-$(G_4S)_3$-$VP2_{mut}$ constructs indicate the deletion of the VP2 start codon and the introduced point mutations to abolish binding to HSPG, respectively. Genetic coupling approach: HEK-293T cells produce AAV particles displaying DARPins as high-affinity ligands on the capsid surface. Biochemical coupling approach: HEK-293T cells produce AAV particles displaying the NpuC intein on the capsid surface. For the generation of cell-type-specific AAV vectors, targeting ligands (DARPins (shown) or scFvs (not shown)) are covalently coupled to assembled capsids via split-intein-mediated protein *trans*-splicing

Here we provide protocols for both types of covalent targeting-ligand coupling, the genetic approach (genetic coupling) and the split-intein approach (biochemical coupling) (Fig. 1). Both procedures start with the production of recombinant AAV2 particles, mutated for HSPG binding and displaying the targeting ligand (genetic coupling) or the NpuC intein (biochemical coupling). In each case, four plasmids are co-transfected into the producer cell line HEK-293T. The AAV transfer vector plasmid carries the gene of interest under control of a suitable promoter and the inverted terminal repeats (ITRs). The AAV helper plasmid pRC-VP2_{mut} encodes the AAV enzymatic and structural proteins. In this plasmid, as well as in a third plasmid called pDARPin-VP2_{mut} (genetic coupling) or pNpuC-$(G_4S)_3$-VP2_{mut} (biochemical coupling), two codons (R585A, R588A) in the VP2 reading frame are mutated to abolish natural receptor binding to HSPG. Furthermore, expression of unmodified VP2 proteins from both plasmids is prevented by deletion of the VP2 start codon. Eventually, adenoviral helper functions are provided by co-transfection of the adenoviral helper plasmid pXX6-80 delivering the adenoviral genes E2a, E4 and VA RNA. Subsequently, the procedure for the purification of generated AAV particles from crude cell lysates by iodixanol density gradient centrifugation enabling the separation of empty and full AAV vector particles are explained. Finally, the protocol for the production and purification of targeting ligands (DARPins or scFvs) fused to a split intein domain NpuN and their coupling to the AAV capsid surface is described.

2 Materials

2.1 Transient Transfection of HEK-293T Cells

1. Medium for cultivation of HEK-293T cells: Dulbecco's Modified Eagle Medium (DMEM) supplemented with 10% FCS and 2 mM L-glutamine. Store at 4 °C.
2. 0.25% trypsin ($PBS^{w/oCa2+,w/oMg2+}$/1 mM EDTA). Store at 4 °C.
3. Medium for preparation of the transfection mix: DMEM without additives ($DMEM^{w/o}$). Store at 4 °C.
4. Medium for medium exchange prior to transfection: DMEM supplemented with 15% FCS and 2 mM L-glutamine (*see* **Note 1**). Store at 4 °C.
5. AAV2 derived transfer vector plasmid containing the transgene expression cassette with the gene of interest under the control of a desired promoter (ITR-promoter-gene of interest-polyA signal-ITR). Store at −20 °C.
6. pRC-VP2_{mut}: plasmid originally described by Münch and colleagues [3]. AAV2 helper plasmid encoding the AAV2 enzymatic and structural proteins required for particle production. Two point mutations are introduced to abolish binding to HSPG (R585A, R588A) and VP2 start codon is deleted (mutation of the unusual VP2 start codon ACG into ACC). Store at −20 °C.
7. pDARPin-VP2_{mut}: plasmid originally described by Münch and colleagues [3]. Coding sequence of the DARPin is N-terminally fused to that of mutated VP2. The plasmid harbors the two point mutations R585A, R588A and is deleted for the VP2 start codon (ACG is mutated into ACC) as described for pRC-VP2_{mut}. Store at −20 °C.

8. pNpuC-$(G_4S)_3$-$VP2_{mut}$: plasmid originally described by Muik and colleagues [9]. The coding sequence of the intein NpuC domain is connected via a $(G_4S)_3$ linker to the N-terminus of mutated VP2. The plasmid harbors the two point mutations R585A, R588A and is deleted for the VP2 start codon (ACG is mutated into ACC) as described for pRC-VP2mut. Store at −20 °C.
9. pxx6-80 [11]: adenoviral helper plasmid encoding the adenoviral helper genes E2a, E4 and the virus-associated RNA (VA RNA) (*see* **Note 2**). Store at −20 °C.
10. pCR3.1-His_6-TL-$(G_4S)_3$-NpuN-HA: plasmid originally described by Muik and colleagues [9]. Coding sequence of the targeting ligand (TL; e.g., a DARPin or a scFv) connected via a $(G_4S)_3$ linker to the N-terminus of the intein NpuN domain. For purification and detection purposes, a hexahistidine (His_6) tag is fused to the very N-terminus and a human influenza hemagglutinin (HA) tag is fused to the very C-terminus of the construct. Store at −20 °C.
11. Polyethylenimine (PEI) (branched, Mw: 25,000 Da): Prepare 0.92 μg/μL PEI solution with cell culture grade water. Adjust to pH 7 using 1 N HCl. Sterilize the PEI solution by filtration. Store aliquots at −20 °C.
12. Serum-substitute required for the production of targeting ligands: serum-free Panexin NTA (PanBiotech). Store at −20 °C.
13. Medium for medium exchange for the production of targeting ligands: DMEM supplemented with 5% Panexin NTA and 2 mM L-glutamine. Store at 4 °C.
14. Filters for cell debris/particle removal for the targeting ligand production: Sartorius Minisart®-Plus, pore size 0.45 μm, membrane material: surfactant-free cellulose acetate.

2.2 Harvest AAV Vector Particles and Benzonase Digestion

1. Cell spatula (length 195 mm, width of blade: 14 mm).
2. PBS M/K: PBS, 2.5 mM KCl, 1 mM Mg_2Cl.
3. Lysis buffer: ultrapure H_2O, 50 mM Tris/HCl, 150 mM NaCl. Adjust to pH 8.5. Sterilize the buffer by filtration. Store at 4 °C.
4. Centrifuge: Heraeus™ Multifuge™ X3R centrifuge.
5. Benzonase® Nuclease (≥250 units/μL).

2.3 AAV Purification via Iodixanol Gradient Centrifugation

1. Solutions for the iodixanol gradient (Table 1).
2. 18G cannula (1.20 × 100 mm).

Table 1
Solutions for iodixanol gradient centrifugation of AAV particles

	Stock	Needed	15% iod. (μL)	25% iod. (μL)	40% iod. (μL)	60% iod. (μL)
PBS	10×	1×	2000	1600	1600	0
NaCl	5 M	1 M	4000	0	0	0
$MgCl_2$	1 M	0.001 M	20	16	16	16
KCl	1 M	0.003 M	50	40	40	40
Phenol red	0.5%		30	24	0	24
Ultrapure H_2O			8900	7653	3677	0
Optiprep™	60%		5000	6667	10,667	16,000
mL total			20	16	16	16
mL per gradient			9	6	5	Up to 7.5

3. Quick-Seal®, Ultra-Clear™ ultracentrifugation tubes, dimensions: 25 × 89 mm, nominal volume: 39.0 mL (Beckman coulter).
4. Metal tube connector (inner diameter: 1.5 mm luer-lock male (LLM)).
5. 23G cannula (0.6 mm × 25 mm).
6. Quick-Seal Cordless Tube Topper (Beckman coulter).
7. Seal former (Beckman coulter).
8. Seal guide (Beckman coulter).
9. Heat sink (Beckman coulter).
10. Type 70 Ti rotor, fixed angle (Beckman coulter).
11. Optima™ L-70K Ultracentrifuge (Beckman coulter).
12. Laborboy (miniature lifting ramp).
13. OptiSeal™ Tube Rack (Beckman coulter).

2.4 Purification of Targeting Ligands via Immobilized Metal Affinity Chromatography and Gel Filtration

1. Ni-NTA Agarose (QIAgen).
2. Bottle/Tube Roller.
3. Econo-Pac® Chromatography Columns (Bio-Rad).
4. Poly Column Rack (Bio-Rad).
5. Storage buffer: 30% ethanol.
6. 10× binding buffer: ultrapure H_2O, 500 mM NaH_2PO_4, 1.5 M NaCl, 100 mM Imidazole (*see* **Note 3**). Adjust to pH 8.0 with 1 N NaOH. Sterilize the buffer by filtration. Store at ambient temperature.

7. Equilibration buffer: ultrapure H_2O, 50 mM NaH_2PO_4, 300 mM NaCl, 10 mM Imidazole. Adjust to pH 8.0 with 1 N NaOH. Sterilize the buffer by filtration. Store at 4 °C.
8. Wash buffer: ultrapure H_2O, 50 mM NaH_2PO_4, 300 mM NaCl, 20 mM Imidazole. Adjust to pH 8.0 with 1 N NaOH. Sterilize the buffer by filtration. Store at 4 °C.
9. Elution buffer: ultrapure H_2O, 50 mM NaH_2PO_4, 300 mM NaCl, 250 mM Imidazole. Adjust to pH 8.0 with 1 N NaOH. Sterilize the buffer by filtration. Store at 4 °C.
10. Disposable PD-10 desalting columns (GE Healthcare).
11. Splice & Store buffer: ultrapure H_2O, 50 mM Tris/HCl, 250 mM NaCl, 1 mM EDTA, 10% (v/v) glycerol. Adjust to pH 8.5 with 1 N NaOH. Store at 4 °C.

2.5 Coupling of Targeting Ligands to the Surface of AAV Capsids by Protein trans-*Splicing*

1. Splice & Store buffer: ultrapure H_2O, 50 mM Tris/HCl, 250 mM NaCl, 1 mM EDTA, 10% (v/v) glycerol. Adjust to pH 8.5 with 1 N NaOH. Store at 4 °C.
2. 10× DTT: ultrapure H_2O, 20 mM Dithiothreitol, 50 mM Tris/HCl, 250 mM NaCl, 1 mM EDTA, 10% (v/v) glycerol. Adjust to pH 8.5 with 1 N NaOH. Store in 100 μL aliquots at −20 °C.

3 Methods

3.1 Transient Transfection of HEK-293T Cells

HEK-293T cells are transfected using PEI as transfection reagent. The transfection procedure described here is for 20 cell culture plates each 150 mm in diameter.

3.1.1 Production of VP2-Modified AAV2 Vectors

1. HEK-293T cells are cultivated in DMEM containing 10% FCS and 2 mM L-glutamine. For passaging they are detached with 0.25% trypsin. Maximal number of passage is 20 before a new aliquot from the cell bank is taken in culture.
2. On the day prior to transfection, seed 1.8–2 × 10^7 HEK-293T cells in 18 mL of DMEM containing 10% FCS and 2 mM L-glutamine per 150 mm cell culture plate (growth surface 147.8 cm^2) to achieve 75–90% cell confluency for transfection.
3. On the day of transfection, prepare the DNA mix and PEI mix required for transfection. Use conical-bottom, sterile polypropylene tubes to prepare the solutions. Prepare DNA mix in a 225 mL tube first and then prepare PEI mix in a 50 mL tube. The following amounts are for transfection of twenty150 mm cell culture plates:
   (a) DNA mix: Combine 39.17 mL of $DMEM^{w/o}$ with 98.6 μg of transfer vector plasmid, 98.6 μg of pRC-$VP2_{mut}$, 98.6 μg of pDARPin-$VP2_{mut}$ or pNpuC-

$(G_4S)_3$-$VP2_{mut}$ (*see* **Note 4**), respectively and 295.6 μg of the helper plasmid pXX6-80.

(b) PEI mix: Combine 2.37 mL of PEI with 39.17 mL of $DMEM^{w/o}$.

The total amount of plasmid DNA per plate is 29.6 μg with a ratio of 1:1:1:3 between transfer vector plasmid, pRC-$VP2_{mut}$, pDARPin-$VP2_{mut}$/pNpuC-$(G_4S)_3$-VP2 $_{mut}$ and pXX6-80. The amount of PEI per μg of DNA is 3.6 μg.

4. Vortex both tubes briefly (15–30 s).
5. Add PEI mix quickly to DNA mix and vortex again (30–60 s).
6. Incubate the transfection mix at room temperature for 20 min (*see* **Note 5**).
7. Meanwhile change the medium of the cells to 14 mL of DMEM supplemented with 15% FCS and 2 mM L-glutamine.
8. After the incubation time, pipette the transfection mix from **step 6** (~4 mL per plate) to the medium in the cell culture plates from **step** 7. Swirl the plates carefully so that the medium is dispersed evenly. Incubate the plates in a cell culture incubator with a humidified atmosphere containing 5% CO_2 at 37 °C.
9. On the next morning (15–18 h post transfection) (*see* **Note 6**), aspirate the medium carefully from the cells and add 18 mL of DMEM containing 10% FCS and 2 mM L-glutamine. Incubate the plates in a cell culture incubator with a humidified atmosphere containing 5% CO_2 at 37 °C.
10. Two days after transfection AAV vector particles are harvested.

3.1.2 Production of Targeting Ligands

1. HEK-293T cells are cultivated in DMEM containing 10% FCS and 2 mM L-glutamine. For passaging they are detached with 0.25% trypsin.
2. On the day prior to transfection, seed 2×10^7 HEK-293T cells in 18 mL of DMEM containing 10% FCS and 2 mM L-glutamine per 175 cm^2 cell culture flask to achieve 75–90% cell confluency for transfection.
3. On the day of transfection, prepare the DNA mix and PEI mix required for transfection. Use two conical-bottom, sterile polypropylene tubes to prepare the solutions. Prepare DNA mix in a 15 mL tube first and then prepare PEI mix in the second 15 mL tube. The amounts are given for the transfection of one 175 cm^2 cell culture flask:
   (a) DNA mix: Combine 2.3 mL of $DMEM^{w/o}$ with 35 μg of pCR3.1-derived expression plasmid.
   (b) PEI mix: Combine 140 μL of PEI with 2.2 mL of $DMEM^{w/o}$.

The total DNA amount per flask is 35 μg and the amount of PEI per μg of DNA is 3.6 μg.

4. Vortex both tubes briefly (15–30 s).
5. Add PEI mix quickly to DNA mix and vortex again (30–60 s).
6. Incubate the transfection mix at room temperature for 20 min (*see* **Note 5**).
7. Meanwhile change the medium of the cells to 12 mL of DMEM supplemented with 15% FCS and 2 mM L-glutamine.
8. After the incubation time, pipette the transfection mix from **step 6** (4.5 mL) to the medium in the cell culture flask from **step** 7. Swirl the flask carefully so that the medium is dispersed evenly. Incubate the flask in a cell culture incubator with a humidified atmosphere containing 5% CO_2 at 37 °C.
9. On the next morning (15–18 h post transfection) (*see* **Note 6**), aspirate the medium carefully off the cells, wash with PBS, and add DMEM containing 5% Panexin NTA and 2 mM L-glutamine. Incubate the flask in a cell culture incubator with a humidified atmosphere containing 5% CO_2 at 37 °C.
10. Supernatants are harvested 48 and 72 h post transfection, passed through a 0.45 μm filter, and stored at 4 °C until used for affinity purification.

3.2 Harvest AAV Vector Particles and Benzonase Digestion

1. Scrape the AAV vector particle producing HEK-293T cells (**step 10** in Subheading 3.1.1) off the culture plates with a cell spatula. Medium remains on the cells during scraping.
2. Transfer the cell suspension into sterile 50 mL polypropylene tubes.
3. Keep the tubes on ice while proceeding with the other cell culture plates.
4. After cells have been scraped off and transferred to the tubes, rinse each plate with PBS M/K to collect the remaining cells. For ten 150 mm cell culture plates use 8 mL of PBS M/K in total: Rinse the first plate with 8 mL of PBS M/K, take this cell suspension and rinse the second cell culture plate and thereby pool cell suspension, take this cell suspension and rinse the third cell culture plate. Proceed until ten plates are rinsed and transfer the cell suspension to the tubes. If required, take fresh 8 mL of PBS M/K and proceed as described for any further ten dishes.
5. Distribute cell suspension evenly over all tubes.
6. Centrifuge the cell suspension for 7 min at 218 × *g* and at 4 °C.
7. Discard supernatant.

8. Add 500 μL of AAV lysis buffer per plate to the cell pellet and resuspend the pellet properly but quickly by pipetting up and down (*see* **Notes** 7 and **8**).
9. Freeze the resuspended cells in liquid nitrogen. The pellet can be stored at −80 °C at this step or can be used for cell lysis immediately.
10. To proceed with cell lysis, thaw the cell pellet at 37 °C in a water bath.
11. Thereafter repeat this freeze-thaw step twice so that in total three freeze-thaw cycles are performed (*see* **Note 9**).
12. Add benzonase to a final concentration of 50 U/mL to the lysate.
13. Incubate the benzonase cell lysate mixture for 30 min at 37 °C in a water bath. Shake every 10 min.
14. Centrifuge the lysate for 20 min at 3700 × *g* and at 4 °C.
15. Transfer the vector particle containing supernatant to a sterile 50 mL polypropylene tube and thereby measure the volume.
16. Add PBS M/K to adjust to a final volume of 17 mL.
17. Store the supernatant on ice until used for AAV purification via iodixanol gradient centrifugation.

3.3 AAV Purification via Iodixanol Gradient Centrifugation

1. Prepare the gradient solutions required for density centrifugation (*see* **Note 10**) (Table 1).
2. Use a long cannula (18 G) and a 20 mL syringe to transfer the AAV particle-containing supernatant (**step 17** in Subheading 3.2) into the ultracentrifugation tube. Avoid the formation of bubbles.
3. The supernatant is under layered with following iodixanol solutions (*see* **Note 11**): First, 9 mL of 15%, then 6 mL of 25%, followed by 5 mL of 40% and finally 60% iodixanol solution until the tube is almost completely filled (up to 7.5 mL). Avoid the formation of bubbles while layering the gradient.
4. Layering of the gradient can be performed using a peristaltic pump (*see* **Note 12**). Both ends of the hose of the peristaltic pump are connected to a tube connector and long cannulas (18 G).
5. Clean and equilibrate the hose with 15 mL of 70% ethanol and 15 mL of PBS M/K.
6. Transfer the appropriate volume of the respective iodixanol solution into a 15 mL tube (9 mL of 15%, 6 mL of 25%, 5 mL of 40%, and 7.5 mL of 60% iodixanol solution) (Table 1).

7. Begin with the 15% iodixanol solution. Put one of the cannulas into the 15% iodixanol solution, start the pump, and let the pump run until two drops have come out.
8. Stop the pump and put the other cannula into the ultracentrifugation tube and start the pump again so that the AAV containing supernatant in the ultracentrifugation tube is under layered with the 15% iodixanol solution.
9. Stop the pump when the 15% iodixanol solution in the 15 mL tube is almost empty.
10. Move the cannula from the 15% iodixanol solution to the 25% iodixanol solution and let the pump run again. Stop the pump when the 25% iodixanol solution in the 15 mL tube is almost empty and repeat these steps for the 40% iodixanol solution.
11. Finally, the gradient is under layered with the 60% iodixanol solution until the ultracentrifugation tube is almost completely filled.
12. Remove the long cannula from the ultracentrifugation tube.
13. Clean the hose with 15 mL of PBS M/K and 15 mL of 70% ethanol.
14. Slowly inject PBS M/K with a syringe and a cannula (23 G) into the ultracentrifugation tube so that the tube is completely filled (*see* **Note 13**).
15. Seal the ultracentrifugation tube using the Quick-Seal cordless tube topper (*see* **Notes 14–16**).
16. Centrifuge the gradient for 2 h at 292,055 × *g* (70 Ti rotor) and at 4 °C.
17. Remove the ultracentrifugation tube from the rotor and put it into an appropriate rack.
18. Use a cannula (23 G) to pierce a hole into the tube next to the neck (*see* **Note 17**).
19. Transfer the ultracentrifugation tube into an OptiSeal™ tube rack and put on a laborboy.
20. Adjust the height of the laborboy so that you can clearly distinguish the different colored layers of the iodixanol gradient.
21. Take a 3 mL syringe with a cannula (23 G), pierce a hole through the side of the tube with a rotating movement into the clear 40% iodixanol layer (just above the yellow phase) and take out the clear phase.
22. Remove the syringe but leave the cannula in the ultracentrifugation tube.
23. Aliquot the AAV preparation into 1.5 mL low-binding tubes and store at −80 °C.

24. Determine the vector genome (vg) number of the preparation by standard quantitative PCR techniques and the number of AAV particles by ELISA (*see* Chapter 4).

3.4 Purification of Targeting Ligands via Immobilized Metal Affinity Chromatography and Gel Filtration

1. Dilute the protein containing supernatant (**step 10** in Subheading 3.1.2) 1:10 with a 1:1:8 ratio of supernatant to 10× binding buffer to water to obtain a final concentration of 1× binding buffer (e.g., 100 mL of supernatant + 800 mL of ultrapure H_2O + 100 mL of 10× binding buffer) (*see* **Note 18**).
2. Prepare a homogeneous Ni-NTA Agarose solution by carefully shaking the Ni-NTA Slurry containing flask (Slurry is 1:1 (v/v) Ni-NTA matrix and ethanol).
3. Transfer the appropriate amount of Ni-NTA Agarose into a 1.5 mL reaction tube (binding capacity is 500–1000 μg of protein per 100 μL of settled matrix) and collect by centrifugation for 5 min at 1000 × *g*.
4. Carefully discard the supernatant with a pipette and wash three times with 1 mL ultrapure H_2O (between washing steps collect the Ni-NTA Agarose by centrifugation for 5 min at 1000 × *g*).
5. Equilibrate the Ni-NTA Agarose two times with 1 mL of equilibration buffer (between equilibration steps collect the Ni-NTA Agarose by centrifugation for 5 min at 1000 × *g*) and resuspend the matrix in 1 mL of equilibration buffer.
6. Add the equilibrated Ni-NTA Agarose to the diluted protein containing supernatant and incubate overnight at 4 °C on a Bottle/Tube Roller (*see* **Note 19**).
7. Set up an empty Econo-Pac® Chromatography Column in a Poly Column Rack (*see* **Note 20**).
8. Equilibrate the column with 10 column volumes (CV) of equilibration buffer.
9. Pour the Ni-NTA Agarose containing protein solution on the column to collect the Ni-NTA Agarose matrix on the porous frit by gravity flow (*see* **Note 21**). Repeat until the whole volume is processed (*see* **Note 22**).
10. Wash the Ni-NTA Agarose matrix two times with 10 CV of wash buffer (*see* **Note 23**).
11. Elute the protein with 10 CV (or at least 1 mL) of elution buffer and collect the eluate. Add 2 mL of elution buffer to remove any remaining bound proteins and combine the flow-through with the first eluate.
12. Set up a disposable PD-10 desalting column in a Poly Column Rack; cut off the bottom cap, remove the top cap, and pour off the excess storage buffer.

13. Equilibrate the PD-10 column with 25 mL of Splice & Store buffer.
14. Add 2.5 mL of eluate of the Ni-NTA Agarose purification step on top of the equilibrated PD-10 desalting column. If the sample is less than 2.5 mL, fill up with elution buffer.
15. Elute the purified targeting ligand with 3.5 mL of Splice & Store buffer and collect the eluate (*see* **Note 24**).
16. Determine the protein concentration of the eluate by standard UV-Vis spectrometry techniques (OD280) using the calculated extinction coefficient ε of the produced targeting ligand.
17. Aliquot the eluate and snap-freeze the purified targeting ligand in liquid nitrogen. Store at −80 °C until further use.

3.5 Coupling of Targeting Ligands to the Surface of AAV Capsids by Protein trans-Splicing

1. Thaw one aliquot of purified targeting ligand (Subheading 3.4) and one aliquot of AAV-NpuC (Subheading 3.3) on ice.
2. Combine the required AAV-NpuC particle number with an 800-fold molar excess of targeting ligand in Splice & Store buffer (e.g., 5×10^{10} AAV-NpuC particles + 4×10^{13} targeting ligands).
3. Start the protein *trans*-splice reaction by adding 1/10 volume of 10× DTT and mix thoroughly with a vortexer.
4. Incubate the coupling reaction for 24 h at 37 °C (*see* **Notes 25–27**).

4 Notes

1. Medium contains 15% FCS to compensate dilution with the transfection mix (containing $DMEM^{w/o}$).
2. The adenoviral genes E1a and E1b are constitutively expressed by HEK-293T cells. Thus, all adenoviral helper functions required for AAV production are provided.
3. The optimal imidazole concentration may vary with each targeting ligand, since (a) the targeting ligand sequence might comprise additional His residues that strengthen the binding to the Ni-NTA Agarose matrix, and (b) depending on the structure of the targeting ligand accessibility of the N-terminal His_6-tag might be impaired. Hence, it may be necessary to adapt the composition of the binding buffer, equilibration buffer, and wash buffer.
4. For the production of genetically coupled AAV2 use pDARPin-$VP2_{mut}$ and for the production of biochemically coupled AAV2 use pNpuC-$(G_4S)_3$-$VP2_{mut}$.

5. The incubation time of the transfection mix should be between 15 and 30 min.
6. Alternatively, medium can be exchanged 4–6 h after transfection on the same day.
7. For 20 plates 10 mL lysis buffer is needed. Distribute the lysis buffer to the tubes (for 20 cell culture plates eight 50 mL tubes are required): add 4 mL of lysis buffer to the first tube, resuspend the pellet, transfer the cell suspension to the next tube, and resuspend second pellet and thereby pool both pellets. Transfer the cell suspension to the third tube and resuspend the third pellet and thereby pool again. Take fresh 3 mL of lysis buffer and resuspend and pool the next three tubes as described above and finally transfer the cell suspension to the first tube containing the cell suspension of the first round. Finally, resuspend and pool the remaining cell pellets in again 3 mL of lysis buffer. At the end, one 50 mL tube contains the complete cell lysate.
8. Avoid multiple up and down pipetting.
9. After the third freezing step, store the cell pellet at −80 °C. The third thaw step is performed on the day of purification via iodixanol gradient centrifugation.
10. Due to the addition of different amounts of phenol red to the gradient solutions, the different iodixanol layers have a different color. The 40% iodixanol layer contains no phenol red. Accordingly, the 40% layer containing the genome-containing AAV vector particles can be identified by its clear color.
11. For layering the gradient, put the ultracentrifugation tube into an appropriate tube rack.
12. Alternatively, manual layering of the gradient can be performed by using a three-way valve and a syringe.
13. If necessary, use a small piece of tissue to dry the neck of the ultracentrifugation tube before proceeding with sealing.
14. To seal the ultracentrifugation tube put the seal former and the seal guide onto the tube. Seal the tube with the tube topper, and when the seal guide is almost touching the tube, remove the tube topper and keep the seal former with the heat sink in position. Finally, remove the seal former.
15. To check whether the tube is sealed wear safety eyeglasses and squeeze the ultracentrifugation tube very carefully.
16. Keep the gradient on ice until the second gradient is ready for centrifugation.
17. Be careful, it might splash out of the hole. This step is important to reduce the pressure of the ultracentrifugation tube.

18. Amino acids such as glutamine compete with the His_6-tagged protein for binding sites on Ni-NTA matrices. Moreover, serum proteins bind weakly to the Ni-NTA matrix and also compete for binding sites. Accordingly, the L-glutamine and serum concentration in the protein containing supernatant needs to be reduced to allow for most effective purification of targeting ligands.
19. Try to omit a classical magnetic stirrer, because friction will harm the Ni-NTA Agarose integrity.
20. Chromatography columns can be reused after washing with H_2O and ethanol.
21. Collect the flow-through for classical SDS-PAGE analysis to control the purification success.
22. Do not let the column run dry at any time.
23. Retain an aliquot from the first and last wash fraction to control the purification success by classical SDS-PAGE analysis.
24. To check the success of the purification strategy, the crude supernatant, the flow-through, wash, as well as the elution fraction can be loaded onto a polyacrylamide gel and analyzed by standard SDS-PAGE and subsequent Coomassie staining.
25. The optimal coupling conditions can vary among the targeting ligands used. To find the optimal condition, test different ionic strength, different pH, different coupling duration, and different ratios of targeting ligand to intact AAV particle.
26. Selective gene transfer can be analyzed using target-receptor expressing AAV-susceptible cells.
27. The coupling efficiency can be assessed by classical SDS-PAGE and subsequent Western Blot using an anti-AAV-VP1/2/3 antibody. The share of coupled VP2 compared to total VP2 can be analyzed by densitometry.

5 Acknowledgement

This work was supported by a grant from the European Union (Horizon 2020 Framework Programme (H2020), CARAT [667980]) to C.J.B

References

1. Santiago-Ortiz JL, Schaffer DV (2016) Adeno-associated virus (AAV) vectors in cancer gene therapy. J Control Release 240:287–301. https://doi.org/10.1016/j.jconrel.2016.01.001
2. Buchholz CJ, Friedel T, Büning H (2015) Surface-engineered viral vectors for selective and cell type-specific gene delivery. Trends Biotechnol 33(12):777–790. https://doi.org/10.1016/j.tibtech.2015.09.008

3. Münch RC, Janicki H, Völker I et al (2013) Displaying high-affinity ligands on adeno-associated viral vectors enables tumor cell-specific and safe gene transfer. Mol Ther Nucleic Acids 21(1):109–118. https://doi.org/10.1038/mt.2012.186
4. Liu Y, Siriwon N, Rohrs JA et al (2015) Generation of targeted adeno-associated virus (AAV) vectors for human gene therapy. Curr Pharm Des 21(22):3248–3256
5. Büning H, Huber A, Zhang L et al (2015) Engineering the AAV capsid to optimize vector-host-interactions. Curr Opin Pharmacol 24:94–104. https://doi.org/10.1016/j.coph.2015.08.002
6. Plückthun A (2015) Designed ankyrin repeat proteins (DARPins): binding proteins for research, diagnostics, and therapy. Annu Rev Pharmacol Toxicol 55:489–511. https://doi.org/10.1146/annurev-pharmtox-010611-134654
7. Dreier B, Plückthun A (2012) Rapid selection of high-affinity binders using ribosome display. Methods Mol Biol 805:261–286. https://doi.org/10.1007/978-1-61779-379-0_15
8. Münch RC, Muth A, Muik A et al (2015) Off-target-free gene delivery by affinity-purified receptor-targeted viral vectors. Nat Commun 6:6246. https://doi.org/10.1038/ncomms7246
9. Muik A, Reul J, Friedel T et al (2017) Covalent coupling of high-affinity ligands to the surface of viral vector particles by protein trans-splicing mediates cell type-specific gene transfer. Biomaterials 144:84–94. https://doi.org/10.1016/j.biomaterials.2017.07.032
10. Zettler J, Schutz V, Mootz HD (2009) The naturally split Npu DnaE intein exhibits an extraordinarily high rate in the protein trans-splicing reaction. FEBS Lett 583(5):909–914. https://doi.org/10.1016/j.febslet.2009.02.003
11. Xiao X, Li J, Samulski RJ (1998) Production of high-titer recombinant adeno-associated virus vectors in the absence of helper adenovirus. J Virol 72(3):2224–2232

Chapter 4

Quantitative and Digital Droplet-Based AAV Genome Titration

Julio Sanmiguel, Guangping Gao, and Luk H. Vandenberghe

Abstract

The adeno-associated viral vector (AAV) platform has developed into a primary modality for efficient in vivo, and in more limited settings, in vitro or ex vivo gene transfer. Its applications range from a tool for experimental purposes to preclinical and clinical gene therapy. The ability to accurately and reproducibly quantify vector concentration is critical for any of these applications. While several quantification assays are available, here we outline a detailed protocol for the quantification of DNase-I protected vector genomes reliant on the polymerase chain reaction (PCR) as a measure of the *active* component of the vector, namely its transgene cargo. With the emergence of droplet digital PCR (ddPCR), we provide side-by-side protocols for traditional TaqMan™ real-time, quantitative PCR (qPCR) and ddPCR, as well as comparative data generated with both methods. Lastly, we discuss the importance of the use of surfactant (here, Pluronic® F-68) in the execution of the assay to limit DNA and AAV adherence to various carriers during the titration, particularly at low concentrations. We believe these protocols can lead to reduced variability and increased comparability between AAV studies.

Key words AAV, Adeno-associated virus, Vector, Titration, Genome, Quantitative PCR, qPCR, Real-time PCR, Droplet digital PCR, ddPCR

1 Introduction

Recombinant AAV is a complex biologic derived from a nonpathogenic virus that is rendered replication-defective through elimination of all viral open reading frames [1]. It is composed of a proteinaceous viral capsid and a single-stranded DNA (ssDNA) vector genome. The vector genome is fully user-defined to deliver transgenic content to a target cell, although the transgene cargo must be flanked by inverted terminal repeats (ITRs), the only element retained from the native virus in the vector genome. In most applications, these ITRs are derived from AAV serotype 2 [2]. These AAV2-based ITR genomes can be packaged into a plethora of natural or engineered AAV capsids [3–5]. Several methods to produce AAV are established that result in vector particles of

Michael J. Castle (ed.), *Adeno-Associated Virus Vectors: Design and Delivery*, Methods in Molecular Biology, vol. 1950,
https://doi.org/10.1007/978-1-4939-9139-6_4,

high concentration and purity, as described in Chapters 3, 7, 19, 21, 22, and 23 [6].

Following production, AAV preparations traditionally undergo various assessments of quantification and qualification, such as vector titration. Titration is a process aimed at quantifying the number of vector particles in a given volume. For AAV, as for many viruses, various assays are available, many of which rely on different principles or measure distinct subpopulations. Specifically, infectious titers determine the concentration of infectious units, which is dependent on the specific conditions of the infectivity assay. Physical titers are analytical measures of the number of physical particles within a set volume. Physical titers for AAV can differ based on the principle that is relied on or the component of the particle that is probed. These include (1) measures of the number of genome containing particles determined by probing for components of the vector genome, (2) measures of the number of assembled capsids as determined by probing for the viral capsid protein, (3) calibrated methods identifying unique physical properties of AAV particles [7–9], and (4) automated imaging methodologies relying on a particle-by-particle assessment and counts. These methods each have their own benefits and limitations, and consequently each may be appropriate depending on the question asked. For example, an assessment of the total amount of assembled capsids is relevant for evaluating safety, and thus in this case measurements of viral capsid protein are preferred.

For most applications, however, the active agent in AAV is considered to be the vector genome, irrespective of its use in experimental laboratory or preclinical, translational, or clinical settings. For this reason, most dosing is based on titration methods probing the viral genome, leading to a determination of viral genomes (vg) or genome containing particles (GC) within a given volume. Genome titration is therefore the focus of the protocols below, which leverage the polymerase chain reaction (PCR). Here, we outline and detail a workflow schematically depicted in Fig. 1, leading to the determination of titer using TaqMan™ quantitative PCR (qPCR) or droplet digital PCR (ddPCR) [10]. These methods have been described in various levels of detail previously [11, 12]; however, here we further optimize and compare these different methods. Previous studies have highlighted the potential for large discrepancies among the results of different laboratories using analogous protocols [13]. This protocol aims to provide minute detail to ensure a greater level of standardization across the AAV field.

2 Materials

1. Nuclease-free water (*see* **Note 1**).
2. GeneAmp® 10× PCR Buffer I (Thermo Fisher).

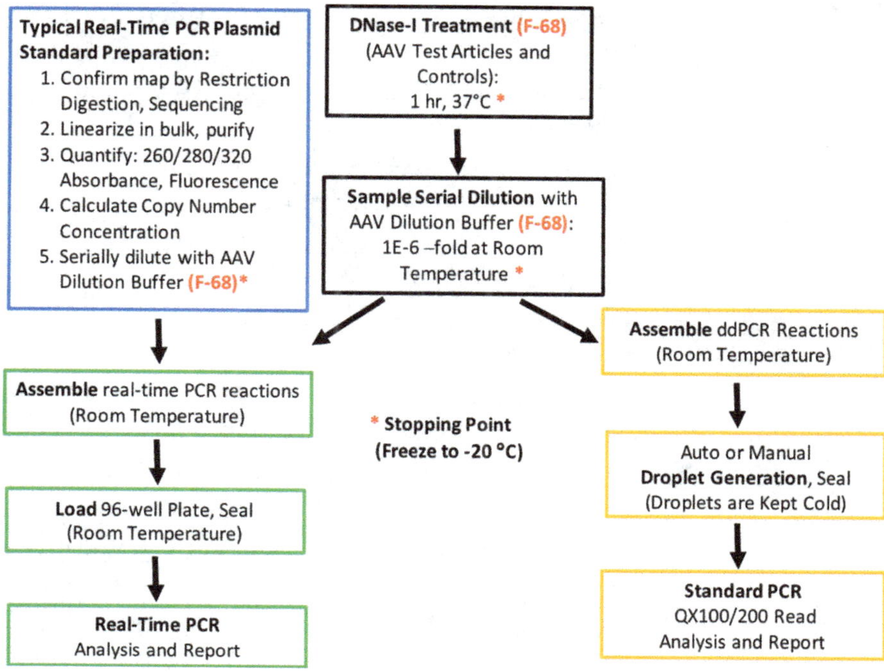

Fig. 1 Protocol schematic and workflow

3. Sheared Salmon Sperm (SSS) DNA (Thermo Fisher) (*see* **Note 2**).
4. Certified DNase- and RNase-free 0.65 mL microcentrifuge tubes, clear.
5. Certified DNase- and RNase-free 0.65 mL microcentrifuge tubes, colored.
6. Certified DNase- and RNase-free 1.7 mL microcentrifuge tubes, clear (*see* **Notes 3** and **4**).
7. L20, L200, and L1000 manual micropipettes.
8. E3-20 and E3-200 LTS electronic pipettes (Rainin Instruments) or electronic equivalent.
9. Electronic pipette rapid charge stand.
10. Sterile low-retention filter micropipette tips.
11. Pipet-Lite™ Multi Pipette L8- or L12-50XLS+ (Rainin Instruments) or electronic equivalent (*see* **Note 5**).
12. Compact PCR Racks for 0.2 mL tubes and strips.
13. 0.2 mL PCR 8-tube strips with attached clear flat caps.
14. Microcentrifuge tube racks.
15. Dedicated Microcentrifuge.
16. Dedicated 8-well PCR tube strip mini centrifuge.

17. DNase-I: RNase-free with 10× incubation buffer (Roche) (*see* **Note 6**).
18. 100× Pluronic® F-68 surfactant (10% solution) (Thermo Fisher) (*see* **Notes 7–9**).
19. 10× Pluronic F-68 (1% solution) diluted in nuclease-free water, prepared fresh each time.
20. 100× SSS DNA (200 ng/μL solution) in 1× PCR Buffer I and 0.1% Pluronic® F-68.
21. 10× SSS DNA (20 ng/μL solution) in 1× PCR Buffer I and 0.1% Pluronic® F-68.
22. AAV sample dilution buffer: 1× PCR Buffer I, 2 ng/μL SSS DNA, 0.1% Pluronic® F-68 (*see* **Note 10**).
23. Blank buffer (Mock negative control): PBS, 35 mM NaCl, 0.001% Pluronic F-68, or nuclease-free water.
24. AAV validation sample for quantitative positive control (*see* **Note 26**).
25. DNase-I linear plasmid control (1E+8 copies/5 μL), serially diluted from stock using AAV sample dilution buffer.
26. AAV experimental samples.
27. Forward sequence detection primer working stock solution: 9 μM for ddPCR, 3 μM for qPCR (*see* **Notes 11** and **12**).
28. Reverse sequence detection primer working stock solution: 9 μM for ddPCR, 3 μM for qPCR (*see* **Notes 11** and **12**).
29. TaqMan™ fluorescent probe working stock solution: 2.5 μM for ddPCR, 2 μM for qPCR (*see* **Note 13**).
30. Applied Biosystems SDS7500 Sequence Detector (Thermo Fisher), or equivalent.
31. Restriction enzymes with reaction buffer.
32. Agarose gel with Ethidium Bromide or equivalent.
33. QIAquick PCR Purification Kit (QIAGEN) with EB Buffer or Plasmid Purification Kit.
34. Linearized plasmid standard set, 1E+8 to 1E+1 copies/5 μL, serially diluted from stock in AAV dilution buffer.
35. Calibrated spectrophotometer (*see* **Note 14**).
36. Qubit™ Fluorometer (Invitrogen), or equivalent.
37. Qubit™ dsDNA HS Assay Kit (Invitrogen), or equivalent.
38. 2× ddPCR™ Supermix for Probes, no dUTP (Bio-Rad).
39. 2× TaqMan™ Universal Master Mix for qPCR (Thermo Fisher).
40. Auto DG™ Automated Droplet Generator (Bio-Rad) [14].
41. Automated Droplet Generation Oil for Probes (Bio-Rad).

42. DG32™ AutoDG™ Cartridges (Bio-Rad).
43. AutoDG™ droplet cooling box.
44. QX100™or QX200™ ddPCR System (Droplet Reader) and QuantaSoft™ Software (Bio-Rad) [15, 16].
45. C1000 Touch™ Thermocycler (Bio-Rad), or equivalent.
46. PX1 PCR Plate Sealer and corresponding Plate Support Block (Bio-Rad).
47. Pierceable Foil Heat Seal (Bio-Rad).
48. ddPCR 96-well plates (Bio-Rad) or equivalent.
49. Optical 96-Well Reaction Plate for qPCR.
50. Optical Adhesive Film Kit for qPCR (Thermo Fisher).
51. Splash-Free 96-Well Base (Thermo Fisher).
52. TempPlate® sealing film, non-sterile (USA Scientific).
53. Corning™ Costar™ 50 mL Sterile Disposable Reagent Reservoirs (Fisher Scientific), or equivalent.
54. Powder-free Nitrile Gloves.
55. Clean Lab Coat.
56. BSL-2 Biological Safety Cabinet with UV light.
57. 15 mL conical tubes.
58. Kimwipes™ Delicate Task Wipers, or equivalent.
59. Paper towels.
60. Wet ice and ice bucket.
61. Timer.
62. Vortex Mixer.
63. Lysol.
64. 70% Ethanol.
65. Lab notebook and pen.
66. Adhesive tape.
67. Color markers, fine and extra-fine point.

3 Methods

3.1 Preparation of the Work Area (See Note 15)

1. Work in a biosafety cabinet or HEPA-filtered PCR station with UV light capability, preferably a dedicated unit for quantification of high copy number targets.
2. Wear gloves at all times and change them often, particularly after leaving (and returning to) the work area.

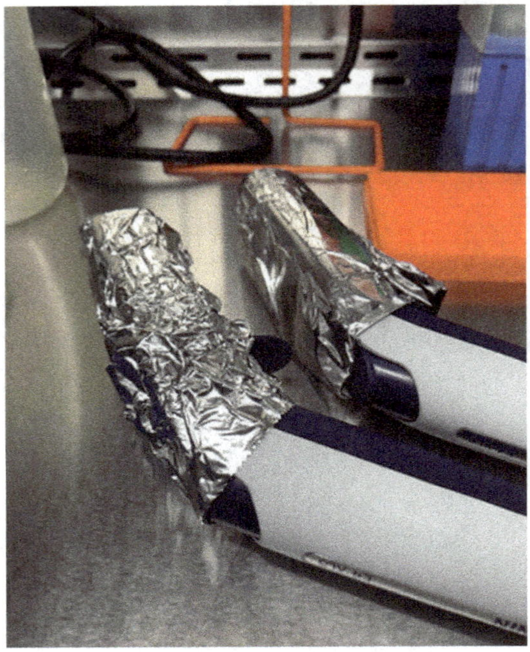

Fig. 2 Protect the LED screens of electronic pipettes from UV light damage

3. Apply a liberal amount of Lysol or equivalent to the work surface area inside the cabinet and wipe it clean with a paper towel. Do the same again with 70% Ethanol.
4. Apply a small amount of Lysol or equivalent to a paper towel and wipe clean the microcentrifuge tube racks that will be used. Repeat with 70% Ethanol. Place the clean racks well inside the clean biosafety cabinet.
5. Apply a liberal amount of Lysol or equivalent to pipette shafts (manual or electronic) over a biohazard trash bag. Repeat with 70% ethanol and wipe the pipette shafts dry with a lint-free Kimwipes™. Place the pipettes in the clean biohazard cabinet.
6. Protect the LED screen of electronic pipettes with aluminum foil while the units are exposed to UV light. Exposure to UV light will crack LED screens over time (Fig. 2).
7. Turn on the UV light for 15 min (*see* **Note 16**).

3.2 Preparation of the Real-Time PCR (qPCR) Standard

1. Prepare the real-time PCR standard in a biological safety cabinet or PCR station, rather than on the bench, where reagents or standards are more likely to become contaminated. Use dedicated reagents.
2. The standard is usually a cis-plasmid (i.e., ITR-flanked transgene construct) used in the manufacturing of the virus. There are two conditions that need to be met in advance of standard preparation: the plasmid must be a clean, clonally pure species

(OD 260/280 between 1.80 and 1.90) and an accurate electronic map must be available (*see* **Note 17**).

3. Confirm the clonal purity of the plasmid by performing an analytical restriction digestion assay of the DNA. Ideally, the assay will target both the recombinant AAV genome embedded in the plasmid and the plasmid backbone. Set up several reactions and choose enzymes that will target several components of the recombinant genome.
4. Linearize the plasmid with 10–20 units of enzyme in the appropriate reaction buffer, preferably overnight. Digest 10 μg of plasmid per 50 μL reaction. For long-term use, 40 μg of linearized plasmid will last for months or years if properly stored (*see* **Note 18**).
5. Use 1% of each reaction (0.5 μL) to verify the efficiency of linearization in separate wells of an agarose gel. There must be a single band per lane in the gel. Run appropriate controls, as needed, side by side.
6. Clean up the linearization reactions using QIAGEN's QIAquick® PCR Purification Kit (or a plasmid purification kit). Elute twice with 50 μL of EB Buffer in a DNase- and RNase-free 1.7 mL tube. Pool the eluates into a single tube and label (*see* **Note 19**).
7. Quantify the pooled linearized plasmid using spectrophotometry or fluorometry, but preferably both (*see* **Note 20**).
8. Determine the copy number concentration of the linearized plasmid and the volume needed for the first standard dilution (Fig. 3). Round off all the estimated volumes for pipetting to two decimal points. A final concentration of 0.1% Pluronic® F-68 must be included at the outset in the first dilution of the standard (1E+10 or 1E+09 copies per 5 μL, depending on the yields from **steps 6** and 7). The "1E+10" and/or the "1E+09" dilution constitute the working standard stock solutions of the linearized plasmid (*see* **Notes 21–23**).
9. Serially dilute the working stock of linearized plasmid standard (1E+10 or 1E+09 copies/5 μL) in tenfold dilutions to 1E+01 copies/5 μL in AAV sample dilution buffer. Vigorously vortex each previous dilution before moving on to the next, and always rinse the tip carrying the plasmid by pipetting up and down 5–10 times after addition of plasmid to diluent (Fig. 4) (*see* **Note 24**).
10. Discard 10 μL from the "1E+01 copies/5 μL" dilution to keep the volumes of the standard set equal across all tubes. The "1E+08 copies/5 μL" through the "1E+01 copies/5 μL" dilutions comprise the standard set to be used downstream for quantification purposes in real-time PCR.

Formula Weight = Plasmid Length (bp) x 662 $\frac{\text{gram}}{\text{mol}} \cdot \frac{1}{\text{bp}}$ = ______ gram/mol

Molar Concentration = $\frac{\text{Measured Plasmid Concentration (gram/Liter)*}}{\text{Formula Weight (gram/mol)}}$ = ______ mol/Liter

Plasmid Concentration = $\frac{\text{6.022E+23 (copies/mol) x Molar Concentration (mol/Liter)}}{\text{1E+6 (μL/Liter)}}$ = ______ copies/μL

Plasmid μL Needed = $\frac{\text{2E+09 (copies/μL) x 100μL}}{\text{Plasmid Concentration (copies/μL)}}$ ** = ______ μL

"1E+10" Stock Dilution Formulation (in order) : Nuclease-Free water to 100μL
10μL 10X PCR Buffer I
10μL 10X (1%) Pluronic® F-68
10μL 10X SSS DNA Solution ***
Plasmid μL Needed

Fig. 3 Workflow for determining linear plasmid copy number and dilution. Calculate the copy number concentration of the linearized plasmid and the volume needed for a dilution of 2E+9 copies per μL (1E+10 copies per 5 μL), based on the known length of the plasmid, "Plasmid Length (bp)," and the mass concentration of the linearized plasmid as determined in Subheading 3.2, **steps 6** and **7**, "measured mass concentration (g)." Choose pipettes that maximize the accuracy and precision of reagent delivery. **See* Subheading 3.2, **steps 6** and **7** and **Notes 19** and **20**. **The final concentration of the stock linear plasmid dilution may be adjusted according to the linear plasmid yields obtained. For example, this value may be reduced from 2E+09 to 2E+08 copies/μL to create a stock of 1E+09 copies/5 μL, rather than 1E+10 copies/5 μL, if the yield of linear plasmid is low. ****See* Subheading 2 for exact formulation

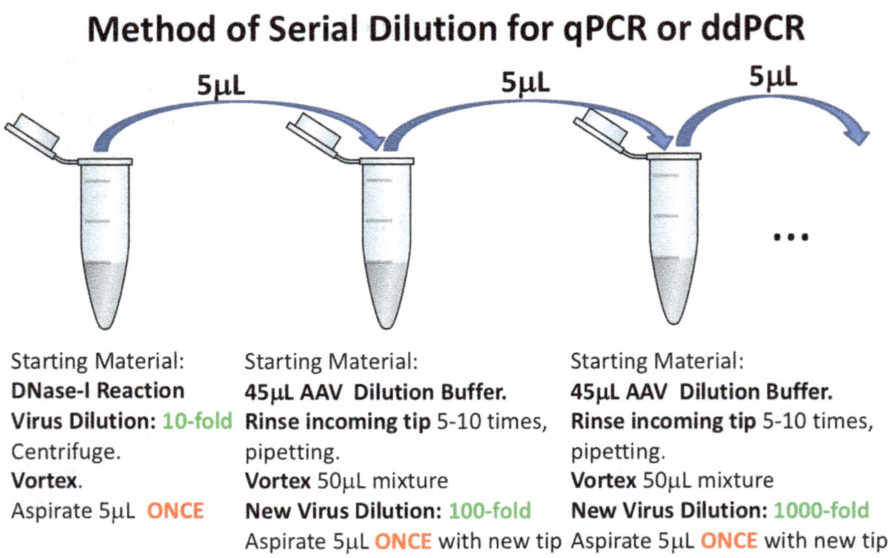

Fig. 4 Serial dilution for AAV quantification. Follow the same protocol whether diluting virus or plasmid

11. From the "1E+09" working stock dilution, prepare an additional "1E+08 copies/5 μL" dilution of the linear standard in AAV dilution buffer to serve as the DNase-I plasmid control in all future quantification runs. Store the linearized plasmid standard dilutions and stocks at −20 °C (*see* **Note 25**).

3.3 Sample Preparation: DNase-I Treatment

1. Prepare the surface area of a dedicated biosafety cabinet as specified in Subheading 3.1, above. In addition, clean a compact PCR rack with Lysol and 70% ethanol and add to the work space.
2. Wear a lab coat and gloves for this procedure. Change gloves often, especially after leaving (and returning to) the work area.
3. Thaw and equilibrate the AAV sample dilution buffer to room temperature.
4. Pulse-centrifuge the AAV test samples and controls to collect them at the bottom of the tube (*see* **Note 26**).
5. Transfer as many 0.2 mL 8-well tube strips as needed to the compact PCR rack in the biohazard cabinet for the DNase-I treatment step. Close all the lids. The DNase-I incubation will be carried out in a programmable thermocycler that can accommodate the strips.
6. Organize the AAV test samples in ascending order followed by the controls and assign a number to them. Label the flat caps of the 0.2 mL tube strips in advance with the assigned number (*see* **Note 27**).
7. The DNase-I treatment step will be the first tenfold dilution of the AAV test samples and controls (*see* **steps 3** and **4** in Subheading 3.4). All dilutions made going forward will be tenfold serial dilutions.
8. Prepare 10× Pluronic® F-68 (1%) reagent (5 μL per reaction). Make enough for all samples and controls, with sufficient extra volume to offset pipetting errors.
9. Prepare the DNase-I reaction mix for all samples and controls in a 1.7 mL microcentrifuge tube as follows (45 μL per reaction):

Reagent	Volume per sample
Nuclease-free Water	33 μL
10× DNase Buffer	5 μL
10× (1%) Pluronic F-68	5 μL
DNase I, RNase-free	2 μL (20 units)
Sample (AAV)	5 μL
Total Volume	50 μL

Multiply the "volume per sample" from above by the number of samples (AAV test samples and controls) to determine the total volume of reagents that will be needed for this step. Include two additional volumes in the calculation to offset pipetting errors (*see* **Note 28**).

10. Before adding DNase-I to the reaction mix, dispense 43 μL of the DNase-free reaction mix into the DNase-I plasmid control tube that will not include the enzyme. Bring to 45 μL with 2 μL of nuclease-free water (*see* **Note 29**).
11. Add the total volume of DNase-I needed for the remaining reactions to the 1.7 mL tube with the reaction mix, plus two extra volumes (*see* **Note 30**).
12. Invert the (complete) DNase-I reaction mix several times and only gently vortex it once. Dispense 45 μL to the remaining tubes in the strips, one at a time, using the multi-dispense function of an electronic 200 μL pipette. Immediately close the lid of each tube after dispensing the mix before moving on to the next one. Centrifuge the strips.
13. Add the test samples and controls to the DNase-I reactions (*see* **Note 21**). Vigorously vortex one sample or control at a time and add 5 μL to the corresponding reaction in the 8-tube strips in the compact PCR rack. Pipet up and down 5–10 times in the reaction mix to rinse the pipette tip (*see* **Note 31**).
14. Place the tube strips in the C1000 Touch thermocycler block and incubate the DNase-I reactions at 37 °C for 1 h (*see* **Note 32**).
15. Return all original samples and controls to the refrigerator or freezer (*see* **Note 33**).

3.4 Sample Preparation: Serial Dilution

1. Prepare the work area as described in Subheading 3.1.
2. Allow the DNase-I reactions to equilibrate to room temperature after taking them out of the thermocycler. Centrifuge the strips and vigorously vortex to mix.
3. AAV test samples, Mock controls, and Validation controls will be serially diluted 100,000-fold more (1,000,000-fold total) with AAV sample dilution buffer. Prepare five 0.65 mL microcentrifuge tubes for each sample or control for the serial dilution step.
4. The DNase-I plasmid controls will be diluted 10,000-fold more (100,000-fold total) with AAV sample dilution buffer. Prepare four 0.65 mL tubes for each plasmid control for the serial dilution step.
5. Set up the 0.65 mL microcentrifuge tubes so that they follow the same organization as the samples in the 0.2 mL strip tubes along the *bottom* length of a microcentrifuge tube rack(s).
6. For brevity, label only the last 0.65 mL tube in each dilution series (the dilution that will be used to deliver the sample to the PCR reaction).

7. Dispense 45 μL of the room temperature-equilibrated AAV sample dilution buffer to each 0.65 mL dilution tube using the multi-dispense mode of an electronic 200 μL pipette (*see* **Note 34**).
8. Serially dilute each DNase-I reaction (Fig. 4).
9. Because most of the 0.65 mL tubes in the rack will not be labeled, use the empty slots in the microcentrifuge tube racks to track your progress, avoid confusion during pipetting, and minimize errors due to skipped tubes. Move each tube up a slot in the rack after it has been used to track progress until completion (*see* **Note 27**).
10. Place the most dilute samples (the labeled tubes) on ice until the next step. Repeat until all the samples or controls are diluted. Discard all unlabeled tubes. Return the initial DNase-I reactions to the freezer and do not discard them until results are obtained (*see* **Notes 35** and **36**).
11. For ddPCR, proceed to Subheadings 3.5 and 3.6. For qPCR, proceed to Subheading 3.7.

3.5 ddPCR Reaction Setup and Droplet Generation (See Note 37)

1. Prepare the work area as described in Subheading 3.1. Only one clean microcentrifuge tube rack will be needed to hold the serially diluted samples and controls, as well as the bulk reaction mixture preparations. Also include a splash-free 96-well base to hold a Bio-Rad 96-well plate (or equivalent) where the reactions will be ultimately assembled.
2. While the interior of the biohazard cabinet is exposed to UV light, create a map of the 96-well plate indicating the position of all samples and controls in each TaqMan™ assay that will be included (Fig. 5).

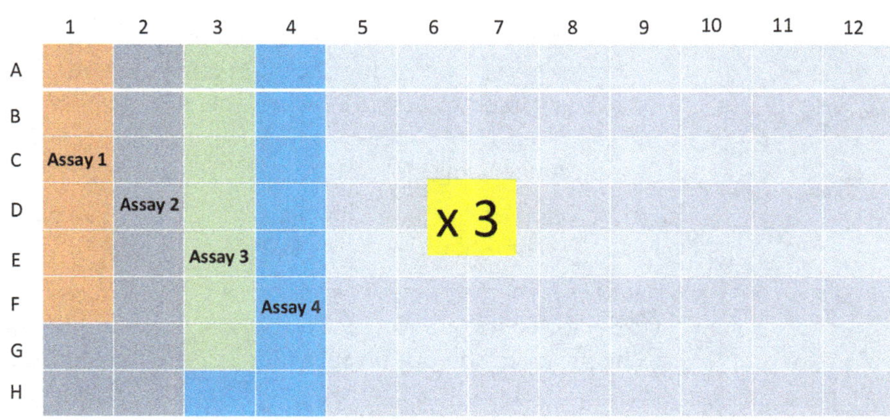

Fig. 5 Schematic view of a 96-well plate layout for the Bio-Rad QX200® Quanta Soft™ ddPCR software. Sample and TaqMan™ assay locations are mapped in advance to assist in the calculation of reagent volumes and determination of loading strategy

3. Set aside a 1.7 mL microcentrifuge tube for each TaqMan™ assay. Prepare all ddPCR reactions for each assay in a master mix as follows (18 μL per reaction):

Reagent	Vol. per reaction (μL)
2× ddPCR™ Supermix for Probes	10
Forward Primer (9 μM)	2
Reverse Primer (9 μM)	2
TaqMan™ Fluorescent Probe (2.5 μM)	2
Nuclease-free water	2
Sample (AAV or control)	2
Final Volume	20

Calculate the number of reactions that are needed to quantify all samples and controls *in triplicate*. Include additional 6–8 reactions per assay to offset pipetting error. The suggested sample order for each TaqMan™ assay is as follows: Test samples (in ascending order), DNase I (−) Plasmid Control, DNase I (+) Plasmid Control, Mock Control, AAV Validation Control, and NTC (No Template Control). A maximum of 32 unique samples may be analyzed in triplicate per plate (*see* **Note 38**).

4. Mix the equilibrated 2× ddPCR Supermix (No dUTP) by inversion and prepare the ddPCR master mix(es) according to the instructions above. Vortex gently to mix.
5. Carefully transfer 18 μL of master mix to the bottom of each reaction well following the map generated in **step 2**; use the multi-dispense function of a 200 μL electronic pipette (single or multichannel) (*see* **Note 39**).
6. Transfer 2 μL of each DNA sample or control to the side of each designated well using the multi-dispense function of a p20 electronic pipette and following the map generated in **step 2** (*see* **Note 21**). Open one tube at a time while dispensing and keep tube lids closed when not in use.
7. After completion, seal the plate with lightly applied Temp-Plate® sealing film (*see* **Note 40**).
8. Start the heat-sealing program in a PX1 PCR Plate Sealer (180 °C for 5 s) and wait for it to reach maximum temperature. The program will not be able to start until maximum temperature is reached. The program will hold until the start button is pressed in **step 11**.
9. Spin down the plate in a dedicated PCR 96-well plate centrifuge at 226 × g for 30 s.

10. Transfer the plate to the plate support block in the PX1 PCR Plate Sealer.
11. Carefully remove the TempPlate® sealing film while holding the plate down, and replace it with a sheet of pierceable foil heat seal (with red marker line up toward the top row of the plate). Close the PX1 Plate Sealer and start the heat-sealing program that has been holding since **step 8**.
12. Remove the plate and briefly vortex to mix the reactions. Centrifuge the plate again.
13. Program the AutoDG™ by matching the position and number of columns on the plate that hold reactions to their exact location in the droplet generation map provided by the instrument. The AutoDG™ will calculate the amount of materials and the volume of droplet generation oil that will be needed for the plate you have programmed in the system.
14. Set the plate inside the AutoDG™ system in the designated location for droplet generation.
15. Populate the AutoDG™ with the necessary materials requested by the instrument to complete the task: DG32™cartridges (up to three), boxes of tips (up to two), droplet cooling box, and a new 96-well plate to contain the newly formed droplets.
16. Once everything is in place, lower the AutoDG™ door hatch and start the instrument. For a full plate of reactions, the AutoDG™ may take up to 45 min to complete.
17. Droplet formation is affected by temperature. Ideally, the room where the AutoDG™ is kept should be between 20 and 25 °C. If the temperature is lower than this, the AutoDG™ may generate an error during the execution of the droplet generation program. To date, in our experience such errors have not impacted results (*see* **Notes 41** and **42**).

3.6 Droplet Digital PCR and Plate Reading

1. Droplets are collected in a new plate over a cooling block. The plate must be heat-sealed promptly without disrupting the droplets, then transferred to a thermocycler for PCR. This should be done within an hour of droplet formation. Keep the plate on the cooling box at all times. The droplets are fragile at this stage, but will be stabilized during PCR.
2. Use the following PCR Reaction conditions:
   (a) 95 °C, 10 min
   (b) 40 cycles of 94 °C, 30 s and 60 °C, 1 min
   (c) 98 °C, 10 min
   (d) 4 °C, indefinitely

The PCR ramp rate should be between 2 °C/3 °C per s. Set the final reaction volume to 40 μL (*see* **Note 43**).

3. Open the QuantaSoft™ application and use the map generated in **step 2** of Subheading 3.5 to create a droplet readout file. Assign the quantification protocol to be used (ABS), supermix type, sample names, and TaqMan™ assay to all wells.
4. Insert the plate in the QX200™ reader and initiate the droplet count. It will take an additional 2.5 h to read a full plate.
5. After completion, click OK and analyze the run. Select the 1D amplitude chart from the available options under the "Analyze" tab, and manually assign a threshold to all wells with the same TaqMan™ assay. The "Analyze" tab allows the operator to choose whether to assign thresholds on an individual reaction basis or as a group. Set the threshold value such that it falls just above the negative droplets at the bottom of the chart. The software will automatically calculate a target concentration per μL of reaction.
6. After the threshold values have been assigned, save the document and export the .csv file to a USB drive for further analysis and final calculations.
7. From the "Concentration" column in the .csv file (Copies/μL), calculate AAV Titer (Genome Copies/mL) using the following formula (*see* **Note 44**):

$$\textbf{AAV Titer} = \text{Concentration} \times 10 \times 1000\,\mu L/\text{mL} \times 1{,}000{,}000$$

8. For the plasmid DNase-I controls, data should be reported as "Copies per DNase-I reaction" rather than as a concentration, in order to determine the recovery of total plasmid input. Calculate using the following formula (*see* **Note 45**):

$$\textbf{Plasmid Copies per DNase-I Reaction} \\ = \text{“Concentration”} \times 20\,\mu L \times 100{,}000 \times 2.5$$

9. The background signal from the "No Template Control" (NTC) wells should also be reported as "Copies per reaction." Calculate using the following formula:

$$\textbf{NTC Copies/reaction} = \text{Concentration} \times 20\,\mu L$$

10. Generate a report from the .csv data and save the file as an Excel sheet or workbook (Table 1). Distribute the report to the AAV production team and quality assurance supervisors for review (*see* **Note 46**).

3.7 Real-Time PCR (qPCR) (See Note 47)

1. Prepare the work area as described in Subheading 3.1.
2. Thaw the qPCR standards.

Table 1
Example ddPCR Report

Sample ID	Titer #1	Titer#2	Titer #3	Average	Stnd Dev	Notes
Sample A TFF	2.52E+12	2.43E+12	2.14E+12	2.36E+12	1.99E+11	
Sample A Final	2.45E+13	2.17E+13	2.03E+13	2.22E+13	2.16E+12	
1E8 1E-5 (Plasmid DNase-I Control)	9.25E+07	8.85E+07	6.35E+07	8.15E+07	1.57E+07	Copies per reaction
1E8D 1E-5 (Plasmid DNase-I Control)	6.00E+05	1.55E+06	1.50E+06	1.22E+06	5.35E+05	Copies per reaction
M/B (Mock Control)	3.70E+09	2.30E+09	2.40E+09	2.80E+09	7.81E+08	
XXXX (Validation Control)	1.20E+13	1.08E+13	9.71E+12	1.08E+13	1.16E+12	
NTC	2.40E+00	8.00E+00	3.60E+00	4.67E+00	2.95E+00	Copies per reaction

In-process samples (TFF) and final concentrates are routinely quantified in our labs, as well as crude lysates. The latter require less dilution. The validation control samples are reproducible to within 10–30% with very low background each run (<0.01%). Intra-assay reproducibility is within 10–30% among different operators

3. Thaw the working TaqMan™ assay stock for qPCR in advance. The oligos should amplify the same target sequence that was used to quantify test samples and controls by ddPCR.
4. Remove the 2× TaqMan™ Universal master mix from storage. Equilibrate all reagents to room temperature as before.
5. Prepare 0.65 mL colored tubes for the real-time PCR reactions: one for NTC, eight for standards, and one for each test sample or control.
6. All real-time PCR reactions will be carried out in a 50 μL final reaction volume in triplicates for each standard, sample, or control (*see* **Note 48**).
7. Prepare the real-time PCR master mix as follows:

Reagent	Vol. per reaction (μL)
TaqMan™ Universal PCR Mix (2×)	26.25
Forward Primer (3 μM)	5.25
Reverse Primer (3 μM)	5.25
Fluorescent Probe (2 μM)	5.25
Nuclease-free water	5.25
Test Sample or control	5.25
Total	52.50

8. Multiply the volume of each reagent by the total number of NTCs, standards, and samples and controls in triplicate. Include an additional two volumes to offset pipetting errors. Prepare the master mix in a 1.7 mL microcentrifuge tube or 15 mL conical tube. Mix by vortexing.
9. Take the same precautions for qPCR as for ddPCR when dispensing reagents and samples. Keep all lids closed (tubes, tip boxes) when not in use. Avoid aerosol formation by limiting blowouts and change gloves often, especially after leaving (and returning to) the work area.
10. Transfer 141.75 μL of PCR master mix to each 0.65 mL tube. Add 15.75 μL of nuclease-free water to the NTC reaction, then add 15.75 μL of each standard, sample, or control DNA to the predesignated tube. Mix each tube by vortexing.
11. Dispense 3 × 50 μL of the reactions to designated wells in the qPCR plate using the multi-dispense function of an electronic pipette, starting with the standards, followed by samples or controls, and ending with the NTC.
12. Seal the plate with optical film and spin it down at 226 × *g* for 30 s.
13. Set up a dedicated file for the run using the SDS software and load the plate into the qPCR instrument.
14. Use the following PCR Reaction conditions:
    (a) 10 min at 95 °C
    (b) 40 cycles of 95 °C, 15 s and 60 °C, 1 min
15. Allow the instrument to establish the threshold automatically.
16. Calculate the final titers (GC/mL) using the following formulas.

    For ssAAV and mock control (*see* **Note 49**):

$$\text{Titer (GC/mL)} = (\text{Qty Mean} \div 5\ \mu L) \times 2 \times 1{,}000{,}000 \times 1000\ \mu L/\text{mL}$$

    For scAAV:

$$\text{Titer (GC/mL)} = (\text{Qty Mean} \div 5\ \mu L) \times 1{,}000{,}000 \times 1000\ \mu L/\text{mL}$$

    For plasmid controls:

$$\text{Total Copies} = \text{Qty Mean} \times 100{,}000$$

    For NTC, no "Qty Mean" is provided in the SDS7500 system. The Ct values are used as an estimate of background signal. Alternatively, a "Qty Mean" value may be calculated independently from the standard curve data if desired.

3.8 Limitations of the Protocol

1. This protocol relies on the denaturation temperature and incubation time of the PCR to break open vector capsids and make AAV genomes therein accessible for quantification. AAV serotypes are differentially stable and we may well be underestimating the titers of the most stable capsids [8, 17].
2. Some AAV genomes or target sequences therein may be refractory to amplification despite our best efforts for efficient quantification, due to entrenched secondary structures, potential methylation of CpG islands at oligonucleotide binding sites, repeat sequences, or our inability to adequately visualize the AAV genome in solution [18]. Unknown, distant interactions may lead to incorrect placement of TaqMan™ assay oligonucleotides. In this case, the combination of qPCR and ddPCR for the same target sequence may yield valuable insights regarding amplification efficiency for the improvement of AAV genome quantification by qPCR. Alternatively, methods that rely on AAV capsid (protein) quantification (semiquantitative SDS-PAGE, ELISA, and AUC) are needed to complement AAV genome quantification by PCR and achieve an analytical understanding of vector lot composition. In the absence of empty capsids and spurious DNA encapsidation products, the successful assembly of recombinant virions is expected to yield a 1:1 capsid-to-genome ratio, and this ratio should serve as a benchmark for the appraisal of quantification methods and packaging efficiency of the recombinant vectors on a lot-to-lot basis. The use of surfactant during genome quantification is crucial for this assessment.
3. Limitations may be imposed by the biology of AAV production itself, where heterogeneity of packaged genomes can make it difficult to confidently assess the concentration of the genome sequence of interest [19]. In this case, neither PCR method nor AAV capsid quantification can reliably assess the integrity of AAV preps; modifications to the biology of AAV production must be implemented for these highly specialized constructs with subsequent evaluation by next generation sequencing (*see* Chapter 5).

3.9 Conclusion

Our data indicate that the adaptation of Pluronic® F-68 from an AAV manufacturing setting [20] to vector quantification work markedly increases quantification accuracy, and moves us closer to a true representation of vector concentrations in recombinant AAV preparations. It is now possible to offer critical quality control support to vector manufacturing facilities from the perspective of genome quantification, as it is possible to arithmetically track AAV genome content in a highly specific manner and with minimal loss of sample. This detailed protocol for AAV quantification by qPCR and ddPCR has been presented in the spirit of current guidelines for publication [21, 22] and to address concerns by the gene therapy community [23]. Our data also suggest that ddPCR and qPCR are equivalent technologies for accurate AAV quantification if the physical properties of the viral capsid are observed. Such considerations may need to be extended to other quantitative assays that are currently hindered by the adherence of the AAV capsid to plastic surfaces, most notably the semiquantitative SDS-PAGE assay [24], which together with PCR provide an overall view of AAV genome packaging efficiency and product quality. Further optimization work is in progress.

4 Notes

1. The use of in-house reagents is strongly discouraged as these are a common source of PCR contamination and variability.
2. SSS DNA serves as a blocking agent and nucleic acid carrier.
3. The microcentrifuge tubes used in this protocol are intentionally not autoclaved. This is to avoid debris or film deposits from accumulating on the internal surface of the tubes during sterilization, which may carry nucleic acid contaminants, potential PCR inhibitors, or active nucleases. Clean buffers or diluents placed in autoclaved tubes that have been heated above 37 °C may exhibit a marked increase in O.D. 260 readings compared to buffers or diluents placed in non-autoclaved tubes that are otherwise identically treated, suggesting the presence of contaminants whose identity cannot be confirmed [25].
4. Use of low-retention or siliconized tubes in conjunction with Pluronic® F-68 has a slight negative effect on ddPCR quantification due to an unknown interaction. Use of these treated plastics is, therefore, discouraged.
5. Electronic and multichannel pipettes will facilitate and speed up liquid handling, particularly for the distribution of vector diluent and sample replicates. However, they are expensive and are not an absolute requirement. AAV quantification will not be significantly affected by the use of calibrated manual pipettes. Manual LTS™ pipettes are our preferred choice due to their ergonomic piston plunging and demonstrated performance.
6. DNase-I is needed to digest un-encapsidated viral genomic DNA or residual plasmid DNA that may have escaped Benzonase® nuclease treatment during the recombinant AAV manufacturing process. Typical efficiency of digestion with DNase-I is 95% or greater, and should be confirmed with an appropriate plasmid control each run (Table 2).

Table 2
Plasmid control samples for qPCR or ddPCR

Sample ID	CMV2 Titer #1	CMV2 Titer#2	CMV2 Titer #3	CMV2 Average	CMV2 Stnd Dev	Notes
1E8 1E-5 (Plasmid DNase-I [−] Control)	8.45E+07	8.85E+07	8.00E+07	**8.43E+07**	4.25E+06	Copies per reaction
1E8 1E-5 (Plasmid DNase-I [+] Control)	3.25E+06	4.70E+06	5.50E+06	4.48E+06	1.14E+06	Copies per reaction

DNase-I efficiency is >95%. Note that contrary to observations made previously [10], we are able to quantify plasmid DNA very accurately

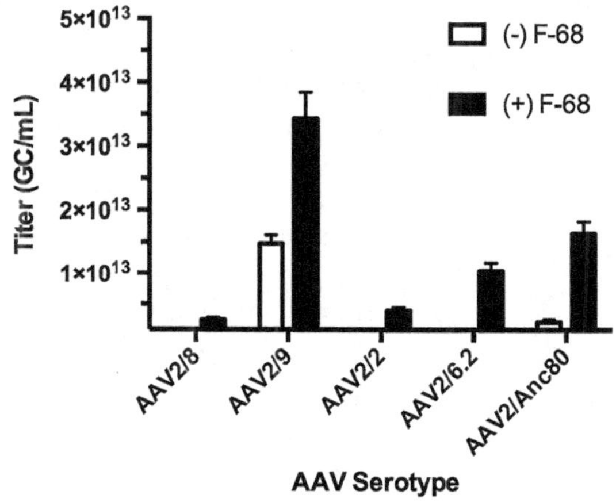

Fig. 6 ddPCR titration is improved by inclusion of Pluronic® F-68 according to AAV serotype. The addition of F-68 during AAV quantification confers to a significant increase in signal that improves vector quantification. The degree of gain ranges from modest (two- to threefold) for AAV2/9 and AAV2/8, to large (five- to tenfold) for AAV2/Anc80 and AAV2/2. The same TaqMan™ assay was used for these measurements, and all five serotypes carried the same single-stranded genome. Based on these observations, it appears that AAV capsids differentially attach to plastic

7. Pluronic® F-68 solution, a.k.a. poloxamer 188 solution, is a BASF registered trademark of a 10% solution of polyoxyethylene-polyoxypropylene block copolymer, a non-ionic surfactant that serves to prevent AAV attachment to plastic [20, 26]. Other surfactants or detergents may be used as well, but these will need to be properly validated. Most importantly, these reagents must not be detrimental to the function of *Taq* polymerase during PCR.
8. When F-68 is not used during AAV quantification there is a two- to tenfold drop in concentration estimates. The extent of the underestimation of titers varies according to serotype (Fig. 6). Also, in the absence of F-68, ddPCR yields higher

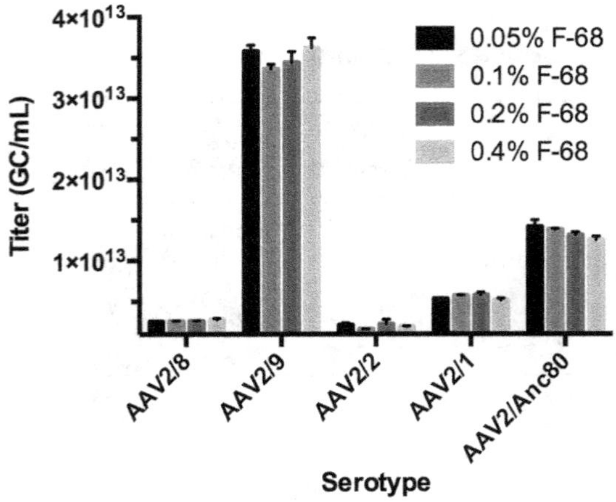

Fig. 7 Pluronic® F-68 optimization for ddPCR. Increasing the amount of F-68 does not significantly improve AAV titration of several serotypes. The same TaqMan™ assay was used for all measurements, and all vectors carried the same single-stranded genome. In light of these results, it was decided to abide by the manufacturer's recommendations for future measurements, and to use a final F-68 concentration of 0.1% (1×) for untested AAV capsid variants

titers than qPCR. This is most likely due to surfactants present in the droplet generation oil, the emulsification of the PCR reaction itself, or both.

9. The recommended final concentration of F-68 is 0.1%. We evaluated different concentrations of F-68 and found that increasing the amount of surfactant yielded variable results but did not significantly improve AAV quantification (Fig. 7). While a final F-68 concentration of 0.05% appears to be appropriate for quantification, we use 0.1% in anticipation of novel AAV variants with unique properties that may make them challenging to quantify. Diluting F-68 beyond 0.1% affects AAV serotypes differentially. The use of F-68 has consequences beyond vector production and quantification [27].
10. SSS DNA in the buffer will degrade over time. The sample diluent may be stored at 4 °C for no more than a week.
11. These solutions follow Bio-Rad's recommended ddPCR primer and probe concentrations. The final oligo concentrations in droplets may be different than in the original 20 μL reaction, as the total volume in the reaction wells increases following droplet formation and only a fraction of this volume is composed of actual droplets. Note that if qPCR primers have already been designed for use at a lower final concentration, their T_m's will likely increase if they are used for ddPCR at a higher concentration and the same annealing temperature. In order to keep the T_m constant at a higher oligo concentration, one would have to shorten the oligo or increase the annealing

temperature. Simply increasing the concentration of qPCR primers is therefore likely to make ddPCR slightly less specific. However, we have not seen an impact on ddPCR quantification when using qPCR primers at higher concentrations. During oligo design, we generally fix the final primer, salt, and magnesium concentrations at the annealing temperature recommended by the instrument manufacturer (Bio-Rad) or set as default (Thermo Fisher), then extend or trim the oligo sequence accordingly using the Two-State Hybridization simulator of the DINAMelt application within the UNAfold Web Server [28, 29]. We seek to design stringent oligos whose T_ms are not lower than 1 °C below the annealing temperature. Since the default annealing temperature for qPCR or ddPCR is 60 °C, the T_ms of our oligos are not lower than 59 °C for specific amplification. Oligo T_ms lower than 59 °C risk that the primer will not bind at the desired temperature, while much higher T_ms risk off-target hybridization. The design results in an exponential overrepresentation of oligos in the reaction mixture compared to target sequences. We have found that commercial algorithms at oligo synthesis websites tend to underestimate oligo T_ms when compared to the DINAMelt application. The Two-State Hybridization simulator consistently yields the shortest primers at the standard qPCR buffer composition of 50 mM Na^+ or K^+ and 1.5 mM Mg^{2+}.

12. Primer design may be fine-tuned for specificity by performing annealing gradient PCR, but this may overlook the fact that the TaqMan™ probe will also have a large contribution to specificity. By the time the PCR cycle reaches the annealing temperature from the denaturation step, the TaqMan™ probe is already firmly bound to the target due to its 10 °C higher T_m. Other considerations for primer design apply also, depending on the complexity of the sample and the background material, which may call for more sophisticated computational resources for quality assurance. However, despite their importance, proper handling of these variables alone will not result in accurate quantification of AAV.

13. TaqMan™ probes are susceptible to photobleaching; therefore, shield these reagents from light as much as possible.

14. If unsure about the state of calibration of your instrument, check it for accuracy against a sample of known concentration such as SSS DNA, or seek professional calibration from the manufacturer. Always validate your spectrophotometer with a different instrument, different samples, and a different method of quantification. It is imperative that the spectrophotometer be properly blanked with exactly the same buffer at the same dilution used to prepare the plasmid samples that are quantified. For example, if your plasmid sample is in QIAGEN EB

buffer and you dilute the plasmid sample tenfold for the spectrophotometer, use a tenfold dilution of naive QIAGEN EB buffer in the same diluent to blank prior to quantification.

15. qPCR and ddPCR are extremely sensitive technologies [30]. Because of this, prepare the work area anew as specified between different steps in the protocol. When working with high copy number targets, aerosol formation increases the probability of cross-contamination. Contamination of negative controls is routinely encountered even with the use of brand new reagents.
16. The UV light should not be on longer than 15 min. The bulbs have a limited life span and a prolonged exposure will not be more effective.
17. It is well established that the quantification of double-stranded (ds) or self-complementary AAV (scAAV) genomes by qPCR is problematic [10, 31, 32]. While the focus of this protocol is not scAAV genome quantification, it has been suggested that a careful choice of standard should resolve most issues for scAAV quantification by qPCR [33]. This, in addition to proper standard handling, purification, and quantification, makes qPCR a competitive technology for quantification in the absence of the more costly ddPCR platform. The standard does not need to be of identical sequence and structure as the experimental vector genome in most cases, but this choice must be considered for very difficult targets.
18. For real-time PCR, it is important that the sequence of interest (or target) be accessible for amplification. That is, the sequence must be linear and unencumbered by secondary structures that may limit its access by Taq polymerase. Most plasmids do not need linearization, but some do. Since it is difficult to predict which plasmids will require linearization, we linearize all plasmid standards used for quantification. Supercoiled plasmids that need linearization may show a shift toward higher Ct values (less sensitivity) for the same initial copy number compared to their identical linear counterparts, and may not be detected by qPCR at all past a certain concentration. These events introduce systemic quantification errors. For TaqMan™ qPCR assay design, always check the plasmid standard sequence for secondary structures using a nucleic acid folding program at the annealing temperature [28, 29]. These technologies generate data that are highly reproducible but not necessarily accurate, and thus statistical metrics of the standards alone cannot guarantee accurate quantification. We hope to provide here a complete picture of the necessary requirements.
19. Do not pre-warm elution buffers. Keep all reagents, and perform all cleanup steps, at room temperature. Pre-warming elution buffers may introduce systemic quantification errors [25].

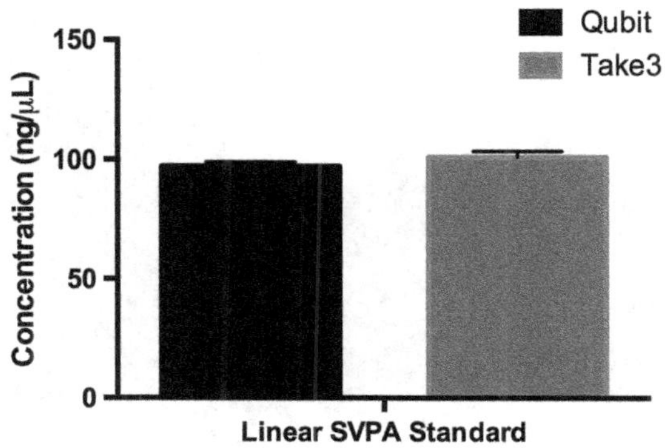

Fig. 8 Quantification of a linearized qPCR standard plasmid. Fluorometry (Qubit™) was compared against Spectrophotometry (BioTek® Take3™). When the DNA being measured is clonally pure and clean, these two types of measurements closely agree. Fluorometry measurements may be lower than spectrophotometer O.D. measurements, because the former is highly specific for double-stranded DNA

20. The two methods offer a cross-reference check for sample purity and accuracy of quantification. That is, for a pure species of high concentration plasmid, quantification by the two methods should significantly agree (Fig. 8). This formally links the physical measurement of the qPCR standard to the results of downstream fluorometry-based PCR methods of quantification, either qPCR or ddPCR. In the event of significant disagreement between these two methods of qPCR plasmid standard quantification, do not continue. Retrace previous steps and evaluate workflow for errors.
21. AAV PCR-based quantification is a sampling exercise. This means that the integrity of the sample as a unit is paramount. Whether the sample to be quantified is plasmid or virus, collect it at the bottom of the tube by centrifugation before pipetting to avoid changes in concentration due to condensation or loose sample spotting inside the tubes. Then vigorously vortex samples to ensure homogeneity. Finally, aspirate the sample by pipetting only once with proper technique. Never reintroduce the pipette tip into the sample during aspiration (or sampling) and re-aspirate, because there is a risk of enriching the pipette tip with extra material that will be detected, quantified, and increase error.
22. The dials of manual LTS pipettes can be set to accommodate up to three decimal points, depending on the unit. Pipette the exact intended volume by taking advantage of this feature for increased accuracy: lock the pipette dials and work within the linear range of the units when preparing standards for real-time PCR. We avoid pipetting less than 2 μL for any step.

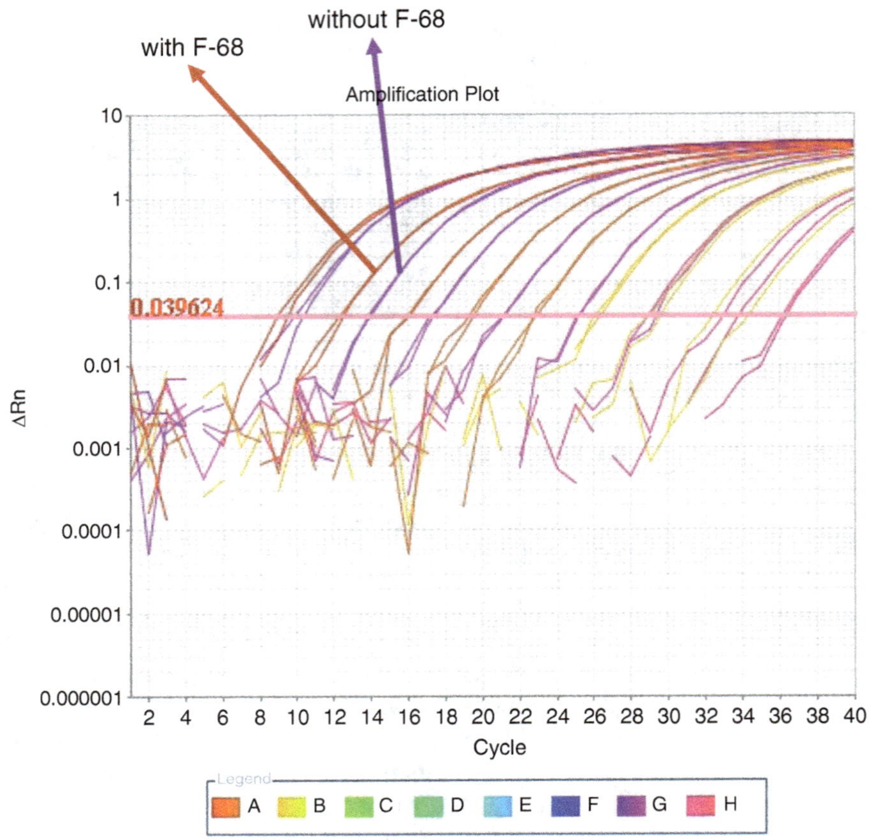

Fig. 9 qPCR amplification plot showing the effect of 0.1% Pluronic® F-68 on qPCR standard performance. Standards of the same copy number concentration, serially diluted from the same source material, demonstrate lower Ct values (a "shift to the left") in the presence of F-68, indicating their higher abundance in solution due to reduced attachment to plastic. This shift translates to higher titers for AAV quantification

23. The use of F-68 for dilution of the standard lowers the Ct values for plasmid DNA during qPCR, demonstrating that pure DNA attaches to plastic in its absence. F-68 maximizes the number of target sequences available for real-time PCR quantification (Fig. 9). The use of this excipient in the standard dilution is justified because it will also be included in all dilutions of AAV experimental samples or controls, where it will serve the same function. Lowering the Ct value of the standards by addition of F-68 will cause the experimental sample titers to decrease, but at the same time, addition of F-68 will cause more experimental sample to become available for quantification in a serotype-dependent manner. F-68, therefore, introduces a correction to quantification which predominantly results in increased titers. In real-time PCR, it is imperative that "apples-to-apples" comparisons be made. Optimal quantification is achieved with real-time PCR when standards and test

samples are identical sequences and are handled within the same or nearly the same environment.

24. Vortexing the tube and rinsing the pipette tip are intended to keep the dilutions homogeneous during the serial dilution step. The same discipline must be applied to the dilution of AAV experimental samples or controls. We make 100 μL serial dilutions of linear standard plasmid at a time (10 μL plasmid + 90 μL AAV dilution buffer), which ensures accurate pipetting and reduced freeze-thawing cycles.
25. The use of a color-coded tube system is highly recommended here, as it will help organize the standards for future use, especially if there are several standards in the freezer.
26. The controls are as follows: DNase-I plasmid control (two reactions, with and without DNase-I), Mock Control (one reaction), and AAV Validation Control (one reaction). The no-template control (NTC) is added at the end, during the PCR plate setup. For the AAV validation control, an in-house vector from previously manufactured sample lots is sufficient. This vector can be any serotype, but its genome should carry the intended target sequence for PCR. For convenience, we recommended using a vector that includes several commonly used target sequences within a standard single-stranded recombinant AAV genome containing ITRs, promoter, transgene, and polyadenylation signal within the packaging limit of the virion. It should be a vector that may be used in an experimental setting, carrying no unusual features. Do not use an artificial composite of target sequences for PCR quantification, which may jeopardize efficient genome packaging by AAV and equal detection by PCR technologies. This control serves to measure inter-assay reproducibility and assess operator performance. We typically use a relatively high titer vector (>5E+12 GC/mL), but this is not absolutely necessary, as quantitative PCR has a wide dynamic range and vector loss is minimal. Several such controls may be needed, depending on the particular sequence being measured.
27. A common source of error in these assays is inadvertent sample switching. Researchers should be vigilant at all times.
28. Do not add DNase-I to the reaction mix at this time. This is because one of the two DNase-I plasmid control reactions will not include the enzyme. If more than one control set is to be included in the plate for a different TaqMan™ assay, adjust the estimation of reagent volumes accordingly.
29. Make sure to pick up one pipette tip at a time, and to keep the tip boxes closed whenever possible.

30. Keep the DNase-I cold at all times and never vigorously vortex to mix. Rather, invert the tube several times and only tap or gently pulse vortex once for mixing.
31. Do not vigorously vortex the PCR tube strips after adding samples or controls to the reaction mixture. DNase-I activity may be diminished.
32. Program the C1000 Touch thermocycler as follows: 25 °C, 2 min; 37 °C, 1 h; 4 °C, ad infinitum. We do not recommend heat-inactivating DNase-I at this time, as vector genomes will be digested if they are exposed to the enzyme after capsid breakdown, which is likely to occur at temperatures used for DNase-I inactivation. The danger is exacerbated because the temperature cools to 4 °C, briefly returning the reaction to DNase-I's optimal temperature range after the genomes have potentially been exposed. Rather, we first dilute the DNase-I reactions (thereby diluting the enzyme), then inactivate DNase-I during PCR. DNase-I does not return to its optimal incubation temperature once thermocycling commences. We have read droplets soon after PCR, 2 h later, or the following day and have observed no significant difference in our positive control titers. Similarly, for qPCR, fluorescence is read in real-time and data collection takes place at 60 °C, well above the optimum temperature of DNase-I. This contributes to AAV titers that are equivalent between qPCR and ddPCR and highlights the importance of complementary methods of AAV quantification. Another important consideration is the length of time and the temperature at which the quantification plate is allowed to stand prior to the start of PCR. For instance, if the plate is left at room temperature for 2 h or more, the titers may significantly decrease not because AAV genomes are degraded, but because DNase-I will slowly degrade primers and probes in the reaction mixture. High-throughput qPCR may require such wait times for automated plate stacking and sequential loading, and these dangers may reduce the accuracy of titration.
33. After the reaction is complete, the protocol may be stopped if needed. The samples may be stored in the thermocycler overnight at 4 °C or placed on ice for transport and stored in a −20 °C freezer until the next steps can be completed. Note that the viral capsids are assumed to be intact after the DNase-I incubation. Unless otherwise noted, most AAV capsids are stable at 37 °C and the encapsidated genomes are, therefore, protected from further action by DNase-I [8, 17]. A simple check may be implemented whereby AAV samples are incubated in the presence or absence of DNase-I for quality assurance.

34. It is strongly recommended to use the electronic pipette that is most appropriate for the task at hand. For instance, dispensing 45 μL volumes with a 1000 μL electronic unit, although convenient, risks introducing large pipetting errors. Avoid shortcuts, if possible. Any changes must yield the same dilutions as before.
35. This is another stopping point. The final dilutions may be transferred to the refrigerator or freezer until the next step. Pluronic® F-68 will prevent AAV particles from adhering to the inner wall of the 0.65 mL tubes under these storage conditions [26], preserving the quantification integrity of the assay. The movement of the virions in solution will be slowed down in cold temperatures as well, with less particles reaching the inner wall of the polypropylene tube.
36. Next, proceed to either qPCR or ddPCR. The same dilutions may be used for either technology and inputs from both technologies will provide clues about the general quality of quantification. Successful quantification by both technologies should result in an equivalent measurement of AAV genome concentration (Fig. 10).
37. The steps that follow are intended for the AutoDG™ droplet generation method. If droplets are to be generated "manually," eight reactions at a time, a manual droplet generator will be needed. Consult with Bio-Rad for details on all aspects of the droplet digital PCR system [15, 16, 34].

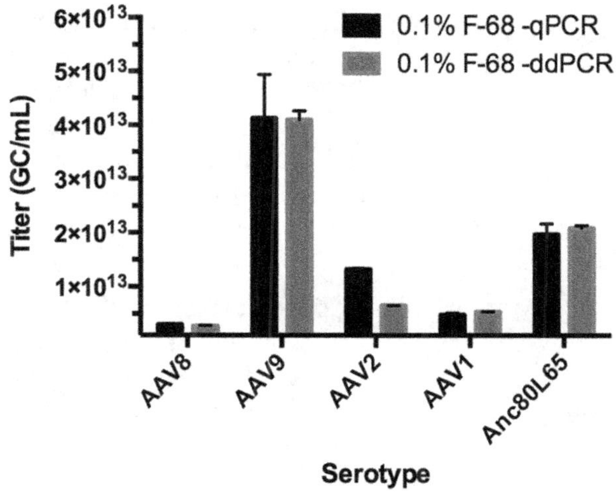

Fig. 10 ddPCR vs. qPCR comparison across AAV serotypes carrying the same genome. The cis-plasmid used to manufacture these vectors was used for the qPCR standard. The same TaqMan™ assay was used for all measurements and exactly the same sample dilutions were assayed by both technologies. Pluronic® F-68 improves qPCR-based AAV genome quantification to a level comparable to ddPCR by making more target available for quantification and correcting qPCR standard performance

38. It may not be possible to include all controls on the plate for any one TaqMan™ assay due to space limitations. In this case, adjust the reaction map accordingly. It is imperative, however, that unassigned wells in any one column be filled with an actual (mock) ddPCR reaction, as opposed to just water or PBS, for efficient droplet generation.
39. A Rainin 200 μL electronic pipet will accurately deliver 18 μL of reagent, despite the fact it nominally dispenses between 20 and 200 μL accurately. Make sure this is the case before routinely implementing it in your quantification workflow.
40. This non-optically rated film is used to protect the reaction wells from contamination while transporting and centrifuging the plate. The film should come off easily after centrifugation.
41. DG32™cartridges (microfluidic channels sufficient for four columns of PCR reactions) come in a single block. If a full plate's worth of droplets are not generated, a portion of the DG32™ cartridge may be unused (up to three columns). To prevent contamination and structural issues with the instrument, it is suggested that no attempt be made to salvage unused cartridges in a DG32™ block.
42. Similarly, not all tips within a tip box may be used in any one instance of droplet generation. Save the unused tips and reassemble them into full boxes inside a biosafety cabinet that has been cleaned as described in Subheading 3.1. Transfer unused tips using sterile forceps or a multichannel pipette to populate the new box of tips.
43. Bio-Rad recommends that reaction plates be kept at 4 °C in the thermocycler or refrigerator for at least 2 h after completion of PCR. This allows droplets to reach a proper size for subsequent reading and counting. During plate readout, as many as 85% of the droplets may be rejected in any one well. Droplet loss, however, does not appear to result in substandard quantification. Nevertheless, this recommendation should be followed, as it appears to maintain a consistent droplet count.
44. "Concentration" in the QuantaSoft™ .csv file refers to copies per μL in the 20 μL reaction. This number is derived from the Poisson distribution and correction that is a feature of the methodology and software, and does not yet represent concentration of the actual sample. To obtain final titers, multiply "Concentration" by 20 μL to obtain the total number of copies per reaction well, and then divide by 2 μL, the input sample volume, to obtain the concentration of the diluted sample (this is represented by the factor "10" in the formula). Multiply by the dilution factor of 1,000,000, and then multiply by 1000 to convert from GC/μL to GC/mL. For AAV experimental samples, this generally means multiplying the "Concentration"

value by 1E+10 unless a different dilution factor was used during preparation. Refer to Bio-Rad's Droplet Digital PCR Application Guide for more information [16].

45. This formula measures the number of plasmid copies that were added to the initial DNase-I reaction. A nominal amount of 1E +08 copies in a volume of 5 μL was added. If the spectrophotometry and fluorometry measurements of the stock linear plasmid were accurate, if all subsequent serial dilutions were executed correctly, and if the surfactant F-68 performed as expected, you should obtain a value very close to 1E+08 copies. Once again, first multiply "Concentration" by 20 μL to obtain the total number of copies per reaction well. Then multiply by the dilution factor of 100,000. Finally, multiply by 2.5, as 5 μL of the linear plasmid was used in the initial DNase-I reaction, but only 2 μL of the 100,000-fold dilution was used for the ddPCR reaction. The expected number of plasmid copies in the 20 μL ddPCR reaction of the DNase-I control (without DNase-I) is 400 copies. A "Concentration" value of 20 copies/μL thus represents nearly 100% recovery of the input plasmid.

46. Adequate training of laboratory personnel is a necessity, irrespective of past experience. Intra-laboratory assessment of AAV quantification must be performed for protocol validation (Fig. 11).

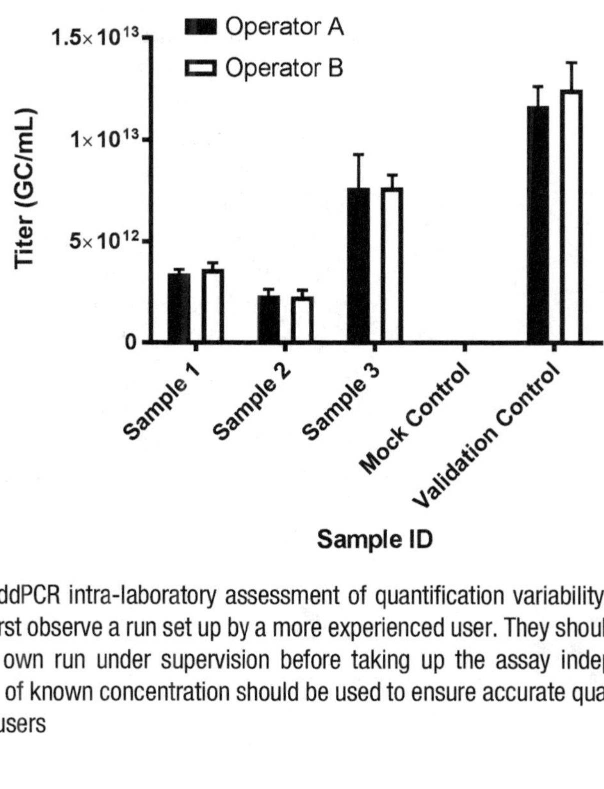

Fig. 11 ddPCR intra-laboratory assessment of quantification variability. Trainees should first observe a run set up by a more experienced user. They should then set up their own run under supervision before taking up the assay independently. Samples of known concentration should be used to ensure accurate quantification by new users

47. For quality control purposes, it is strongly recommended that ddPCR results for a given TaqMan™ assay be compared against standard-based qPCR quantification during the development and validation of new assays, or for standard troubleshooting if both instruments are available. Preferably, real-time PCR validation of new assays should be carried out prior to or concomitantly with ddPCR validation runs, using the same samples to avoid introducing inadvertent biases. This will ensure that future results from one platform do not wrongly influence changes to the other. If the results do not match, revisit the entire procedure and troubleshoot as needed (Fig. 10).
48. All of these reactions will be run in triplicate, but for convenience we assemble one large reaction and split it into three wells of the plate using an electronic pipette. Although this is a shortcut, our experience suggests that if all reactions are set up and pipetted correctly, the outcome will be similar to performing an individual reaction for each well. For ddPCR we add any one sample or control reaction to the plate in triplicate by distributing 2 μL per well using the multi-dispense function of an electronic p20 pipette and a 6 μL aspiration. The qPCR plate is thus set up in a manner equivalent to the ddPCR plate.
49. In qPCR the total copies per reaction ("Qty Mean") are provided directly from the standard curve, so no back calculation is needed as in ddPCR. To obtain GC/μL of the diluted sample, simply divide "Qty Mean" by the input volume per well: 5 μL. ssAAVs are known to package plus (+) and minus (−) genome strands in a 1:1 ratio [35], and the TaqMan™ probe will bind only one of these strands. Because the PCR reaction will generate a detectable AAV amplicon from an undetectable AAV genome strand, and this amplicon will amplify exponentially in subsequent cycles, there is a one cycle differential of detection between the two strands, which corresponds to a twofold reduction compared to the double-stranded linearized plasmid standard. Hence, the final titer is multiplied by 2. For scAAVs this is not needed because the two complementary AAV genome strands are combined in a single particle. Therefore, barring the issue of first-order kinetics of re-annealing, the Ct values of the standard can be compared directly with the Ct values of the scAAV genome in qPCR [33] (*see* **Note 17**).

Acknowledgments

We wish to extend our appreciation to Qin Su and Ru Xiao, directors of Vector Production at the Horae Gene Therapy Center, UMass Medical School and the Gene Transfer Vector Core (vector.meei.harvard.edu) at the Grousbeck Gene Therapy Center, Schepens Eye Research Institute, Mass Eye and Ear, respectively, for providing valuable vectors for the comparisons presented here. This work was supported by 1P01AI100263-05, 1R01NS076991-05, R01 HL097088, and 4P01HL131471-01 (GPG) and Giving/Grousbeck (LHV).

Conflict of Interest: *LHV is an inventor on gene therapy technologies licensed to various biopharmaceutical companies, a founder of Akouos and Gensight, a consultant to multiple entities with gene and genome editing therapeutic interest, and receives sponsored research from Selecta and Solid Biosciences and Lonza Houston.*

References

1. Samulski RJ, Chang LS, Shenk T (1987) A recombinant plasmid from which an infectious adeno-associated virus genome can be excised in vitro and its use to study viral replication. J Virol 61(10):3096–3101
2. Grimm D, Pandey K, Nakai H, Storm TA, Kay MA (2006) Liver transduction with recombinant adeno-associated virus is primarily restricted by capsid serotype not vector genotype. J Virol 80(1):426–439. https://doi.org/10.1128/JVI.80.1.426-439.2006
3. Rabinowitz JE, Rolling F, Li C, Conrath H, Xiao W, Xiao X, Samulski RJ (2002) Cross-packaging of a single adeno-associated virus (AAV) type 2 vector genome into multiple AAV serotypes enables transduction with broad specificity. J Virol 76(2):791–801
4. Hildinger M, Auricchio A, Gao G, Wang L, Chirmule N, Wilson JM (2001) Hybrid vectors based on adeno-associated virus serotypes 2 and 5 for muscle-directed gene transfer. J Virol 75(13):6199–6203. https://doi.org/10.1128/JVI.75.13.6199-6203.2001
5. Gao G, Vandenberghe LH, Wilson JM (2005) New recombinant serotypes of AAV vectors. Curr Gene Ther 5(3):285–297
6. Clement N, Grieger JC (2016) Manufacturing of recombinant adeno-associated viral vectors for clinical trials. Mol Ther Methods Clin Dev 3:16002. https://doi.org/10.1038/mtm.2016.2
7. Sommer JM, Smith PH, Parthasarathy S, Isaacs J, Vijay S, Kieran J, Powell SK, McClelland A, Wright JF (2003) Quantification of adeno-associated virus particles and empty capsids by optical density measurement. Mol Ther 7(1):122–128
8. Pacouret S, Bouzelha M, Shelke R, Andres-Mateos E, Xiao R, Maurer A, Mevel M, Turunen H, Barungi T, Penaud-Budloo M, Broucque F, Blouin V, Moullier P, Ayuso E, Vandenberghe LH (2017) AAV-ID: a rapid and robust assay for batch-to-batch consistency evaluation of AAV preparations. Mol Ther 25(6):1375–1386. https://doi.org/10.1016/j.ymthe.2017.04.001
9. Burnham B, Nass S, Kong E, Mattingly M, Woodcock D, Song A, Wadsworth S, Cheng SH, Scaria A, O'Riordan CR (2015) Analytical ultracentrifugation as an approach to characterize recombinant adeno-associated viral vectors. Hum Gene Ther Methods 26(6):228–242. https://doi.org/10.1089/hgtb.2015.048
10. Lock M, Alvira MR, Chen SJ, Wilson JM (2014) Absolute determination of single-stranded and self-complementary adeno-associated viral vector genome titers by droplet digital PCR. Hum Gene Ther Methods 25(2):115–125. https://doi.org/10.1089/hgtb.2013.131
11. Werling NJ, Satkunanathan S, Thorpe R, Zhao Y (2015) Systematic comparison and validation of quantitative real-time PCR methods for the quantitation of adeno-associated viral products. Hum Gene Ther Methods 26(3):82–92. https://doi.org/10.1089/hgtb.2015.013

12. Rohr UP, Wulf MA, Stahn S, Steidl U, Haas R, Kronenwett R (2002) Fast and reliable titration of recombinant adeno-associated virus type-2 using quantitative real-time PCR. J Virol Methods 106(1):81–88
13. Lock M, McGorray S, Auricchio A, Ayuso E, Beecham EJ, Blouin-Tavel V, Bosch F, Bose M, Byrne BJ, Caton T, Chiorini JA, Chtarto A, Clark KR, Conlon T, Darmon C, Doria M, Douar A, Flotte TR, Francis JD, Francois A, Giacca M, Korn MT, Korytov I, Leon X, Leuchs B, Lux G, Melas C, Mizukami H, Moullier P, Muller M, Ozawa K, Philipsberg T, Poulard K, Raupp C, Riviere C, Roosendaal SD, Samulski RJ, Soltys SM, Surosky R, Tenenbaum L, Thomas DL, van Montfort B, Veres G, Wright JF, Xu Y, Zelenaia O, Zentilin L, Snyder RO (2010) Characterization of a recombinant adeno-associated virus type 2 reference standard material. Hum Gene Ther 21(10):1273–1285. https://doi.org/10.1089/hum.2009.223
14. Bio-Rad (2017) Automated droplet generator instruction manual. Bio-Rad, Hercules, CA. http://www.bio-rad.com/webroot/web/pdf/lsr/literature/10043138.pdf
15. Bio-Rad (2017) QX200™ droplet reader and QuantaSoft™ software instruction manual. http://www.bio-rad.com/webroot/web/pdf/lsr/literature/10031906.pdf
16. Bio-Rad (2017) Droplet Digital™ PCR applications guide. Bio-Rad, Hercules, CA. http://www.bio-rad.com/webroot/web/pdf/lsr/literature/Bulletin_6407.pdf
17. Rayaprolu V, Kruse S, Kant R, Venkatakrishnan B, Movahed N, Brooke D, Lins B, Bennett A, Potter T, McKenna R, Agbandje-McKenna M, Bothner B (2013) Comparative analysis of adeno-associated virus capsid stability and dynamics. J Virol 87(24):13150–13160. https://doi.org/10.1128/JVI.01415-13
18. Plumridge A, Meisburger SP, Pollack L (2017) Visualizing single-stranded nucleic acids in solution. Nucleic Acids Res 45(9):e66. https://doi.org/10.1093/nar/gkw1297
19. Xie J, Mao Q, Tai PWL, He R, Ai J, Su Q, Zhu Y, Ma H, Li J, Gong S, Wang D, Gao Z, Li M, Zhong L, Zhou H, Gao G (2017) Short DNA hairpins compromise recombinant adeno-associated virus genome homogeneity. Mol Ther 25(6):1363–1374. https://doi.org/10.1016/j.ymthe.2017.03.028
20. Bennicelli J, Wright JF, Komaromy A, Jacobs JB, Hauck B, Zelenaia O, Mingozzi F, Hui D, Chung D, Rex TS, Wei Z, Qu G, Zhou S, Zeiss C, Arruda VR, Acland GM, Dell'Osso LF, High KA, Maguire AM, Bennett J (2008) Reversal of blindness in animal models of leber congenital amaurosis using optimized AAV2-mediated gene transfer. Mol Ther 16(3):458–465. https://doi.org/10.1038/sj.mt.6300389
21. Bustin SA, Benes V, Garson JA, Hellemans J, Huggett J, Kubista M, Mueller R, Nolan T, Pfaffl MW, Shipley GL, Vandesompele J, Wittwer CT (2009) The MIQE guidelines: minimum information for publication of quantitative real-time PCR experiments. Clin Chem 55(4):611–622. https://doi.org/10.1373/clinchem.2008.112797
22. Huggett JF, Foy CA, Benes V, Emslie K, Garson JA, Haynes R, Hellemans J, Kubista M, Mueller RD, Nolan T, Pfaffl MW, Shipley GL, Vandesompele J, Wittwer CT, Bustin SA (2013) The digital MIQE guidelines: minimum information for publication of quantitative digital PCR experiments. Clin Chem 59(6):892–902. https://doi.org/10.1373/clinchem.2013.206375
23. Wilson JM (2015) A call to arms for improved vector analytics! Hum Gene Ther Methods 26(1):1–2. https://doi.org/10.1089/hgtb.2015.1502
24. Mueller C, Ratner D, Zhong L, Esteves-Sena M, Gao G (2012) Production and discovery of novel recombinant adeno-associated viral vectors. Curr Protoc Microbiol Chapter 14:Unit14D 11. https://doi.org/10.1002/9780471729259.mc14d01s26
25. Lewis LK, Robson M, Vecherkina Y, Ji C, Beall G (2010) Interference with spectrophotometric analysis of nucleic acids and proteins by leaching of chemicals from plastic tubes. BioTechniques 48(4):297–302. https://doi.org/10.2144/000113387
26. Torcello-Gomez A, Wulff-Perez M, Galvez-Ruiz MJ, Martin-Rodriguez A, Cabrerizo-Vilchez M, Maldonado-Valderrama J (2014) Block copolymers at interfaces: interactions with physiological media. Adv Colloid Interface Sci 206:414–427. https://doi.org/10.1016/j.cis.2013.10.027
27. Mason JB, Gurda BL, Van Wettere A, Engiles JB, Wilson JM, Richardson DW (2017) Delivery and evaluation of recombinant adeno-associated viral vectors in the equine distal extremity for the treatment of laminitis. Equine Vet J 49(1):79–86. https://doi.org/10.1111/evj.12547
28. Markham NR, Zuker M (2005) DINAMelt web server for nucleic acid melting prediction. Nucleic Acids Res 33(Web Server issue):W577–W581. https://doi.org/10.1093/nar/gki591

29. Markham NR, Zuker M (2008) UNAFold: software for nucleic acid folding and hybridization. Methods Mol Biol 453:3–31. https://doi.org/10.1007/978-1-60327-429-6_1

30. Hindson BJ, Ness KD, Masquelier DA, Belgrader P, Heredia NJ, Makarewicz AJ, Bright IJ, Lucero MY, Hiddessen AL, Legler TC, Kitano TK, Hodel MR, Petersen JF, Wyatt PW, Steenblock ER, Shah PH, Bousse LJ, Troup CB, Mellen JC, Wittmann DK, Erndt NG, Cauley TH, Koehler RT, So AP, Dube S, Rose KA, Montesclaros L, Wang S, Stumbo DP, Hodges SP, Romine S, Milanovich FP, White HE, Regan JF, Karlin-Neumann GA, Hindson CM, Saxonov S, Colston BW (2011) High-throughput droplet digital PCR system for absolute quantitation of DNA copy number. Anal Chem 83(22):8604–8610. https://doi.org/10.1021/ac202028g

31. Fagone P, Wright JF, Nathwani AC, Nienhuis AW, Davidoff AM, Gray JT (2012) Systemic errors in quantitative polymerase chain reaction titration of self-complementary adeno-associated viral vectors and improved alternative methods. Hum Gene Ther Methods 23 (1):1–7. https://doi.org/10.1089/hgtb.2011.104

32. Wang Y, Ling C, Song L, Wang L, Aslanidi GV, Tan M, Ling C, Srivastava A (2012) Limitations of encapsidation of recombinant self-complementary adeno-associated viral genomes in different serotype capsids and their quantitation. Hum Gene Ther Methods 23 (4):225–233. https://doi.org/10.1089/hgtb.2012.090

33. Wagner A, Rohrs V, Kedzierski R, Fechner H, Kurreck J (2013) A novel method for the quantification of adeno-associated virus vectors for RNA interference applications using quantitative polymerase chain reaction and purified genomic adeno-associated virus DNA as a standard. Hum Gene Ther Methods 24 (6):355–363. https://doi.org/10.1089/hgtb.2013.095

34. Bio-Rad (2017) QX200™ Droplet Generator Instruction Manual. Bio-Rad, Hercules, CA. http://www.bio-rad.com/webroot/web/pdf/lsr/literature/10031907.pdf

35. Rose JA, Berns KI, Hoggan MD, Koczot FJ (1969) Evidence for a single-stranded adenovirus-associated virus genome: formation of a DNA density hybrid on release of viral DNA. Proc Natl Acad Sci U S A 64(3):863–869

Chapter 5

Single-Stranded DNA Virus Sequencing (SSV-Seq) for Characterization of Residual DNA and AAV Vector Genomes

Emilie Lecomte, Adrien Leger, Magalie Penaud-Budloo, and Eduard Ayuso

Abstract

With the success of clinical trials using recombinant adeno-associated viral vectors (rAAV), regulatory agencies ask for a more comprehensive characterization of process- and product- related impurities found in rAAV stocks in order to assess the potential risks for patients. During production, rAAV capsids are known to internalize illegitimate DNA fragments in addition to their recombinant genome. These contaminants can come from plasmid or helper virus DNA as well as from the producer host cell. Here, we describe a method based on high-throughput sequencing to identify and quantify residual DNA in rAAV vector lots. Contrary to qPCR, SSV-Seq (Single-Stranded DNA Virus Sequencing) offers a nonselective approach to determine the percentage of each DNA contaminant and analyze rAAV vector genome identity.

Key words AAV, Viral vectors, Quality control, Residual DNA, Identity, Sequencing

1 Introduction

Positive outcomes of clinical trials using recombinant adeno-associated virus (rAAV), together with the market authorization of Glybera and Luxturna, have significantly boosted the gene therapy field. To ensure the quality and the purity of clinical grade rAAV batches the Food and Drug Administration (FDA) and the European Pharmacopoeia published a series of compulsory quality control assays. The presence of residual DNA in the final product is a significant concern. These fragments could be derived from the host cell genome, the plasmid DNA, or helper virus sequences used for production. Together, these sequences imply potential risks including the synthesis of viral protein, immunogenicity, and insertional mutagenesis. To limit these risks, the FDA recommends the level of host cell DNA to be lower than 10 ng per parenteral dose

Magalie Penaud-Budloo and Eduard Ayuso contributed equally to this work.

Michael J. Castle (ed.), *Adeno-Associated Virus Vectors: Design and Delivery*, Methods in Molecular Biology, vol. 1950, https://doi.org/10.1007/978-1-4939-9139-6_5, © Springer Science+Business Media, LLC, part of Springer Nature 2019

[1]. However, no instruction is given regarding the method to quantify DNA contaminants. Currently, quantitative qPCR is used by most manufacturers to monitor host cell DNA levels. Although quick and simple, qPCR has many pitfalls and is subject to intra- and inter-laboratory variability [2]. This strategy also requires selecting representative targets for which there is no consensus, in particular for the cellular DNA. Here, we propose a novel technique based on Illumina sequencing to quantify DNA contaminants comprehensively in rAAV vector batches. The Single-Stranded DNA Virus Sequencing (SSV-Seq) protocol consists of four successive experimental steps (Fig. 1): (1) DNA extraction from single-stranded AAV vector batches, (2) second-strand synthesis, (3) preparation of the sequencing library, and (4) high-throughput sequencing based on Illumina technology. This is followed by data analysis using our dedicated bioinformatics pipeline (Fig. 3). SSV-Seq is a significant improvement over qPCR, as it can detect, identify, and quantify all DNA species. Furthermore, SSV-Seq is a more comprehensive method than qPCR, allowing a precise quantification of the host cell residual DNA. Using SSV-Seq, we have shown that the vector plasmid is the major DNA impurity found in rAAV produced in mammalian HEK293 cells [3]. Similarly, the baculovirus genome is the main contaminant found in the insect Sf9 cell production system [4]. As an additional advantage, SSV-Seq can identify single nucleotide variants (SNV) and indels in the rAAV genome.

2 Materials

Decontaminate laboratory surfaces and equipment by wiping with RNase AWAY™ reagent. Irradiate pipettes, racks, and reagents that are not sensitive to UV with a 15 min exposure to UV light. To prevent environmental and cross contamination, aliquot all reagents for single use. Diligently follow all waste disposal regulations.

2.1 Negative Control and Internal Normalizer

1. Ultrapure DNase/RNase free distilled water (dH_2O).
2. TE 10:1: 10 mM Tris–HCl, 1 mM EDTA, pH 8.0.
3. λ Phage DNA in TE 10:1 buffer.
4. g-TUBE™ (Covaris).
5. NucleoSpin® Gel and PCR Clean-up kit (Macherey-Nagel).
6. Bioruptor (Diagenode) (*see* **Note 1**).
7. 0.5 mL Bioruptor® plus microtubes (Diagenode).
8. Agarose gel.
9. Microcentrifuge.

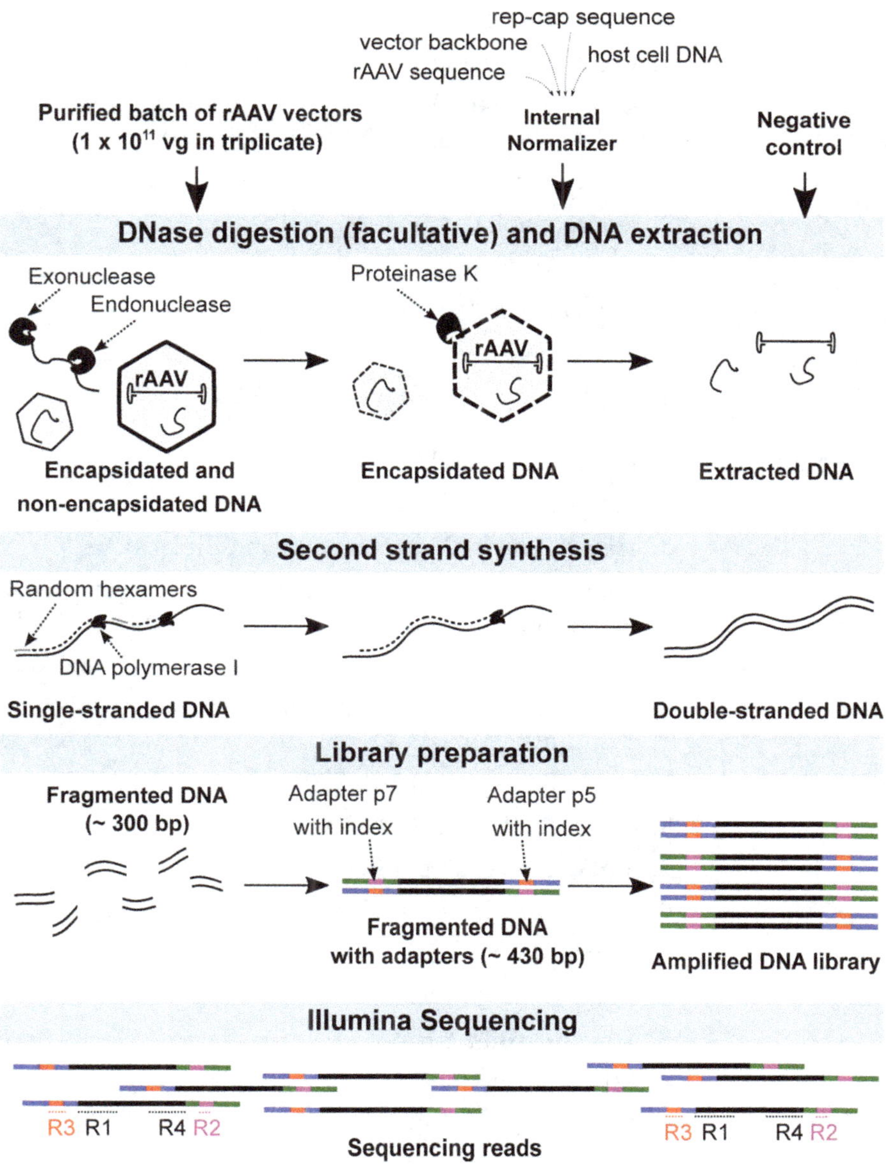

Fig. 1 SSV-Seq workflow. 3×10^{11} vector genomes (1×10^{11} vg in triplicate) are required for SSV-Seq analysis of DNA species in a recombinant AAV vector stock. DNase pretreatment can be performed to specifically identify and quantify DNA impurities that are encapsidated in rAAV capsids. After DNA extraction and second-strand synthesis, a sequencing library is prepared based on the TruSeq Illumina dual-index strategy [7]. The sequencing reads obtained by high-throughput sequencing are then processed through the SSV-Conta bioinformatics pipeline (Fig. 3)

2.2 DNase Digestion and DNA Extraction

1. Baseline-ZERO™ DNase at 1 U/μL with 10× Reaction buffer and 10× Stop solution (Epicentre Biotechnologies).
2. Plasmid-Safe™ ATP-Dependent DNase at 10 U/μL supplied with 10× Reaction buffer and 25 mM ATP solution (Epicentre Biotechnologies).

3. 20 mg/mL Proteinase K.
4. Cell lysis solution (*see* **Note 2**) (Qiagen).
5. 4 mg/mL and 100 U/mL exonuclease- and endonuclease-free RNase A solution (*see* **Note 2**) (Qiagen).
6. Protein precipitation solution (*see* **Note 2**) (Qiagen).
7. Isopropanol.
8. RNA-grade Glycogen.
9. 70% and 100% ethanol.
10. Microcentrifuge.
11. High-speed centrifuge.

2.3 Second-Strand Synthesis

1. PCR thermal cycler.
2. 0.2 mL PCR tubes and caps.
3. 10× NEBuffer™ 2 (New England Biolabs).
4. 18.6 nmol of Random Primer 6 (New England Biolabs) dissolved in 80 μL of ultrapure DNase/RNase free distilled water.
5. 10 U/μL DNA Polymerase I (*E. coli*) (New England Biolabs).
6. 2 mM nucleotide mix: 2 mM dATP, 2 mM dTTP, 2 mM dGTP, 2 mM dCTP in dH_2O, pH 7.0.
7. 1 mM fluorescein-12-dUTP.
8. 0.5 M EDTA, pH 8.0.
9. 5 mM Tris–HCl, pH 8.5.
10. Protein precipitation solution (*see* **Note 2**) (Qiagen).
11. Isopropanol.
12. RNA-grade Glycogen.
13. 70% and 100% ethanol.
14. Microcentrifuge.
15. High-speed centrifuge.

2.4 Illumina Sequencing Library Preparation

1. PCR tubes.
2. SPRIselect® (Beckman Coulter).
3. Bioruptor (Diagenode).
4. 0.5 mL Bioruptor® plus microtubes (Diagenode).
5. TE 10:1: 10 mM Tris–HCl, 1 mM EDTA, pH 8.0.
6. Magnetic rack.
7. Freshly prepared 80% ethanol.
8. 10× T4 DNA ligase Reaction buffer (New England Biolabs).
9. 10 mM deoxynucleotide (dNTP) mix: 10 mM dATP, 10 mM dTTP, 10 mM dGTP, 10 mM dCTP.

10. 100 mM dATP.
11. 3 U/μL T4 DNA polymerase (New England Biolabs).
12. 5 U/μL DNA Polymerase I, Large (Klenow) Fragment (New England Biolabs).
13. 10 U/μL T4 Polynucleotide Kinase (New England Biolabs).
14. 5 U/μL Klenow fragment lacking 3′ to 5′ exonuclease activity (New England Biolabs).
15. 2000 U/μL Quick ligase (New England Biolabs).
16. 2× Quick ligase buffer (New England Biolabs).
17. 100 μM oligonucleotides P5-iX and P7-iX in 5 mM Tris, pH 8.5.
    (a) P5-iX oligonucleotide: 5′-AATGATACGGCGACCACC GAGATCTACACNNNNNNACACTCTTTCCCTACA CGACGCTCTTCCGATC*T.
    (b) P7-iX oligonucleotide 5′-P-GATCGGAAGAGCACACG TCTGAACTCCAGTCACNNNNNNATCTCGTATG CCGTCTTCTGCTTG.

    The asterisk (*) represents a phosphorothioate linkage, P represents a phosphorylated 5′-end, and NNNNNN represents a TruSeq Illumina-compatible six base-index (iX) used to identify samples after multiplexing.
18. PfuUltra II Fusion HS DNA Polymerase (Agilent Technologies).
19. 10× PfuUltra II Reaction Buffer (Agilent Technologies).
20. 25 mM PCR-grade deoxynucleotide mix: 25 mM dATP, 25 mM dTTP, 25 mM dGTP, 25 mM dCTP (Agilent Technologies).
21. 100 μM Primer P5-F 5′-AATGATACGGCGACCACCG.
22. 100 μM Primer P7-R 5′-CAAGCAGAAGACGGCATAC.
23. Spectrophotometer (*see* **Note 3**).
24. High Sensitivity DNA kit including microfluidic chips, reagents, and consumables (Agilent Technologies).
25. 2100 Bioanalyzer (Agilent Technologies).

2.5 High-Throughput Sequencing

1. KAPA Library Quantification Kit (Kapa Biosystems).
2. PhiX Control v3 DNA (Illumina).
3. Second-generation sequencer compatible with TruSeq Illumina adaptors (*see* **Note 4**).

2.6 Bioinformatics

The SSV-Conta package is used to quantify and characterize DNA contaminants in AAV vector batches after Illumina-based sequencing. The package requires a UNIX environment. The latest version of the package is freely available under a GPL License (*see* **Note 5**) with extensive documentation detailing principles, installation, and usage at https://github.com/emlec/SSV-Conta.

2.6.1 Software

1. bcl2fastq conversion software (Illumina) is used to convert BCL files generated by Illumina sequencing systems to standard FASTQ files (*see* **Note 6**). bcl2fastq can be downloaded from the Illumina website (https://support.illumina.com/downloads.html). Installation instructions are provided in the software guide.
2. Samtools [5] or the python module pysam (https://github.com/pysam-developers/pysam) are used to determine sequencing coverage at each base of the reference sequence and for variant calling analysis.

   The following programs are included in the SSV-Conta package. The program versions listed below may not correspond to the latest versions hosted by github since each program may include new features:
3. Quade (v0.3.2, python2.7 object oriented script) is used to demultiplex samples from mixed FASTQ files (*see* **Note 7**).
4. Sekator (v0.2.1, python2.7/Cython/C) is used to perform FASTQ quality trimming and adapter trimming (*see* **Note 8**).
5. RefMasker (v0.1, python2.7 object oriented script) is used to mask homologies between fasta reference sequences identified by Blastn.
6. ContaVect (v0.2.1, python2.7 object oriented script) is used to identify and quantify DNA species in rAAV batches (*see* **Note 9**).
7. fastq_control_sampler (C program) is used to generate the paired-end reads of the in silico control from FASTA reference sequences and a randomly generated sequence (*see* **Note 10**).

2.6.2 Reference Sequences

The following reference sequences in FASTA format (*see* **Note 11**) are required for the SSV-Seq analysis:

1. Recombinant AAV genome sequence.
2. Vector shuttle backbone sequence from the vector plasmid or the vector bacmid.
3. Helper shuttle sequence containing the rep-cap sequence from the helper plasmid or the rep-cap bacmid.
4. Producer cell genome sequence: for characterization of rAAV vectors produced in HEK293 mammalian cells, use the adenovirus 5 sequence (nucleotides 1 to 4344 of the Ad5 complete genome, AC_000008) (*see* **Note 12**) and the human genome (GRCh38 primary assembly). For characterization of rAAVs

produced in Sf9 insect cells, use either the closely related reference genome Sf21 (number GCA_000753635.2, WGS number JQCY02.1) or the recently published Sf9 cell line genome (number GCA_002213285.1) (*see* **Note 13**).

5. The λ phage sequence (J02459.1).
6. The coliphage phi-X174 complete genome sequence (J02482.1).

3 Methods

3.1 Negative Control and Internal Normalizer

1. High-throughput sequencing is a highly sensitive technology, and thus a negative control composed of fragmented λ phage DNA is included in the SSV-Seq protocol in order to control for environmental contamination during sample handling (Fig. 1).
2. For λ phage DNA fragmentation, dilute 2 μg of λ phage DNA in 150 μL of TE 10:1, load onto one g-TUBE™, and centrifuge at 13,000 × *g* for 30 s. Invert the tube and repeat the centrifugation. Determine the fragment distribution of the sheared DNA by agarose gel electrophoresis. Fragment size must be centered on 5–6 kb (ranging from 3 to 10 kb). Store at −20 °C.
3. For normalization of SSV-Seq analyses, prepare an internal normalizer control in parallel to the rAAV sample (Fig. 1). The internal normalizer consists of a mix of all DNA molecules that are expected to be found in rAAV stocks.
4. For producer cell DNA fragmentation, dilute 6 μg of genomic DNA (gDNA) extracted from HEK293 or Sf9 cells in 150 μL of TE 10:1. Load into one g-TUBE™. Follow the same protocol as for λ phage DNA (**step 2**).
5. For the internal normalizer of HEK293-derived vectors, digest the vector plasmid using restriction enzyme(s) to release the rAAV genome (including ITRs) from the plasmid backbone. Separate both fragments, the rAAV genome and the plasmid backbone, by agarose gel electrophoresis and purify them using the NucleoSpin® Gel and PCR Clean-up kit. Linearize the helper rep-cap plasmid by restriction endonuclease and purify using the NucleoSpin® Gel and PCR Clean-up kit.
6. For the internal normalizer of Sf9-derived vectors, prepare the rAAV genome fragment as above (**step 5**), and shear the vector bacmid DNA and the rep-cap bacmid DNA as follows. Prepare 2 μg of bacmid DNA in 100 μL of TE 10:1 and transfer to a 0.5 ml Bioruptor microtube. Sonicate 1 × 15 s at 160 W (LOW power on Bioruptor UCD-200) to fragment DNA at an average size of 1.5 kb as verified by agarose gel electrophoresis.

7. Prepare an internal normalizer control for each rAAV vector batch by mixing all DNA species that can theoretically be found in the rAAV lot in a final volume of 100 μL of ultrapure DNase/RNase free distilled water (dH_2O). Quantify each DNA fragment isolated in **steps 5** and **6** by qPCR using the appropriated target sequence. For the analysis of rAAV vectors generated in HEK293 cells by dual-plasmid transfection, prepare the internal control by mixing 1×10^{11} copies of the rAAV vector genome fragment (**step 5**) [6], 1×10^{10} copies of the vector plasmid backbone fragment (**step 5**), 4×10^{9} copies of linearized helper plasmid (**step 5**), and 0.4 ng of HEK293 sheared gDNA (**step 4**) (*see* **Note 14**). For the analysis of rAAV vectors produced in Sf9 insect cells by dual-baculovirus infection, mix 1×10^{11} copies of the rAAV vector genome fragment (**step 6**) [6], 2×10^{7} copies of sonicated vector bacmid DNA (**step 6**), 2×10^{7} copies of sonicated rep-cap bacmid DNA (**step 6**), and 22 ng of Sf9 sheared gDNA (**step 4**) (*see* **Note 14**).

3.2 DNase Digestion and DNA Extraction

1. For the analysis of rAAV vector lots with DNase pretreatment (+DNase) (*see* **Note 15**), prepare the following mixture: 10^{11} copies of rAAV vector genome as determined by free ITR qPCR [6], 24.2 ng of sheared λ phage DNA, 20 μL of baseline-ZERO 10× reaction buffer, 8 μL of 25 mM ATP, 10 μL (10 U) of baseline-ZERO DNase, 4 μL (40 U) of plasmid-safe DNase, and dH_2O to a final volume of 200 μL. The reaction is performed in triplicate (*see* **Note 16**). Incubate for 2 h at 37 °C. Stop the reaction with 20 μL of baseline-ZERO 10× stop solution and 30 min incubation at 75 °C. For the analysis of rAAV vector lots without DNase pretreatment (-DNase) (*see* **Note 15**), substitute baseline-ZERO and plasmid-safe DNases with dH_2O. Perform the reaction in triplicate.
2. For the environmental negative control, add only 484 ng of sheared λ phage DNA to the "-DNase" reaction mixture.
3. For the internal normalizer, replace rAAV and λ phage DNA in the "-DNase" reaction mixture with the 100 μL normalizer library prepared as described in **step 5** of Subheading 3.1.
4. Add 300 μL of cell lysis solution to all samples and vortex vigorously for 10 s.
5. Add 20 μL of proteinase K to all samples, mix by inversion, and incubate for 3 h at 55 °C.
6. Add 1.5 μL of RNase A solution to all samples, mix by inversion, and incubate for 15 min at 37 °C.

7. Cool the tubes on ice for 1 min, add 100 μL of protein precipitation solution, and vortex vigorously for 20 s at high speed.
8. Incubate on ice for 5 min and centrifuge at 16,000 × *g* for 5 min at 4 °C.
9. Carefully transfer the supernatant to a clean tube. Add 300 μL of isopropanol and 2 μL of glycogen and mix by inverting gently ten times. Place the tubes at −20 °C overnight.
10. Centrifuge at 25,000 × *g* for 45 min at 4 °C. Place on ice for approximately 5 min (until the pellet becomes visible). Carefully discard the supernatant.
11. Add 300 μL of 70% ethanol and centrifuge at 25,000 × *g* for 10 min at 4 °C.
12. Carefully discard the supernatant and air-dry the pellet for 1 h before resuspension in 20 μL of dH_2O.
13. Incubate at 65 °C for 1 h (homogenize by pipetting after 30 min of incubation), then incubate at room temperature for 1 h. Store extracted DNA at 4 °C, or at −20 °C for extended storage.

3.3 Second-Strand Synthesis

1. Transfer all samples into PCR tubes. For rAAV samples, pool the three replicate tubes and distribute 20 μL into three tubes.
2. Prepare the random hexanucleotides mix by combining 5 μL of 10× NEBuffer 2, 12.5 μL of random primers, and 4.5 μL of dH_2O.
3. Prepare the DNA polymerase mix by combining 2 μL of 2 mM nucleotide mix, 1 μL (10 U) of DNA polymerase I, and 8 μL of dH_2O (*see* **Note 17**).
4. Add 22 μL of the random hexanucleotides mix to each sample, heat 5 min at 95 °C in a PCR machine, and place immediately on ice.
5. Add 11 μL of the DNA polymerase mix to each tube and place the tubes in a precooled PCR machine at 4 °C.
6. Use a program for second-strand synthesis consisting of a temperature increase from 4 °C to 37 °C with a ramp of 0.1 °C/s for random primer hybridization, followed by DNA polymerization at 37 °C for 1 h. Stop the reaction by adding 1 μL of 0.5 M EDTA.
7. Add 300 μL of dH_2O and 100 μL of protein precipitation solution. Vortex for 20 s at high speed.
8. Incubate for 5 min on ice and centrifuge for 5 min at 16,000 × *g* at 4 °C.
9. Transfer the supernatant in a new tube and add 300 μL of isopropanol and 2 μL of glycogen. Homogenize by inverting the tube ten times.

10. Place at −20 °C overnight.
11. Centrifuge at 25,000 × *g* for 45 min at 4 °C. Place the tube on ice for 5 min.
12. Carefully discard the supernatant.
13. Wash the pellet with 300 μL of 70% ethanol and centrifuge for 10 min at 25,000 × *g* at 4 °C.
14. Carefully discard the supernatant and air-dry for approximately 1 h.
15. Resuspend the pellet in 30 μL of 5 mM Tris–HCl pH 8.5 and leave the tube 1 h at room temperature. Pool the replicates for rAAV samples and determine DNA concentration by UV spectrophotometry (*see* **Note 18**).
16. Store at 4 °C if processed the next day or at −20 °C for longer time.

3.4 Illumina Sequencing Library Preparation

The next-generation sequencing (NGS) library is prepared according to a protocol adapted from Kozarewa and Turner [7].

1. Anneal the 40 μM adapters for the adaptor ligation step as follows. In PCR tubes, mix 20 μL of 100 μM oligonucleotide P5-iX, 20 μL of 100 μM oligonucleotide P7-iX, 5 μL of ultrapure DNase/RNase free water, and 5 μL of 10× T4 DNA ligase reaction buffer. In a thermal cycler, denature for 2.5 min at 97.5 °C, then cool to 20 °C with a ramp of 0.1 °C every 5 s. Place the tubes immediately at 4 °C after oligonucleotide annealing. Aliquot adapters and store at −80 °C.
2. In 0.5 mL Bioruptor microtube, dilute 200 ng of each sample in 100 μL of TE 10:1. For the environmental control, add a volume corresponding to the maximal sample volume added for the rAAV conditions.
3. Perform DNA fragmentation using a Bioruptor sonicator following the conditions: 12 cycles of 30 s ON at LOW power (160 W) followed by 90 s OFF, with short spin centrifugation every 2 cycles (*see* **Note 19**). Place tubes on ice.
4. Thoroughly shake the SPRIselect bottle to resuspend SPRI beads and add 160 μL of the bead slurry to each tube. Vortex 30 s to 1 min until homogenous, then incubate for 5 min at room temperature.
5. Place the tubes on a magnetic rack and allow the beads to settle to the magnet. Carefully remove and discard the clear supernatant.
6. Add 360 μL of freshly prepared 80% ethanol and incubate at room temperature for 30 s. Carefully remove and discard the ethanol supernatant. Repeat the ethanol wash a second time.

7. Air-dry the beads for 10–15 min, leaving the tube on the magnetic rack.
8. To elute the sample, remove the tube from the magnet and add 45 μL of dH_2O. Pipet ten times to resuspend the beads, then incubate for 2 min at room temperature.
9. Place the tube on the magnetic rack and allow the beads to settle the magnet. Transfer the clear eluate to a new tube. Take a 2 μL aliquot ("aliquot #1").
10. After purification of the fragmented DNA, repair ends as follows: add 30 μL of dH_2O, 10 μL of 10× T4 DNA ligase reaction buffer, 4 μL of 10 mM dNTP mix, 5 μL (15 U) of T4 DNA polymerase, 1 μL (5 U) of DNA polymerase I large (Klenow) fragment, and 5 μL of T4 polynucleotide kinase to each tube. Incubate for 30 min at room temperature.
11. Repeat SPRIselect purification **steps 4–9**, but elute in 32 μL of dH_2O.
12. For A-tailing of the 3′ ends, add 5 μL of 10× NEBuffer 2, 10 μL of 1 mM dATP (freshly diluted in dH_2O from the 100 mM dATP solution), and 3 μL (15 U) of Klenow fragment to each tube. Incubate for 30 min at 37 °C.
13. Repeat purification **steps 4–9** with the following SPRIselect parameters: 80 μL of SPRIselect beads (**step 4**), two washes with 180 μL of 80% ethanol (**step 6**), and elution with 15 μL of dH_2O (**step 8**).
14. For ligation of adapters, combine 3 μL of dH_2O, 25 μL of 2× Quick ligase buffer, 2 μL of 4 μM adapters (diluted in dH_2O from the 40 μM adapters prepared in **step 1**), and 5 μL (10,000 U) of Quick ligase. Incubate for 15 min at room temperature.
15. Repeat purification **steps 4–9** with the following SPRIselect parameters: 50 μL of SPRIselect beads (**step 4**), two washing steps with 180 μL of 80% ethanol (**step 6**), and elution with 50 μL of dH_2O (**step 8**). Repeat this same purification a second time, but elute with 30 μL of dH_2O.
16. Perform PCR amplification from 5 μL of each sample prior to sequencing (*see* **Note 20**). To make the PCR mix, combine 36 μL of dH_2O, 5 μL of 10× PfuUltra II reaction buffer, 1.25 μL of 10 μM primer P5-F, 1.25 μL of 10 μM primer P7-F, 0.5 μL of 25 mM PCR-grade deoxynucleotide mix, and 1 μL of PfuUltra II Fusion DNA Polymerase. Use the following amplification program: 1 cycle of initial denaturation at 95 °C for 2 min; 15 cycles of denaturation at 95 °C for 20 s, annealing at 60 °C for 20 s, and elongation at 72 °C for 15 s; final elongation at 72 °C for 3 min.

17. Repeat SPRIselect purification twice, as described in **step 15**. Take a 2 μL aliquot ("aliquot #2").
18. Verify the distribution of DNA fragment size post-sonication (aliquot #1) and post-PCR (aliquot #2) on an Agilent 2100 Bioanalyzer system using the High Sensitivity DNA kit according to the manufacturer's instructions (*see* **Note 19**). The average size should be centered on approximately 300 bp after sonication (Fig. 2a) and 400 bp after adapter ligation and PCR (Fig. 2b).

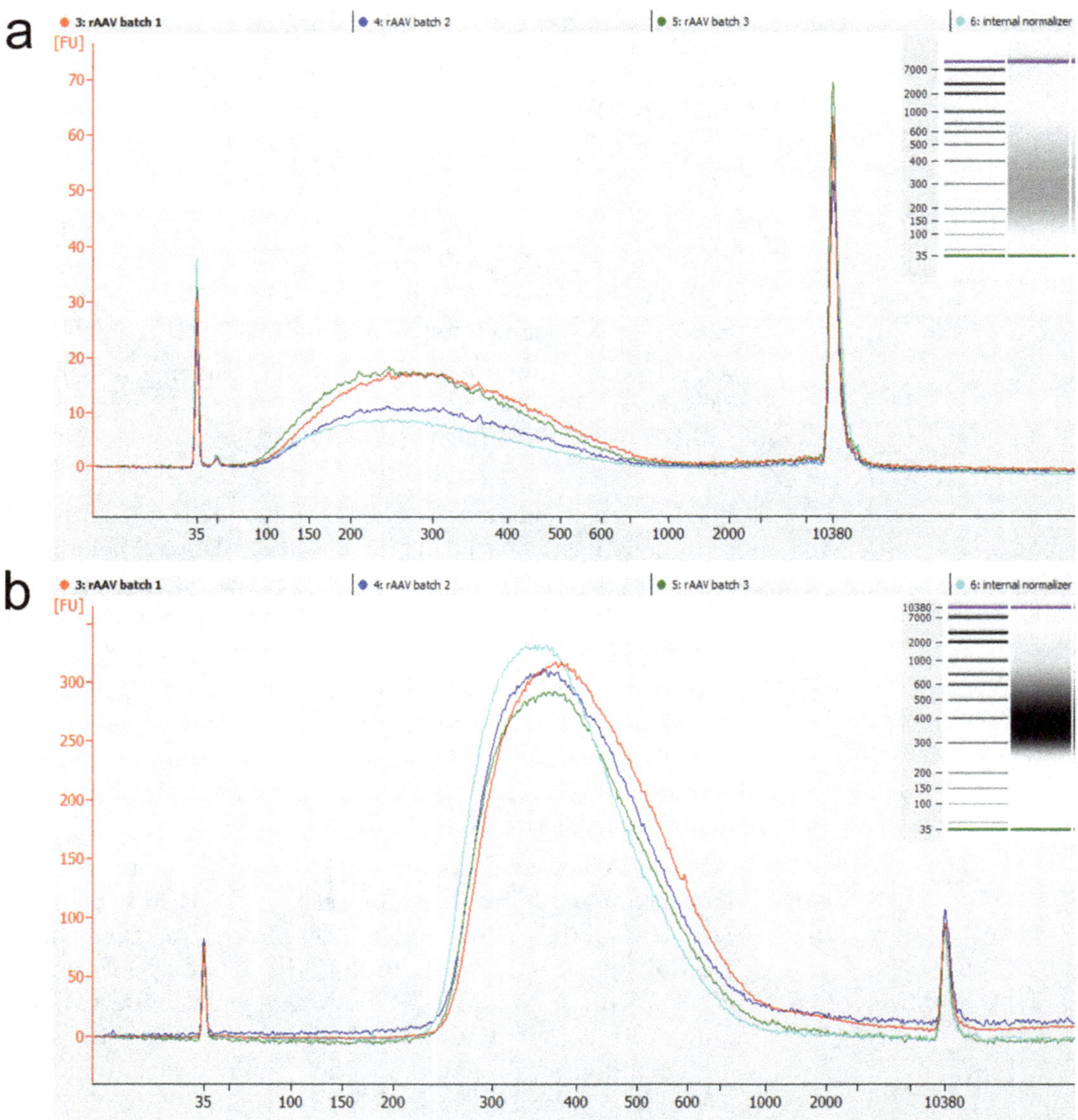

Fig. 2 DNA fragment size. The distribution of DNA fragment size in post-sonication (**a**) and post-PCR (**b**) samples (aliquot #1 and #2, respectively) is determined using an Agilent 2100 Bioanalyzer instrument

3.5 High-Throughput Sequencing

1. Quantify samples using a KAPA Library Quantification kit according to the manufacturer's instructions. DNA concentration must be higher than 4 nM (10 nM recommended) to perform Illumina sequencing. For multiplexing (optional), pool samples in equimolar quantities (*see* **Note 21**).
2. Spike 1% of PhiX Control v3 DNA into the samples before sequencing with a HiSeq 2500 system (Illumina) using the rapid-run paired-end mode (2 × 99 bp).

3.6 Bioinformatics

Perform data analysis using a dedicated bioinformatics pipeline as shown in Fig. 3. All steps are provided as UNIX bash shell command lines. As the pipeline involves the chained execution of several command lines, it is recommended to automate this pipeline using a workflow engine such as Snakemake.

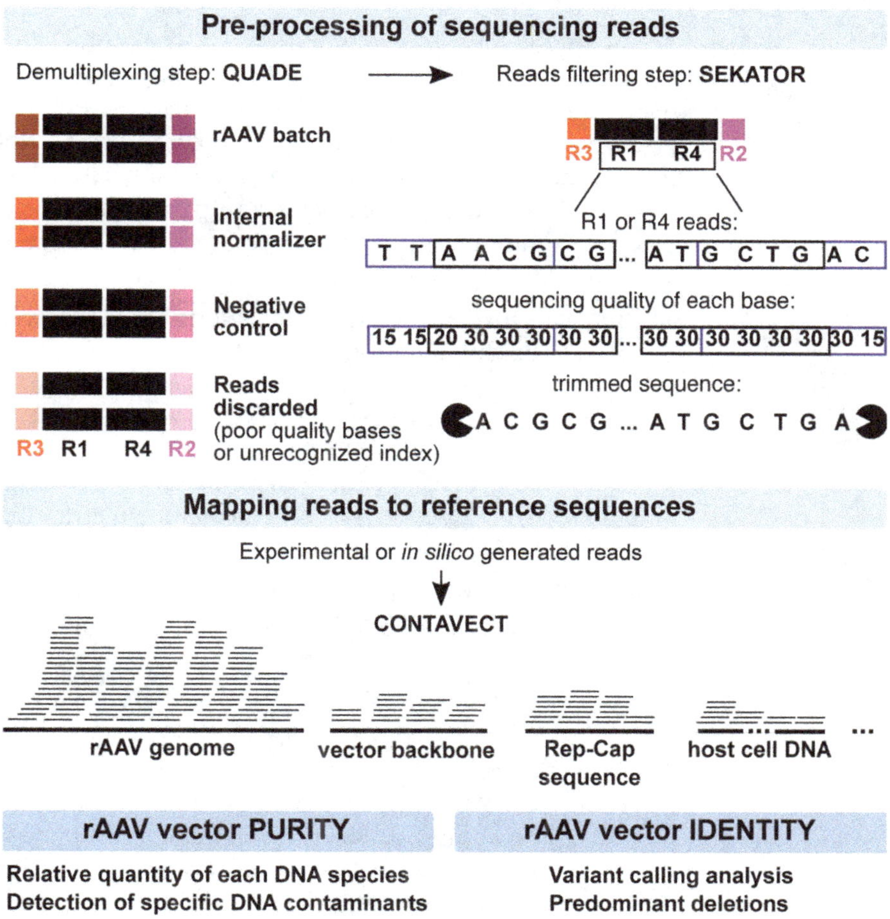

Fig. 3 SSV-Conta bioinformatics pipeline. The sequencing reads are preprocessed using the programs Quade (FASTQ files demultiplexer) and Sekator (quality and adapter trimmer) to ensure the reliability of the analysis. Next, ContaVect attributes reads to reference sequences, allowing the relative quantification of each DNA species in the rAAV vector stock. Additional analyses such as variant calling can be performed on the BAM files

3.6.1 Preprocessing of Sequencing Reads

1. Generate FASTQ files using the bcl2fastq program (*see* **Note 22**). The following command is used for insert sequencing reads of 99 bases (Rl and R4) and index reads of 6 bases (dual-indexing, R2 and R3) and must be modified with your specific read lengths:

```
configureBclToFastq.pl --input-dir <BaseCalls_dir> --out-
put-dir <Fastq_dir>
--sample-sheet <SamplesSheet.csv> --force --use-bases-
mask Y99,Y6,Y6,Y99
```

2. Run Quade with the option -i to create an example configuration file containing all program parameters:

```
Quade.py -i
```

3. Fill in the Quade configuration file indicating all program parameters: the path to the non-demultiplexed FASTQ files, the start and end positions of each index (e.g., 1 and 6, respectively, for index reads composed of 6 bases), the index sequences specific to each sample, the minimal quality for one base of the index to consider a read pair valid (25 is the value most commonly used by our laboratory), and the expected output files (*see* **Note 23**).
4. Run the Quade program with the option -c followed by the path to the configuration file:

```
Quade.py -c <Quade_conf_file.txt>
```

   Quade generates demultiplexed FASTQ files and a tabulated report (Quade_report.csv) containing metrics that can be used for quality control.
5. Run the Sekator program with the option -i to create an example configuration file containing all program parameters:

```
Sekator.py -i
```

6. Fill in the Sekator configuration file indicating all program parameters. We typically use the following: minimal size of read after trimming = 30; perform quality trimming from the left and/or extremities = True (for both left_trim and right_-trim); size of the sliding window in which quality will be computed = 6; step of sliding window during trimming = 2; minimal quality in a given window to be retained during trimming = 28; perform a step of adapter trimming = True; minimal fraction of the length of the adapter matching on the

read = 0.3; minimal SSW score/base of the adapter matching on the read = 1; ssw_match = 2, ssw_mismatch = 2; ssw_gapO = 3; ssw_gapE = 1.

7. Run the Sekator program with the option -c followed by the path to the configuration file:

```
Sekator.py -c <Sekator_conf_file.txt>
```

Similar to Quade, filtered FASTQ files and a tabulated report containing metrics are generated.

3.6.2 Creation of FASTQ Files for the In Silico Control

1. Run fastq_control_sampler using a command based on the following:

```
fastq_control_sampler -m -n 1000 -e 0.001 -s 99 -l 250 -u
1000 -r 1000000 <rAAV_genome_sequence.fa> 6500000
<helper_shuttle_sequence.fa>
5000 <vector_shuttle_backbone_sequence.fa> 105000
```

The above command uses these example parameters: number of pairs generated for the random reference = 1000; frequency of errors per base in FASTQ = 0.001; size of reads to be selected in FASTA references = 99; lower size of fragments obtained after sonication = 250; upper size of fragments obtained after sonication = 1000; size of random reference to be generated = 1,000,000; number of pairs generated for the rAAV genome = 6,500,000; number of pairs generated for the shuttle helper = 5000; number of pairs generated for the backbone of the shuttle vector = 105,000.

2. Adjust the above parameters according to the characteristics of the rAAV library and sequencing run. Determine the lower and upper fragment sizes using the Agilent 2100 Bioanalyzer system and High Sensitivity DNA kit as described in **step 18** of Subheading 3.4. The number of in silico reads generated for each reference sequence should be a rough estimate of the expected proportion of each DNA species in the rAAV batch. The size of the in silico reads must be equal to that of the experimental dataset. The frequency of errors per base should reflect the error rate based on the sequencing technology. fastq_control_sampler generates paired-end reads in FASTQ files.

3.6.3 Mapping of Reads to Reference Sequences

1. Copy and fill in the configuration file with your data files and settings as indicated in the template "Conf.txt," provided in the ContaVect source files. The parameters commonly used are: reference masking = True; blastn evalue = 0.1; quality filtering = False; adapter trimming = False; bwa mem option

= M; minimal quality mapping of a read to be considered as valid = 20; minimal size of mapping = 25.

2. Run the ContaVect program:

```
ContaVect.py <Conf.txt>
```

3. Sequence alignments are saved in separate BAM files depending on the sequence they aligned to, such as "rAAV vector genome" and "producer cell genome sequence". The file "Unmapped.bam" contains unmapped reads and alignments with length below the minimum mapping size. The file "LowMapq.bam" contains sequence alignments with mapping quality below the minimum mapping quality. The file "Secondary.bam" contains additional reads alignments (*see* **Note 24**). The file "Reference_distribution.csv" summarizes the number of reads aligned to each reference as well as the sum of the unmapped and LowMapq reads.

3.6.4 Determine Relative Quantities of rAAV Genomes and DNA Contaminants

1. Calculate the proportion of each DNA contaminant as a percentage of the total mapped reads. Exclude reads assigned to the λ phage genome and the PhiX genome, as well as unmapped and LowMapq reads (*see* **Note 25**).
2. Generate read coverage plots to identify and visualize overrepresentation of a particular region in the reference sequences (*see* **Note 26**). For reference sequences composed of a unique element, such as the rAAV genome and vector shuttle, first generate a file containing the sequencing coverage along each base of the reference sequence with samtools depth [5] or the python module pysam. Then, at each position, divide the sequencing coverage by the total number of sequenced nucleotides. Finally, represent the normalized sequencing coverages obtained from the rAAV vector lot and its internal normalizer on the same graphic. For reference sequences composed of several elements such as the human or *Spodoptera frugiperda* genome, calculate the number of reads mapped to each element, e.g., chromosomes, mitochondrial DNA, or contigs. For the rAAV batch, normalize to the read count of the internal normalizer and visualize each element in one or more graphics, depending of the number of elements. A value of 1 indicates a random distribution, a value of 2 indicates twofold enrichment, and a value of 0.5 indicates twofold depletion. The read coverage along each element, such as the read distribution on each chromosome, can also be visualized as described above for reference sequences composed of a unique element.

3.6.5 Variant Calling Analysis

1. Run the samtools command line:

```
samtools mpileup -d 50000000 -L 50000000 -BAugf ref.fa aln.
bam | bcftools call -vmO z -o out.vcf.gz
```

The -d (--max-depth) and -L (--max-idepth) parameters must be adapted to the dataset. A compressed VCF file will be generated with sequence variations (*see* **Notes 27** and **28**).

4 Notes

1. Sonication steps are adapted for the Bioruptor UCD-200 sonicator. If another water bath sonicator is used, adapt sonication cycle (number of cycles, OFF/ON time and power) to reach the same target size for fragmented DNA.
2. Cell lysis solution, RNase A solution, and protein precipitation solution are sold together as Gentra Puregene tissue kits (Qiagen).
3. A fluorescence or UV spectrophotometer such as Qubit™ or Nanodrop™ is typically used for DNA quantitation and quality assessment.
4. Only the Illumina HiSeq 2500 system has been used for SSV-Seq analysis.
5. The source codes are available under the GNU General Public License v2.0 or v3.0 (GNU GPL). GNU GPL is a widely used free software license which means that users have freedom to run, copy, study, share, and modify the software. As the GNU GPL is a copy left license, the derivative work can only be distributed under the same license terms.
6. bcl2fastq conversion software can both demultiplex data and convert BCL files generated by Illumina sequencing systems to standard FASTQ files. For this analysis, bcl2fastq is only used to convert BCL files to FASTQ files, as the demultiplexing step is performed by Quade. Quade demultiplexes the reads by taking into account the quality of each base of index sequences, thus reducing read misattributions. The program bcl2fastq converts BCL files generated on Illumina sequencing systems for RTA 1.18.54 version or earlier. When using a more recent version of RTA, install bcl2fastq2 conversion software and follow the instructions provided in the software guide.
7. Quade is a FASTQ file demultiplexer with the ability to exclude reads containing poorly identified index sequences. The quality of each base constituting the index reads is checked and should be above a defined minimal value. According to this criterion, reads are classified as "pass" or "fail." Although simple and

double indexing are both supported by Quade, double indexing combined with index sequence quality filtering is recommended to reduce reads misattribution [4].

8. Sekator performs paired-end FASTQ quality trimming and adapter trimming to ensure reliability of the results. As shown in Fig. 3, the quality trimming step is performed from both ends of reads with an adjustable sliding window. Reads that are too short after trimming are discarded along with their paired mate. An adapter trimming step is also performed, keeping the read without the adapter sequence. The short reads are finally discarded with their paired mate.
9. Align sequencing reads against the reference sequences using ContaVect, which is based on the program BWA [8]. If there is sequence homology between two reference sequences, the RefMasker program (used as a preprocessing module in ContaVect) masks this region in the lowest represented sequence and, thus, the reads are attributed to the highest represented DNA species. The masking process is performed before the alignment stage. The use of RefMasker increases the sensitivity (more true positives) and specificity (more true negatives) of the read alignment against a reference.

   ContaVect generates standard genomic files (e.g., BAM, BED, and BedGraph) as well as detailed text reports including the number of reads assigned for each reference. The binary BAM format is commonly used to store sequence alignments, both from aligned and unaligned reads. The SAM format contains the same information but is text-based. The BED file generated using ContaVect reports the coverage at all positions of the reference sequence. BedGraph is more concise, because uncovered regions are not reported and contiguous positions of coverage are reported only once. Use the python module pysam to read and manipulate mapped reads that are stored in the SAM/BAM files. From the version pysam v0.13.0, the header information is separated from the AlignmentFile and placed into the AlignmentHeader. Contavect does not currently work with this version. Thus, install versions of pysam earlier than v0.13.0 (version 0.11.2.2 works with Contavect).
10. The program fastq_control_sampler generates two FASTQ files (R1 and R2) from FASTA reference sequences that are provided by the user, as well as from a randomly generated reference sequence. The in silico control files are useful to mimic a real next-generation sequencing library, and thus to determine the prediction accuracy of the ContaVect mapping process. fastq_control_sampler needs reference sequences in uppercase characters, otherwise the part of the sequence in lowercase characters is excluded. Lowercase characters are usually reserved for “soft masking” in repeat regions.

11. The FASTA format begins with a ">" character and specifies the sequence name. It is followed by line(s) of sequence data (nucleotide sequences). For genomes that are composed of several chromosomes (e.g., the human genome) and/or several scaffolds (e.g., the *Sf* genome), the multiple sequences must be combined into a single FASTA file.
12. An ~4.3 kb fragment from the left arm of the adenovirus type 5 genome was introduced into chromosome 19 during immortalization of HEK293 cells [9]. This fragment provides adenoviral E1A/E1B cofactors that are necessary for rAAV genome replication.
13. A whole-genome sequence of the *Spodoptera frugiperda* Sf9 insect cell line has been recently published under GenBank assembly accession number GCA_002213285.1 [10]. This sequence was obtained using the Pacific Biosciences long-read sequencing technology and Canu assembly. Consisting of 514 Mb (2396 scaffolds), it is larger than the *Sf*21 cell line genome (358 Mb, 37,235 scaffolds) [11] used in the SSV-Seq pipeline described in this chapter.
14. Each DNA species is added in a proportion comparable to the typical percentage detected by qPCR in rAAV batches (*reviewed in* [12]). Nevertheless, the proportion of each DNA species varies between vector lots, and also depends on the variability of the qPCR [2]. For the host cell DNA, use a number of copies close to the limit of quantification (LOQ) of qPCR detection of Sf9 or HEK293 gDNA. For example, use 1×10^3 copies (22 ng) of fragmented Sf9 gDNA in 100 µL of internal normalizer, because the LOQ of qPCR targeting the *Spodoptera frugiperda* ribosomal protein L37A gene [4] is 2×10^4 copies per mL.
15. The DNase digestion step is facultative. An optimized nuclease treatment removes free DNA molecules in the solution and enriches DNA that is encapsidated in (or protected by) rAAV capsids. For the analysis of total DNA, do not digest with DNase prior to DNA extraction.
16. For rAAV samples, three replicates are included in the SSV-Seq protocol in order to minimize experimental errors. One of these replicates can also be used to control the efficiency of critical steps such as the second-strand synthesis (*see* **Note 17**).
17. One tube of the rAAV sample replicates can be used as a control for the second-strand synthesis step. In this case, the DNA polymerase mix is composed of 2 µL of modified nucleotide mix (2 mM dATP, 2 mM dCTP, 2 mM dGTP, and 1.5 mM dTTP diluted in dH_2O), 0.25 µL of 1 mM fluorescein-12-dUTP solution, 1 µL (10 U) of DNA polymerase I, and 7.75 µL of dH_2O. After the second-strand synthesis reaction,

bring to 100 μL with dH_2O and purify using the NucleoSpin Gel and PCR Clean-up kit following the manufacturer's instructions. Fluorescein can be measured and compared to a fluorescein-labeled DNA probe with a known concentration.

18. The ratio of the absorbance at 260 and 280 nm (A260/280) must be between 1.7 and 2.1. DNA concentration is usually between 10 and 60 ng/μL.

19. The distribution of DNA fragment size after sonication (aliquot #1) can be verified immediately after sonication or at the end of the library preparation (**step 18** in Subheading 3.4).

20. A PCR-free protocol can be implemented for preparation of the Illumina sequencing library to avoid coverage bias due to the PCR amplification [13].

21. Multiple samples can be loaded onto the same flow cell for simultaneous analysis, following the manufacturer's instructions. Sample multiplexing reduces the sequencing cost but decreases the sequencing depth. Adapt the number of multiplexed samples to the sequencing depth that will be required for the bioinformatics analyses.

22. The FASTQ format is the standard format to store biological nucleotide sequences and the corresponding quality scores from high-throughput sequencing instruments. The first line begins with an "@" character and specifies the name of the reads. A "+" character in the third line separates the sequence of the read (second line) and the Phred quality score of the sequence (fourth line). The Phred quality score (Q score) indicates the probability that a given base is called incorrectly by the sequencer. Q scores are logarithmically related to the base calling error probabilities: $Q = -10 \log_{10} P$. For example, if a Phred quality score of 30 (Q30) is assigned to a base, the probability of an incorrect base call is 1 in 1000. These Q score values are used in our pipeline in order to ensure reliable results.

23. Combining dual-indexing and an index quality filter of 25 reduces read misattribution during the demultiplexing step. For dual-indexing, the FASTQ files containing the insert sequencing reads (R1 and R4) are named seq_R1 and seq_R2, while the index_R1 and index_R2 correspond to index reads (R2 and R3). Using an index quality filter of 25 typically results in 75% passed reads (20% of the reads fail the quality filter and 5% are not assigned). The program can generate three FASTQ files containing the reads that passed the quality filter, the "failed" reads, and the reads that are not assigned.

24. The correct mapping of a read may be ambiguous, and thus multiple read alignments can occur for the same read. In this case, one of those alignments (typically the best alignment) is

considered primary and all the other alignments are considered secondary.

25. Even though they are excluded from the final percentages, the number of unmapped and LowMapq reads are important to consider, because a high proportion of them may indicate the presence of inaccurate reference sequence. A high number of unmapped and LowMapq reads may also indicate DNA contamination during library preparation.
26. Graphic representations of the sequencing coverage can be generated using the python library matplotlib (https://github.com/matplotlib/matplotlib).
27. Samtools mpileup provides the coverage of mapped reads on a reference sequence at a single base pair resolution. The output of samtools mpileup can be piped into BCFtools to call genomic variants (SNPs and Indels).
28. When a variant is detected in the rAAV genome for an rAAV vector batch, single nucleotide variant (SNV) analysis of the vector plasmid or bacmid in the corresponding internal normalizer can help to determine if the mutation occurred during rAAV production (if the SNV is not in the vector plasmid or bacmid sequence) or during the generation of the biological raw material (if the SNV is present in the plasmid or bacmid sequence).

References

1. Food and Drug Administration (FDA) (2012) Vaccines and related biological products advisory committee meeting: Cell lines derived from human tumors for vaccine manufacture. FDA briefing Document
2. Ayuso E, Blouin V, Lock M, McGorray S, Leon X, Alvira MR, Auricchio A, Bucher S, Chtarto A, Clark KR, Darmon C, Doria M, Fountain W, Gao G, Gao K, Giacca M, Kleinschmidt J, Leuchs B, Melas C, Mizukami H, Müller M, Noordman Y, Bockstael O, Ozawa K, Pythoud C, Sumaroka M, Surosky R, Tenenbaum L, van der Linden I, Weins B, Wright JF, Zhang X, Zentilin L, Bosch F, Snyder RO, Moullier P (2014) Manufacturing and characterization of a recombinant adeno-associated virus type 8 reference standard material. Hum Gene Ther 25(11):977–987. https://doi.org/10.1089/hum.2014.057
3. Lecomte E, Tournaire B, Cogné B, Dupont J-B, Lindenbaum P, Martin-Fontaine M, Broucque F, Robin C, Hebben M, Merten O-W, Blouin V, François A, Redon R, Moullier P, Léger A (2015) Advanced characterization of DNA molecules in rAAV vector preparations by single-stranded virus next-generation sequencing. Mol Ther Nucleic Acids 4:e260. https://doi.org/10.1038/mtna.2015.32
4. Penaud-Budloo M, Lecomte E, Guy-Duché A, Saleun S, Roulet A, Lopez-Roques C, Tournaire B, Cogné B, Léger A, Blouin V, Lindenbaum P, Moullier P, Ayuso E (2017) Accurate identification and quantification of DNA species by next-generation sequencing in adeno-associated viral vectors produced in insect cells. Hum Gene Ther Methods 28 (3):148–162. https://doi.org/10.1089/hgtb.2016.185
5. Li H, Handsaker B, Wysoker A, Fennell T, Ruan J, Homer N, Marth G, Abecasis G, Durbin R, Genome Project Data Processing Subgroup (2009) The sequence alignment/map format and SAMtools. Bioinformatics 25 (16):2078–2079. https://doi.org/10.1093/bioinformatics/btp352
6. D'Costa S, Blouin V, Broucque F, Penaud-Budloo M, François A, Perez IC, Le Bec C, Moullier P, Snyder RO, Ayuso E (2016)

Practical utilization of recombinant AAV vector reference standards: focus on vector genomes titration by free ITR qPCR. Mol Ther Methods Clin Dev 5:16019. https://doi.org/10.1038/mtm.2016.19

7. Kozarewa I, Turner DJ (2011) Amplification-free library preparation for paired-end Illumina sequencing. Methods Mol Biol 733:257–266. https://doi.org/10.1007/978-1-61779-089-8_18
8. Li H, Durbin R (2009) Fast and accurate short read alignment with Burrows-Wheeler transform. Bioinformatics 25(14):1754–1760. https://doi.org/10.1093/bioinformatics/btp324
9. Louis N, Evelegh C, Graham FL (1997) Cloning and sequencing of the cellular-viral junctions from the human adenovirus type 5 transformed 293 cell line. Virology 233 (2):423–429. https://doi.org/10.1006/viro.1997.8597
10. Nandakumar S, Ma H, Khan AS (2017) Whole-genome sequence of the Spodoptera frugiperdaSf9 insect cell line. Genome Announc 5(34). https://doi.org/10.1128/genomeA.00829-17
11. Kakumani PK, Malhotra P, Mukherjee SK, Bhatnagar RK (2014) A draft genome assembly of the army worm, Spodoptera frugiperda. Genomics 104(2):134–143. https://doi.org/10.1016/j.ygeno.2014.06.005
12. Penaud-Budloo M, François A, Clément N, Ayuso E (2018) Pharmacology of recombinant adeno-associated virus production. Mol Ther Methods Clin Dev 8:166–180. https://doi.org/10.1016/j.omtm.2018.01.002
13. Kondratov O, Marsic D, Crosson SM, Mendez-Gomez HR, Moskalenko O, Mietzsch M, Heilbronn R, Allison JR, Green KB, Agbandje-McKenna M, Zolotukhin S (2017) Direct head-to-head evaluation of recombinant adeno-associated viral vectors manufactured in human versus insect cells. Mol Ther 25(12):2661–2675. https://doi.org/10.1016/j.ymthe.2017.08.003

Chapter 6

In Situ Hybridization for Detection of AAV-Mediated Gene Expression

Jacqueline E. Hunter, Brittney L. Gurda, Sea Young Yoon, Michael J. Castle, and John H. Wolfe

Abstract

Techniques to localize vector transgenes in cells and tissues are essential in order to fully characterize gene therapy outcomes. In situ hybridization (ISH) uses synthesized complementary RNA or DNA nucleotide probes to localize and detect sequences of interest in fixed cells, tissue sections, or whole tissue mounts. Variations in techniques include adding labels to probes, such as fluorophores, which can allow for the simultaneous visualization of multiple targets. Here we provide the steps necessary to: (1) label probes for colorimetric visualization and (2) perform ISH on OCT cryo-preserved fixed frozen tissues.

Key words Riboprobes, In situ hybridization, ISH, Fluorescence in situ hybridization, FISH, Cell-specific labeling, Hybridization, DIG labeling

1 Introduction

In situ hybridization (ISH) of RNA-DNA hybrid probes in cytological specimens was first reported in the late 1960s [1]. A decade later, the use of fluorescent in situ hybridization (FISH) was published [2]. The introduction of various tags and non-isotopic visualization methods, such as fluorophores and colorimetric development, allowed for advances in resolution while enhancing speed and safety of the technique, and eventually gave rise to simultaneous visualization of multiple targets and live-cell imaging [3]. Today, ISH, and various alterations of the original technique, is among the most common protocols of any laboratory needing to localize DNA or messenger (m)RNA in cytological preparations of fixed cells, histologic tissue sections, or even whole mounts. This technique is invaluable for assessing the distribution of transduced cells after gene delivery, and can detect adeno-associated virus (AAV)-mediated gene expression with greater accuracy and sensitivity than standard immunolabeling of the expressed protein,

Michael J. Castle (ed.), *Adeno-Associated Virus Vectors: Design and Delivery*, Methods in Molecular Biology, vol. 1950, https://doi.org/10.1007/978-1-4939-9139-6_6,

which can vary depending on the intracellular localization of the protein, the efficiency of antibody binding to the protein, the half-life of the protein, and other factors. ISH is particularly important when AAV is used to deliver a protein that is secreted or axonally transported, because in this case the presence of protein within a cell does not indicate that it was transduced by the AAV vector, and thus protein immunolabeling cannot be used to examine cellular transduction. The protocol outlined in this chapter has been used extensively in our laboratory and others for the past 20 years to assess various therapeutic methods, including viral vector-mediated gene therapy, whole cell delivery, and ex vivo gene therapy [4–23].

The design of RNA probes (riboprobes) for ISH has been commercialized, and a range of optimized riboprobes that efficiently bind to common target sequences can be purchased. The design of probe sequences will thus not be discussed in this chapter, although methods for generating and purifying digoxigenin (DIG)-UTP labeled riboprobes by in vitro transcription of template DNA are provided. The ISH protocol outlined here has three components: (1) antigen retrieval; (2) denaturation and hybridization of labeled RNA probes to the corresponding DNA followed by stringent washes; and (3) detection by antibody staining. This protocol focuses on ISH methods for cryo-preserved, paraformaldehyde (PFA)-fixed, OCT-embedded tissue preparations, but techniques using paraffin-embedded whole mount tissues and embryos have been established and can be found elsewhere [24].

Fluorescent ISH (FISH) can also be combined with immunofluorescent antibody (IFA) detection of markers for cell type or other functions. However, the reader is cautioned that the conditions of fixation and preparation of cells or tissues for IFA and ISH can work against each other, thus it is important to establish specific use of reagents for each combination of FISH and IFA reagents. Since FISH requires permeabilization for the probes to access mRNA, the sensitivity of detection of cell surface proteins may be reduced. In our experience, the detection of cytosolic proteins usually works well, such as combining FISH with IFA for tyrosine hydroxylase or neuron-specific enolase [9, 14].

2 Materials

Use sterile, ultrapure water for all solutions, unless otherwise noted.

2.1 DIG Labeling of Riboprobes

1. DIG RNA Labeling Kit (SP6/T7), containing:
   (a) 10× NTP labeling mixture
   (b) 10× Transcription buffer
   (c) 20 U/μL Protector RNase A Inhibitor

   (d) 20 U/μL SP6 RNA polymerase
   (e) 20 U/μL T7 RNA polymerase
   (f) 20 U/μL T3 RNA polymerase
   (g) 10 U/μL DNase I, RNase-free
2. Riboprobe template DNA.
3. 0.5 M EDTA, pH 8.0.
4. 20 mg/mL glycogen.
5. 4 M LiCl.
6. 100% Ethanol.
7. 5× MOPS gel running buffer: 0.1 M MOPS, 0.04 M sodium acetate, 0.005 M EDTA (pH 8). Dissolve MOPS and sodium acetate in DEPC-treated water. Adjust to pH 7 with NaOH. Add EDTA. Filter-sterilize through a 0.2 μm filter. Wrap the bottle in aluminum foil to protect it from light. Store at room temperature.
8. 10× RNA gel loading buffer: 50% glycerol, 0.25% bromophenol blue, 0.25% xylene cyanol FF, 0.01 M EDTA (pH 8). Store at 4 °C.
9. DEPC-treated water: fill large 2 L glass bottles with water (almost to top). In fume hood, add 2 mL of diethyl pyrocarbonate (DEPC) to each jug, seal tightly, and shake. USE A GLASS PIPETTE, as DEPC can melt plastic. Add nitrogen gas to the DEPC bottle, cap, and store at 4 °C. Do not store longer than 9 months. Loosen water caps and incubate at 37 °C for at least 30 min (overnight is usually used). Autoclave to degrade the remaining DEPC, leaving cap loosened. Tighten cap and store at 4 °C.
10. Agarose.
11. 12.3 M formaldehyde.
12. 100% formamide, deionized.
13. 0.4 mg/mL ethidium bromide.
14. Spectrophotometer.

2.2 In Situ Hybridization

1. RNase Zap (RNase decontamination solution).
2. 10 mg/mL Proteinase K enzyme: dissolve 100 mg of Proteinase K in 10 mL of nuclease-free water. Store at −20 °C in 200 μL aliquots.
3. 10× PBS: dissolve 80 g of NaCl, 2 g of KCl, 14.4 g of Na_2HPO_4, and 2.4 g of KH_2PO_4 in 800 mL of nuclease-free water. Adjust pH to 7.4 with HCl and bring volume to 1 L with nuclease-free water. To make 1× PBS, dilute 100 mL of 10× PBS with 900 mL of nuclease-free water.

4. 3% Hydrogen Peroxide: add 20 mL of 30% hydrogen peroxide and 20 mL of 10× PBS to 160 mL of nuclease-free water.
5. 1× Proteinase K buffer: 0.10 M Tris–HCl (pH 8), 0.05 M EDTA (pH 8). Combine 20 mL of 1.0 M Tris–HCl (pH 8) and 20 mL of 0.5 M EDTA (pH 8). Bring to 200 mL with nuclease-free water.
6. 0.1 M Triethanolamine (TEA) buffer, pH 8: weigh 7.43 g of TEA and add to 400 mL of DEPC-treated water. Add 11.5 mL of 2 N NaOH (*see* **Note 1**).
7. 0.1 M TEA/0.25% Acetic Anhydride wash: add 500 μL of Acetic Anhydride to 200 mL of 0.1 M TEA buffer. Recap acetic anhydride immediately after taking aliquot (*see* **Note 2**).
8. 2× SSC: 20× SSC diluted in nuclease-free water.
9. 1.0 M Tris–HCl, pH 7.5: dissolve 121.1 g of Trizma base in 800 mL of nuclease-free water. Adjust pH to 7.5 with ~68 mL of 12 N HCl. Bring volume to 1 L with nuclease-free water. Filter-sterilize.
10. Tris-EDTA-NaCl buffer: 4.200 M NaCl, 0.140 M Tris–HCl (pH 7.5), 0.014 M EDTA (pH 8). Dissolve 24.55 g NaCl in 70 mL of nuclease-free water. Add 14 mL of 1 M Tris–HCl (pH 7.5) and 2.8 mL of 0.5 M EDTA (pH 8). Bring volume to 100 mL with nuclease-free water. Filter-sterilize. Make 1 mL aliquots and store at −20 °C.
11. 100% Deionized Formamide: store 5 mL aliquots at −20 °C.
12. 40% Dextran Sulfate: heat 10 mL of Nuclease-Free water to 65 °C in a water bath for 15–20 min. Dissolve 5 g of dextran sulfate while stirring under low heat. Final volume will be 12 mL. Store in 3–4 mL aliquots at 4 °C (*see* **Note 3**).
13. 5× Maleic Acid buffer, pH 7.5: 0.50 M Maleic acid, 0.75 M NaCl. Dissolve 58.05 g of Maleic acid, 43.83 g of NaCl, and 40 g of NaOH pellets in 800 mL of nuclease-free water. Adjust pH to 7.5 with ~16 mL of 2 N NaOH. Bring volume to 1 L with nuclease-free water. Filter-sterilize.
14. 10% Blocking solution: dissolve 50 g of Blocking Reagent (Roche) in 500 mL of 1× maleic acid buffer by microwaving for 5 min. Check every 30 s after the first 3 min to prevent boil-over. Allow to cool on benchtop. Add 500 μL of DEPC and stir at room temperature for 10 min, then incubate in a 37 °C oven overnight. On the next day, autoclave for 45 min. Make 100 aliquots of 500 μL and store at −20 °C. These 500 μL aliquots are to be used in the hybridization solution. Aliquot the remaining solution into 40 mL aliquots and store at −20 °C. These 40 mL aliquots are used to make the colorimetric blocking buffer.

15. 5× RNase buffer: 2.50 M NaCl, 0.05 M Tris (pH 7.5), 5 mM EDTA (pH 8). Dissolve 146.1 g of NaCl in 800 mL of water (*see* **Note 4**). Add 50 mL of 1 M Tris (pH 7.5) and 5 mL of 0.5 M EDTA. Bring volume to 1 L with water.
16. Formamide Wash buffer: 50% deionized formamide, 2× SSC.
17. TSA Cyanine 3 System (Perkin Elmer). Contains the blocking powder, the streptavidin-HRP conjugate, the Cy3 substrate, and the Cy3 dilution buffer.
18. Cy3 Substrate: The Cy3 substrate comes lyophilized. Reconstitute the pellet by adding 300 μL of water. Store the substrate at 4 °C in the dark (wrap the tube in aluminum foil for protection).
19. 1× TNT Wash Buffer: 0.100 M Tris–HCl (pH 7.5), 0.150 M NaCl, 0.05% Tween 20. Dissolve 8.7 g of NaCl in 850 mL of water, then add 100 mL of 1 M Tris–HCl (pH 7.5) and 5 mL of 10% Tween 20. Bring final volume to 1 L with water.
20. 1× TNB Block (for immunofluorescent ISH): 0.10 M Tris–HCl (pH 7.5), 0.15 M NaCl, 0.5% blocking solution. Dissolve 5.26 g of NaCl in 450 mL of water and add 60 mL of 1 M Tris–HCl (pH 7.5). Heat solution to 60 °C in a water bath for approximately 30 min. Slowly add 3 g of blocking powder (*see* **item 17**) to the warm solution while stirring. Bring final volume to 600 mL with water. Prepare 10–15 mL aliquots and store at −20 °C.
21. Blocking buffer (for colorimetric ISH): 2× SSC, 0.05% Triton X-100, 2% blocking solution (*see* **item 14**).
22. Anti-Digoxigenin alkaline phosphatase primary antibody (Roche #1093274). Use at a 1:2500 dilution for colorimetric ISH, or 1:150 for immunofluorescent ISH.
23. Biotinylated anti-sheep IgG secondary antibody.
24. Antibody Dilution buffer: 1× Maleic Acid buffer, 1% blocking solution, 0.3% Triton X-100.
25. 10× Tris/NaCl: 1 M Tris–HCl (pH 9.5), 1 M NaCl. Dissolve 121.1 g of Trizma Base and 58.44 g of NaCl in 800 mL of water, then adjust pH to 9.5 with HCl. Bring volume to 1 L with water. Filter-sterilize.
26. 10× $MgCl_2$: 0.5 M $MgCl_2$. Dissolve 101.7 g of $MgCl_2$ in 800 mL of water. Bring volume to 1 L and filter-sterilize.
27. BCIP/NBT reaction Buffer: 100 mM Tris (pH 9.5), 100 mM NaCl, 50 mM $MgCl_2$. Add 20 mL of 10× Tris/NaCl and 20 mL of 10× $MgCl_2$ to a graduated cylinder, then bring to 100 mL with water.
28. BCIP/NBT tablets.

29. 10× Stop Buffer: 0.10 M Tris–HCl (pH 7.5), 0.01 M EDTA (pH 8), 1.50 M NaCl. Dissolve 88 g of NaCl in 800 mL of water. Add 10 mL of 1 M Tris (pH 7.5) and 2 mL of 0.5 M EDTA. Bring volume to 1 L with water and filter-sterilize.
30. 4% Paraformaldehyde.
31. HybriSlip Hybridization Covers (Grace Bio-Labs).
32. CoverWell Incubation Chambers (Grace Bio-Labs).
33. Aqua Poly/Mount mounting medium (Polysciences, Inc.).
34. Vectashield Antifade Mounting Medium with DAPI (Vector Labs).
35. 24 × 60 mm, no. 1.5 glass coverslips.
36. Humid chamber (*see* **Note 5**).
37. Hybridization oven.
38. Slide holders and solution containers.
39. 100% Ethanol.
40. DEPC-treated water (*see* **item 9** in Subheading 2.1).

3 Methods

3.1 DIG Labeling of Riboprobes

3.1.1 RNA Labeling Reaction

1. Add 1 μg of riboprobe template DNA or 4 μL of control DNA to a sterile, RNase-free reaction vial. Bring the total sample volume to 13 μL with water (sterile, RNase-free, DEPC-treated).
2. Place the reaction vial on ice, then add each reagent as shown in Table 1.
3. Mix and centrifuge briefly.
4. Incubate for 2 h at 37 °C.
5. Pulse spin, then add 2 μL of DNase I, RNase-free (10 U/μL) to remove template DNA.
6. Incubate for 15 min at 37 °C.
7. Stop the reaction by adding 2 μL of 0.5 M EDTA (pH 8.0).
8. Ethanol precipitate probes overnight at −80 °C as shown in Table 2.
9. Centrifuge the RNA precipitate at 4 °C for 20 min in a microfuge at full speed.
10. Wash the RNA pellet once with 1 mL of 70% ethanol, centrifuge for 15 min, then air-dry.
11. Resuspend the RNA pellet in 75 μL of DEPC-treated water.
12. Visualize the labeled RNA by running 5 μL on a 1% agarose/formaldehyde gel at 80 V (*see* Subheading 3.1.2).

Table 1
Preparation of riboprobe DIG labeling reaction

Reagent	Volume	Control #1	Control #2	Template (1×) Sense	Template (1×) Antisense	Template (2×) Antisense	Template (5×) Antisense
10× NTP mixture	2 μL	2 μL	2 μL	2 μL	2 μL	4 μL	10 μL
10× Transcription buffer	2 μL	2 μL	2 μL	2 μL	2 μL	4 μL	10 μL
Protector RNase A Inhibitor (20 units/μL)	1 μL	1 μL	1 μL	1 μL	1 μL	2 μL	5 μL
SP6 RNA polymerase (20 units/μL), OR	2 μL	–	2 μL	–	–	–	–
T7 RNA polymerase (20 units/μL), OR	2 μL	2 μL	–	2 μL	–	–	–
T3 RNA polymerase (20 units/μL)	2 μL	–	–	–	2 μL	4 μL	10 μL
Total reaction volume	20 μL	20 μL	20 μL	20 μL	20 μL	40 μL	100 μL

Table 2
Preparation of ethanol precipitation

Riboprobe labeling reaction mixture	24 μL	44 μL	144 μL
Glycogen (20 mg/mL) (optional)	1 μL	1 μL	1 μL
4 M LiCl	2.5 μL	5 μL	12.5 μL
100% EtOH (2.5× volume to be precipitated)	75 μL	125 μL	400 μL

13. Quantify RNA using a spectrophotometer.
14. Aliquot riboprobe at 4 μg/tube and store at −80 °C.

3.1.2 Visualization of Probes Using a Formaldehyde Gel

1. Prepare a 1% agarose gel by microwaving 1 g of agarose in 62 mL of DEPC-treated water, then cooling to 60 °C. Add 20 mL of 5× MOPS gel running buffer and 18 mL of 12.3 M formaldehyde (2.2 M final concentration) to the dissolved agarose. Pour the mixture into a gel caster.
2. Prepare the sample for denaturation of RNA as shown in Table 3.

Table 3
Preparation of RNA denaturation reaction

RNA	5 μL
5× MOPS Gel running buffer	2 μL
12.3 M Formaldehyde	3.5 μL
100% Formamide, deionized	10 μL
0.4 mg/mL ethidium bromide	1 μL
Total reaction volume	21.5 μL

3. Vortex and spin samples, then incubate at 65 °C for 15 min. Remove samples from water bath and place on ice. Spin samples briefly on a microfuge and return to ice.
4. Add 2.4 μL of 10× RNA gel loading buffer, then vortex and spin. Place samples on ice.
5. While samples are incubating, pre-run the gel at 80 V (5 V/cm gel) for 30 min.
6. Load samples onto gel along with control RNA and RNA marker on separate lanes.
7. Run until dye is halfway through the gel.
8. View under UV light.

3.2 *In Situ* Hybridization

3.2.1 Hybridization

1. Before beginning, clean solution containers with RNase Zap and rinse with DEPC-treated water to eliminate all RNase. Be careful not to introduce RNase throughout Day 1. Clean gloves with RNase Zap periodically and conduct all procedures with RNase-free reagents on a dedicated and clean RNase-free bench.
2. Warm 200 mL of Proteinase K buffer (without Proteinase K enzyme) to 37 °C in a water bath (*see* **Note 6**). Be sure to use autoclaved, RNase-free glassware.
3. Remove slides from −80 °C and allow to dry on lab bench for *at least* 1 h (*see* **Note 7**). Load slides in a slide holder after they are dry. Take out two slides that are known to be positive for your target sequence and label as controls: one positive control (label "Antisense" and treat the same as all other slides) and one negative control (label "Sense" and use a sense riboprobe in **steps 13** and **14**).
4. Dip the slides for 1 min into 200 mL of each of the following solutions in the given order to rehydrate the tissue sections (*see* **Note 8**):
   (a) 100% Ethanol
   (b) 100% Ethanol

Table 4
Preparation of hybridization solution

Reagent	Amount (μL)
digoxigenin (DIG)-labeled riboprobe, 4 μg	X
Nuclease-free water	440–X
Tris-EDTA-NaCl buffer	360
100% deionized Formamide	2500
40% Dextran sulfate	1200
10% Blocking solution	500
Total	5000

(c) 95% Ethanol

(d) 70% Ethanol

(e) 50% Ethanol

(f) 2× SSC

5. For colorimetric ISH, skip to **step 7**. For fluorescent ISH, place slides in 200 mL of 3% hydrogen peroxide in 1× PBS for 25 min at room temperature.
6. Wash slides twice in 1× PBS for 5 min each.
7. Add 200 μL of 10 mg/mL Proteinase K enzyme to the pre-warmed 200 mL Proteinase K buffer and mix. Transfer slides immediately after adding Proteinase K. Dip the slide holder 2 quick times and then incubate at 37 °C for 3 min for tissue sections less than 5 μm thick, 4 min for 6–10 μm, or 5 min for 11–20 μm.
8. Dip slides in 200 mL of RNase-free water twice for 10 s each time.
9. Incubate the slides in 200 mL of freshly made 0.1 M TEA buffer (pH 8) for 3 min.
10. Incubate the slides in 0.1 M TEA/Acetic Anhydride solution for 10 min. DO NOT add acetic anhydride until immediately before use. Dip slide holders up and down to mix.
11. Dip the slides for 1 min into 200 mL of each of the following solutions in the given order to dehydrate the tissue sections (use the same solutions used for rehydration in **step 4**):

    (a) 2× SSC

    (b) 50% Ethanol

    (c) 70% Ethanol

    (d) 95% Ethanol

(e) 100% Ethanol

(f) 100% Ethanol

12. Air-dry the slides for at least 1 h at room temperature.
13. Prepare the hybridization solution as shown in Table 4. Remember to prepare a second hybridization solution containing a sense riboprobe for negative control. While the slides are air-drying, combine the riboprobe and nuclease-free water in a 1.7 mL microcentrifuge tube, then place the tube in boiling water for 10 min. Place the tube directly onto ice after boiling. While on ice, combine with the remaining reagents as shown in Table 4 in a 15 mL tube (*see* **Note 9**). Probe aliquots are 4 μg, so the final concentration will be 0.8 μg/mL. After combining all reagents, the mixture can be vortexed and returned to boiling water for 2–5 min to remove bubbles (this is not necessary if no bubbles are present) (*see* **Note 10**).
14. Prepare HybriSlip hybridization coverslips (coverslips with peel-off backing). Pipette 90 μL of hybridization solution onto each coverslip.
15. Carefully roll the slide over the coverslip so that all sections are in contact with the hybridization solution. This works best if a long line of solution is laid across the coverslip, then rolled onto the slide gently without pushing. The slide can then be picked up and the coverslip adjusted so that all sections are covered. Apply gentle pressure to the coverslip to remove any remaining bubbles and affix it to the slide.
16. Place the slides in a humid chamber along with 1 wet paper towel and 50 mL of dH_2O. Do not allow slides to touch the paper towel. Incubate in a hybridization oven at 63 °C overnight (at least 8 h). Place a beaker filled with water in the oven (*see* **Note 11**).

3.2.2 Post-hybridization

An RNase-free environment is no longer required from this stage onward. Do not use the dedicated RNase-free bench for any of the below steps.

1. Pre-warm 200 mL of the following solutions at the temperatures indicated below:

Formamide wash solution	65 °C
RNase buffer #1	37 °C
RNase buffer #2	65 °C
2× SSC #1	RT
2× SSC #2	37 °C
2× SSC #3	37 °C

2. Remove the slides from the humid chamber and carefully remove the hybridization coverslip. Place the slides onto a slide holder that is immersed in SSC (to prevent drying while loading slides). Wash in the following order:
   (a) 2× SSC for 30 min with low-speed shaking at room temperature.
   (b) Formamide wash solution for 30 min at 65 °C. Gently agitate the slides (move up and down in container) periodically during the first 5 min of incubation.
   (c) 2× SSC #2 for 10 min at 37 °C.
   (d) 2× SSC #3 for 10 min at 37 °C.
3. Incubate the slides in RNase buffer #1 containing 400 μL (10 mg/mL) of RNase A for 30 min at 37 °C. Add slides immediately after adding RNase to buffer.
4. Wash slides in RNase buffer #2 (no RNase A added) for 30 min at 65 °C. For colorimetric immunohistochemical detection, continue to Subheading 3.2.3. For immunofluorescent detection, skip to Subheading 3.2.4.

3.2.3 Colorimetric Immunohistochemical Detection

1. Place slides in 200 mL of blocking buffer while shaking at low speed for 2 h at room temperature. If desired, slides can remain in blocking buffer at 4 °C overnight (no shaking) (*see* **Note 12**).
2. Prepare antibody dilution buffer as described in Table 5 (*see* **Note 13**). Add anti-digoxigenin AP antibody last, then mix gently.
3. Using CoverWell coverslips (rubber side up), pipette 220 μL of antibody solution in a line from the left to the right edges of the coverslip. Remove any bubbles with a micropipette.
4. Press the bottom edge of the slide to the middle of the bottom rubber edge of the CoverWell, and gently roll the CoverWell onto the slide, allowing solution to fill the entire CoverWell from the bottom. Keep the slide and CoverWell in contact with the bench to prevent the coverslip from sliding off. The coverslip can then be moved gently to an optimal position (*see* **Note 14**). If bubbles form, they can be moved to a corner of the coverslip by tilting the slide, or can be pressed out by gently easing the corner of the CoverWell that contains bubbles over the edge of the slide, releasing the bubbles (be careful not to lose solution).
5. Place the slides in a humid chamber along with 1 wet paper towel and 50 mL of dH_2O. Do not allow slides to touch the paper towel. Close humid chamber and incubate for 3 h at RT or overnight at 4 °C. Quickly check all slides for infiltration of

Table 5
Preparation of antibody dilution buffer

Stock solutions	Vol. stock solution	Final concentration (antibody solution)
dd water	6.7/10.05/13.4/16.75 mL	
5× Maleic acid buffer	2/3/4/5 mL	1× Maleate
10% Blocking solution	1.0/1.5/2.0/2.5 mL	1% Blocking solution
10% Triton X-100	0.3/0.45/0.6/0.75 mL	0.3% Triton X-100
Anti-Digoxigenin AP (ADD LAST)	4/6/8/10 μL	1:2500 dilution
Total volume	10/15/20/25 mL	

bubbles before incubating, as bubbles can stick to the edges initially but move to the middle once slides are laid flat in the chambers.

6. Carefully remove the CoverWell coverslips and load slides onto a slide holder immersed in 1× Maleate buffer (to prevent drying while slides are loaded). After removing, place CoverWell coverslips in deionized water with soap (*see* **Note 15**). While the slides are washed in **steps** 7 and **8**, clean the coverslips by rinsing off soap thoroughly, first with warm tap water, then with deionized water. Gently rub the plastic under water to remove all residual antibody, which can cause background. Dry CoverWell coverslips with paper towels and lay out in a dark room for **step 10**.
7. Wash slides twice in 200 mL of 1× Maleate buffer at room temperature for 10 min each, with low-speed shaking.
8. Wash in BCIP/NBT reaction buffer at room temperature for 10 min with low-speed shaking.
9. Dissolve 1 BCIP/NBT tablet in 10 mL of sterile distilled water.
10. Place 220 μL of dissolved BCIP/NBT onto a CoverWell coverslip, then roll slide onto coverslip as described in **step 4**. Follow any notes on the slide labels to ensure that the same sections are coverslipped (*see* **Note 14**).
11. As described in **step 5**, place slides in humid chamber and check all slides for infiltration of bubbles. Incubate for 12–14 h in the dark at room temperature (*see* **Note 16**).
12. If 4% paraformaldehyde has already been made and is stored at 4 °C, pour into solution container(s) before beginning so that it can warm to room temperature during **steps 13** and **14**.
13. Carefully remove the CoverWell coverslips and load slides onto a slide holder immersed in stop buffer to prevent drying.

14. Wash slides in Stop Buffer for at least 30 min at room temperature with low-speed shaking.
15. Fix in 200 mL of 4% paraformaldehyde for 20 min at room temperature with low-speed shaking.
16. Wash 3× with 200 mL of Stop Buffer for 1 min each to remove paraformaldehyde.
17. Place in fresh Stop Buffer for up to 24 h until ready to coverslip.
18. Coverslip in Aqua Polymount. Take a rectangular glass coverslip (24 × 60 mm, no. 1.5) and place an even line of Polymount along the entire bottom edge of the coverslip, leaving no space between the Polymount and the edge (space can cause bubbles to form). Remove any bubbles with a micropipette. Take the slide out of stop buffer, dry the sides and back, and place the bottom edge against the bottom edge of the coverslip with Polymount. Allow the Polymount to slowly and evenly move up the slide so that no bubbles form. Wipe off any excess, press out any bubbles, and air-dry. Once dried, excess Polymount can be cleaned off with 95% ethanol.

3.2.4 Immunofluorescent Detection

1. Using CoverWell coverslips, incubate in 1× TNB for 1 h at room temperature with shaking at low speed. Use 200 μL per slide.
2. Dilute the anti-Digoxigenin primary antibody 1:150 in 1× TNB, cover each slide with 200 μL (use CoverWell coverslips), and incubate for 30 min at room temperature.
3. Wash 3× with TNT for 5 min each.
4. Dilute the biotinylated anti-sheep IgG secondary antibody 1:100 in 1× TNB. Cover each slide with 200 μL (use CoverWell coverslips) and incubate for 30 min at room temperature.
5. Wash 3× with TNT for 5 min each.
6. Dilute the streptavidin-HRP conjugate 1:100 in 1× TNB. Cover each slide with 200 μL (use CoverWell coverslips) and incubate for 30 min at room temperature.
7. Wash 3× with TNT for 5 min each.
8. Dilute the Cy3 substrate 1:50 in Cy3 dilution buffer. Cover each slide with 200 μL (use CoverWell coverslips) and incubate for 5 min in the dark at room temperature. Do not allow the reaction to proceed for more than 5 min.
9. Wash 3× with TNT for 5 min each.
10. Place slides in 1× PBS and store in the dark at 4 °C until ready to coverslip. Alternately, cell type-specific immunofluorescent staining may be performed at this point.
11. Coverslip using Vectashield mounting medium with DAPI.

4 Notes

1. Always prepare TEA fresh.
2. Acetic anhydride is rapidly hydrolyzed to acetic acid upon exposure to air. Keep closed tightly when not in use.
3. Dextran sulfate aliquots are good for up to 1 year. This solution can also be prepared by vortexing dextran sulfate with heated water in a 50 mL tube.
4. Do not use DEPC-treated water.
5. Use a commercially available humid chamber. Alternatively, a humid chamber can be constructed by gluing cut plastic pipettes to the base of a plastic chamber (such as the Corning 245 mm Square BioAssay Dish). Glue pipettes in pairs spaced 1 inch apart. Before adding slides, add a wet paper towel and 50 mL of dH_2O to the chamber. Place slides on the pipettes, ensuring that slides do not touch paper towels.
6. If using a full slide holder (48 slides), 160 mL of solution can be used instead of 200 mL for all steps. 180 mL can be used for a half-full slide holder.
7. Tissue sections should be mounted on RNase-free adhesion microscope slides. To avoid introduction of RNase during adhesive coating, we recommend purchasing precoated adhesion slides that are certified RNase-free.
8. For **steps 4–10** (Subheading 3.2.1), if using multiple slide holders, run both slide holders simultaneously: proceed through each wash with the second slide holder following one container behind the first. The same ethanol and SSC solutions (**step 4**) and RNase-free water washes (**step 8**) can be used for both slide holders. However, a fresh 3% hydrogen peroxide solution (**step 5**, if needed), Proteinase K solution (**step 7**), TEA buffer (**step 9**), and TEA/acetic anhydride wash (**step 10**) should be used for each slide holder.
9. Cut off the end of the pipette used to transfer dextran sulfate, as it is typically too thick to transfer with a standard pipette tip.
10. Adjust the total volume according to the number of slides needed (90 μL per slide). Extra can be stored at −20 °C. When using next, vortex and boil for 10 min, then put on ice and allow to cool.
11. The incubation temperature may need to be optimized for the specific riboprobe that is used.
12. If you have more than 48 slides, it is *strongly recommended* that you leave them at 4 °C overnight and continue the next day, as coverslipping can be very time consuming.

13. Use 220 μL per slide. CoverWells can be difficult to use, so prepare extra volume. 15 mL is recommended for one full slide holder (48 slides), 25 mL for two full slide holders (96 slides).
14. CoverWell coverslips may not cover all tissue sections on the slide. To ensure that the same tissue sections are coverslipped in each step, we recommend always coverslipping the sections that are closest to the frosted slide label. If these sections are damaged or folded, coverslip the sections furthest from the label instead, and clearly label the slide to ensure that these same sections are coverslipped in all subsequent steps.
15. Do not use soap that contains lotion, such as hand soap.
16. Try not to exceed 16 h, as areas with strong signal can be overexposed.

Acknowledgments

This work has been supported by NIH research grants R01-NS038690, R01-NS088667, R01-DK063973, and U01-HD079066. BLG was supported by NIH training grant T32-NS007413.

References

1. Gall JG, Pardue ML (1969) Formation and detection of RNA-DNA hybrid molecules in cytological preparations. Proc Natl Acad Sci U S A 63(2):378–383
2. Bauman JG, Wiegant J, Borst P, van Duijn P (1980) A new method for fluorescence microscopical localization of specific DNA sequences by in situ hybridization of fluorochromelabelled RNA. Exp Cell Res 128(2):485–490
3. Levsky JM, Singer RH (2003) Fluorescence in situ hybridization: past, present and future. J Cell Sci 116(Pt 14):2833–2838. https://doi.org/10.1242/jcs.00633
4. Berges BK, Wolfe JH, Fraser NW (2005) Stable levels of long-term transgene expression driven by the latency-associated transcript promoter in a herpes simplex virus type 1 vector. Mol Ther 12(6):1111–1119. https://doi.org/10.1016/j.ymthe.2005.06.478
5. Berges BK, Yellayi S, Karolewski BA, Miselis RR, Wolfe JH, Fraser NW (2006) Widespread correction of lysosomal storage in the mucopolysaccharidosis type VII mouse brain with a herpes simplex virus type 1 vector expressing beta-glucuronidase. Mol Ther 13(5):859–869. https://doi.org/10.1016/j.ymthe.2005.12.017
6. Castle MJ, Gershenson ZT, Giles AR, Holzbaur EL, Wolfe JH (2014) Adeno-associated virus serotypes 1, 8, and 9 share conserved mechanisms for anterograde and retrograde axonal transport. Hum Gene Ther 25(8):705–720. https://doi.org/10.1089/hum.2013.189
7. Cearley CN, Vandenberghe LH, Parente MK, Carnish ER, Wilson JM, Wolfe JH (2008) Expanded repertoire of AAV vector serotypes mediate unique patterns of transduction in mouse brain. Mol Ther 16(10):1710–1718. https://doi.org/10.1038/mt.2008.166
8. Cearley CN, Wolfe JH (2006) Transduction characteristics of adeno-associated virus vectors expressing cap serotypes 7, 8, 9, and Rh10 in the mouse brain. Mol Ther 13(3):528–537. https://doi.org/10.1016/j.ymthe.2005.11.015
9. Cearley CN, Wolfe JH (2007) A single injection of an adeno-associated virus vector into nuclei with divergent connections results in widespread vector distribution in the brain and global correction of a neurogenetic disease. J Neurosci 27(37):9928–9940. https://doi.org/10.1523/JNEUROSCI.2185-07.2007

10. Husain T, Passini MA, Parente MK, Fraser NW, Wolfe JH (2009) Long-term AAV vector gene and protein expression in mouse brain from a small pan-cellular promoter is similar to neural cell promoters. Gene Ther 16 (7):927–932. https://doi.org/10.1038/gt.2009.52
11. Karolewski BA, Wolfe JH (2006) Genetic correction of the fetal brain increases the lifespan of mice with the severe multisystemic disease mucopolysaccharidosis type VII. Mol Ther 14 (1):14–24. https://doi.org/10.1016/j.ymthe.2006.02.012
12. McFadden K, Griffin TA, Levy V, Wolfe JH, Valentino RJ (2012) Overexpression of corticotropin-releasing factor in Barrington's nucleus neurons by adeno-associated viral transduction: effects on bladder function and behavior. Eur J Neurosci 36(10):3356–3364. https://doi.org/10.1111/j.1460-9568.2012.08250.x
13. Passini MA, Lee EB, Heuer GG, Wolfe JH (2002) Distribution of a lysosomal enzyme in the adult brain by axonal transport and by cells of the rostral migratory stream. J Neurosci 22 (15):6437–6446. doi: 20026665
14. Passini MA, Watson DJ, Vite CH, Landsburg DJ, Feigenbaum AL, Wolfe JH (2003) Intraventricular brain injection of adeno-associated virus type 1 (AAV1) in neonatal mice results in complementary patterns of neuronal transduction to AAV2 and total long-term correction of storage lesions in the brains of beta-glucuronidase-deficient mice. J Virol 77 (12):7034–7040
15. Passini MA, Wolfe JH (2001) Widespread gene delivery and structure-specific patterns of expression in the brain after intraventricular injections of neonatal mice with an adeno-associated virus vector. J Virol 75 (24):12382–12392. https://doi.org/10.1128/JVI.75.24.12382-12392.2001
16. Vite CH, McGowan JC, Niogi SN, Passini MA, Drobatz KJ, Haskins ME, Wolfe JH (2005) Effective gene therapy for an inherited CNS disease in a large animal model. Ann Neurol 57(3):355–364. https://doi.org/10.1002/ana.20392
17. Vite CH, Passini MA, Haskins ME, Wolfe JH (2003) Adeno-associated virus vector-mediated transduction in the cat brain. Gene Ther 10(22):1874–1881. https://doi.org/10.1038/sj.gt.3302087
18. Watson DJ, Longhi L, Lee EB, Fulp CT, Fujimoto S, Royo NC, Passini MA, Trojanowski JQ, Lee VM, McIntosh TK, Wolfe JH (2003) Genetically modified NT2N human neuronal cells mediate long-term gene expression as CNS grafts in vivo and improve functional cognitive outcome following experimental traumatic brain injury. J Neuropathol Exp Neurol 62(4):368–380
19. Watson DJ, Passini MA, Wolfe JH (2005) Transduction of the choroid plexus and ependyma in neonatal mouse brain by vesicular stomatitis virus glycoprotein-pseudotyped lentivirus and adeno-associated virus type 5 vectors. Hum Gene Ther 16(1):49–56. https://doi.org/10.1089/hum.2005.16.49
20. Wood SK, McFadden K, Griffin T, Wolfe JH, Zderic S, Valentino RJ (2013) A corticotropin-releasing factor receptor antagonist improves urodynamic dysfunction produced by social stress or partial bladder outlet obstruction in male rats. Am J Physiol Regul Integr Comp Physiol 304(11):R940–R950. https://doi.org/10.1152/ajpregu.00257.2012
21. Yoon SY, Bagel JH, O'Donnell PA, Vite CH, Wolfe JH (2016) Clinical improvement of alpha-mannosidosis cat following a single cisterna magna infusion of AAV1. Mol Ther 24 (1):26–33. https://doi.org/10.1038/mt.2015.168
22. Yoon SY, Gay-Antaki C, Ponde DE, Poptani H, Vite CH, Wolfe JH (2014) Quantitative, noninvasive, in vivo longitudinal monitoring of gene expression in the brain by co-AAV transduction with a PET reporter gene. Mol Ther Methods Clin Dev 1:14016. https://doi.org/10.1038/mtm.2014.16
23. Zhu J, Kang W, Wolfe JH, Fraser NW (2000) Significantly increased expression of beta-glucuronidase in the central nervous system of mucopolysaccharidosis type VII mice from the latency-associated transcript promoter in a nonpathogenic herpes simplex virus type 1 vector. Mol Ther 2(1):82–94. https://doi.org/10.1006/mthe.2000.0093
24. Swain GP, Jacobs AJ, Frei E, Selzer ME (1994) A method for in situ hybridization in whole-mounted lamprey brain: neurofilament expression in larvae and adults. Exp Neurol 126 (2):256–269. https://doi.org/10.1006/exnr.1994.1063

Chapter 7

Use of AAV Vectors for CRISPR-Mediated In Vivo Genome Editing in the Retina

Wenhan Yu and Zhijian Wu

Abstract

Degenerative retinal diseases such as retinitis pigmentosa (RP) and Leber's congenital amaurosis (LCA) may lead to blindness without effective treatment. With the rapid advancement of the CRISPR/Cas9 genome editing technology, in vivo application of CRISPR/Cas9 holds immense potential for treatment of these diseases. Adeno-associated virus (AAV) vectors are an ideal gene transfer tool for delivery of CRISPR components to the retina. Here, we describe a protocol for utilizing an AAV-based CRISPR/Cas9 system for in vivo genome editing in the retina.

Key words CRISPR/Cas9, AAV vector, Genome editing, Gene therapy, Retina, Retinal degeneration

1 Introduction

The rapidly advancing CRISPR/Cas9 genome editing technology has brought revolutionary changes in life sciences. In the CRISPR/Cas9 system, the Cas9 DNA endonuclease creates double-strand DNA breaks (DSBs) at desired locations in the genome, directed by a short guide RNA (sgRNA) comprised of 20 nucleotides that recognizes the target DNA via Watson-Crick base paring. Endogenous DNA repair mechanisms such as non-homologous end joining (NHEJ) or homology directed repair (HDR) are activated to repair the DSBs. NHEJ is the major DNA repair pathway in which insertion or deletion (indel) of random nucleotides usually occurs, and is used to achieve gene disruption or gene deletion for genome editing. In contrast, HDR leads to an error-free repair based on a template DNA with homologous sequences, though with much lower efficiency than NHEJ [1].

Numerous studies have shown the great potential of the CRISPR/Cas9 technique for disease treatment [2]. To develop therapies for various degenerative retinal diseases such as retinitis pigmentosa (RP), Leber's congenital amaurosis (LCA), and age-related macular degeneration (AMD), the CRISPR components

Michael J. Castle (ed.), *Adeno-Associated Virus Vectors: Design and Delivery*, Methods in Molecular Biology, vol. 1950, https://doi.org/10.1007/978-1-4939-9139-6_7,

need to be efficiently delivered to the retina. To this end, three gene transfer methods have been reported, including electroporation of plasmids carrying CRISPR components [3–6], lipid-mediated transfection of CRISPR ribonucleoprotein (RNP) complexes [7], and viral vector (AAV and lentivirus) mediated CRISPR transfer [8–15]. Among them, the AAV vector approach appears to be most efficient for in vivo retinal gene delivery. Here we describe a protocol for the generation of AAV vectors carrying the CRISPR components and the delivery of these AAV vectors to the mouse retina.

2 Materials

2.1 Vector Construction and In Vitro sgRNA Selection

2.1.1 Design and Assembly of sgRNA

1. CRISPR design tool (http://crispr.mit.edu/).
2. Benchling (https://benchling.com/).
3. pV-Cas9 plasmid carrying the Cas9 expression cassette [14].
4. pV-sgRNA-tdTomato plasmid carrying the U6 promoter-driven sgRNA and a tdTomato expression cassette [14].
5. T4 polynucleotide kinase.
6. 10× T4 PNK reaction buffer.
7. 10 mM adenosine triphosphate (ATP).
8. UltraPure DNase/RNase-free distilled water.
9. Taq buffer.
10. Thermal cycler.
11. SapI restriction enzyme (New England Biolabs).
12. 10× CutSmart buffer.
13. Alkaline phosphatase, calf intestinal (CIP).
14. Tris-acetate-EDTA (TAE) buffer.
15. Agarose gel.
16. Gel Extraction kit.
17. Quick Ligase (New England Biolabs).
18. Quick Ligase buffer.
19. DH10B competent cells (New England Biolabs).
20. SOC medium.
21. LB agar plates.
22. LB medium supplemented with 100 μg/mL ampicillin.
23. Plasmid miniprep kit.
24. Plasmid maxiprep kit.

2.1.2 In Vitro sgRNA Selection

1. HEK293 cells (ATCC® #CRL-1573).
2. HEK293-GFP cells (GenTarget #SC001).
3. Dulbecco's modified Eagle's medium (DMEM).
4. Fetal bovine serum (FBS).
5. Lipofectamine 2000.
6. Gel Extraction kit.
7. Opti-MEM® I reduced serum medium.
8. Phosphate buffered saline (PBS).
9. Trypsin-EDTA.
10. DNeasy Blood & Tissue Kit (QIAGEN).
11. SURVEYOR assay kit (Integrated DNA Technologies):
    (a) 0.15 M $MgCl_2$ stock solution.
    (b) SURVEYOR nuclease S.
    (c) SURVEYOR enhancer S.
    (d) Stop Solution.
12. High-fidelity DNA polymerase: PrimeSTAR polymerase (Clontech) or similar.
13. DNA polymerase buffer (2× PrimeSTAR GC buffer or similar).
14. 25 mM Deoxy-ribonucleoside triphosphate (dNTP).

2.2 AAV Production

2.2.1 AAV Generation by Transient Transfection

1. Helper plasmid pHLP19-AAV8 carrying AAV2 Rep and AAV8 Cap genes.
2. Helper plasmid pLAdeno5 carrying adenovirus helper genes.
3. Roller bottles.
4. 1 M HEPES buffer.
5. 2× HBS buffer: 50 mM HEPES, 280 mM NaCl, and 1.5 mM Na_2HPO_4, pH 7.3.
6. 0.3 M $CaCl_2$ solution.
7. 500 mL conical tubes.
8. TSM buffer: 50 mM TrisCl, 150 mM NaCl, and 2 mM $MgCl_2$, pH 8.0.
9. Benchtop centrifuge.

2.2.2 AAV Isolation and Purification

1. Microfluidizer.
2. 1 M $CaCl_2$ solution.
3. Benzonase.
4. 40% PEG8000/2.5 N NaCl solution.

5. HSSE-RNase A buffer: 50 mM HEPES, 150 mM NaCl, 1% Sarcosyl, 20 mM EDTA, and 10 μg/mL RNase A, pH 8.0.
6. 1.5 g/mL CsCl: mix 60 g CsCl with approximately 80 mL of 25 mM Tris–HCl, pH 8.0. Determine density by weighing 1 mL solution.
7. 1.4 g/mL CsCl: mix 60 g CsCl with approximately 100 mL of 25 mM Tris–HCl, pH 8.0.
8. 1.3 g/mL CsCl: mix 60 g CsCl with approximately 125 mL of 25 mM Tris–HCl, pH 8.0.
9. Ultracentrifuge with SW32Ti and SW40Ti rotors (Beckman).
10. 38.5 mL centrifuge tubes.
11. 14 mL centrifuge tubes.
12. Halogen beam illuminator.
13. 5 mL syringes.
14. 18G needles.
15. Slide-A-Lyzer dialysis cassettes.
16. Tris-buffered saline: 10 mM Tris–HCl, 180 mM NaCl, pH 7.4.

2.2.3 AAV Titration by qPCR

1. Linearized AAV vector standard DNA.
2. Titration primers.
3. FAM-conjugated probes.
4. qPCR dilution buffer: 10 mM Tris–HCl pH 8.0, 1 mM EDTA, 10 μg/mL yeast tRNA, 0.01% Tween 80.
5. DNase buffer: 10 mM Tris–HCl, pH 8.0, and 10 mM $MgCl_2$.
6. 200 mM EDTA, pH 8.0.
7. 2× PCR Master Mix.
8. LoBind microcentrifuge tubes.
9. 96-well qPCR plate
10. qPCR optical adhesive film.
11. Real-Time PCR System (QuantStudio™ 3 or similar).

2.3 Delivery of AAV to the Retina by Subretinal Injection

1. 100 mg/mL fluorescein (Alcon).
2. 12.5 mg/mL ketamine.
3. 2.5 mg/mL xylazine.
4. 1% atropine eye drops.
5. 0.5% tropicamide eye drops.
6. 18G needles.
7. Ophthalmic surgical microscope.
8. Hamilton glass syringes.

9. 34-gauge blunt-tip needles.
10. Antibiotic ointment.

2.4 Validation of In Vivo Genome Editing

2.4.1 Retina Whole Mount

1. Micro scissors.
2. Micro forceps.
3. 4% paraformaldehyde (PFA).
4. PBS (prechilled or room temperature).
5. Microscope slides.
6. Micro cover glasses.
7. Mounting medium: Fluoromount-G or similar.
8. Fluorescence microscope or confocal scanning microscope.

2.4.2 Dissociation of Retinal Cells for FACS

1. StemPro™ accutase cell dissociation reagent (Thermo Fisher).
2. Thermal Shaker.
3. Test tubes with cell strainer snap cap and regular cap.
4. Hank's balanced salt solution (HBSS).
5. Fluorescence-activated cell sorting (FACS) system (FACSAria or similar).

2.4.3 Targeted DNA High-Throughput Sequencing

1. Nanodrop spectrophotometer.
2. MiSeq Reagent Nano kit, v.2 and v.3 (Illumina).
3. Real-Time Analysis (RTA) software 1.18.54 (Illumina).
4. CASAVA sequencing analysis software v1.8.2 (Illumina).
5. Bedtools software package [16].
6. CRISPResso software package [17].
7. Gel Extraction kit.

2.5 Assessment After In Vivo Genome Editing

1. Espion E2 Electroretinography Console.
2. 12.5 mg/mL ketamine.
3. 2.5 mg/mL xylazine.
4. 1% atropine eye drops.
5. 0.5% tropicamide eye drops.
6. Ophthalmic gel (Gonak™ or similar).
7. OptoMotry VR system, version 1.4.0 (CerebralMechanics, Inc.).
8. 10% sucrose in PBS.
9. 20% sucrose in PBS.
10. PBST solution: 5% Donkey serum and 0.1% Triton X-100 in PBS.
11. Tissue-Tek™ OCT compound.

12. Cryostat microtome.
13. Primary antibodies and fluorochrome-conjugated secondary antibodies.
14. 0.2 μg/mL 4,6-diamidino-2-phenylindole (DAPI) in PBST.
15. Mounting medium: Fluoromount-G or similar.

3 Methods

Carry out all procedures at room temperature unless otherwise specified.

3.1 Vector Construction and In Vitro sgRNA Selection

3.1.1 Design and Assembly of sgRNA

1. Input the DNA sequence of interest into the online programs CRISPR design tool and Benchling. Select five sgRNA target sequences based on the ranking of potential off-target and on-target activities.
2. Purchase synthesized oligonucleotides (oligos) of each sense and antisense target sequence with appropriate overhangs for assembly. For sgRNA assembly to pV-sgRNA-tdTomato, add 5′-ACCG-3′ to the 5′-end of the sense oligos, and add 5′-AAC-3′ to the 5′-end and a "C" to the 3′-end of the antisense oligos.
3. Phosphorylate the oligos in a 20 μL reaction containing: 1× T4 PNK reaction buffer, 200 μM oligonucleotide, 2 μL 10 mM ATP, and 10 U T4 PNK. Incubate at 37 °C for 30 min followed by 75 °C for 10 min. Dilute the reaction to 200 μL by adding 180 μL UltraPure water.
4. Anneal the sense and antisense oligos in a 20 μL reaction containing: 4 μM sense oligos, 4 μM antisense oligos, and 1× standard Taq buffer. Incubate at 95 °C for 5 min and ramp down to 25 °C at a rate of 1 °C per 20 s.
5. Digest 2 μg of pV-sgRNA-tdTomato in a 50 μL reaction containing 1× CutSmart buffer and 20 U SapI restriction enzyme. Incubate at 37 °C for 2 h. Add 10 U CIP to the reaction and incubate for 30 min to dephosphorylate the digested ends.
6. Separate the digested plasmid on a 1% agarose gel. Purify the digested plasmid using the Gel Extraction kit following manufacturer's instructions.
7. Ligate the annealed oligos with the purified digested plasmid in a 20 μL reaction containing: 1× Quick Ligase buffer, 50 ng purified digested plasmid, 2 μL annealed oligos and 1 μL Quick Ligase. Incubate for 5 min.
8. Transform the ligation mixture into DH10B competent bacterial cells following manufacturer's instructions. Briefly, add 4 μL of the ligation mixture to 100 μL competent cells and

incubate on ice for 30 min, followed by incubation at 42 °C for 30 s. Immediately place the bacteria on ice for 2 min. Add 800 μL of SOC medium to the bacteria and shake at the speed of 220 rpm at 37 °C for 1 h. Centrifuge at 1500 × *g* for 3 min and discard the supernatant. Resuspend the bacteria in 100 μL of SOC medium and spread on LB agar plates with 100 μg/mL ampicillin. Incubate the plate at 37 °C overnight.

9. Pick several bacterial colonies for cultures in 3 mL of LB medium with 100 μg/mL ampicillin and incubate at 37 °C overnight with shaking at 220 rpm.
10. Isolate the plasmid DNA using a miniprep kit following manufacturer's instructions. Elute plasmid DNA with 50 μL Tris–HCl solution, pH 8.0.
11. Validate the assembly of sgRNA sequences by redigestion using SapI restriction enzyme and/or by sanger sequencing.

3.1.2 In Vitro sgRNA Selection

1. Seed 4 × 10^5 HEK293-GFP cells in a 6-well plate filled with 2 mL of DMEM medium supplemented with 10% FBS. Culture cells in a 37 °C incubator supplemented with 10% CO_2.
2. When the cells are approximately 80% confluent, transfect the cells in each well with 1 μg of pV-Cas9 plasmid and 1 μg of pV-sgRNA-tdTomato plasmid with verified insertion of the sgRNA target sequence, using Lipofectamine 2000 reagent and following manufacturer's instructions.
3. Harvest the cells 48 h after transfection. Extract genomic DNA using the DNeasy Blood & Tissue Kit following manufacturer's instructions.
4. Amplify fragments containing target sequences by PCR using high-fidelity DNA polymerase. Purify the PCR amplicons using a Gel Extraction kit following manufacturer's instructions (*see* **Note 1**).
5. Detect and determine cleavage activity using the SURVEYOR assay, following manufacturer's instructions and the previously published protocol [18]. Briefly, first set up a 20 μL reaction containing approximately 500 ng purified amplicon and 1× standard Taq buffer. Denature and reanneal the amplicons to allow heteroduplex formation by incubating at 95 °C for 5 min and ramping down to 25 °C at a rate of 5 °C per min. Add 2.5 μL of 0.15 M $MgCl_2$ stock solution, 1 μL of SURVEYOR nuclease S, 1 μL of SURVEYOR enhancer S, and 0.5 μL of UltraPure water to the annealed heteroduplex solution to prepare a 25 μL reaction for SURVEYOR digestion. Incubate at 42 °C for 30 min, then add 2 μL of Stop Solution to terminate the reaction.

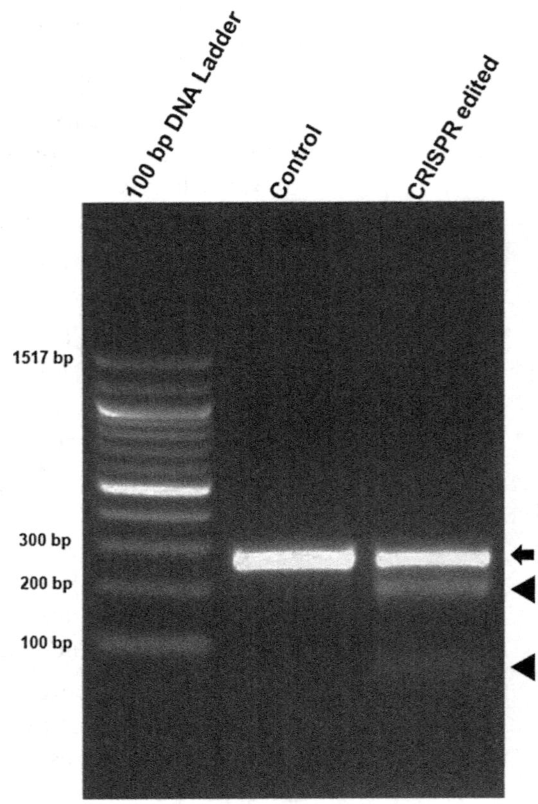

Fig. 1 Sample SURVEYOR assay result. The SURVEYOR nuclease recognizes the mismatch (if any) in the reannealed heteroduplex, which results from indels introduced into the predicted cleavage site in the CRISPR edited cells. Black arrow: undigested amplicons; black triangles: cleavage products of the SURVEYOR nuclease

6. Visualize and image the DNA fragments on a 2% agarose gel.
7. Estimate the integrated intensity of the PCR amplicon and cleaved bands using ImageJ or other software for gel quantification. Calculate the estimated SURVEYOR cleavage (*fcut*) using the following formula [18]:

$$fcut = (b + c)/(a + b + c),$$

   where a is the integrated intensity of the undigested amplicons and b and c are the integrated intensities of each cleavage product (Fig. 1).
8. The sgRNA with higher value of *fcut* usually has higher on-target activity. Select one or two sgRNAs with the highest *fcut* for further studies.

3.2 AAV Vector Production

3.2.1 AAV Generation by Transient Transfection

1. Seed the HEK293 cells in 5 roller bottles at a density of 3×10^7 cells per roller bottle. Culture cells in 300 mL of DMEM medium with 10% FBS in 37 °C incubator supplemented with 10% CO_2.
2. When the cells are approximately 80% confluent, add 5 mL of 1 M HEPES buffer to each roller bottle before transfection to stabilize the pH of the medium. To set up the transfection solution, for each roller bottle, mix 150 μg vector plasmid (either pV-Cas9 or pV-sgRNA-tdTomato containing the selected sgRNA sequence) with 150 μg pHLP19-AAV8 and 150 μg pLAdeno5 helper plasmids, and add the plasmid mixture to 15 mL of 0.3 M $CaCl_2$. Add 15 mL of 2× HBS buffer and gently mix the solution by pipetting up and down five times. Immediately add the transfection solution to the roller bottle. Incubate the cells in a 37 °C incubator supplemented with 3% CO_2 for 6 h to overnight and replace the medium with 100 mL of DMEM serum-free medium.
3. Incubate the cells in a 37 °C incubator supplemented with 5% CO_2.
4. 48 h after transfection, harvest cells from the roller bottles by vigorous swirling without excessive splashing or foaming.
5. Transfer the dislodged cells and medium from the 5 roller bottles to one 500 mL conical tube and centrifuge at 3000 × *g* for 30 min at 4 °C. Discard the supernatant and resuspend the pellets with 200 mL of TSM buffer. The harvested cells can be used for AAV purification or stored at −80 °C. If they are stored, thaw in a 37 °C water bath before proceeding to the next step.

3.2.2 AAV Isolation and Purification

1. Homogenize and break the cells using a microfluidizer (*see* **Note 2**) according to manufacturer's instructions.
2. Centrifuge at 3000 × *g* for 30 min and pour supernatant into a new 500 mL centrifuge tube to eliminate cell debris.
3. Add 1 M $CaCl_2$ to a final concentration of 25 mM. Mix thoroughly and incubate at 4 °C for 10 min. Centrifuge at 3000 × *g* for 1 h. Collect the supernatant and transfer it to a new 500 mL centrifuge tube.
4. To digest residual-free DNA, add Benzonase (100 U/mL) to the supernatant and incubate at 37 °C for 1 h.
5. To precipitate the AAV particles, add 40% PEG8000/2.5 N NaCl to a final concentration of 8% PEG, mix thoroughly, and incubate on ice for 2 h. Centrifuge at 3000 × *g* for 30 min at 4 °C. Discard the supernatant.
6. Resuspend the PEG pellet in 25 mL of HSSE-RNase A buffer. Incubate at 37 °C for 30 min.

7. Prepare the CsCl step gradient by adding 5 mL of 1.5 g/mL CsCl to the bottom of a new 38.5 mL centrifuge tube, followed by 8 mL of 1.3 g/mL CsCl as the middle layer. Add the viral suspension to the top. Carefully balance the tubes and centrifuge in a SW32Ti rotor at 28,000 rpm for 18 h.
8. After step gradient ultracentrifugation, carefully place the tube above a halogen beam illuminator, collect the viral bands using a 5 mL syringe and 18G needle, and transfer to a new 14 mL centrifuge tube. Add 1.4 g/mL CsCl to fill the tube and centrifuge in a SW40Ti rotor at 38,000 rpm for 72 h (*see* **Note 3**).
9. After linear gradient ultracentrifugation, carefully place the tube above a halogen beam illuminator, collect the viral band with 5 mL syringe and 18-gauge needle. Dialyze overnight using Slide-A-Lyzer cassettes in Tris-buffered saline. Store the virus solution at −80 °C (*see* **Note 3**).

3.2.3 AAV Titration by qPCR

1. Use LoBind tubes for all steps in this section. Prepare standard DNA for qPCR using 1 μg purified linearized AAV vector plasmid. Because 1 μg of a 1 kb plasmid contains 9.26×10^{11} plasmid molecules, 1 μg of the standard plasmid will contain a number of copies equal to:

$$9.26 \times 10^{11}/N,$$

where N is the length of the standard DNA plasmid in kb. Make serial fourfold dilutions of this digested DNA in qPCR dilution buffer, starting at 10^7 copies/μL and preparing a total of eight dilutions.
2. For each vector sample, set up two independent initial dilutions (100-fold and 300-fold) in qPCR dilution buffer.
3. To eliminate residual plasmid DNA and AAV replication intermediates from the vector samples, add 10 μL of each diluted vector sample to 40 μL of DNase buffer and 10 U DNase. 1 μg plasmid DNA with a volume of 10 μL was used as a control for DNase activity, and DNA negative sample was used as a control for contamination. Incubate all samples at 37 °C for 1 h. Add 50 μL of 200 mM EDTA (pH 8.0) and incubate at 95 °C for 30 min to terminate the reaction. This step further dilutes the samples by tenfold.
4. Dilute each sample 100-fold by adding 5 μL of sample to 495 μL qPCR dilution buffer and vortexing thoroughly. The vector samples are now diluted 10^5-fold or 3×10^5-fold.
5. For each sample prepare qPCR reactions in triplicate containing: 1× PCR Master Mix, 900 nM sense primer, 900 nM antisense primer, 250 nM FAM-conjugated probe, and 5 μL of diluted sample or standard DNA in a final volume of 25 μL

Table 1
Program for real-time qPCR

Step	Temperature (°C)	Time (s)	Fluorescent collection
1	95	20	No
2	95	1	No
3	60	20	Yes

Repeat step 2 and 3 for an additional 39 cycles

qPCR dilution buffer. Load the reaction solution on a 96-well qPCR plate. Seal the plate with optical adhesive film and run the qPCR reaction in a real-time PCR machine following the program in Table 1.

6. Calculate the titer in vector genomes (vg) per mL using the formula:

$$\text{Titer(vg/mL)} = \left(\left(N_1 \times 4 \times 10^7\right) + \left(N_2 \times 1.2 \times 10^8\right)\right)/2,$$

where N_1 and N_2 are the average number of copies in the 5 μL sample from the 10^5-fold and 3×10^5-fold AAV vector dilution wells, respectively (*see* **Note 4**).

3.3 Delivery of AAV to the Retina by Subretinal Injection

1. To prepare the working solution, dilute AAV vectors to the desired concentration using sterilized Tris-buffered saline containing 0.1% fluorescein.
2. Anesthetize the animal by intraperitoneal injection of a mixture of ketamine (80 mg/kg) and xylazine (8 mg/kg) and dilate the pupils with topical atropine (1%) and tropicamide (0.5%).
3. Position the animal on its side. Expose the eye that will be injected under an ophthalmic surgical microscope. Some fundus-vessels should be visible.
4. Gently make an incision through the cornea adjacent to the limbus at the nasal side using a sharp 18G needle.
5. Use a Hamilton syringe with a 0.5 in 34G blunt-tip needle to perform subretinal injection. Load 1 μL of vector working solution. Carefully insert the needle through the incision while avoiding the lens, and push through the retina. Depress the plunger to inject the solution into the subretinal space. The spread of the vector can be visualized by fluorescein.
6. After injection, apply antibiotic ointment to prevent potential infection.

3.4 Validation of In Vivo Genome Editing

3.4.1 Retina Whole Mount

1. Euthanize the mouse. Enucleate the eyeballs and incubate in chilled PBS for 15 min (*see* **Note 5**).
2. Gently squeeze the eyeballs several times to detach the retina.
3. Fix the eyeballs in 4% PFA for 1 h.
4. Remove the cornea using micro scissors. Gently pull out the lens using fine forceps.
5. Carefully separate the retina from the sclera and dissect it from the eye cup at the optic nerve head. Wash the retina with PBS.
6. Place the retina on a microscope slide with the vitreous side facing up. Make three or four cuts from the periphery half way to the center and flatten the retina using a fine brush.
7. The retina whole mount can be used for immunofluorescent staining or other treatments. To observe expression and knockdown of EGFP, coverslip the retina with mounting reagent and observe under a fluorescence microscope.

3.4.2 Dissociation of Retinal Cells for FACS

1. Euthanize the mouse and enucleate the eyes. Remove the cornea and lens and carefully separate the retina from other tissue.
2. Place the retina in 1 mL of accutase solution for dissociation. Incubate at 37 °C for 10 min with shaking at 1000 rpm in a thermal shaker.
3. Prepare test tubes with cell strainer caps. Rinse cell strainers with 0.5 mL of HBSS.
4. Pipette the retina-accutase solution up and down several times to completely dissociate the tissue.
5. Add the retina-accutase solution through the cell strainer cap.
6. Add 3 mL of HBSS through the cap to wash it. Replace the cell strainer cap with a regular cap.
7. Centrifuge at 450 × *g* for 4 min.
8. Discard the supernatant.
9. Resuspend the cells in 1 mL of HBSS. The suspension is ready for FACS.
10. Use FACS to analyze the number of EGFP-expressing cells. The AAV-transduced photoreceptor cells indicated by tdTomato expression can also be analyzed and sorted by FACS.

3.4.3 Targeted DNA High-Throughput Sequencing

1. Predicted on- and off-target sites are indicated by the online programs during sgRNA selection (*see* Subheading 3.1.1). Based on this information, design primers for amplifying DNA fragments containing each predicted on- and off-target site. The length of the amplicons should not exceed 600 bp.

2. Perform PCR amplification as described above in **step 4** in Subheading 3.1.2. Gel purify the amplicons using a Gel Extraction Kit (*see* **Note 1**). Determine the quantity of each amplicon using a Nanodrop spectrophotometer.
3. For each treatment group, pool all amplicons containing each predicted on- and off-target site in equal ratio as one sample.
4. Run an aliquot of the pooled amplicons on an Illumina MiSeq device using MiSeq Reagent Nano kit v.2. Use 25 cycles followed by two index reads for quality control. Rebalance the pool based on the percentage of reads seen for each amplicon's indexes.
5. Sequence the rebalanced pool on the MiSeq using MiSeq Reagent Nano Kit v.3.
6. Process the sequencing data using RTA 1.18.54 and CASAVA (v1.8.2.).
7. Convert the Aligned Binary Alignment/Map (BAM) files to FASTQ format using the Bedtools software package [16].
8. To obtain the mutation rate and pattern of the predicted on- and off-target sites, analyze the FASTQ data using the CRISPResso software package [17] or similar software.

3.5 Assessment After In Vivo Genome Editing

3.5.1 Electroretinography (ERG)

The ERG is used to measure retinal function. It records the electrical response of neuronal and nonneuronal cells in the retina to a light stimulus. The primary outcomes of the ERG test usually consist of the a-wave, b-wave, latency of response, etc. The a-wave is a negative deflection derived from rod (dark-adapted) and cone (light-adapted) photoreceptors. The b-wave is a positive deflection derived mainly from Müller and ON-bipolar cells.

1. Dark-adapt the mice overnight.
2. Conduct the test in a dark room with only dim red light.
3. Anesthetize the mouse by intraperitoneal injection of a mixture of ketamine (80 mg/kg) and xylazine (8 mg/kg). Dilate the pupils with topical atropine (1%) and tropicamide (0.5%).
4. Place the mouse on a warm pad. Gently place the reference electrode in its mouth and the recording electrodes on the cornea of each eye. Apply a drop of Gonak™ or similar contact gel on the electrodes or eyes to improve electrical contact and maintain corneal hydration before placing the recording electrodes.
5. Record the dark-adapted ERG at a light intensity of 10^{-4}, 10^{-3}, 10^{-2}, 1, 10, 100, and 1000 cd.s/m^2. The main ERG parameters of each step are listed in Table 2.
6. Light-adapt the mouse for 2 min.

Table 2
Main parameters of ERG

		Light intensity (cd s/m^2)	Total sweeps	Internal time (s)
Dark-adapted ERG	Step 1	10^{-4}	5	5
	Step 2	10^{-3}	3	10
	Step 3	10^{-2}	3	20
	Step 4	1	3	30
	Step 5	10	2	45
	Step 6	10^2	2	60
	Step 7	10^3	1	60
Light-adapted ERG	Step 1	0.3	20	0.2
	Step 2	1	20	0.2
	Step 3	3	20	0.2
	Step 4	10	20	0.2
	Step 5	30	20	0.2
	Step 6	100	20	0.2
	Flicker	100	20	0

7. Record the light-adapted ERG at a light intensity of 0.3, 1, 3, 10, 30, and 100 cd.s/m^2. The main ERG parameters of each step are listed in Table 2.

3.5.2 Optomotor Test

The measurement of the optomotor response (compensatory head/body movement) is a well-established method to evaluate visual acuity in the context of impaired visual function in disease models [19, 20]. The animals do not need training for this visual behavior test.

1. Place the mouse on the platform surrounded with LCD screens. Close the lid of the instrument. The mouse should be allowed to move freely on the platform.
2. When starting the test, rotating images of virtual walls of cylinders with vertical sine wave grating will be displayed on the LCD screens at a spatial frequency controlled by the OptoMotry VR software (v.1.4).
3. During the test, the display of the images can be reset by inputting the location of the mouse's head on the video image using a crosshair superimposed on the computer display. The software will adjust the projection of images to center the rotation of the cylinder at the mouse's viewing position.
4. The test starts at a spatial frequency of 0.042 cycles/degree. When you observe unambiguous mouse head movements tracking the cylinder rotation, accept the current spatial frequency and the software will proceed to the next test at an increased frequency. In contrast, if the mouse fails to track the cylinder rotation in 10 min, reject the current frequency and

the software will proceed to the next test at a slightly decreased frequency. Visual behavior at serial frequency settings from 0.03 cycles/degree to 0.642 cycles/degree will be repeatedly tested to determine visual acuity.

5. Tracking clockwise rotation determines the visual acuity of the left eye and tracking counterclockwise rotation determines the visual acuity of the right eye [21].

3.5.3 Retinal Cryosection Preparation

1. Euthanize the mouse and enucleate the eyes after marking the orientation (*see* **Note 5**).
2. Fix the eyes in 4% PFA for 1 h.
3. Remove the cornea and lens. Place the eye cup in 10% sucrose and incubate at 4 °C for 30 min.
4. Place the eye cup in 20% sucrose and incubate at 4 °C for 30 min or overnight, until it completely sinks.
5. Embed the eye cup in OCT compound. Carefully adjust the position of the eye cup in the OCT, then perform rapid freezing using dry ice or liquid nitrogen. Store in −80 °C.
6. Cut retinal cryosections at 10 μm thickness using a cryostat microtome.

3.5.4 Immunofluorescent Staining

1. Completely air-dry the sections before beginning. Once the procedure starts, prevent sections from drying out.
2. Use PBS to rinse sections, and block with PBST containing 5% donkey serum for 1 h.
3. Incubate sections with primary antibodies diluted in 2% donkey serum in PBS for 2 h to overnight.
4. Wash sections with PBST three times for 5 min each.
5. Incubate sections with fluorochrome-conjugated secondary antibodies diluted in 2% donkey serum for 1 h.
6. Incubate the sections with 0.2 μg/mL 4,6-diamidino-2-phenylindole (DAPI) in PBST for 10 min.
7. Wash the sections with PBST four times for 10 min each.
8. Mount the sections in Fluoromount-G reagent. Keep away from light.
9. Once the mounting medium has dried, observe and take images using a fluorescence microscope.

4 Notes

1. The integrity and fidelity of the specific amplicon that contains the predicted CRISPR cleavage site is critical for obtaining accurate results in the SURVEYOR assay. Gel purification is

necessary when nonspecific amplicons are present. To minimize potential DNA damage, UV exposure should be limited.

2. A microfluidizer is a high-pressure homogenizing device for large-scale homogenization. Alternatively, for small-scale production, sonication can be utilized for cell homogenization.
3. With halogen beam illumination, two bands can be visualized after step gradient ultracentrifugation. The lower/heavier band contains full AAV particles, while the upper/lighter band contains empty particles. Only the lower band is collected and subjected to linear gradient ultracentrifugation. After the second round of ultracentrifugation, only one band containing full AAV particles is visualized.
4. Because a double-stranded DNA plasmid is used for the standard wells and a single-stranded AAV genome is measured in the unknown wells, this formula multiplies the average value of the unknown wells by 2. If a self-complementary (double-stranded) AAV genome is measured in the unknown wells, use the formula:

$$\text{Titer(vg/mL)} = \left(\left(N_1 \times 2 \times 10^7\right) + \left(N_2 \times 6 \times 10^7\right)\right)/2$$

5. When enucleating the eyes, keep the nictitating membrane intact to indicate the orientation. Alternatively, mark the orientation with tissue-marking dye before enucleating the eyeball.

References

1. Mali P, Yang L, Esvelt KM, Aach J, Guell M, DiCarlo JE, Norville JE, Church GM (2013) RNA-guided human genome engineering via Cas9. Science 339(6121):823–826. https://doi.org/10.1126/science.1232033
2. Lau CH, Suh Y (2017) In vivo genome editing in animals using AAV-CRISPR system: applications to translational research of human disease. F1000Res 6:2153. https://doi.org/10.12688/f1000research.11243.1
3. Wang S, Sengel C, Emerson MM, Cepko CL (2014) A gene regulatory network controls the binary fate decision of rod and bipolar cells in the vertebrate retina. Dev Cell 30(5):513–527. https://doi.org/10.1016/j.devcel.2014.07.018
4. Bakondi B, Lv W, Lu B, Jones MK, Tsai Y, Kim KJ, Levy R, Akhtar AA, Breunig JJ, Svendsen CN, Wang S (2016) In vivo CRISPR/Cas9 gene editing corrects retinal dystrophy in the S334ter-3 rat model of autosomal dominant retinitis pigmentosa. Mol Ther 24(3):556–563. https://doi.org/10.1038/mt.2015.220
5. Latella MC, Di Salvo MT, Cocchiarella F, Benati D, Grisendi G, Comitato A, Marigo V, Recchia A (2016) In vivo editing of the human mutant rhodopsin gene by electroporation of plasmid-based CRISPR/Cas9 in the mouse retina. Mol Ther Nucleic Acids 5(11):e389. https://doi.org/10.1038/mtna.2016.92
6. Li P, Kleinstiver BP, Leon MY, Prew MS, Navarro-Gomez D, Greenwald SH, Pierce EA, Joung JK, Liu Q (2018) Allele-specific CRISPR/Cas9 genome editing of the single-base P23H mutation for rhodopsin associated dominant retinitis pigmentosa. bioRxiv. https://doi.org/10.1101/197962
7. Kim K, Park SW, Kim JH, Lee SH, Kim D, Koo T, Kim KE, Kim JH, Kim JS (2017) Genome surgery using Cas9 ribonucleoproteins for the treatment of age-related macular degeneration. Genome Res 27(3):419–426. https://doi.org/10.1101/gr.219089.116
8. Hung SS, Chrysostomou V, Li F, Lim JK, Wang JH, Powell JE, Tu L, Daniszewski M, Lo C, Wong RC, Crowston JG, Pebay A, King AE, Bui BV, Liu GS, Hewitt AW (2016) AAV-mediated CRISPR/Cas gene editing of

retinal cells in vivo. Invest Ophthalmol Vis Sci 57(7):3470–3476. https://doi.org/10.1167/iovs.16-19316

9. Giannelli SG, Luoni M, Castoldi V, Massimino L, Cabassi T, Angeloni D, Demontis G, Leocani L, Andreazzoli M, Broccoli V (2018) Cas9/sgRNA selective targeting of the P23H Rhodopsin mutant allele for treating retinitis pigmentosa by intravitreal AAV9. PHP.B-based delivery. Hum Mol Genet 27 (5):761–779. https://doi.org/10.1093/hmg/ddx438
10. Kim E, Koo T, Park SW, Kim D, Kim K, Cho HY, Song DW, Lee KJ, Jung MH, Kim S, Kim JH, Kim JH, Kim JS (2017) In vivo genome editing with a small Cas9 orthologue derived from Campylobacter jejuni. Nat Commun 8:14500. https://doi.org/10.1038/ncomms14500
11. Huang X, Zhou G, Wu W, Duan Y, Ma G, Song J, Xiao R, Vandenberghe L, Zhang F, D'Amore PA, Lei H (2017) Genome editing abrogates angiogenesis in vivo. Nat Commun 8 (1):112. https://doi.org/10.1038/s41467-017-00140-3
12. Ruan GX, Barry E, Yu D, Lukason M, Cheng SH, Scaria A (2017) CRISPR/Cas9-mediated genome editing as a therapeutic approach for leber congenital amaurosis 10. Mol Ther 25 (2):331–341. https://doi.org/10.1016/j.ymthe.2016.12.006
13. Suzuki K, Tsunekawa Y, Hernandez-Benitez R, Wu J, Zhu J, Kim EJ, Hatanaka F, Yamamoto M, Araoka T, Li Z, Kurita M, Hishida T, Li M, Aizawa E, Guo S, Chen S, Goebl A, Soligalla RD, Qu J, Jiang T, Fu X, Jafari M, Esteban CR, Berggren WT, Lajara J, Nunez-Delicado E, Guillen P, Campistol JM, Matsuzaki F, Liu GH, Magistretti P, Zhang K, Callaway EM, Zhang K, Belmonte JC (2016) In vivo genome editing via CRISPR/Cas9 mediated homology-independent targeted integration. Nature 540(7631):144–149. https://doi.org/10.1038/nature20565
14. Yu W, Mookherjee S, Chaitankar V, Hiriyanna S, Kim JW, Brooks M, Ataeijannati Y, Sun X, Dong L, Li T, Swaroop A, Wu Z (2017) Nrl knockdown by AAV-delivered CRISPR/Cas9 prevents retinal degeneration in mice. Nat Commun 8:14716. https://doi.org/10.1038/ncomms14716
15. Holmgaard A, Askou AL, Benckendorff JNE, Thomsen EA, Cai Y, Bek T, Mikkelsen JG, Corydon TJ (2017) In vivo knockout of the vegfa gene by lentiviral delivery of CRISPR/Cas9 in mouse retinal pigment epithelium cells. Mol Ther Nucleic Acids 9:89–99. https://doi.org/10.1016/j.omtn.2017.08.016
16. Quinlan AR, Hall IM (2010) BEDTools: a flexible suite of utilities for comparing genomic features. Bioinformatics 26(6):841–842. https://doi.org/10.1093/bioinformatics/btq033
17. Pinello L, Canver MC, Hoban MD, Orkin SH, Kohn DB, Bauer DE, Yuan GC (2016) Analyzing CRISPR genome-editing experiments with CRISPResso. Nat Biotechnol 34(7):695–697. https://doi.org/10.1038/nbt.3583
18. Ran FA, Hsu PD, Wright J, Agarwala V, Scott DA, Zhang F (2013) Genome engineering using the CRISPR-Cas9 system. Nat Protoc 8 (11):2281–2308. https://doi.org/10.1038/nprot.2013.143
19. Prusky GT, Alam NM, Beekman S, Douglas RM (2004) Rapid quantification of adult and developing mouse spatial vision using a virtual optomotor system. Invest Ophthalmol Vis Sci 45(12):4611–4616. https://doi.org/10.1167/iovs.04-0541
20. Kretschmer F, Sajgo S, Kretschmer V, Badea TC (2015) A system to measure the optokinetic and optomotor response in mice. J Neurosci Methods 256:91–105. https://doi.org/10.1016/j.jneumeth.2015.08.007
21. Douglas RM, Alam NM, Silver BD, McGill TJ, Tschetter WW, Prusky GT (2005) Independent visual threshold measurements in the two eyes of freely moving rats and mice using a virtual-reality optokinetic system. Vis Neurosci 22(5):677–684. https://doi.org/10.1017/S0952523805225166

Part II

AAV Delivery to the Central Nervous System

Chapter 8

Intravenous Infusion of AAV for Widespread Gene Delivery to the Nervous System

Dominic J. Gessler, Phillip W. L. Tai, Jia Li, and Guangping Gao

Abstract

The central nervous system (CNS) is a fascinating and intricate set of biological structures that we have yet to fully understand. Studying the in vivo function of the CNS and finding novel methods for treating neurological disorders have been particularly challenging. One difficulty is correcting genetic disorders afflicting the CNS in a targeted manner. Recombinant adeno-associated viruses (rAAVs) have emerged as promising therapeutic tools for treating genetic defects of the CNS, due to their excellent safety profile and ability to cross the blood-brain barrier (BBB). While stereotactic injection of AAV is promising for localized gene delivery, it is less desirable for some applications because of the technique's invasiveness and limited intraparenchymal spread. Alternatively, intravascular administration can achieve widespread delivery of rAAV to the CNS. In this chapter, we will discuss the prevalent routes of administration to deliver rAAV to the CNS via intravenous (IV) injection in mice. We will highlight key considerations for using rAAV, and the advantages and disadvantages of each administration method. We will also briefly discuss intravenous delivery in larger animal models, factors that may impact experimental interpretations, and outlooks for clinical translation.

Key words CNS, Mouse, Adeno-associated virus, Blood-brain barrier, Neutralizing antibodies, AAV vector design

1 Introduction

1.1 General Introduction

The central nervous system (CNS), which consists of the brain and spinal cord, is arguably the most intricate organ in vertebrates. Its diversity of subanatomical structures and variety of specialized and subspecialized cell types define the complex nature of this system of tissues. The high-level organization of the CNS presents many difficulties in studying CNS function and understanding neurological diseases. While certain in vitro models may recapitulate some aspects of CNS function, full understanding of the CNS requires study in vivo. This notion is particularly true for investigating

Dominic J. Gessler and Phillip W. L. Tai contributed equally to this work.

Michael J. Castle (ed.), *Adeno-Associated Virus Vectors: Design and Delivery*, Methods in Molecular Biology, vol. 1950, https://doi.org/10.1007/978-1-4939-9139-6_8,

intercellular mechanisms of the CNS and the development of therapies for neurologic diseases.

Recombinant adeno-associated viruses (rAAVs), which are derived from apathogenic, non-genotoxic parvoviruses that are gutted of their natural DNA genomes, have not only become a powerful tool for studying the brain, but are also among the most promising therapeutic vehicles for treating genetic diseases that affect the nervous system. A clinical trial using systemic delivery of rAAV to treat spinal muscular atrophy (SMA) in infants has shown remarkable safety and therapeutic efficacy [1]. Although there is substantial recent excitement surrounding the potential for rAAVs and their use as biotherapies, there are still challenges associated with their use in treating neurological disease. In addition, there are many factors that investigators must consider in the design of vectors and their mode of delivery in order to efficiently transduce tissues of the CNS.

Stereotactic intracerebroventricular (ICV) and intrathecal (IT) injections to deliver rAAV via the cerebrospinal fluid (CSF) for widespread distribution to the CNS, or direct intraparenchymal stereotactic administration, are the preferred methods of delivering rAAV to specific brain regions. Although these approaches may be advantageous for diseases with local CNS pathology, intraparenchymal and ICV injections are invasive, and it may not be desirable to only transduce a restricted area of the brain. For example, in disease models where rAAV requires targeting of both CNS and peripheral tissues (e.g., liver, spleen, kidney, skeletal muscle), single intravenous (IV) infusions under a ubiquitous promoter may be preferred. Notably, many disorders, such as spinal muscular atrophies, neuronal ceroid lipofuscinoses, and spinocerebellar degenerations, require widespread treatment of the entire CNS [2–4]. Additionally, neurological disorders may display a primary CNS phenotype, but also present a secondary etiology that affects the entire organism (CNS and peripheral organs). Intravascular delivery in these cases, as well as in their preclinical model counterparts, has substantial advantages. Utilizing the naturally existing vasculature, IV delivery of rAAV can achieve widespread CNS transduction and deliver therapeutic vectors to the entire CNS.

Although IV administration is minimally invasive and simultaneously targets both peripheral organs and tissues of the CNS, intravascular delivery unfortunately also has its own set of challenges. Namely, the ability of intravenously administered drugs to efficiently reach the CNS is hindered by the blood-brain barrier (BBB). This is less of a challenge for certain drugs (e.g., small molecules and compounds with ideal hydrophobicity and lipophilicity profiles [5]), but is a substantial barrier for large hydrophilic molecules (e.g., oligonucleotides, genes, proteins, and viruses) [6, 7]. Furthermore, attempting to target the CNS through intravascular delivery leads to a wider distribution of rAAV, which may carry the risk of off-target gene delivery to the tissues of peripheral organs, inducing potential immunotoxicities.

A chief advantage of rAAVs as therapeutic vectors is that some AAV serotypes are capable of crossing the BBB and transducing the CNS after IV delivery with relatively high efficiencies, especially when administered to neonatal animals [8–10]. We will discuss three modes of IV rAAV delivery to target the CNS: facial vein, retro-orbital, and lateral tail vein. Importantly, we will outline considerations prior to any in vivo experiment. We will primarily focus on experimental considerations in mouse models, as the mouse is the most commonly used animal model for in vivo gene transfer via rAAV administration. It should be noted that CNS gene transfer by some rAAV vector serotypes is largely consistent between rodents and large animals, but enough interspecies differences in transduction efficiency among serotypes have been reported to warrant careful consideration [11].

1.2 Natural and Engineered AAV Capsids

This section, along with Subheadings 1.3 and 1.4, will cover vector design considerations that must be tailored for the investigator's specific experimental aims. In addition, the researcher should take note that the rAAV gene therapy field is rapidly evolving. Below, we highlight concepts that stand out as critical factors for vector design for the explicit purpose of transducing the CNS via IV administration. The details herein are best used as a general starting guide for researchers that are new to this field.

One valuable property of AAV is the inherent capsid diversity observed in naturally occurring AAV. This advantage has led to the discovery and design of new AAV capsids with improved tissue tropism and potentially reduced immunogenicity [12]. New capsids can be developed by three existing strategies: (1) natural discovery of AAV capsids from pro-viral DNA isolated from host species [13, 14], (2) directed evolution [15, 16], and (3) rational capsid design, which includes the conversion of residues to paralogous motifs that are unique to other serotypes to gain different tropism profiles [17, 18], and the fusion of peptide sequences to the capsid structure to alter tissue tropism [19–22] (*see* Chapter 3).

Multiple AAV serotypes and variants have been extensively evaluated under a variety of administration parameters. The first widely used AAV serotype, AAV2, does not efficiently cross the BBB. Early studies utilized direct injection of rAAV into the parenchyma or ventricles to target the CNS, resulting in limited CNS distribution and regional transgene expression [23, 24].

The discovery of new serotypes has opened the door to vectors with improved ability to cross the BBB, providing an opportunity to treat CNS diseases via intravenous delivery of therapeutic rAAVs that achieve long-term expression in large areas of the brain. Several AAV serotypes have been shown by different research groups to cross the BBB [13, 19, 25–28]. In fact, with the exception of rAAV2 and rAAV5, it has been demonstrated that the capacity to cross the adult BBB is a common feature among natural AAV serotype isolates [26, 27]. After crossing the BBB, serotypes can display different tropisms [27]. Certain capsids stand out for efficient CNS transduction following systemic IV injection. Notably, rAAV9, rAAVrh.8, rAAVrh.10, rAAVrh.39, and rAAVrh.43 demonstrate robust global transduction of glial and neuronal cell types within the hypothalamus, medulla, striatum, hippocampus, choroid plexus, caudate-putamen, corpus callosum, and substantia nigra [26, 27, 29], with variable transduction in the cortex, habenular nucleus, cornu ammonis, dentate gyrus, thalamus, cerebellum, and olfactory bulb.

In adult animals, rAAV9 transduces mostly astrocytes and brain endothelial cells [10, 26]. rAAVrh.10 vectors can also efficiently transduce a range of cell types of the CNS, including neurons. rAAV1 transduces primarily granule cells in the cerebellum, while rAAV6 and rAAV6.2 transduce mostly Purkinje cells [27]. Of note, rAAVrh.10, rAAV9, rAAV7, and rAAVrh.39 show high transduction efficiencies in motor neurons [27]. However, these vectors have very broad tropisms that extend beyond neuronal cell types. In contrast, rAAVrh.8 exhibits naturally lower peripheral tissue dissemination than other AAV serotypes, particularly in the liver, enhancing the safety profile of systemic gene delivery to the CNS [26].

Several groups have taken a directed evolution approach to develop capsids that can cross the BBB with efficiencies greater than natural isolates. By applying selective pressure to a library of artificially created AAV capsid variants, novel vectors that cross the BBB and transduce specific cell types of the murine brain can be developed. Using such methods, Deverman et al. engineered the nonnatural AAV capsid AAV-PHP.B, which has the capacity to transduce astrocytes, neurons, oligodendrocytes, cerebellar Purkinje cells, and several interneuron populations across multiple brain regions after IV administration, and demonstrates greater CNS transduction and reduced peripheral transduction relative to AAV9 [19]. Interestingly, Hordeaux et al. recently demonstrated that AAV-PHP.B underperforms relative to AAV9 in another commonly used mouse strain, BALB/cJ, as well as in nonhuman primates [30]. The authors argue that a directed evolution approach that selects high performance capsids in the C57BL/6 J mouse strain may introduce strain-specific properties.

Extra caution should be considered when utilizing capsids selected or tested exclusively in one model system. All rAAVs delivered intravascularly to target the CNS should first be tested with standard gene reporter systems, such as EGFP or luciferase, before proceeding with the transgene of interest.

1.3 Regulatory Elements for Controlling Gene Expression

A major drawback among current rAAVs that have strong tropism for CNS is their lack of CNS-restricted tropism. For example, rAAV9 also confers robust transduction of peripheral tissues such as liver, heart, lungs, muscle, and pancreas [31–33]. To tailor rAAV-delivered gene expression for the intended therapeutic target, the use of tissue- or cell type-specific promoters, as well as tissue-specific miRNA-mediated de-targeting elements, are attractive vector design strategies.

Adaptation of native promoter sequences for tissue- or cell type-specific expression of vector transgenes may be preferred over ubiquitous and constitutively active promoter cassettes. Both muscle- or liver-specific expression have strongly relied on the use of natively derived promoters such as muscle creatine kinase (CKM) or thyroxine-binding globulin (TBG) promoters [34, 35]. Multiple CNS cell type-specific promoters are available and have been utilized to limit transgene expression to certain cell types (Table 1). However, further testing of promoters is often required. Common challenges for promoter and rAAV compatibility are: (1) the limited availability of well-described regulatory cassettes for CNS-specific expression, (2) the packaging capacity of AAV, which requires that the transgene expression cassette, including promoter, transgene, polyA sequence, and other auxiliary elements, be below a limit of 4.7 kb, (3) changes in the metabolic state of the cell and environmental dynamics that may alter promoter activities, and (4) the fact that promoters which appear specific for one CNS cell type after brain injection may also express in a subset of peripheral tissues after IV delivery. Optimization of CNS-specific rAAV regulatory cassettes is complicated by the diversity of cell types in the CNS, with new cell types continuing to be identified [36–38]. The CNS can be divided into four main cell types: (1) neurons, (2) astrocytes, (3) oligodendrocytes, and (4) microglia. Current knowledge of

Table 1
Lists of CNS promoters and cell targets

Promoter	Gene name	Cell type	Citations
NSE	Neuron specific enolase	Neuron	[64, 65]
SYN1	Synapsin 1	Neuron	[66–68]
GFAP	Glial fibrillary acidic protein	Astrocyte	[69, 70]
MBP/MAG	Myelin basic protein/Myelin-associated glycoprotein	Oligodendrocyte	[71]
OLIG2	Oligodendrocyte transcription factor 2	Oligodendrocyte	[72]
POMC	Proopiomelanocortin	Neurons/hypothalamus	RIKEN
SOD1	Superoxide dismutase 1	Neurons/spinal cord	RIKEN

gene regulation in the diverse cell populations of the CNS is incomplete, and the design and selection of promoters that can truly confer cell type-specific transgene expression is an ongoing challenge.

Many cell- or tissue-specific promoters fail to achieve experimental or therapeutic levels of transgene expression. In such cases, strong ubiquitous and constitutively active regulatory cassettes, such as the CMV-enhancer/chicken β-actin (CBA) promoter, are popular in rAAV vector design. Additionally, no cell- or tissue-specific promoter is guaranteed to be devoid of leaky off-target activation. The suppression of peripheral organ expression in these situations can be achieved by miRNA-mediated de-targeting elements [39]. This strategy involves modification of the rAAV transgene by incorporating miRNA binding sites (miR-BS) into the 3′-UTR to recruit endogenous cell type-specific, highly expressed miRNAs to inhibit expression. For example, to mitigate potential protein toxicity from unintentional transduction of peripheral organs and from leaky promoter activity, miR-BSs designed to recruit miR1 or miR122 to exclude expression from striated muscle and liver, respectively [40], can be introduced. Such designs have shown remarkable ability to specifically de-target expression from peripheral organs while maintaining expression in the CNS [8, 40].

Importantly, to circumvent innate immunity, vector design incorporating miR-BSs that utilize miRNAs present in antigen presenting cells (APCs) such as miR-142 have also been used with success (*see* **Note 1**) [41, 42].

1.4 Dose and Timing

Intravenous administration of rAAVs circulates the virus through the vascular system. Therefore, rAAV is disseminated throughout the peripheral organs, substantially reducing the number of rAAV particles that can reach and cross the BBB. Transduction efficiency of the CNS is therefore intrinsically dose dependent [29, 32]. Intravenous delivery at high doses might lead to transgene expression in undesirable cell types, potentially causing unwanted side effects. Dose also seems to influence the transduction pattern of subanatomical CNS regions. While a high dose such as 2×10^{14} vector genomes per kilogram (vg/kg) is efficiently able to achieve widespread transduction and has been successfully translated into humans, a lower dose (2×10^{12} vg/kg) may only reach CNS regions close to blood vessels, such as choroid plexus [1].

Permeability for certain molecules across the BBB seems to be age-independent, and differences between the neonatal and the adult BBB have been suggested to influence IV delivery of AAV. The transduction efficiency of rAAVs is dependent on the interplay between cellular and humoral host factors and capsid properties. During growth and development, the physiological profile of the BBB changes, and expression of molecules facilitating or preventing

AAV from crossing the BBB may change as well [43]. Age-dependent differences in CNS transduction efficiency and cellular tropism [10, 26, 27, 44] observed after IV administration of rAAVs support this notion. Notably, pre-infusion with mannitol has been shown to increase permeability of the BBB for AAV2 in adult mice [45]. Additionally, certain diseases require early intervention and thus neonatal administration might be preferred. However, because AAV genomes predominantly persist in the nucleus as nonreplicating episomes [46], growth of the host tissue may dilute the transgene and reduce vector potency during normal aging. This is a problem for neonatal targeting of organs that consist of cells with high replicative rates, such as liver, but is less of an issue for the CNS.

With these considerations in mind, it is particularly important to conduct systematic dose assessments at suitable ages, to determine which conditions confer the desired strength and distribution of gene expression for each rAAV vector.

2 Materials

1. Alcohol prep pads.
2. 70% isopropyl alcohol.
3. Absorbent gauze pads.
4. Isoflurane.
5. Anesthesia machine.
6. Anesthesia chamber and nose cone.
7. 50 mL conical tube.
8. Mouse Tail Illuminator restrainer.
9. Ice.
10. rAAV vector.
11. Filter-sterilized PBS.
12. Neonatal (0–2 days) or adult mice.
13. 0.3 mL 31G insulin syringes.
14. 0.5 mL 28G insulin syringes.
15. Padded injection platform.
16. Heating pads or hand warmers.
17. Eye ointment.

3 Methods

In this section, we will focus specifically on techniques utilized for mouse models, since they are currently the standard and first-pass in vivo system for preclinical investigation of therapeutic rAAVs. There are three standard approaches for IV delivery to target rAAV to the mouse CNS: facial vein (Subheading 3.2), retro-orbital (Subheadings 3.3 and 3.4), and lateral tail vein (Subheading 3.5). Each method has advantages and disadvantages, and preference is primarily dependent on the age of mice at treatment. For neonatal treatments, the superficial temporal/facial vein is visible and accessible posterior to the eye. In adults, lateral tail vein administration is more convenient. Retro-orbital injection can be performed at all ages, and is required in adult disease animal models that must be treated at young ages when tail vein administration is not possible.

Translation to larger animals requires additional considerations (*see* **Note 2**).

Several aspects hinder extrapolation of rAAV function from mice to larger animal models: (1) preexisting immunity in large animals that necessitate prescreening for neutralizing antibodies (NAbs) (*see* **Note 3**), (2) inherent differences in humoral or cellular immunity between species (*see* **Note 1**), (3) differences in brain size, (4) longevity of transduced vectors, (5) faithful translation of disease models, and (6) genomic diversity of outbred strains common with large animal models. These factors should be carefully considered before translation to larger animals, and steps taken to minimize the risk of immune response and toxicity (*see* **Notes 4** and **5**).

3.1 Preparation of rAAV Vectors for Injection

Recombinant AAV vectors can be produced in-house by triple transfection of human embryonic kidney cells (HEK293) [47] (*see* Chapters 3, 7, 19, 21, 22, and 23). Alternatively, vectors can be obtained from a local vector core or commercial manufacturer. High-quality rAAV vector formulations for injection should be free of contaminants that might harm animals, such as bacteria, endotoxins, or cellular proteins. Low abundance of empty virions in preparations is also desirable. Analysis of rAAV purity by silver-stained SDS-PAGE and assessment of empty particles by electron microscopy is recommended before sample preparation for injection. In addition, newer methods for assessing vector heterogeneity such as *A*deno-*A*ssociated-*V*irus *G*enome *P*opulation sequencing (AAV-GPseq) and sequencing of single-stranded DNA virus preparations (SSV-Seq) (*see* Chapter 5), which employ high-throughput sequencing to quantify and identify the abundance of contaminating DNAs or truncated genomes, are becoming the standard for many quality control pipelines to ensure genomic homogeneity in vector preps [48, 49].

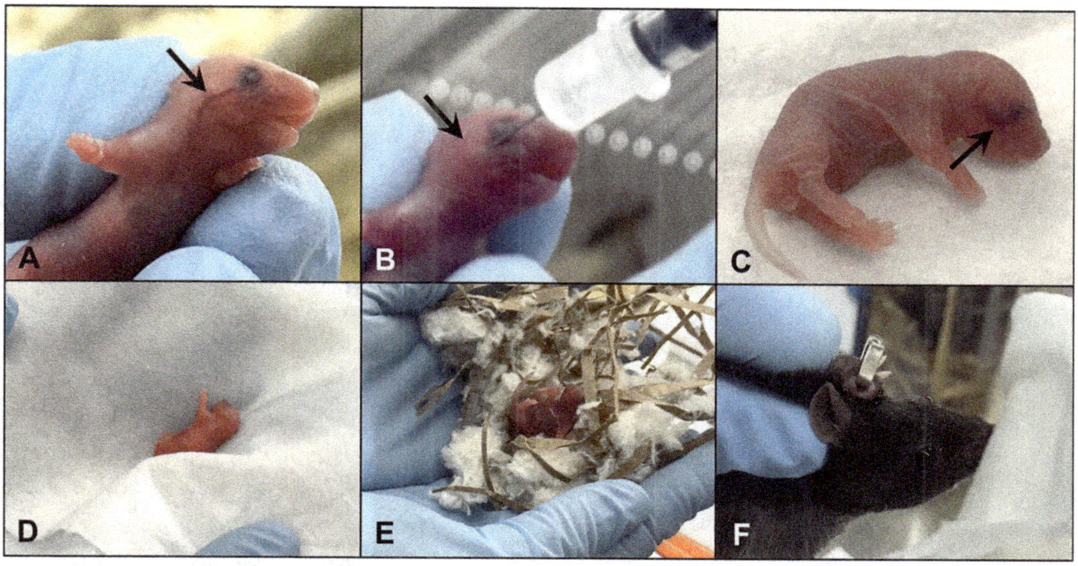

Fig. 1 Facial vein injection of neonatal mice. (**a**) Neonatal mouse preinjection. Arrow indicates visible facial vein. (**b**) Image of injection. Note angle of needle and blanching of facial vein during injection (arrow), indicating successful delivery of vector. (**c**) Black blemish (arrow) at the site of injection post-administration is another indication of successful delivery. (**d**) The pup is rolled in alcohol-soaked tissue, followed by (**e**) rolling in dirty bedding before returning to the dam. (**f**) The dam is also dampened on the nose with alcohol to mask foreign odors

3.2 Facial Vein Injection (Fig. 1)

1. Thaw rAAV vector(s) (if previously frozen) and keep on ice.
2. Calculate the total amount of vector needed for experiments and prepare enough volume for 1 extra pup. 25–50 μL of vector can be injected per pup. Use filter-sterilized 1× PBS to dilute to the desired dose, if needed (*see* **Note 6**).
3. Load 25–50 μL of AAV into a 0.3 mL 31G insulin syringe.
4. Separate neonates from the dam.
5. Working with one pup at a time, immobilize neonates by placing them on a chilled paper towel on top of ice for 30–60 s. Wait until the pup's color takes on a bluish hue and the animal exhibits less movement. Avoid prolonged exposure to ice to avoid mortality.
6. Alternatively, add 0.5 mL of isoflurane to tissue paper in an anesthesia chamber (50 mL conical tube) and tighten the cap to allow isoflurane to vaporize for 1 min. Hold the chamber horizontally, place a pup inside, and tighten the cap. Remove the pup immediately once it has completely stopped moving. Before proceeding, check that the pup is alive by checking for rise and fall of the rib cage.
7. Hold the pup by the thumb and index finger (of nondominant hand) and gently spread the skin of the area of the injection site

to visualize the facial vein (Fig. 1a). Avoid exerting excessive pressure on the pup to avoid crushing the animal.

8. Sterilize the injection site with an alcohol pad.
9. Support elbows on a stable surface. Grip the syringe with the thumb, middle finger, and ring finger. Poise the index finger over the plunger.
10. Position the needle tip at a ~10° angle over the superficial temporal vein, partially insert the needle tip into the vein, adjust the angle of the needle to 0–5°, slowly insert the needle approximately 3 mm into the vein track, and slowly inject the 25–50 μL vector volume (Fig. 1b).
11. Wait 5 s, slowly withdraw the needle, and immediately compress the injection site using an absorbent gauze to stop any bleeding. A successful injection will show a black blemish at the point of injection (Fig. 1c).
12. Ensure the pup's body temperature and color return to normal. Dampen tissue paper with 70% EtOH and gently rub the pup (Fig. 1d). Then rub the pup with used bedding to mask foreign odors that may lead to rejection of the pup (Fig. 1f).
13. Place the cage on warm heating pad to aid neonate recovery. Rub the dam's nose with tissue paper dampened with 70% EtOH (Fig. 1f) and return the pup to the dam.

3.3 Retro-Orbital Injection of Neonates (Fig. 2) (See Note 7)

1. Thaw rAAV vector(s) (if previously frozen) and keep on ice.
2. Calculate the total amount of vector needed for experiments and prepare enough volume for 1 extra pup. 50 μL of vector can be injected per pup (use filter-sterilized 1× PBS to dilute to the proper dose).
3. Load 50 μL of vector into a 0.3 mL 31G insulin syringe.
4. Separate neonates from the dam.
5. Working with one pup at a time, immobilize neonates by placing them on a chilled paper towel on top of ice for 30–60 s. Wait until the pup's color takes on a bluish hue and the animal exhibits less movement. Avoid prolonged exposure to ice to avoid mortality.
6. Alternatively, add 1 mL of isoflurane to tissue paper in an anesthesia chamber (50 mL conical tube) and tighten the cap to allow isoflurane to vaporize for 1 min. Hold the chamber horizontally, place a pup inside, and tighten the cap. Remove the pup immediately once it has completely stopped moving. Before proceeding, check that the pup is alive by checking for rise and fall of the rib cage.
7. Construct a padded injection platform to position the pup (Fig. 2a, b).

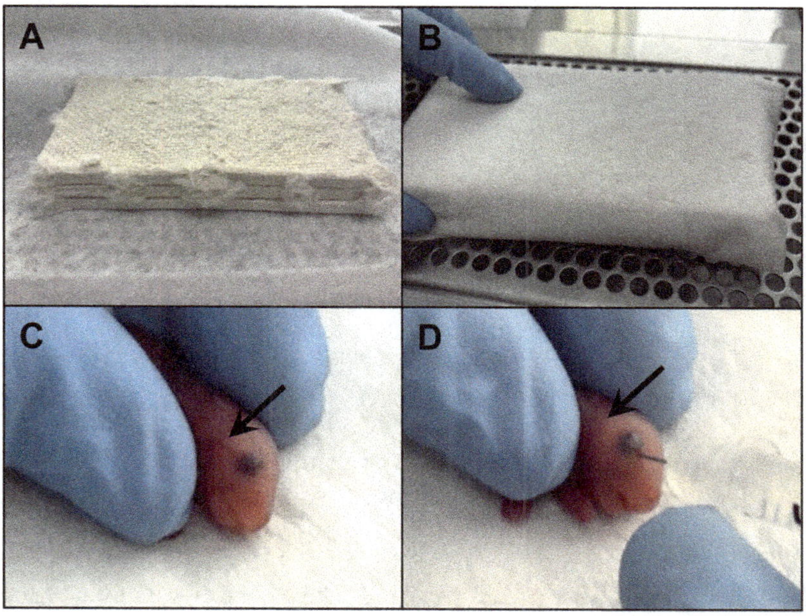

Fig. 2 Retro-orbital injection of neonatal mice. (**a**, **b**) Construction of a padded injection platform (10 cm × 5 cm × 3 cm [L × W × H]). Pads are stacked and wrapped with a paper towel. (**c**) The pup is placed in a lateral position and grasped at the head with the thumb and index finger. Arrow indicates visible superficial temporal vein. (**d**) The needle is inserted with bevel down, at the 3 o'clock position, into the eye approximately 3 mm at a 30° angle. Blanching of the vein (arrow) indicates successful injection

8. Place the pup in a lateral position and grasp the head with the thumb and index finger of the nondominant hand. Avoid exerting excessive pressure on the pup to avoid crushing the animal (Fig. 2c).
9. Sterilize the injection site with an alcohol pad.
10. Support wrists on a stable surface, and grip the syringe with the thumb, middle finger, and ring finger, with index finger poised over the plunger.
11. Place the needle bevel on the eye at a 3 o'clock position, insert approximately 3 mm at a 30° angle into the center of the retro-orbital sinus, and slowly inject the 50 μL vector volume (Fig. 2d).
12. Wait 5 s and slowly withdraw the needle, keeping the bevel outward to protect the eye from being scratched. Immediately compress the injection site using absorbent gauze to stop any bleeding.
13. Ensure the pup's body temperature and color return to normal. Dampen tissue paper with 70% EtOH and gently rub the pup. Then rub the pup with used bedding to mask foreign odors that may lead to rejection of the pup.

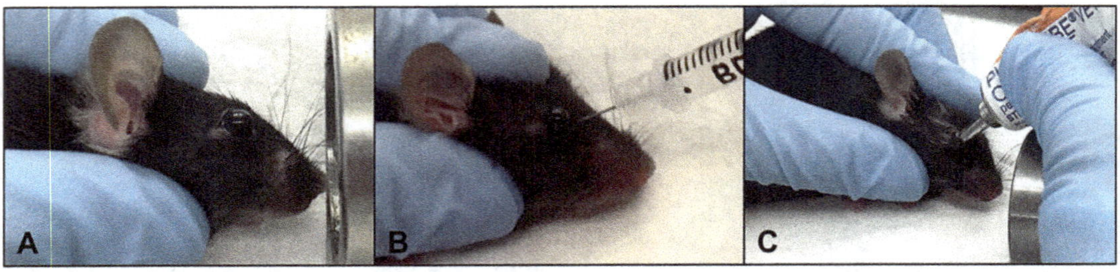

Fig. 3 Retro-orbital injection of adult mice. (**a**) Manipulation of the mouse to cause slight bulge of the eye. The thumb is placed just lateral to the animal's trachea and the fur on the head is drawn toward the palm of the hand. (**b**) The needle is inserted approximately 5 mm at a 45° angle into the center of the retro-orbital sinus. (**c**) Ointment is applied liberally to the eye to aid recovery

14. Place the cage on a warm heating pad to aid neonate recovery. Rub the dam's nose with tissue paper dampened with 70% EtOH and return the pup back to the dam.

3.4 Retro-Orbital Injection of Adults (Fig. 3) (See Note 7)

1. Thaw rAAV vector(s) (if previously frozen) and keep on ice.
2. Calculate the total amount of vector needed for experiments and prepare enough volume for 1 extra mouse. 150–200 μL of vector can be injected per mouse (use filter-sterilized 1× PBS to dilute to the proper dose).
3. Load 150–200 μL of vector into a 0.5 mL 28G insulin syringe.
4. Sedate the mouse with isoflurane in an anesthesia chamber. Wait until breathing slows, then retrieve the mouse from the chamber.
5. Keep the mouse under general anesthesia by using a nose cone to administer 2% isoflurane in medical-grade 21% oxygen at a flow of 2 L/min. Monitor movement and ensure there is no limb withdraw reflex to toe pinch. Grasp the mouse so that its back rests on the palm of the nondominant hand with its head toward the thumb. The thumb is placed just lateral to the animal's trachea and the fur on the animals' head is drawn toward the palm. This causes the animal's eye to bulge slightly (Fig. 3a).
6. Support wrists on a stable surface. Grip the syringe with the thumb, middle finger, and ring finger. Poise index finger over the plunger.
7. Under gentle pressure to bulge the eye, direct the needle into the medial canthus (junction of eyelids closest to the animal's nose) with the bevel facing outward (lateral toward globe). Insert the needle approximately 5 mm at a 45° angle into the center of the retro-orbital sinus and slowly inject the 150–200 μL volume (Fig. 3b).

Fig. 4 Lateral tail vein injection. (**a**) The animal is positioned securely in the restrainer. The tail is fitted through the aperture and held straight and taut. (**b**) Two lateral veins are visible on each side of the tail. Arrow indicates a single vein. (**c**) The needle tip is partially inserted at a 30° angle into the vein. Once the tip has penetrated the vein, readjust the needle so that it is in-line with the length with the vein. Blanching of the tail vein indicates successful injection

8. Withdraw the needle carefully and slowly. Ensure the needle does not scratch cornea. Use sterile gauze to stop any bleeding and apply ointment liberally to the eye to aid recovery (Fig. 3c).
9. After the injection, return the mouse to its cage. It will take approximately 1 min to recover from isoflurane.

3.5 Lateral Tail Vein Injection (Fig. 4)

1. Thaw rAAV vector(s) (if previously frozen) and keep on ice.
2. Calculate the total amount of vector needed for experiments and prepare enough volume for 1 extra mouse. 300 μL of vector can be injected per mouse (use filter-sterilized 1× PBS to dilute to the proper dose).
3. Load 300 μL of vector into a 0.5 mL 28G insulin syringe.
4. Place the mouse into the mouse tail illuminator restrainer apparatus (Fig. 4a). Do not over-restrain the mouse as this will constrict the animal from breathing.
5. Spray 70% isopropyl alcohol on the tail.
6. Hold the tail straight and taut (Fig. 4b). There are two lateral veins on each side of the tail (*see* **Note 8**).
7. Partially insert the needle tip at a 30° angle into the vein. Once the tip has penetrated the vein halfway up the bevel, readjust the needle so that it is in-line with the length of the vein (Fig. 4c). If the needle is inserted properly, there should be no pressure when the syringe plunger is depressed.
8. Inject the 300 μL volume slowly. The vein will blanch. After approximately 4 s the vein color will turn to normal.
9. Carefully withdraw the needle and compress the injection site with absorbent gauze for 10 s to stop any bleeding.
10. Return the mouse to its cage.

4 Notes

1. Long-term expression of transgenes can be blocked by the host immune response. The transgene product can be recognized as nonself, triggering the activation of both humoral and cell-mediated immune responses. This is both problematic for completely foreign gene products, as well as those that are mutated forms of endogenously expressed proteins. Transgene cassettes that carry miR-BS to recruit the antigen presenting cell (APC)-enriched miR-142, thereby blocking expression in dendritic cells, have shown great promise [50, 51]. Such constructs are within the size limitations of rAAV and are compatible with ubiquitous or CNS-specific promoters that can achieve high-level expression.
2. Although rAAV tropism profiles are often recapitulated across strains of mice and different species, differences in tropism between mammalian species do exist. Even strain-to-strain differences can impact experimental interpretation. For example, AAVPHP.B demonstrated robust CNS transduction in the C57/BL6 strain after IV administration, but comparatively low CNS transduction in the BALB/cJ strain [30]. Additionally, PHP.B also demonstrates weak BBB permeability in nonhuman primates (NHPs), including marmosets and rhesus macaques [30, 52]. Tropism differences are often explained by differences in immunity (*see* **Notes 1** and **3**). Of note, the impact of NAbs against current AAV serotypes is lowest in rat and sheep [53]. Consequently, studies to validate the translational potential of rAAVs warrant evaluation in model systems that best recapitulate the human patient, such as nonhuman primates. However, it should be noted that transgene-directed immune responses can be mounted against rAAV-delivered transgenes that produce foreign or human proteins in NHPs [54].
3. There is strong evidence that many standard animal models harbor preexisting antibodies against AAV serotypes AAV1, AAV2, AAV6, and AAV9 [53]. Neutralizing antibodies (NAbs) against AAVs are present in the sera of mice, rats, rabbits, dogs, sheep, and pigs. The CNS has traditionally been considered an immune privileged organ. However, autoimmune diseases such as multiple sclerosis suggest that the CNS is not completely protected from inflammation [55]. Because targeting the CNS via IV injection will expose animals to vector on a systemic level, immune response should be heavily considered. Therefore, prescreening of animals for NAbs is recommended to assess vector effectiveness in all animal subjects, regardless of species. Of note, we and others have

found that even NAb titers between 1:2 and 1:5 can prevent in vivo transduction by intravascular AAV9 [53]. Several studies have focused on developing novel capsids that can efficiently evade host immunity. High-throughput human IgG screening via directed evolution of capsids, in combination with the development of humanized liver mouse models, has identified capsids that can transduce human liver as well as exhibit low immunogenicity [56]. However, a capsid that can both bypass the BBB and evade the human host immune system has yet to be developed.

4. Immune tolerance can be achieved by delivery of rAAV to the liver to harness its tolerogenic nature [57]. Hepatocyte-directed targeting induces antigen-specific regulatory T cells (Tregs) to modulate immune tolerance [58–60]. Although the exact mechanism underpinning immune tolerance by the liver is not fully understood, this method has been shown to be highly effective [57, 61]. Liver APCs, which play roles in the induction and maintenance of tolerance, consist of nonconventional hepatic APCs that include Kupffer cells (KCs), liver sinusoidal endothelial cells (LSEC), and hepatic stellate cells that express low levels of major histocompatibility complex (MHC)-I/II and co-stimulatory molecules [57]. T lymphocytes are also found within the liver, but are much different than those found in lymph nodes, spleen, and peripheral blood. For CNS transduction, stimulation of immune tolerance has many practical applications. For example, in the mouse model of Pompe disease, investigators showed that liver-directed AAV prevented production of antibodies against the transgene, acid alpha-glucosidase (GAA) [62]. Transduction of both liver and CNS to simultaneously induce tolerance and express the enzyme in neuronal tissue may perhaps be the most effective method to ensure sustained transgene expression in the CNS after IV delivery [57], so long as the transgene is not toxic to the liver. Investigators interested in exploring immune tolerance should be cautioned that strain-to-strain differences can impact results. For example, C57BL/6 mice demonstrate transgene tolerance after systemic AAV delivery, but the same is not true for BALB/c mice, which show impaired tolerance. Studies conducted in dogs and nonhuman primates (NHP) show that AAV-mediated liver-induced tolerance is dependent on the transgene and its species of origin [57].

5. Systemic AAV9 administration has achieved successful motor neuron transduction in a variety of animal models. However, recent studies testing AAV9 variants in NHPs and piglets have raised concerns for potential toxicity. Hinderer et al. demonstrated that administration of 2×10^{14} vector genome copies per kilogram body weight of an AAV9 variant (clade F) vector

resulted in severe toxicity with marked transaminase elevations [63]. Histological lesions within the central and peripheral nervous system were also observed. Interestingly, the authors found no evidence for T-cell responses to the vector capsid or transgene product in NHPs, suggesting that the observed toxicity was independent of an immune response to the capsid or transgene product. However, it was surmised that vector-related activation of cellular stress led to hepatocyte damage and subsequent activation of systemic inflammation. Piglets demonstrated no evidence of hepatic toxicity but exhibited proprioceptive deficits and ataxia [63]. The mechanistic cause for vector-related adverse effects was not entirely explored, but these findings highlight concerns that the risks of high-dose IV administration to target the CNS is still not fully understood. Researchers are encouraged to explore the extent that cellular or systemic toxicity may impact experimental interpretations regardless of the AAV capsids or species utilized.

6. One major drawback for rAAV administration to neonatal mice via facial vein or retro-orbital injection is that the total delivery volume should never exceed 100 μL (25–50 μL is recommended), thus relatively high-titer vector preparations ($\sim 10^{13}$ vg/mL) are required for the delivery of a high dose.
7. There are several considerations for retro-orbital injections. First, anesthesia is required. Second, procedural frequency should be limited to one injection per eye per day, and there should be at least a 1-day interval between injections. The injection volume should never exceed 100 μL per eye in neonates (50 μL is recommended). Trauma to the eye and swelling around the injection site may be caused by incorrect placement of the needle. Once inserted into the retro-orbital sinus, it is essential that minimal movement of the needle occurs, in order to prevent vessels from rupturing, causing bleeding and loss of vector. Juvenile and adult mice can also be subjected to retro-orbital injections by following procedures outlined for neonates.
8. For tail vein injections, the lateral veins are very superficial and easily accessible. The total injection volume should not exceed 0.1–0.3 mL, depending on the size of the mouse (20–40 g). Thus, high-titer vectors ($\sim 10^{13}$ vg/mL) are required for the delivery of high doses. Adult mice 4 weeks or older are recommended.

References

1. Mendell JR, Al-Zaidy S, Shell R, Arnold WD, Rodino-Klapac LR, Prior TW, Lowes L, Alfano L, Berry K, Church K, Kissel JT, Nagendran S, L'Italien J, Sproule DM, Wells C, Cardenas JA, Heitzer MD, Kaspar A, Corcoran S, Braun L, Likhite S, Miranda C, Meyer K, Foust KD, Burghes AHM, Kaspar BK (2017) Single-dose gene-replacement

therapy for spinal muscular atrophy. N Engl J Med 377(18):1713–1722. https://doi.org/10.1056/NEJMoa1706198

2. Hebert MD, Whittom AA (2007) Gene-based approaches toward Friedreich ataxia therapeutics. Cell Mol Life Sci 64(23):3034–3043. https://doi.org/10.1007/s00018-007-7293-6
3. Hobert JA, Dawson G (2006) Neuronal ceroid lipofuscinoses therapeutic strategies: past, present and future. Biochim Biophys Acta 1762(10):945–953. https://doi.org/10.1016/j.bbadis.2006.08.004
4. Suzuki K, Kastuno M, Banno H, Sobue G (2009) Pathogenesis-targeting therapeutics for spinal and bulbar muscular atrophy (SBMA). Neuropathology 29(4):509–516. https://doi.org/10.1111/j.1440-1789.2009.01013.x
5. Daneman R, Prat A (2015) The blood-brain barrier. Cold Spring Harb Perspect Biol 7(1): a020412. https://doi.org/10.1101/cshperspect.a020412
6. Banks WA (2009) Characteristics of compounds that cross the blood-brain barrier. BMC Neurol 9(Suppl 1):S3. https://doi.org/10.1186/1471-2377-9-S1-S3
7. Hawkins BT, Davis TP (2005) The blood-brain barrier/neurovascular unit in health and disease. Pharmacol Rev 57(2):173–185. https://doi.org/10.1124/pr.57.2.4
8. Ahmed SS, Li H, Cao C, Sikoglu EM, Denninger AR, Su Q, Eaton S, Liso Navarro AA, Xie J, Szucs S, Zhang H, Moore C, Kirschner DA, Seyfried TN, Flotte TR, Matalon R, Gao G (2013) A single intravenous rAAV injection as late as P20 achieves efficacious and sustained CNS Gene therapy in Canavan mice. Mol Ther 21(12):2136–2147. https://doi.org/10.1038/mt.2013.138
9. Ahmed SS, Schattgen SA, Frakes AE, Sikoglu EM, Su Q, Li J, Hampton TG, Denninger AR, Kirschner DA, Kaspar B, Matalon R, Gao G (2016) rAAV gene therapy in a Canavan's disease mouse model reveals immune impairments and an extended pathology beyond the central nervous system. Mol Ther 24(6):1030–1041. https://doi.org/10.1038/mt.2016.68
10. Foust KD, Nurre E, Montgomery CL, Hernandez A, Chan CM, Kaspar BK (2009) Intravascular AAV9 preferentially targets neonatal neurons and adult astrocytes. Nat Biotechnol 27(1):59–65. https://doi.org/10.1038/nbt.1515
11. Asokan A, Schaffer DV, Samulski RJ (2012) The AAV vector toolkit: poised at the clinical crossroads. Mol Ther 20(4):699–708. https://doi.org/10.1038/mt.2011.287
12. Castle MJ, Turunen HT, Vandenberghe LH, Wolfe JH (2016) Controlling AAV tropism in the nervous system with natural and engineered capsids. Methods Mol Biol 1382:133–149. https://doi.org/10.1007/978-1-4939-3271-9_10
13. Gao G, Alvira MR, Somanathan S, Lu Y, Vandenberghe LH, Rux JJ, Calcedo R, Sanmiguel J, Abbas Z, Wilson JM (2003) Adeno-associated viruses undergo substantial evolution in primates during natural infections. Proc Natl Acad Sci U S A 100(10):6081–6086. https://doi.org/10.1073/pnas.0937739100
14. Gao GP, Alvira MR, Wang L, Calcedo R, Johnston J, Wilson JM (2002) Novel adeno-associated viruses from rhesus monkeys as vectors for human gene therapy. Proc Natl Acad Sci U S A 99(18):11854–11859. https://doi.org/10.1073/pnas.182412299
15. Grimm D, Lee JS, Wang L, Desai T, Akache B, Storm TA, Kay MA (2008) In vitro and in vivo gene therapy vector evolution via multispecies interbreeding and retargeting of adeno-associated viruses. J Virol 82(12):5887–5911. https://doi.org/10.1128/JVI.00254-08
16. Maheshri N, Koerber JT, Kaspar BK, Schaffer DV (2006) Directed evolution of adeno-associated virus yields enhanced gene delivery vectors. Nat Biotechnol 24(2):198–204. https://doi.org/10.1038/nbt1182
17. Asokan A, Conway JC, Phillips JL, Li C, Hegge J, Sinnott R, Yadav S, DiPrimio N, Nam HJ, Agbandje-McKenna M, McPhee S, Wolff J, Samulski RJ (2010) Reengineering a receptor footprint of adeno-associated virus enables selective and systemic gene transfer to muscle. Nat Biotechnol 28(1):79–82. https://doi.org/10.1038/nbt.1599
18. Wang D, Li S, Gessler DJ, Xie J, Zhong L, Li J, Tran K, Van Vliet K, Ren L, Su Q, He R, Goetzmann JE, Flotte TR, Agbandje-McKenna M, Gao G (2018) A Rationally engineered capsid variant of AAV9 for systemic CNS-directed and peripheral tissue-detargeted gene delivery in neonates. Mol Ther Methods Clin Dev 9:234–246. https://doi.org/10.1016/j.omtm.2018.03.004
19. Deverman BE, Pravdo PL, Simpson BP, Kumar SR, Chan KY, Banerjee A, Wu WL, Yang B, Huber N, Pasca SP, Gradinaru V (2016) Cre-dependent selection yields AAV variants for widespread gene transfer to the adult brain. Nat Biotechnol 34(2):204–209. https://doi.org/10.1038/nbt.3440
20. Kotterman MA, Schaffer DV (2014) Engineering adeno-associated viruses for clinical gene

therapy. Nat Rev Genet 15(7):445–451. https://doi.org/10.1038/nrg3742

21. Bartel M, Schaffer D, Buning H (2011) Enhancing the clinical potential of AAV vectors by capsid engineering to evade pre-existing immunity. Front Microbiol 2:204. https://doi.org/10.3389/fmicb.2011.00204
22. Asokan A (2010) Reengineered AAV vectors: old dog, new tricks. Discov Med 9 (48):399–403
23. Su X, Kells AP, Huang EJ, Lee HS, Hadaczek P, Beyer J, Bringas J, Pivirotto P, Penticuff J, Eberling J, Federoff HJ, Forsayeth J, Bankiewicz KS (2009) Safety evaluation of AAV2-GDNF gene transfer into the dopaminergic nigrostriatal pathway in aged and parkinsonian rhesus monkeys. Hum Gene Ther 20(12):1627–1640. https://doi.org/10.1089/hum.2009.103
24. Davidson BL, Stein CS, Heth JA, Martins I, Kotin RM, Derksen TA, Zabner J, Ghodsi A, Chiorini JA (2000) Recombinant adeno-associated virus type 2, 4, and 5 vectors: transduction of variant cell types and regions in the mammalian central nervous system. Proc Natl Acad Sci U S A 97(7):3428–3432. https://doi.org/10.1073/pnas.050581197
25. Gao G, Vandenberghe LH, Alvira MR, Lu Y, Calcedo R, Zhou X, Wilson JM (2004) Clades of Adeno-associated viruses are widely disseminated in human tissues. J Virol 78 (12):6381–6388. https://doi.org/10.1128/JVI.78.12.6381-6388.2004
26. Yang B, Li S, Wang H, Guo Y, Gessler DJ, Cao C, Su Q, Kramer J, Zhong L, Ahmed SS, Zhang H, He R, Desrosiers RC, Brown R, Xu Z, Gao G (2014) Global CNS transduction of adult mice by intravenously delivered rAAVrh.8 and rAAVrh.10 and nonhuman primates by rAAVrh.10. Mol Ther 22 (7):1299–1309. https://doi.org/10.1038/mt.2014.68
27. Zhang H, Yang B, Mu X, Ahmed SS, Su Q, He R, Wang H, Mueller C, Sena-Esteves M, Brown R, Xu Z, Gao G (2011) Several rAAV vectors efficiently cross the blood-brain barrier and transduce neurons and astrocytes in the neonatal mouse central nervous system. Mol Ther 19(8):1440–1448. https://doi.org/10.1038/mt.2011.98
28. Choudhury SR, Fitzpatrick Z, Harris AF, Maitland SA, Ferreira JS, Zhang Y, Ma S, Sharma RB, Gray-Edwards HL, Johnson JA, Johnson AK, Alonso LC, Punzo C, Wagner KR, Maguire CA, Kotin RM, Martin DR, Sena-Esteves M (2016) In vivo selection yields AAV-B1 capsid for central nervous system and muscle gene therapy. Mol Ther 24 (7):1247–1257. https://doi.org/10.1038/mt.2016.84
29. Miyake N, Miyake K, Yamamoto M, Hirai Y, Shimada T (2011) Global gene transfer into the CNS across the BBB after neonatal systemic delivery of single-stranded AAV vectors. Brain Res 1389:19–26. https://doi.org/10.1016/j.brainres.2011.03.014
30. Hordeaux J, Wang Q, Katz N, Buza EL, Bell P, Wilson JM (2018) The neurotropic properties of AAV-PHP.B are limited to C57BL/6J mice. Mol Ther 26(3):664–668. https://doi.org/10.1016/j.ymthe.2018.01.018
31. Bevan AK, Duque S, Foust KD, Morales PR, Braun L, Schmelzer L, Chan CM, McCrate M, Chicoine LG, Coley BD, Porensky PN, Kolb SJ, Mendell JR, Burghes AH, Kaspar BK (2011) Systemic gene delivery in large species for targeting spinal cord, brain, and peripheral tissues for pediatric disorders. Mol Ther 19 (11):1971–1980. https://doi.org/10.1038/mt.2011.157
32. Gray SJ, Matagne V, Bachaboina L, Yadav S, Ojeda SR, Samulski RJ (2011) Preclinical differences of intravascular AAV9 delivery to neurons and glia: a comparative study of adult mice and nonhuman primates. Mol Ther 19 (6):1058–1069. https://doi.org/10.1038/mt.2011.72
33. Zincarelli C, Soltys S, Rengo G, Rabinowitz JE (2008) Analysis of AAV serotypes 1-9 mediated gene expression and tropism in mice after systemic injection. Mol Ther 16(6):1073–1080. https://doi.org/10.1038/mt.2008.76
34. Moscioni D, Morizono H, McCarter RJ, Stern A, Cabrera-Luque J, Hoang A, Sanmiguel J, Wu D, Bell P, Gao GP, Raper SE, Wilson JM, Batshaw ML (2006) Long-term correction of ammonia metabolism and prolonged survival in ornithine transcarbamylase-deficient mice following liver-directed treatment with adeno-associated viral vectors. Mol Ther 14(1):25–33. https://doi.org/10.1016/j.ymthe.2006.03.009
35. Salva MZ, Himeda CL, Tai PW, Nishiuchi E, Gregorevic P, Allen JM, Finn EE, Nguyen QG, Blankinship MJ, Meuse L, Chamberlain JS, Hauschka SD (2007) Design of tissue-specific regulatory cassettes for high-level rAAV-mediated expression in skeletal and cardiac muscle. Mol Ther 15(2):320–329. https://doi.org/10.1038/sj.mt.6300027
36. Marques S, Zeisel A, Codeluppi S, van Bruggen D, Mendanha Falcao A, Xiao L, Li H, Haring M, Hochgerner H, Romanov RA, Gyllborg D, Munoz-Manchado AB, La Manno G, Lonnerberg P, Floriddia EM, Rezayee F, Ernfors P, Arenas E, Hjerling-

Leffler J, Harkany T, Richardson WD, Linnarsson S, Castelo-Branco G (2016) Oligodendrocyte heterogeneity in the mouse juvenile and adult central nervous system. Science 352(6291):1326–1329. https://doi.org/10.1126/science.aaf6463

37. Hochstim C, Deneen B, Lukaszewicz A, Zhou Q, Anderson DJ (2008) Identification of positionally distinct astrocyte subtypes whose identities are specified by a homeodomain code. Cell 133(3):510–522. https://doi.org/10.1016/j.cell.2008.02.046

38. Tasic B, Menon V, Nguyen TN, Kim TK, Jarsky T, Yao Z, Levi B, Gray LT, Sorensen SA, Dolbeare T, Bertagnolli D, Goldy J, Shapovalova N, Parry S, Lee C, Smith K, Bernard A, Madisen L, Sunkin SM, Hawrylycz M, Koch C, Zeng H (2016) Adult mouse cortical cell taxonomy revealed by single cell transcriptomics. Nat Neurosci 19 (2):335–346. https://doi.org/10.1038/nn.4216

39. Xie J, Ameres SL, Friedline R, Hung JH, Zhang Y, Xie Q, Zhong L, Su Q, He R, Li M, Li H, Mu X, Zhang H, Broderick JA, Kim JK, Weng Z, Flotte TR, Zamore PD, Gao G (2012) Long-term, efficient inhibition of microRNA function in mice using rAAV vectors. Nat Methods 9(4):403–409. https://doi.org/10.1038/nmeth.1903

40. Xie J, Xie Q, Zhang H, Ameres SL, Hung JH, Su Q, He R, Mu X, Seher Ahmed S, Park S, Kato H, Li C, Mueller C, Mello CC, Weng Z, Flotte TR, Zamore PD, Gao G (2011) MicroRNA-regulated, systemically delivered rAAV9: a step closer to CNS-restricted transgene expression. Mol Ther 19(3):526–535. https://doi.org/10.1038/mt.2010.279

41. Brown BD, Venneri MA, Zingale A, Sergi Sergi L, Naldini L (2006) Endogenous microRNA regulation suppresses transgene expression in hematopoietic lineages and enables stable gene transfer. Nat Med 12(5):585–591. https://doi.org/10.1038/nm1398

42. Brown BD, Cantore A, Annoni A, Sergi LS, Lombardo A, Della Valle P, D'Angelo A, Naldini L (2007) A microRNA-regulated lentiviral vector mediates stable correction of hemophilia B mice. Blood 110(13):4144–4152. https://doi.org/10.1182/blood-2007-03-078493

43. Zhao Z, Nelson AR, Betsholtz C, Zlokovic BV (2015) Establishment and Dysfunction of the Blood-Brain Barrier. Cell 163(5):1064–1078. https://doi.org/10.1016/j.cell.2015.10.067

44. Duque S, Joussemet B, Riviere C, Marais T, Dubreil L, Douar AM, Fyfe J, Moullier P, Colle MA, Barkats M (2009) Intravenous administration of self-complementary AAV9 enables transgene delivery to adult motor neurons. Mol Ther 17(7):1187–1196. https://doi.org/10.1038/mt.2009.71

45. Fu H, Muenzer J, Samulski RJ, Breese G, Sifford J, Zeng X, McCarty DM (2003) Self-complementary adeno-associated virus serotype 2 vector: global distribution and broad dispersion of AAV-mediated transgene expression in mouse brain. Mol Ther 8(6):911–917

46. Penaud-Budloo M, Le Guiner C, Nowrouzi A, Toromanoff A, Cherel Y, Chenuaud P, Schmidt M, von Kalle C, Rolling F, Moullier P, Snyder RO (2008) Adeno-associated virus vector genomes persist as episomal chromatin in primate muscle. J Virol 82(16):7875–7885. https://doi.org/10.1128/JVI.00649-08

47. Gao G, Sena-Esteves M (2012) Introducing genes into mammalian cells: viral vectors. In: Molecular cloning: a laboratory manual, vol 2. Cold Spring Harbor Laboratory Press, Cold Spring Harbor, pp 1209–1313

48. Lecomte E, Tournaire B, Cogne B, Dupont JB, Lindenbaum P, Martin-Fontaine M, Broucque F, Robin C, Hebben M, Merten OW, Blouin V, Francois A, Redon R, Moullier P, Leger A (2015) Advanced characterization of DNA molecules in rAAV vector preparations by single-stranded virus next-generation sequencing. Mol Ther Nucleic Acids 4:e260. https://doi.org/10.1038/mtna.2015.32

49. Tai PWL, Xie J, Fong K, Seetin M, Heiner C, Su Q, Weiand M, Gao G (2018) Adeno-associated virus genome population sequencing achieves full vector genome resolution and reveals human-vector chimeras. Mol Ther Methods Clin Dev 9:130–141

50. Boisgerault F, Gross DA, Ferrand M, Poupiot J, Darocha S, Richard I, Galy A (2013) Prolonged gene expression in muscle is achieved without active immune tolerance using microrRNA 142.3p-regulated rAAV gene transfer. Hum Gene Ther 24 (4):393–405. https://doi.org/10.1089/hum.2012.208

51. Majowicz A, Maczuga P, Kwikkers KL, van der Marel S, van Logtenstein R, Petry H, van Deventer SJ, Konstantinova P, Ferreira V (2013) Mir-142-3p target sequences reduce transgene-directed immunogenicity following intramuscular adeno-associated virus 1 vector-mediated gene delivery. J Gene Med 15 (6–7):219–232. https://doi.org/10.1002/jgm.2712

52. Matsuzaki Y, Konno A, Mochizuki R, Shinohara Y, Nitta K, Okada Y, Hirai H (2018) Intravenous administration of the

adeno-associated virus-PHP.B capsid fails to upregulate transduction efficiency in the marmoset brain. Neurosci Lett 665:182–188. https://doi.org/10.1016/j.neulet.2017.11.049

53. Rapti K, Louis-Jeune V, Kohlbrenner E, Ishikawa K, Ladage D, Zolotukhin S, Hajjar RJ, Weber T (2012) Neutralizing antibodies against AAV serotypes 1, 2, 6, and 9 in sera of commonly used animal models. Mol Ther 20 (1):73–83. https://doi.org/10.1038/mt.2011.177
54. Samaranch L, Sebastian WS, Kells AP, Salegio EA, Heller G, Bringas JR, Pivirotto P, DeArmond S, Forsayeth J, Bankiewicz KS (2014) AAV9-mediated expression of a non-self protein in nonhuman primate central nervous system triggers widespread neuroinflammation driven by antigen-presenting cell transduction. Mol Ther 22(2):329–337. https://doi.org/10.1038/mt.2013.266
55. Schwartz M, Deczkowska A (2016) Neurological disease as a failure of brain-immune crosstalk: the multiple faces of neuroinflammation. Trends Immunol 37(10):668–679. https://doi.org/10.1016/j.it.2016.08.001
56. Paulk NK, Pekrun K, Zhu E, Nygaard S, Li B, Xu J, Chu K, Leborgne C, Dane AP, Haft A, Zhang Y, Zhang F, Morton C, Valentine MB, Davidoff AM, Nathwani AC, Mingozzi F, Grompe M, Alexander IE, Lisowski L, Kay MA (2018) Bioengineered AAV capsids with combined high human liver transduction in vivo and unique humoral seroreactivity. Mol Ther 26(1):289–303. https://doi.org/10.1016/j.ymthe.2017.09.021
57. Keeler GD, Markusic DM, Hoffman BE (2017) Liver induced transgene tolerance with AAV vectors. Cell Immunol. https://doi.org/10.1016/j.cellimm.2017.12.002
58. Breous E, Somanathan S, Vandenberghe LH, Wilson JM (2009) Hepatic regulatory T cells and Kupffer cells are crucial mediators of systemic T cell tolerance to antigens targeting murine liver. Hepatology 50(2):612–621. https://doi.org/10.1002/hep.23043
59. Cao O, Loduca PA, Herzog RW (2009) Role of regulatory T cells in tolerance to coagulation factors. J Thromb Haemost 7(Suppl 1):88–91. https://doi.org/10.1111/j.1538-7836.2009.03417.x
60. Dobrzynski E, Herzog RW (2005) Tolerance induction by viral in vivo gene transfer. Clin Med Res 3(4):234–240
61. Kumar SRP, Hoffman BE, Terhorst C, de Jong YP, Herzog RW (2017) The balance between CD8(+) T Cell-mediated clearance of AAV-encoded antigen in the liver and tolerance is dependent on the vector dose. Mol Ther 25 (4):880–891. https://doi.org/10.1016/j.ymthe.2017.02.014
62. Sun B, Bird A, Young SP, Kishnani PS, Chen YT, Koeberl DD (2007) Enhanced response to enzyme replacement therapy in Pompe disease after the induction of immune tolerance. Am J Hum Genet 81(5):1042–1049. https://doi.org/10.1086/522236
63. Hinderer C, Katz N, Buza EL, Dyer C, Goode T, Bell P, Richman LK, Wilson JM (2018) Severe toxicity in nonhuman primates and piglets following high-dose intravenous administration of an adeno-associated virus vector expressing human SMN. Hum Gene Ther 29(3):285–298. https://doi.org/10.1089/hum.2018.015
64. Kugler S, Meyn L, Holzmuller H, Gerhardt E, Isenmann S, Schulz JB, Bahr M (2001) Neuron-specific expression of therapeutic proteins: evaluation of different cellular promoters in recombinant adenoviral vectors. Mol Cell Neurosci 17(1):78–96. https://doi.org/10.1006/mcne.2000.0929
65. Shevtsova Z, Malik JM, Michel U, Bahr M, Kugler S (2005) Promoters and serotypes: targeting of adeno-associated virus vectors for gene transfer in the rat central nervous system in vitro and in vivo. Exp Physiol 90(1):53–59. https://doi.org/10.1113/expphysiol.2004.028159
66. Stierl M, Penzkofer A, Kennis JT, Hegemann P, Mathes T (2014) Key residues for the light regulation of the blue light-activated adenylyl cyclase from Beggiatoa sp. Biochemistry 53(31):5121–5130. https://doi.org/10.1021/bi500479v
67. Hedegaard C, Kjaer-Sorensen K, Madsen LB, Henriksen C, Momeni J, Bendixen C, Oxvig C, Larsen K (2013) Porcine synapsin 1: SYN1 gene analysis and functional characterization of the promoter. FEBS Open Bio 3:411–420. https://doi.org/10.1016/j.fob.2013.10.002
68. McLean JR, Smith GA, Rocha EM, Hayes MA, Beagan JA, Hallett PJ, Isacson O (2014) Widespread neuron-specific transgene expression in brain and spinal cord following synapsin promoter-driven AAV9 neonatal intracerebroventricular injection. Neurosci Lett 576:73–78. https://doi.org/10.1016/j.neulet.2014.05.044
69. Gessler DJ, Li D, Xu H, Su Q, Sanmiguel J, Tuncer S, Moore C, King J, Matalon R, Gao G (2017) Redirecting N-acetylaspartate metabolism in the central nervous system normalizes myelination and rescues Canavan disease. JCI Insight 2(3):e90807. https://doi.org/10.1172/jci.insight.90807

70. Cho W, Hagemann TL, Johnson DA, Johnson JA, Messing A (2009) Dual transgenic reporter mice as a tool for monitoring expression of glial fibrillary acidic protein. J Neurochem 110 (1):343–351. https://doi.org/10.1111/j.1471-4159.2009.06146.x

71. von Jonquieres G, Frohlich D, Klugmann CB, Wen X, Harasta AE, Ramkumar R, Spencer ZH, Housley GD, Klugmann M (2016) Recombinant human myelin-associated glycoprotein promoter drives selective AAV-mediated transgene expression in oligodendrocytes. Front Mol Neurosci 9:13. https://doi.org/10.3389/fnmol.2016.00013

72. Chen CT, Gottlieb DI, Cohen BA (2008) Ultraconserved elements in the Olig2 promoter. PLoS One 3(12):e3946. https://doi.org/10.1371/journal.pone.0003946

Chapter 9

Intraspinal and Intracortical Delivery of AAV Vectors for Intersectional Circuit Tracing in Non-transgenic Species

James M. Conner, Greg L. Bain, and Jennifer N. Dulin

Abstract

The mapping of long-range axonal projections is a rapidly growing endeavor in the field of neuroscience. Recent advances in the development of adeno-associated viral vectors that can achieve efficient retrograde transport now enable the characterization and manipulation of specific neuronal subpopulations using Cre-dependent, intersectional approaches. Importantly, these approaches can be applied to non-transgenic animals, making it possible to carry out detailed anatomical studies across a variety of species including nonhuman primates. In this chapter, we demonstrate the utility of such intersectional strategies by describing methods for targeting viral constructs to distinct subsets of corticospinal motor neurons based on their projections to specific spinal cord segments.

Key words Retrograde transport, Cre-dependent gene expression, rAAV2-retro, Corticospinal motor neurons, Projection mapping

1 Introduction

1.1 Intersectional Viral Vector-Based Approaches

As whole-brain connectivity mapping initiatives gain increasing momentum in neuroscience, methods to characterize the axonal projections of defined neuronal populations are becoming of increasing utility to the field. The use of transgenic mice with cell-specific patterns of Cre recombinase expression (Cre driver lines) [1], in conjunction with viral vectors whose genetic payloads are expressed in a Cre-dependent manner, has enabled investigators to target genetically encoded tools (e.g., fluorescent reporters and axonal tracers, optogenetic [2–6] and chemogenetic [7–9] activity modulators, and calcium indicators [10–12]) to neuronal populations on the basis of their molecular expression profiles. However, such an approach fails to distinguish cell subpopulations based on their unique patterns of efferent projections. Moreover, it is impractical to extend transgenic approaches across a wide variety of species, such as nonhuman primates. A much more versatile alternative is to utilize a purely viral vector-based strategy, thereby

Michael J. Castle (ed.), *Adeno-Associated Virus Vectors: Design and Delivery*, Methods in Molecular Biology, vol. 1950, https://doi.org/10.1007/978-1-4939-9139-6_9, © Springer Science+Business Media, LLC, part of Springer Nature 2019

liberating the investigator from reliance on transgenic lines. Importantly, the use of an intersectional viral approach permits selective targeting of cell populations based on their unique axonal projection patterns. In such a system, Cre-encoding viral vectors with retrograde affinity are injected into a target region, taken up at axon terminals, and transported back to cell bodies along efferent projections that innervate the site of Cre vector delivery. Subsequent introduction of a second virus with Cre-***dependent*** gene expression to one (or more) of these efferent projecting populations will result in recombination and genetic manipulation of cells in a *circuit-specific* manner. Although such an intersectional approach is hardly a new concept, efficacy of gene delivery has historically been limited by the low efficiency of retrograde transport when using many viral vectors, such as the canine adenovirus-2 (CAV-2) [13–18] or various serotypes of adeno-associated virus (AAV) [19–23].

1.2 A New Gold Standard for Viral Retrograde Transport

Recently, Karpova and colleagues applied a directed evolution approach to generate an AAV vector capable of robust retrograde transduction [24]. This capsid, rAAV2-retro, undergoes retrograde transport at approximately 10 times greater efficiency than other AAV serotypes or CAV-2 [24]. Indeed, in the short time since AAV2-retro was made commercially available, its high utility for retrograde gene delivery has been validated in multiple projection mapping studies [25–30]. Hence, this capsid may now be considered the "gold standard" for highly efficient retrograde gene delivery to a wide variety of brain circuits. Similar to other AAV serotypes, AAV2-retro can package a wide variety of genetic payloads; a particular advantage is its capacity to efficiently deliver Cre to neurons on the basis of their efferent input to defined brain structures, thereby enabling the manipulation of cell populations with distinct long-range projection patterns via intersectional, Cre-dependent targeting.

1.3 Intersectional Targeting of Corticospinal Neurons

In this chapter we describe how to use an AAV-based intersectional approach to specifically target virally encoded genes to unique subpopulations of corticospinal motor neurons (CSMNs) based on their long-range projections to distinct spinal segments. CSMNs originate in diverse cortical regions and extend descending projections to multiple spinal segments, presumably as a means of influencing distinct aspects of fine motor control. For instance, it has long been appreciated that CSMNs projecting to lumbar spinal cord segments (associated with hind limb motor control) largely reside in a distinct cortical region from those that project to cervical segments (associated with forelimb motor control) [31]. In such cases where two functionally distinct cell populations are topographically isolated, it is possible to selectively target one by restricting the site of virus delivery. However, the distribution of CSMNs that project to distinct segments within the cervical spinal

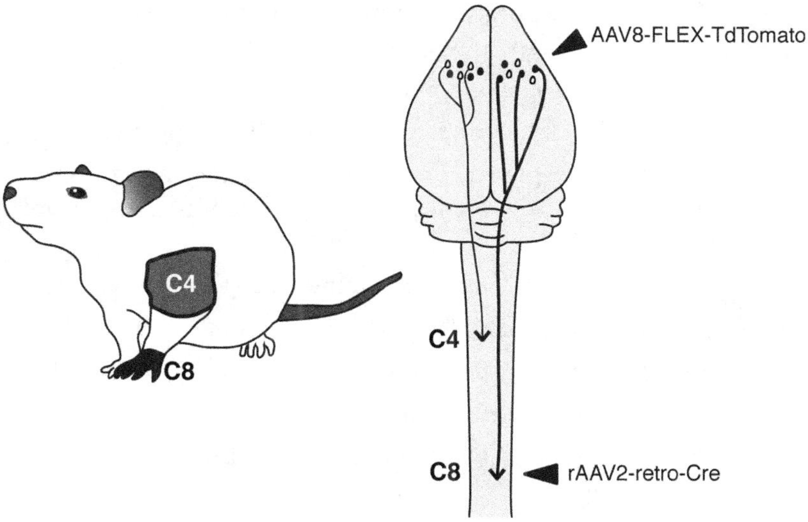

Fig. 1 Schematic representation of intersectional AAV targeting of C8 spinal cord-projecting corticospinal motor neurons

cord, such as those associated with proximal (C4) and distal (C8) forelimb motor control, share a completely overlapping topographical distribution within the primary motor cortex in the rat [32]. As such, selectively targeting these subpopulations requires isolation on the basis of their distinct projections to specific spinal segments (Fig. 1). Here, we will demonstrate how the intersectional viral approach can be used for specific gene delivery to CSMNs that provide efferent input to lower (C7/8) cervical spinal segments, a population associated with distal forelimb motor control. The injection coordinates can be modified for virus delivery to other regions of the central nervous system, and this protocol is equally applicable to studies in both transgenic and non-transgenic animals, including large animals such as nonhuman primates (*see* **Note 1**).

2 Materials

1. AAV2-retro encoding Cre recombinase (*see* **Note 2**).
2. AAV8 encoding a Cre-dependent fluorescent tracer, such as AAV8-FLEX-TdTomato (*see* **Note 3**).
3. Phosphate-buffered saline (PBS) or Hank's Balanced Salt Solution (HBSS) for diluting virus stock to working concentrations.
4. Protein LoBind microcentrifuge tubes.
5. Glass capillary micropipette puller.
6. Glass capillary micropipette beveling device (optional).

7. Picospritzer/pressure-driven microinjection system with glass capillary micropipette holder.
8. Nitrogen gas tank with regulator.
9. Rectal temperature probe and heating pad.
10. Stereotaxic frame with alignment system for small animal brain and spinal cord injections (such as David Kopf Instruments #980).
11. Surgical stereo zoom microscope.
12. Dental drill for performing craniectomy.
13. Tabletop centrifuge.
14. Surgical tools, including:
    (a) Scalpel with #15 scalpel blades.
    (b) Agricola retractor (4 cm).
    (c) Friedman-Pearson rongeurs with curved tip and 1 mm cup.
    (d) Two Dumont #5 forceps.
    (e) Adson forceps.
    (f) Two Bulldog serrefine clamps.
    (g) Wound clip auto-applier with stainless steel wound clips.
15. Tabletop heat sterilizer.
16. 10% bleach solution.
17. 70% ethanol solution.
18. Anesthesia.
19. Hair clippers.
20. Betadine or iodine solution.
21. Sterile 2″ × 2″ surgical gauze pads.
22. Sterile cotton swabs.
23. 5 μL Hamilton syringes with 24G, 2.75″ needles, point style 3 (such as Hamilton #88000) (*see* **Note 4**).
24. 10 mL disposable plastic syringes.
25. 25G ¾″ disposable needles.
26. 30G ½″ insulin syringes.
27. Suture (e.g., polypropylene #4.0).
28. Gelfilm or Avitene absorbable film, cut into 2 cm × 2 cm pieces and rehydrated in PBS.
29. Borosilicate glass capillary tubes.
30. Lactated Ringer's solution.
31. Postoperative solutions (e.g., antibiotics, analgesics).
32. Fine-tipped permanent marker.

3 Methods

3.1 Preparation of Surgical Area

1. Pull glass micropipettes and bevel micropipette tips to a tip thickness of approximately 50 μm at an angle of 20°.
2. Disinfect all work surfaces with 70% ethanol. Lay down a sterile/clean surgical area drape and place stereotaxic frame in the center of the workspace. Ensure that the stereotaxic manipulator components are set at right angles.
3. Position the temperature-controlled heating pad on the stereotaxic frame.
4. Sterilize surgical tools with a tabletop heat sterilizer.
5. Thoroughly wash Hamilton syringe by gently drawing up and expelling 70% ethanol 3–5 times, then repeating with HBSS 3–5 times.
6. Prepare fresh 10% bleach solution for decontamination of surgical tools and surfaces.
7. Place the following items next to the surgical area:
   (a) Sterile cotton swabs.
   (b) Sterile 2″ × 2″ surgical gauze.
   (c) 10 mL syringe with 25G needle containing Lactated Ringer's solution.
   (d) Gelfoam or Avitene collagen sponges.
8. Position stereomicroscope for surgery.

3.2 Preparation of Virus for Injection

1. Remove an AAV aliquot from storage at −80 °C and thaw on ice.
2. Centrifuge the aliquot for 5 min at 13,000 rpm (17,000 × *g*) in a microcentrifuge using a rotor with a sealed containment lid. This will separate virus-containing supernatant from any contaminating particulate that may be present in the stock virus preparation.
3. Carefully remove supernatant, taking care not to draw up solution from the bottom of the tube. Transfer to a new protein LoBind microcentrifuge tube and dilute to a working concentration with PBS or HBSS (*see* **Note 5**).
4. With a clean Hamilton syringe, draw up 3–4 μL of AAV and load into a pulled glass micropipette. Carefully tap the micropipette shaft to remove any bubbles. Place on a piece of gauze and keep on ice.

3.3 Spinal Cord AAV-Cre Injections

In this protocol, we give injection coordinates for (**1**) the cervical (C7/8) spinal cord ventral gray matter and (**2**) the primary forelimb motor cortex in the adult Fischer F344 rat. The specific anatomical sites for AAV injections should be determined by the experimenter ahead of time according to experimental needs. Consult species-specific brain and spinal cord atlases to determine the specific anterior/posterior (A/P), medial/lateral (M/L), and dorsal/ventral (D/V) coordinates to be targeted.

1. Anesthetize experimental subjects using methods approved by the Institutional Animal Care and Use Committee. For example, we use an anesthetic cocktail containing ketamine (50 mg/kg), xylazine (2.6 mg/kg), and acepromazine (0.5 mg/kg).
2. Once the animal is fully anesthetized, shave the upper back between the shoulder blades and neck. Disinfect the skin with betadine and 70% ethanol.
3. Insert the rectal probe into the animal and set heating pad temperature to 37 °C.
4. Position the animal in the stereotaxic frame with its head securely fixed with ear bars. A small cushion may be placed underneath the animal's stomach to position the body parallel to the work surface.
5. Palpate the back of the animal to locate the T2 vertebral process, which is noticeably raised relative to other vertebrae in the cervical cord. Using a #15 scalpel blade, make a 4 cm midline incision through the animal's skin in the rostral-to-caudal direction, extending 3 cm above and 1 cm below the T2 vertebral process. With the Agricola retractor, retract the incised skin and carefully cut through the muscle layers at the midline with the scalpel blade (*see* **Note 6**).
6. Use the retractor to push muscle layers to the side until the cervical dorsal spinous processes are visible. Using the raised T2 vertebral process as a landmark, identify the C7 vertebral segment to be laminectomized. Clean any blood and muscle off of the laminae with cotton swabs, and carefully perform a laminectomy using Friedman-Pearson rongeurs (*see* **Note 7**).
7. Carefully secure the glass micropipette containing AAV2-retro-Cre into the stereotaxic injector alignment system. Adjust the duration of nitrogen gas flow through the microinjection system in order to dispense a volume of approximately 20 nL per pulse, as estimated by carefully monitoring the meniscus level in the micropipette with repeated pulsing.
8. Position the tip of the micropipette just above the spinal cord surface at the midline; record the stereotaxic coordinates and use these as a reference. A total of 2.4 μL of virus should be injected into 12 sites total (6 sites per hemisphere). A/P coordinates: 3 sites spanning the exposed C7/8 spinal segments. M/L coordinates: ±550–600 μm. Depth: 700 μm and 1100 μm.

9. Using the stereotaxic alignment system, move the injection micropipette to the first injection site. At each injection site, use a 30G ½″ insulin syringe to first carefully poke a small hole in the dura, taking care not to damage the underlying spinal cord. Wait until cerebrospinal fluid has stopped flowing out of the hole in the dura, then lower the glass micropipette into the spinal cord to a depth of 1100 μm below the spinal cord surface. Inject 200 nL of virus over 1 min, then raise the micropipette to a depth of 700 μm below the surface and repeat. After the second injection, allow the micropipette to remain in place for 2 min prior to raising it out of the tissue to allow the virus to adequately diffuse into the tissue parenchyma.
10. Continue until all injections are complete. Carefully remove the glass micropipette from the microinjection device and dispose according to institutional biosafety regulations.
11. Carefully suture the animal's muscles (2 layers) and close the incision with wound clips.

3.4 Cortical AAV Injections

Injection of Cre-dependent AAV into the motor cortex can be performed in the same surgical session or in a subsequent surgical session. Cre-dependent reporter gene expression in corticospinal motor neurons requires sufficient time for both retrograde transport of the Cre-expressing virus and for peak expression of the Cre-dependent reporter gene (*see* **Note 8**).

1. Making sure that the animal is still fully anesthetized, shave the dorsal surface of the head and disinfect the skin with betadine and 70% ethanol.
2. Make a 2 cm midline incision in the skin with a #15 scalpel blade to expose the skull. Use the bulldog serrefine clamps to retract the skin flaps to each side.
3. Position the animal's head securely into the stereotaxic instrument using the ear bars. Adjust the head positioning such that the Lambda-Bregma axis is completely flat and parallel to the work surface, and secure in place with the bite bar.
4. Locate Bregma and mark with a fine-tipped black permanent marker.
5. Using a dental drill, carefully drill a window in the skull encompassing the forelimb area of the primary motor cortex (M1) of one hemisphere. Gently remove the bone flap with Dumont #5 forceps, and place a piece of PBS-soaked gelfoam or collagen sponge onto the exposed dura. Repeat for the other hemisphere.
6. Carefully secure the glass micropipette containing Cre-dependent AAV into the stereotaxic injector alignment

system, again ensuring that the dispensing volume is approximately 20 nL per pulse.

7. Position the tip of the micropipette just above Bregma and record the stereotaxic coordinates. Calculate sites for cortical injection: aim to inject 10 sites spaced evenly within the M1 forelimb area. A/P coordinates from Bregma: +1.0 to −1.0 mm. M/L coordinates from Bregma: ±2.5 to 4.2 mm. Depth: 1.0 mm. Inject approximately 200 nL of Cre-dependent virus per site over a period of 1 min. After each injection, leave the glass capillary in place for an additional minute to allow the virus to diffuse into the brain parenchyma. After all sites have been injected, repeat the process for the other hemisphere (*see* **Note 9**).
8. Carefully remove the glass micropipette from the microinjection device and dispose according to institutional biosafety regulations.
9. Cover the exposed cortical surfaces with rehydrated gelfilm, and close the skin with wound clips.
10. Inject animals with postoperative solutions and allow them to recover from anesthesia in a quiet, warm environment. Animals should be closely monitored by lab personnel until fully recovered from anesthesia.
11. Decontaminate all surgical frames and surfaces, surgical instruments, and Hamilton syringes with 10% bleach.

4 Notes

1. It is crucial to obtain institutional approval before performing any of the procedures described within this protocol. Use of the AAV vectors described in this protocol requires Institutional Biosafety Committee approval. All animal surgeries and procedures described herein require approval of the Institutional Animal Care and Use Committee. Always follow institutional guidelines regarding specific details of animal care, and working with and disposing of biohazardous materials.
2. There are several AAV2-retro-Cre viral vector preparations that are available commercially, including Addgene #24593 AAV-pgk-Cre and Addgene #51507 AAV-pmSyn1-EBFP-Cre.
3. We and others have achieved robust local transduction of rodent cortical neurons using the AAV8 capsid [33–37]; AAV2 [38–44] and AAV1 [45] have also been used. A wide variety of Cre-dependent AAV constructs are commercially available.

4. Separate Hamilton syringes should be dedicated for each type of viral vector used. This precaution limits the possibility of cross-contamination between the Cre viruses and the Cre-dependent viruses, which could result in nonspecific (off-target) recombination.
5. Viral vectors for injection should typically be used at a working concentration of approximately $2 \times 10^{12} - 5 \times 10^{12}$ viral genome copies per mL. Keep virus working solutions on ice. Any unused stock virus can be stored at 4 °C for up to 1 month or refrozen for future use; AAV has been reported to withstand up to four freeze-thaw cycles without losing potency [46], but this should be avoided if possible. It is strongly suggested that newly acquired viruses are diluted to their working concentrations and stored as 5–10 μL aliquots at −80 °C. Thus, a suitable amount can be thawed for each surgical session.
6. Take extreme caution not to cut the bilateral fat pads that lie under the superficial muscles in the cervical region of the spinal cord. This tissue is highly vascularized and even a small nick can cause extensive bleeding. In the event that excessive bleeding occurs during surgery, apply gelfoam or collagen foam to the site of bleeding and administer 3–5 mL of Lactated Ringer's solution to animals that experience excessive bleeding during surgery.
7. Large blood vessels course longitudinally beneath the laminae just lateral to the spinal cord; be conservative and do not remove bone more laterally than necessary.
8. Gene expression in corticospinal neurons, including distal axon regions, should plateau in approximately 3–4 weeks following AAV-FLEX-TdTomato injections.
9. If at any time the injection pipette becomes clogged or if you run out of virus, it is possible to fill another micropipette and resume injections after rezeroing the pipette at Bregma.

Acknowledgments

This work was supported by the Craig H. Neilsen Foundation, Paralyzed Veterans of America Research Foundation, and TIRR Foundation (J.N.D.).

References

1. Harris JA, Hirokawa KE, Sorensen SA, Gu H, Mills M, Ng LL, Bohn P, Mortrud M, Ouellette B, Kidney J, Smith KA, Dang C, Sunkin S, Bernard A, Oh SW, Madisen L, Zeng H (2014) Anatomical characterization of Cre driver mice for neural circuit mapping and manipulation. Front Neural Circuits 8:76. https://doi.org/10.3389/fncir.2014.00076
2. Boyden ES, Zhang F, Bamberg E, Nagel G, Deisseroth K (2005) Millisecond-timescale,

genetically targeted optical control of neural activity. Nat Neurosci 8(9):1263–1268. https://doi.org/10.1038/nn1525

3. Yizhar O, Fenno LE, Davidson TJ, Mogri M, Deisseroth K (2011) Optogenetics in neural systems. Neuron 71(1):9–34. https://doi.org/10.1016/j.neuron.2011.06.004
4. Han X, Chow BY, Zhou H, Klapoetke NC, Chuong A, Rajimehr R, Yang A, Baratta MV, Winkle J, Desimone R, Boyden ES (2011) A high-light sensitivity optical neural silencer: development and application to optogenetic control of non-human primate cortex. Front Syst Neurosci 5:18. https://doi.org/10.3389/fnsys.2011.00018
5. Chow BY, Han X, Dobry AS, Qian X, Chuong AS, Li M, Henninger MA, Belfort GM, Lin Y, Monahan PE, Boyden ES (2010) High-performance genetically targetable optical neural silencing by light-driven proton pumps. Nature 463(7277):98–102. https://doi.org/10.1038/nature08652
6. Han X, Boyden ES (2007) Multiple-color optical activation, silencing, and desynchronization of neural activity, with single-spike temporal resolution. PLoS One 2(3):e299. https://doi.org/10.1371/journal.pone.0000299
7. Sternson SM, Roth BL (2014) Chemogenetic tools to interrogate brain functions. Annu Rev Neurosci 37:387–407. https://doi.org/10.1146/annurev-neuro-071013-014048
8. Roth BL (2016) DREADDs for neuroscientists. Neuron 89(4):683–694. https://doi.org/10.1016/j.neuron.2016.01.040
9. Zhu H, Roth BL (2014) Silencing synapses with DREADDs. Neuron 82(4):723–725. https://doi.org/10.1016/j.neuron.2014.05.002
10. Akerboom J, Carreras Calderon N, Tian L, Wabnig S, Prigge M, Tolo J, Gordus A, Orger MB, Severi KE, Macklin JJ, Patel R, Pulver SR, Wardill TJ, Fischer E, Schuler C, Chen TW, Sarkisyan KS, Marvin JS, Bargmann CI, Kim DS, Kugler S, Lagnado L, Hegemann P, Gottschalk A, Schreiter ER, Looger LL (2013) Genetically encoded calcium indicators for multi-color neural activity imaging and combination with optogenetics. Front Mol Neurosci 6:2. https://doi.org/10.3389/fnmol.2013.00002
11. Tian L, Hires SA, Looger LL (2012) Imaging neuronal activity with genetically encoded calcium indicators. Cold Spring Harb Protoc 2012(6):647–656. https://doi.org/10.1101/pdb.top069609
12. Tian L, Hires SA, Mao T, Huber D, Chiappe ME, Chalasani SH, Petreanu L, Akerboom J, McKinney SA, Schreiter ER, Bargmann CI, Jayaraman V, Svoboda K, Looger LL (2009) Imaging neural activity in worms, flies and mice with improved GCaMP calcium indicators. Nat Methods 6(12):875–881. https://doi.org/10.1038/nmeth.1398
13. Gore BB, Soden ME, Zweifel LS (2013) Manipulating gene expression in projection-specific neuronal populations using combinatorial viral approaches. Curr Protoc Neurosci 65:4.35.1–4.3520. https://doi.org/10.1002/0471142301.ns0435s65
14. Boender AJ, de Jong JW, Boekhoudt L, Luijendijk MC, van der Plasse G, Adan RA (2014) Combined use of the canine adenovirus-2 and DREADD-technology to activate specific neural pathways in vivo. PLoS One 9(4):e95392. https://doi.org/10.1371/journal.pone.0095392
15. Hnasko TS, Perez FA, Scouras AD, Stoll EA, Gale SD, Luquet S, Phillips PE, Kremer EJ, Palmiter RD (2006) Cre recombinase-mediated restoration of nigrostriatal dopamine in dopamine-deficient mice reverses hypophagia and bradykinesia. Proc Natl Acad Sci U S A 103(23):8858–8863. https://doi.org/10.1073/pnas.0603081103
16. Hung LW, Neuner S, Polepalli JS, Beier KT, Wright M, Walsh JJ, Lewis EM, Luo L, Deisseroth K, Dolen G, Malenka RC (2017) Gating of social reward by oxytocin in the ventral tegmental area. Science 357 (6358):1406–1411. https://doi.org/10.1126/science.aan4994
17. Schwarz LA, Miyamichi K, Gao XJ, Beier KT, Weissbourd B, DeLoach KE, Ren J, Ibanes S, Malenka RC, Kremer EJ, Luo L (2015) Viral-genetic tracing of the input-output organization of a central noradrenaline circuit. Nature 524(7563):88–92. https://doi.org/10.1038/nature14600
18. Yackle K, Schwarz LA, Kam K, Sorokin JM, Huguenard JR, Feldman JL, Luo L, Krasnow MA (2017) Breathing control center neurons that promote arousal in mice. Science 355 (6332):1411–1415. https://doi.org/10.1126/science.aai7984
19. Kaspar BK, Erickson D, Schaffer D, Hinh L, Gage FH, Peterson DA (2002) Targeted retrograde gene delivery for neuronal protection. Mol Ther 5(1):50–56. https://doi.org/10.1006/mthe.2001.0520
20. Hollis ER 2nd, Kadoya K, Hirsch M, Samulski RJ, Tuszynski MH (2008) Efficient retrograde neuronal transduction utilizing self-complementary AAV1. Mol Ther 16 (2):296–301. https://doi.org/10.1038/sj.mt.6300367

21. Castle MJ, Gershenson ZT, Giles AR, Holzbaur EL, Wolfe JH (2014) Adeno-associated virus serotypes 1, 8, and 9 share conserved mechanisms for anterograde and retrograde axonal transport. Hum Gene Ther 25 (8):705–720. https://doi.org/10.1089/hum.2013.189
22. Kaspar BK, Llado J, Sherkat N, Rothstein JD, Gage FH (2003) Retrograde viral delivery of IGF-1 prolongs survival in a mouse ALS model. Science 301(5634):839–842. https://doi.org/10.1126/science.1086137
23. Rothermel M, Brunert D, Zabawa C, Diaz-Quesada M, Wachowiak M (2013) Transgene expression in target-defined neuron populations mediated by retrograde infection with adeno-associated viral vectors. J Neurosci 33 (38):15195–15206. https://doi.org/10.1523/JNEUROSCI.1618-13.2013
24. Tervo DG, Hwang BY, Viswanathan S, Gaj T, Lavzin M, Ritola KD, Lindo S, Michael S, Kuleshova E, Ojala D, Huang CC, Gerfen CR, Schiller J, Dudman JT, Hantman AW, Looger LL, Schaffer DV, Karpova AY (2016) A designer AAV variant permits efficient retrograde access to projection neurons. Neuron 92 (2):372–382. https://doi.org/10.1016/j.neuron.2016.09.021
25. Chabrat A, Brisson G, Doucet-Beaupre H, Salesse C, Schaan Profes M, Dovonou A, Akitegetse C, Charest J, Lemstra S, Cote D, Pasterkamp RJ, Abrudan MI, Metzakopian E, Ang SL, Levesque M (2017) Transcriptional repression of Plxnc1 by Lmx1a and Lmx1b directs topographic dopaminergic circuit formation. Nat Commun 8(1):933. https://doi.org/10.1038/s41467-017-01042-0
26. Danielson NB, Turi GF, Ladow M, Chavlis S, Petrantonakis PC, Poirazi P, Losonczy A (2017) In Vivo Imaging of Dentate Gyrus Mossy Cells in Behaving Mice. Neuron 93 (3):552–559 e554. https://doi.org/10.1016/j.neuron.2016.12.019
27. Ciabatti E, Gonzalez-Rueda A, Mariotti L, Morgese F, Tripodi M (2017) Life-long genetic and functional access to neural circuits using self-inactivating rabies virus. Cell 170 (2):382–392 e314. https://doi.org/10.1016/j.cell.2017.06.014
28. Wagner MJ, Kim TH, Savall J, Schnitzer MJ, Luo L (2017) Cerebellar granule cells encode the expectation of reward. Nature 544 (7648):96–100. https://doi.org/10.1038/nature21726
29. Zhang GW, Sun WJ, Zingg B, Shen L, He J, Xiong Y, Tao HW, Zhang LI (2018) A non-canonical reticular-limbic central auditory pathway via medial septum contributes to fear conditioning. Neuron 97(2):406–417 e404. https://doi.org/10.1016/j.neuron.2017.12.010
30. Hashimoto M, Yamanaka A, Kato S, Tanifuji M, Kobayashi K, Yaginuma H (2018) Anatomical evidence for a direct projection from purkinje cells in the mouse cerebellar vermis to medial parabrachial nucleus. Front Neural Circuits 12:6. https://doi.org/10.3389/fncir.2018.00006
31. Li XG, Florence SL, Kaas JH (1990) Areal distributions of cortical neurons projecting to different levels of the caudal brain stem and spinal cord in rats. Somatosens Mot Res 7 (3):315–335
32. Wang L, Conner JM, Rickert J, Tuszynski MH (2011) Structural plasticity within highly specific neuronal populations identifies a unique parcellation of motor learning in the adult brain. Proc Natl Acad Sci U S A 108 (6):2545–2550. https://doi.org/10.1073/pnas.1014335108
33. Dulin JN, Adler AF, Kumamaru H, Poplawski GHD, Lee-Kubli C, Strobl H, Gibbs D, Kadoya K, Fawcett JW, Lu P, Tuszynski MH (2018) Injured adult motor and sensory axons regenerate into appropriate organotypic domains of neural progenitor grafts. Nat Commun 9(1):84. https://doi.org/10.1038/s41467-017-02613-x
34. Soderblom C, Lee DH, Dawood A, Carballosa M, Jimena Santamaria A, Benavides FD, Jergova S, Grumbles RM, Thomas CK, Park KK, Guest JD, Lemmon VP, Lee JK, Tsoulfas P (2015) 3D imaging of axons in transparent spinal cords from rodents and nonhuman primates. eNeuro 2(2). https://doi.org/10.1523/ENEURO.0001-15.2015
35. Blackmore MG, Wang Z, Lerch JK, Motti D, Zhang YP, Shields CB, Lee JK, Goldberg JL, Lemmon VP, Bixby JL (2012) Krüppel-like Factor 7 engineered for transcriptional activation promotes axon regeneration in the adult corticospinal tract. Proc Natl Acad Sci U S A 109(19):7517–7522. https://doi.org/10.1073/pnas.1120684109
36. Wang Z, Reynolds A, Kirry A, Nienhaus C, Blackmore MG (2015) Overexpression of sox11 promotes corticospinal tract regeneration after spinal injury while interfering with functional recovery. J Neurosci 35 (7):3139–3145. https://doi.org/10.1523/JNEUROSCI.2832-14.2015
37. Jayaprakash N, Wang Z, Hoeynck B, Krueger N, Kramer A, Balle E, Wheeler DS, Wheeler RA, Blackmore MG (2016) Optogenetic interrogation of functional synapse formation by corticospinal tract axons in the

injured spinal cord. J Neurosci 36 (21):5877–5890. https://doi.org/10.1523/JNEUROSCI.4203-15.2016

38. Hollis ER 2nd, Ishiko N, Yu T, Lu CC, Haimovich A, Tolentino K, Richman A, Tury A, Wang SH, Pessian M, Jo E, Kolodkin A, Zou Y (2016) Ryk controls remapping of motor cortex during functional recovery after spinal cord injury. Nat Neurosci 19(5):697–705. https://doi.org/10.1038/nn.4282

39. Wang X, Liu Y, Li X, Zhang Z, Yang H, Zhang Y, Williams PR, Alwahab NSA, Kapur K, Yu B, Zhang Y, Chen M, Ding H, Gerfen CR, Wang KH, He Z (2017) Deconstruction of corticospinal circuits for goal-directed motor skills. Cell 171(2):440–455 e414. https://doi.org/10.1016/j.cell.2017.08.014

40. Liu Y, Keefe K, Tang X, Lin S, Smith GM (2014) Use of self-complementary adeno-associated virus serotype 2 as a tracer for labeling axons: implications for axon regeneration. PLoS One 9(2):e87447. https://doi.org/10.1371/journal.pone.0087447

41. Jara JH, Stanford MJ, Zhu Y, Tu M, Hauswirth WW, Bohn MC, DeVries SH, Ozdinler PH (2016) Healthy and diseased corticospinal motor neurons are selectively transduced upon direct AAV2-2 injection into the motor cortex. Gene Ther 23(3):272–282. https://doi.org/10.1038/gt.2015.112

42. Hollis ER 2nd, Jamshidi P, Löw K, Blesch A, Tuszynski MH (2009) Induction of corticospinal regeneration by lentiviral trkB-induced Erk activation. Proc Natl Acad Sci U S A 106(17):7215–7220. https://doi.org/10.1073/pnas.0810624106

43. Lee DH, Luo X, Yungher BJ, Bray E, Lee JK, Park KK (2014) Mammalian target of rapamycin's distinct roles and effectiveness in promoting compensatory axonal sprouting in the injured CNS. J Neurosci 34 (46):15347–15355. https://doi.org/10.1523/JNEUROSCI.1935-14.2014

44. Geoffroy CG, Hilton BJ, Tetzlaff W, Zheng B (2016) Evidence for an age-dependent decline in axon regeneration in the adult mammalian central nervous system. Cell Rep 15 (2):238–246. https://doi.org/10.1016/j.celrep.2016.03.028

45. Hilton BJ, Anenberg E, Harrison TC, Boyd JD, Murphy TH, Tetzlaff W (2016) Re-establishment of cortical motor output maps and spontaneous functional recovery via spared dorsolaterally projecting corticospinal neurons after dorsal column spinal cord injury in adult mice. J Neurosci 36(14):4080–4092. https://doi.org/10.1523/JNEUROSCI.3386-15.2016

46. Gruntman AM, Su L, Su Q, Gao G, Mueller C, Flotte TR (2015) Stability and compatibility of recombinant adeno-associated virus under conditions commonly encountered in human gene therapy trials. Hum Gene Ther Methods 26(2):71–76. https://doi.org/10.1089/hgtb.2015.040

Chapter 10

MRI-Guided Focused Ultrasound for Targeted Delivery of rAAV to the Brain

Zeinab Noroozian, Kristiana Xhima, Yuexi Huang, Brian K. Kaspar, Sebastian Kügler, Kullervo Hynynen, and Isabelle Aubert

Abstract

Recombinant adeno-associated viral (rAAV) vectors are a promising tool for therapeutic gene delivery to the brain. However, the delivery of rAAVs across the blood-brain barrier (BBB) and entry into the brain remains a major challenge for rAAV-based gene therapy. To circumvent this limitation, transcranial MRI-guided focused ultrasound (MRIgFUS) combined with intravenously injected microbubbles has been used to transiently and reversibly increase BBB permeability in targeted brain regions. Systemic administration of rAAVs at the time of sonication with focused ultrasound (FUS) facilitates the passage of rAAVs through the BBB and into the brain parenchyma. We and others have demonstrated that FUS-mediated rAAV delivery to the brain results in efficient transduction and transgene expression in vivo. Using this approach, the dose of intravenously injected rAAV variants that can cross the BBB can be reduced by 100 times, achieving significant transgene expression in the brain parenchyma with reduced peripheral transduction. Moreover, this strategy can be used to deliver rAAV variants that do not cross the BBB from the blood to selected brain regions. Here, we provide a detailed protocol for FUS-induced BBB permeability for targeted rAAV delivery to the brain of adult mice and rats.

Key words Adeno-associated virus, Gene therapy, Focused ultrasound, Blood-brain barrier, Gene therapy, Viral vector delivery

1 Introduction

Strategies are being developed to deliver therapeutic genes to the brain with the potential to treat patients affected by neurological disorders [1]. Extensive evidence from preclinical research and clinical studies suggests that targeted gene therapy is a promising approach to enhance treatment efficacy and reduce the toxicity of therapeutic drugs for a wide variety of neurological conditions including neurodegenerative diseases, brain tumors, stroke, and epilepsy [2, 3]. Recombinant adeno-associated virus (rAAV)

Zeinab Noroozian and Kristiana Xhima contributed equally to this work.

Michael J. Castle (ed.), *Adeno-Associated Virus Vectors: Design and Delivery*, Methods in Molecular Biology, vol. 1950, https://doi.org/10.1007/978-1-4939-9139-6_10,

vectors have emerged as an attractive gene delivery vehicle for the brain [1–3]. Recent developments in gene therapy [4–6] include rAAVs that, when given at high dosages, have the capability to cross the blood-brain barrier (BBB) [7–11]. For example, rAAV9 was recently used in the first life-saving treatment of neurodegeneration in infants with spinal muscular atrophy [12]. For neurodegenerative disorders of the brain, such as Alzheimer's disease (AD), the delivery of rAAVs and their genetic cargo has to date been achieved by invasive intraparenchymal injection to desired brain regions [13, 14]. In recent years, several companies have joined forces to develop innovative gene-based immunotherapy approaches for AD. These partenerships could be transformative, surmounting the limitations of current systemic biweekly injections of therapeutic antibodies targeting AD pathology, in which only a small amount penetrates the brain. A single treatment with gene therapy can lead to long-term production of therapeutic transgenes and provide sustained clinical benefits.

Current limitations in the design of rAAV-based gene therapy for diseases of the central nervous system (CNS) include the restricted access of rAAVs to the brain and spinal cord when administered systemically [2]. This can be overcome, when appropriate, with high dosages of rAAVs that are intrinsically capable of crossing the BBB. An alternative approach is to use focused ultrasound (FUS) disruption of the BBB, thereby enhancing delivery of intravenously injected rAAV vectors to the CNS (Fig. 1). Focused ultrasound, combined with microbubbles and guided by magnetic resonance imaging (MRI), has been applied to open the BBB in a safe, noninvasive, and reversible manner in targeted areas of brain in animal models [15] and, most recently, in AD patients [16]. MRI-guided FUS (MRIgFUS)-mediated delivery of rAAVs to targeted areas of the brain and spinal cord can be achieved with systemic doses which are up to 100 times lower than those conventionally used with rAAVs capable of crossing the BBB [11, 15, 17–20]. Intravenous injections of rAAVs, with and without FUS, can lead to peripheral transduction, limiting CNS delivery and treatment efficacy and potentially generating adverse side effects. Strategies for mitigating peripheral rAAVs transduction [21] and expression [22] can be deployed as required.

FUS, combined with intravenously injected microbubble contrast agents circulating in the vasculature, is a noninvasive drug delivery strategy that results in local, reversible, and safe increases in BBB permeability [23]. Under these conditions, circulating microbubbles are stimulated by ultrasonic energy, generating stable oscillations. These oscillations induce mechanical stress and also transient downregulation of the tight junction proteins between endothelial cells that play a role in BBB permeability [24, 25]. MRI-guided FUS is used to achieve precise targeting of focal spots and to visualize the induction of BBB permeability using

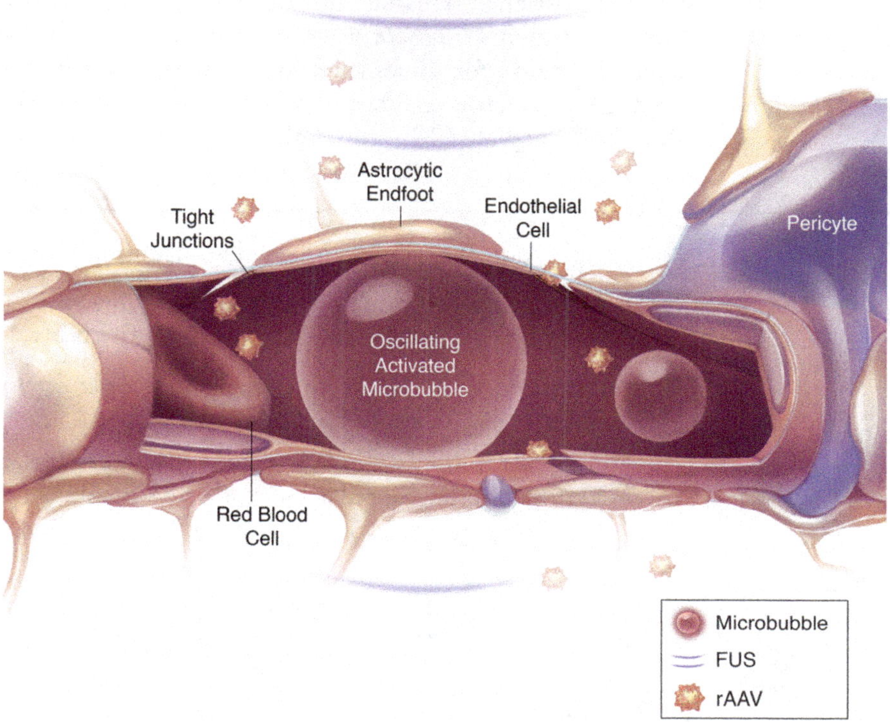

Fig. 1 FUS-mediated delivery of rAAVs across the BBB. Circulating microbubbles in the brain vasculature are activated and oscillate in response to ultrasonic energy delivered by FUS. The interaction of FUS with microbubbles leads to an increase in BBB permeability, which is localized and temporary, allowing rAAVs injected intravenously to enter FUS-targeted brain regions. Reproduced with permission from Sunnybrook Research Institute

MRI contrast agents [26]. FUS-induced BBB permeability in the presence of microbubbles has been used to deliver a wide variety of therapeutic agents in preclinical animal models, such as drugs, antibodies, cells, plasmid DNA, viral vectors, and other drug vehicles [15, 27, 28]. Using MRI-guided FUS, our group has demonstrated targeted gene delivery to the mouse brain after a single systemic injection of rAAV9 vectors. Efficient transduction of neurons and glia in the brain was reported using a low systemic dose of 2.5×10^9 vector genomes (vg) per g [20]. Another study reported efficient rAAV2 brain transduction using a systemic dose of 1×10^9 vg/g [29]. In addition to FUS-facilitated viral vector delivery for gene therapy, this technique has also been developed for optogenetic applications. Recombinant AAVs encoding light-activated channels delivered by this approach allow for noninvasive manipulation of neuronal activity, thereby avoiding damage caused by direct brain infusion. No inflammatory response or long-term tissue damage was reported following FUS-facilitated delivery of rAAV9 encoding the light-sensitive protein Channelrhodopsin-2 (ChR2) to the mouse brain [30].

FUS-mediated gene delivery to the CNS is a promising tool for basic neuroscience research and for providing effective and long-lasting treatments for neurological disorders. Here, we provide a detailed method for the efficient delivery of systemically administered rAAV vectors to targeted brain areas in rodents using MRIgFUS-mediated BBB disruption.

2 Materials

2.1 Animals

Researchers must comply with institutional animal use and other applicable policies, guidelines, and regulations (*see* **Note 1**).

1. Adult mice or rats balanced for sex and weight across experimental groups (*see* **Note 2**).
2. Housing for both mice and rats: individual micro-isolator cages on HEPA filtered vented racks with bed-o-cob bedding and enrichment material including nestlets, paper-based material, and tunnels.
3. Commercially available irradiated laboratory rodent diet, soft food (food pellet placed in water), and Nutra-Gel diet (Bio-Serv).
4. Chlorinated RO water supplied through an automated watering system on the rack or in autoclaved water bottles.
5. A thermal blanket (e.g. circulating water warming pad) or forced warm air (e.g. Bair Hugger animal warming unit) for thermoregulation.

2.2 rAAV Preparation

1. rAAV solution diluted in sterile 0.9% NaCl (*see* **Note 3**).
2. A 250 μL Hamilton syringe with a 22G needle (Hamilton #1725RN).
3. 1% bleach solution for disinfection of materials.
4. Class IIA2 biosafety cabinet.

2.3 FUS-Induced BBB Disruption

1. 22G angiocatheter.
2. 1 mL insulin syringes with 27G × ½″ needle.
3. 0.9% NaCl sterile IV solution.
4. Depilatory cream.
5. Cotton swabs.
6. Alcohol pads.
7. Gauze.
8. Surgical tape.
9. Sterile surgical instruments.
10. Ophthalmic ointment.

11. Small animal hair clipper.
12. An anesthesia machine with a rodent circuit including an induction chamber and anesthesia nose cone for induction and maintenance of anesthesia during catheterization. Use isoflurane with 50% medical air and 50% oxygen as carrier gases.
13. A standard animal 7T MRI scanner (Bruker) is used for imaging. Standard clinical 1.5T or 3T scanners can also be used for these studies.
14. MRI body coil (Bruker, 86 mm Quad SN37).
15. MRI surface coil (Bruker, 86 mm Quad Receive).
16. MR-compatible isoflurane anesthesia machine including an anesthesia nose cone for isoflurane anesthesia during MRI with 100% medical air carrier gas (*see* **Note 4**).
17. MR-Compatible Focused Ultrasound (FUS) system. The FUS Instruments RK-300 system, which is designed to work with the small-bore Bruker 7T MR scanner, includes the following components:
    (a) Spherically focused transducer. In rats, FUS-induced BBB permeability is achieved using a 551.5 kHz transducer (focal number = 0.8, external diameter = 75 mm, internal diameter = 20 mm). A 1.68 MHz transducer is used for experiments carried out in mice (focal number = 0.8, external diameter = 75 mm, internal diameter = 20 mm).
    (b) Function generator (Agilent 33220A).
    (c) Scope card (14 bit Alazar Tech ATS460).
    (d) Radio frequency amplifier (NPTECH NP2519 50W).
    (e) Power meter.
    (f) Transducer matching circuit.
    (g) Three-axis transducer positioning system [31].
    (h) Polyvinylidene difluoride (PVDF) or lead zirconate titanate (PZT) hydrophone.
    (i) Acoustic pressure feedback controller [32].
    (j) Plastic tank filled with degassed, deionized water.
    (k) MR-compatible plastic sled with anesthesia nose cone. The sled is used to transfer the animal between the MR scanner for imaging and the FUS system to induce BBB permeability. The relative position of the sled between the MR and FUS system is registered in the calibration stage (*see* Subheading 3.3.2). The sled contains a fixture, comprising two Kapton polyimide membranes filled with degassed, deionized water, which sits in the degassed, deionized water tank to provide good ultrasound coupling to the animal brain.

18. Ultrasound gel (Wavelength CL, National Therapy Products Inc.).
19. Ultrasound phantom.
20. DEFINITY microbubbles (0.02 mL/kg, Lantheus Medical Imaging) stored at 4 °C. For BBB disruption, similar results can be achieved with other microbubble contrast agents (*see* **Note 5**).
21. Gadolinium-based MRI contrast agent, Gadovist (0.1 mL/kg, Schering AG), stored at room temperature.
22. 1 mL insulin syringes with 27G × ½″ needles and 18G blunt-end needles.
23. Sterile 0.9% NaCl. Mice receive a 0.2 mL saline flush and rats receive a 0.5 mL saline flush after each injection of rAAV, microbubbles, or MRI contrast agent.

2.4 Animal Recovery

1. 1 mL insulin syringes with 27G × ½″ needles.
2. Enrofloxacin antibiotic (2.5 mg/kg Baytril).
3. Opioid analgesic (0.05 mg/kg Buprenorphine).
4. Silver sulfadiazine cream (Flamazine).
5. Recovery cages with paper bedding.
6. Small animal heat lamp.

3 Methods

3.1 Animal Preparation

1. It is recommended that group-housed rodents be separated to individual cages for at least 1 week prior to treatment.
2. Rodents should be handled at least 3 days prior to treatment. Handling twice a day for 5 min decreases anxiety levels and improves the animal's response to the experimental procedure.
3. Inject 1 mL of sterile saline subcutaneously the day before sonication and less than 45 min before sonication to prevent dehydration.

3.2 rAAV Preparation

1. Prepare the viral working solution on the same day or 1 day in advance of FUS treatment. Handle the viral solution on ice or store at 4 °C.
2. rAAV preparation must be carried out in a class IIA2 biosafety cabinet. Clean the workstation with 1% bleach solution prior to all procedures.
3. Stock rAAV solution should be diluted such that the maximum volume of all intravenous injectables does not exceed 25 mL/kg for the heaviest mouse in the experimental cohort. Similar calculations should be performed for dilution of stock rAAV

solution in rat studies using a maximal dose volume of 20 mL/kg of body weight (*see* **Note 6**).

4. Fill the Hamilton syringe with rAAV solution. Use a different syringe for each rAAV vector if multiple vectors are being tested.
5. All materials that come into contact with rAAV solution (e.g., tips, tubes) should be placed in 1% bleach solution then disposed in biohazard containers as appropriate. Clean instruments with 1% bleach solution, rinse with distilled water, then rinse with alcohol (*see* **Note 7**).

3.3 FUS-Induced BBB Disruption

When using the RK-300 FUS system, which is designed to work with the small-bore Bruker 7T MRI, the FUS procedure is performed outside the MR room. An MR-compatible, spatially registered sled is used to transfer the animal between the MR room and the FUS system (*see* **Note 8**).

3.3.1 Ultrasound System Setup

1. The experimental system setup is presented in Fig. 2a.
2. The ultrasound transducer should be placed within a plastic tank of degassed, deionized water, and mounted in a mechanical arm holder. Movement of the transducer arm is controlled by a motorized positioning system.
3. Position the MRI-compatible sled on the plate covering the water tank.

3.3.2 Focus Finding

1. Register the position of the transducer focal spot to the MRI-coordinate system. Align the transducer focal spot directly underneath the polyimide membrane fixture on the MRI-compatible sled using the motorized positioning system.
2. At a low power level (0.5 W), aim the ultrasound beam at the water surface. The focal spot be visible as a fountain in the water.
3. Secure a plastic focus finding marker at the focal spot (directly above the fountain in the water) on the sled.
4. Remove the sled from the ultrasound setup and place on the MRI bed for imaging.
5. Perform a tri-pilot imaging sequence to image the focus finder marker.
6. Record the MRI coordinates of the marker (Left/Right, Anterior/Posterior, and Superior/Inferior). Use this location as a reference for targeting the brain.
7. Using an ultrasound phantom positioned on the sled and coupled with ultrasound gel, ensure that the reflected power is less than 10% on the power meter reading.

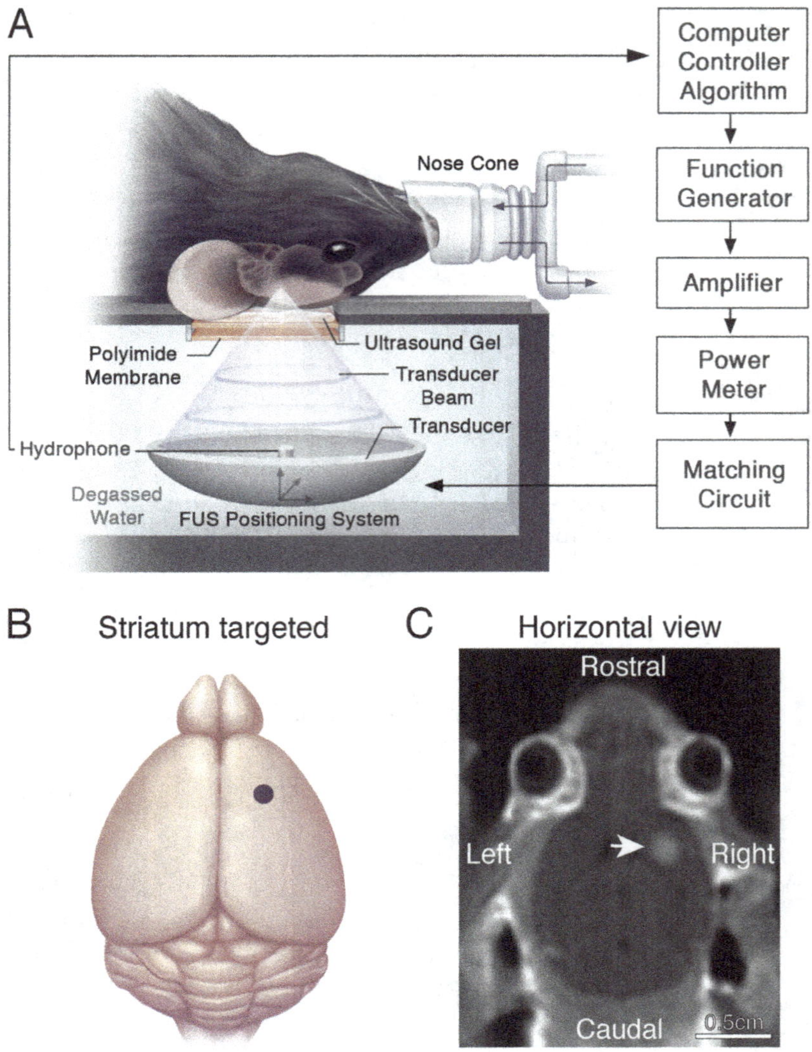

Fig. 2 MRIgFUS-mediated BBB permeability in the unilateral mouse striatum. (**a**) The animal is placed in the supine position on an MRI-compatible sled. The skull is coupled to a polyimide membrane with ultrasound gel. A second polyimide membrane is coupled to a degassed, deionized water tank containing the transducer and hydrophone positioned in the center of the transducer. Sonication parameters and procedures are described in Subheading 3.3. (**b**) Relative neuroanatomical location of the FUS focal spot targeted to the striatum (indicated by a purple dot). Target locations in the brain are selected on axial T2-weighted MR images. (**c**) Post-sonication, a contrast-enhanced T1-weighted MR image is used to visualize the entry of an MRI contrast agent, Gadovist, in the sonicated mouse striatum (indicated by a white arrow) and confirm BBB opening in the targeted location. Increased voxel intensity in the sonicated region is proportional to the extent of BBB permeability and amount of virus delivered into the brain. Reproduced with permission from Sunnybrook Research Institute

8. Analyze the fast Fourier transform (FFT) of the hydrophone data captured at 20 MHz sampling frequency. No significant peaks other than the driving frequency and its harmonics should be visible, and the baseline noise floor should be lower than the pre-calibrated level.

3.3.3 Tail Vein Catheter Insertion

1. Move the animal to a procedure room located adjacent to the MRI suite to reduce transit time for the anesthetized animal. Tail vein catheterization should be performed aseptically in a Class IIA2 biosafety cabinet.
2. Provide external warmth to the animal cages for 30 min before starting the procedure using forced warm air at 40 °C. Cover the cages with a large sheet to help maintain heat inside the cages (*see* **Note 9**).
3. Weigh the animal to determine the volume of injectables that it will receive during the experiment.
4. Place the animal inside the anesthesia induction chamber and adjust the isoflurane concentration to 5%, with 50% medical air and 50% oxygen as carrier gas. When the animal is anesthetized, remove it from the induction chamber.
5. Insert the nose of the animal into the anesthesia nose cone and reduce the isoflurane concentration to 2% or to effect.
6. Place the animal on a circulating warm water blanket to maintain body temperature at 37 °C.
7. Apply ophthalmic ointment to the eyes.
8. Shave the fur on top of the skull, spanning an area from the neck region (between the ears) to the frontal portion of the head (between the eyes), using a small animal clipper followed by application of depilatory cream. Rinse off the depilatory cream thoroughly with sterile saline. Clean the shaved area with gauze and an alcohol wipe (*see* **Note 10**).
9. Swab the tail with an alcohol wipe. This also increases visibility of the tail vein.
10. Locate the tail vein and insert the angiocatheter into the vein. If the needle is in the vein, blood will be visible in the hub of the catheter when the needle is removed.
11. Remove the needle from catheter and leave the plastic hub inside the tail.
12. Secure the catheter hub to the tail using surgical tape. Firmly tape the wood stick of a cotton swab applicator to the tail near the catheter hub to facilitate multiple injections during the experiment.

3.3.4 Animal Positioning

1. Bring the anesthetized animal to the MRI suite, and place the animal on the MRI-compatible sled in a supine position such that head of the animal contacts the Kapton polyimide film.
2. Position the anesthetized animal on the sled by placing its teeth over the incisor bar and sliding the adaptor forward until the nose cone is stabilized on the animal's nose. The nose cone should be attached to tubing that leads to an anesthesia

machine, delivering 2% isoflurane with 100% medical air as carrier gas (*see* **Note 4**). Maintain anesthesia at this level for the duration of the experiment. The nose cone should be connected to tubing in such a way that the entire sled can be easily connected and separated while the head position remains fixed.

3. Apply ultrasonic gel between the head and the membrane on the sled.
4. Firmly attach the head and body of the animal to the sled with surgical tape.
5. Cover the animal with a warming bag.
6. Place the animal (attached to the sled) into the MRI sliding table and connect the nose cone to an isoflurane anesthesia machine. Slide the table slowly into the MRI scanner.
7. Perform a tri-pilot scan to ensure good head positioning and water coupling (*see* **Note 11**). The specific acquisition parameters for all MRI scans are listed in Table 1.
8. Perform an axial T2-weighted FSE sequence.
9. Perform a baseline axial T1-weighted FSE sequence.
10. Remove the animal (attached to the sled) from the MRI scanner and position the sled on top of the ultrasound setup. Reconnect the nose cone to an isoflurane anesthetic machine.
11. Select the ultrasound target locations from the axial T2-weighted FSE scan (*see* **Note 12**). To target a single focal spot in the striatum, we selected coordinates as shown in Fig. 2b, c. For unilateral treatments, ensure that the orientation of the axial MR image used for targeting corresponds to

Table 1
Parameters used for MR imaging

	Rat			Mouse		
MRI Parameters	**Tri-pilot**	**T1w**	**T2w**	**Tri-pilot**	**T1w**	**T2w**
Sequence type	Flash	RARE	RARE	Flash	RARE	RARE
Echo time (ms)	3	10	75	3	10	75
Repetition time (ms)	200	500	4000	200	500	4000
RARE factor	n/a	2	10	n/a	2	10
Averages	1	3	2	1	3	4
Field of view (mm)	60	50	50	60	40	40
Matrix size	256	150	200	256	160	160
Slice thickness (mm)	1	1.5	1.5	1	1.5	1.5

Table 2
Parameters used for FUS-induced BBB permeability

Parameters	Rat	Mouse
Frequency (MHz)	0.55	1.68
Length of sonication (s)	120	120
Duty cycle (%)	1	1
Pulse repetition frequency (Hz)	1	1
Burst duration (ms)	10	10
Microbubble type	DEFINITY	DEFINITY
Microbubble dose (mL/kg)	0.02	0.02
Microbubble size (μm) and dispersion	1.1–3.3	1.1–3.3
Passive cavitation detection	PVDF hydrophone	PZT hydrophone
Sampling rate (MHz)	20	20
Detection frequency (kHz)	825	840
Emissions target (pressure ratio)	0.5	0.5
Starting acoustic pressure (MPa)	0.128	0.25
Acoustic pressure increment (MPa)	0.008	0.025
Contrast enhancement (imaging detection)	Gadovist (0.1 mL/kg)	Gadovist (0.1 mL/kg)

desired brain side to be treated. Multiple sonication spots should be at least 1 mm apart so that the focal spots do not overlap.

3.3.5 Sonication

1. Start the sonication using the following parameters: 10 ms burst length, 1 Hz pulse repetition, and 120 s sonication duration. A complete list of all sonication parameters is provided in Table 2.
2. Increase the applied acoustic pressure in a stepwise manner with each burst. For the 551 kHz transducer, use an initial acoustic pressure of 0.128 MPa and increase incrementally by 0.008 MPa after each burst. For the 1.68 MHz transducer, use a starting pressure of 0.25 MPa and increase incrementally by 0.025 MPa. Acoustic emissions are recorded by the hydrophone.
3. Acoustic emissions recorded during each burst using the hydrophone are processed with a real-time algorithm that detects subharmonic or ultraharmonic frequencies to adjust treatment pressures. Once these emissions are detected, the acoustic pressure is reduced to 50% for the duration of the sonication. Use ultraharmonic emissions as a threshold for

the 551 kHz transducer, and use subharmonic emissions as a threshold for the 1.68 MHz transducer.

3.3.6 Microbubble Administration

1. Immediately prior to sonication, activated DEFINITY microbubbles must be diluted with saline to evenly distribute microspheres. Slowly withdraw activated DEFINITY microbubbles from the stock suspension using an 18G blunt needle attached to a 1 mL syringe. For rats, dilute 0.1 mL of DEFINITY stock with 0.9 mL of saline, and inject 0.2 mL/kg of the diluted solution (equivalent to 0.02 mL/kg of undiluted DEFINITY) via the tail vein. For mice, dilute 0.02 mL of DEFINITY stock in 0.98 mL of saline and inject 1 mL/kg of diluted DEFINITY solution (equivalent to 0.02 mL/kg of undiluted DEFINITY) intravenously.
2. The acoustic feedback controller records baseline emissions for 10 s. Therefore, DEFINITY microbubbles should be injected 10 s after the start of sonication.
3. Inject saline to flush the microbubbles through the catheter (0.2 mL for mice, 0.5 mL for rats).
4. For additional sonications (in another anatomical location), wait 5 min between microbubble injections. This ensures that the previously injected microbubbles are cleared from the bloodstream.

3.3.7 rAAV Delivery

1. Keep the Hamilton syringe with rAAV solution on ice for the duration of the experiment.
2. Administer rAAV immediately after microbubble injection. For multiple sonication locations, we recommend administering rAAV after the first microbubble injection, and with the start of the first sonication.
3. Place Hamilton syringe into the tail vein catheter hub and inject the viral solution slowly.
4. Inject saline to flush rAAV through the catheter (0.2 mL for mice, 0.5 mL for rats).

3.3.8 Confirm FUS-Induced BBB Permeability

1. For rats, inject 0.1 mL/kg of gadolinium-based MRI contrast agent (Gadovist) at stock concentration through the tail vein catheter hub. For mice, dilute 0.1 mL of Gadovist stock in 0.9 mL of saline in a 1 mL syringe and inject 1 mL/kg of the diluted solution (equivalent to 0.1 mL/kg of undiluted Gadovist).
2. Inject saline to flush the contrast agent through the catheter (0.2 mL for mice, 0.5 mL for rats).
3. Move the animal to the MRI scanner and acquire a T1-weighted FSE image.

4. The increase in signal intensity indicates influx of the MRI contrast agent at sonicated regions, and is proportional to both the degree of BBB opening and the amount of rAAV delivered into the brain (Fig. 2c).

3.3.9 Recovery

1. Detach the rodent from the sled and return it to the animal procedure room.
2. Let the animal recover from anesthesia in a recovery cage on paper-based bedding and position the heat lamp nearby. The recovery cages should be placed on a circulating water blanket.
3. Gently remove the tail vein catheter and apply pressure with gauze until blood stops flowing from the puncture site.
4. Reapply ophthalmic ointment, clean the shaved head with an alcohol wipe, and apply Flamazine.
5. Inject the animal subcutaneously with enrofloxacin (2.5 mg/kg, Baytril) and Buprenorphine (0.05 mg/kg).
6. The animal must be continuously monitored until locomotion, breathing rate, and behavioral activities are normal.
7. Once the animal is fully recovered, return the animal to the animal housing room and provide soft food and Nutra-Gel.

3.4 Post-FUS Handling

1. Provide soft food and Nutra-Gel for 2 days posttreatment (*see* **Note 13**).
2. Inject Buprenorphine for 2 days post-treatment to reduce pain and discomfort from the procedure.
3. Inject 1 mL of saline subcutaneously for at least 2 days post--treatment to prevent dehydration.
4. Give antibiotics (enrofloxacin) for at least 2 days post-treatment to prevent infection.
5. Weigh the animals and monitor their nesting activity in their home cage post-experiment. These observations can also provide information about health and welfare of the animals.
6. Assess transgene expression after FUS-mediated rAAV delivery (*see* **Notes 14** and **15**).

4 Notes

1. All animal work was approved by the Sunnybrook Research Institute Animal Care Committee and was performed in compliance with the Canadian Council on Animal Care Policies & Guidelines and the regulatory requirements of the Animals for Research Act of Ontario.

2. It is important to include sex-matched experimental groups in gene therapy studies. Emerging evidence suggests there are sex differences with respect to systemic AAV gene transfer efficiency in the rodent brain [33, 34]. For example, Maguire et al. reported increased transgene expression in adult female mice compared to male mice after intravenous injection of AAV9 encoding the firefly luciferase (Fluc) and green fluorescent protein (GFP) reporter genes [33].
3. It is recommended to prepare a stock solution of rAAV from which three or more concentrations can be tested, in order to establish the dose required for transgene expression in FUS-treated areas [20]. We recommend first testing three doses (a given concentration of rAAV, 10× more, and 10× less), and then testing additional doses as needed.
4. During MRIgFUS BBB disruption, animals are anesthetized using isoflurane gas with 100% medical air. Isoflurane anesthesia with oxygen carrier gas has been shown to significantly reduce the amount of BBB opening, evaluated using contrast-enhanced MRI, when compared with medical air [35, 36]. Previous studies demonstrate that the circulation time for perfluorocarbon microbubbles is greatly decreased when oxygen is used as the carrier gas relative to medical air [37, 38]. As an alternative, intraperitoneal injections of ketamine and xylazine can be used for anesthesia.
5. There are several commercially available microbubble contrast agents approved for clinical use that have been applied for safe and localized induction of BBB permeability. The most widely used microbubble contrast agents are composed of either lipid (DEFINITY) or human serum albumin (Optison, GE Healthcare) shells filled with the perfluorocarbon gas Perflutren. DEFINITY and Optison microbubbles produce nearly equivalent results with respect to the extent of BBB permeability [39]. It is also important to note that the safety profile of FUS-induced BBB permeability is dependent on the microbubble dose administered. A range of DEFINITY doses have been described as safe and effective for BBB opening. For clinical imaging using DEFINITY as a contrast agent, 20 μL/kg is the maximal dose recommended by the manufacturer [40], which has been shown to successfully increase BBB permeability following FUS in rodents [39, 41]. The first clinical trial in AD patients used DEFINITY at a dose of 4 μL/kg for FUS-mediated BBB opening in humans [16]. Using a clinical imaging dose of DEFINITY microbubbles for FUS-mediated BBB permeability, McMahon et al. reported no acute inflammatory response, hemorrhage, or edema following treatment. In contrast, using a higher microbubble dose (tenfold greater than the dose applied for clinical imaging), significant

neuroinflammation and tissue damage including edema, hemorrhage, neutrophil infiltration, and cell death were observed following FUS, as reported in previous studies [42–44].

6. For intravenous injections, the maximum injectable volume for a mouse is 25 mL/kg per day administered slowly [45]. During the MRIgFUS experiment, DEFINITY, Gadovist, and saline injections are all administered through the tail vein. Thus, for a 30 g mouse, this leaves a volume of about 90 μL for virus injection. If the injection volume of the viral working solution is less than 20 μL, catheter-associated variability may affect the delivered dosage in the bloodstream. For example, in a recent study using mice, working viral solutions were calculated based on the following values:
   (a) rAAV stock titer: 1.4×10^{10} vg/μL
   (b) Intravenous injection dose: 3.0×10^{9} vg/g
   (c) Number of mice: 6
   (d) Average weight: 27 g
   (e) Heaviest weight: 30 g

   Injecting the heaviest mouse with the maximum volume (90 μL) at a dose of 3.0×10^{9} vg/g required a virus concentration of 1.0×10^{9} vg/μL, and thus the stock solution was diluted 14-fold in sterile 0.9% NaCl to prepare the working solution. 500 μL of working solution was prepared for this experiment. Similar calculations can be applied to prepare diluted virus solution for rat studies. The maximum injectable volume for a rat is 20 mL/kg daily [45].

7. If exposure to bleach is not recommended for disinfecting instruments by the manufacturer, clean supplies with 100% acetone, rinse with distilled water, and then rinse with 100% ethanol.

8. For FUS systems designed to work with whole-body MR scanners, such as the FUS Instruments RK-100, the FUS procedure can be carried out directly inside the MR scanner. This procedure is described in an earlier published protocol [26].

9. Alternatively, heaters, recirculating warm water blankets, or heat lamps (at a safe distance from the animal) are all effective means of providing warmth. It is important to keep animals warm throughout the procedure. This improves blood circulation, which helps to visualize the tail vein and install the catheter. It also improves tolerance to anesthesia during the procedure and recovery from anesthesia. Increasing blood flow is also important for efficient transport of injectables from the bloodstream to the brain vasculature.

10. Proper animal preparation is a critical aspect of this procedure. The hair on the animal's head must be removed completely to avoid attenuating the ultrasound beam.
11. Reposition the head if necessary and repeat the localizer scan to provide optimal acoustic coupling between the transducer and desired target brain region.
12. Previous studies have demonstrated the feasibility of rAAV delivery using FUS targeted to the cortex, hippocampus, and striatum in the rat and mouse brain. The focal sizes of BBB opening generated by the two transducers typically used in our experiments are described here. In water, for 550 kHz, the full width at half maximum intensity (FWHM) of pressure is approximately 3 mm in diameter laterally and 15 mm along the beam. For 1.6 MHz, FWHM is less than 1 mm laterally and 3–4 mm along the beam. For FUS-induced BBB permeability, we use only peak pressure, and thus the size of the actual focal opening in the lateral dimension is normally smaller than the FWHM. Along the beam, however, FUS-induced BBB disruption tends to be wider due to standing wave formation in close proximity to the rodent skull. The spatial resolution of FUS-induced BBB opening in the axial and lateral dimensions using the 1.6 MHz transducer can be visualized by the distribution of GFP expression after targeted delivery of rAAV9-CMV-GFP to the mouse striatum using MRIgFUS as shown in Fig. 3. In contrast to a single focal spot, FUS has also been applied across several spots of the brain and spinal cord for broad CNS AAV delivery [19, 20].
13. In some instances, this feeding period with soft food and Nutra-Gel may need to be extended, such as for rodent strains that have low body weight and for animals that lose more than 20% of body weight after the procedure. If substantial weight loss is observed, we recommend providing soft food and Nutra-Gel for a week or the entire survival duration after treatment.
14. The survival time of animals after treatment is variable and depends on several factors including the nature of the rAAV vector, the target cells, and the objectives of the study. The duration of time necessary to detect rAAV-mediated gene expression is highly dependent on the type of virus used. Some rAAVs provide fast, robust expression within 1 week, while others require 4 weeks to reach maximum expression in brain [20, 29, 30, 46]. FUS-mediated gene delivery studies detected efficient transgene expression in targeted regions using rAAV9 after 12–15 days [20, 30]. Dose-dependent GFP expression levels following targeted delivery of rAAV9-CMV-GFP to the striatum 12 days after FUS is illustrated in

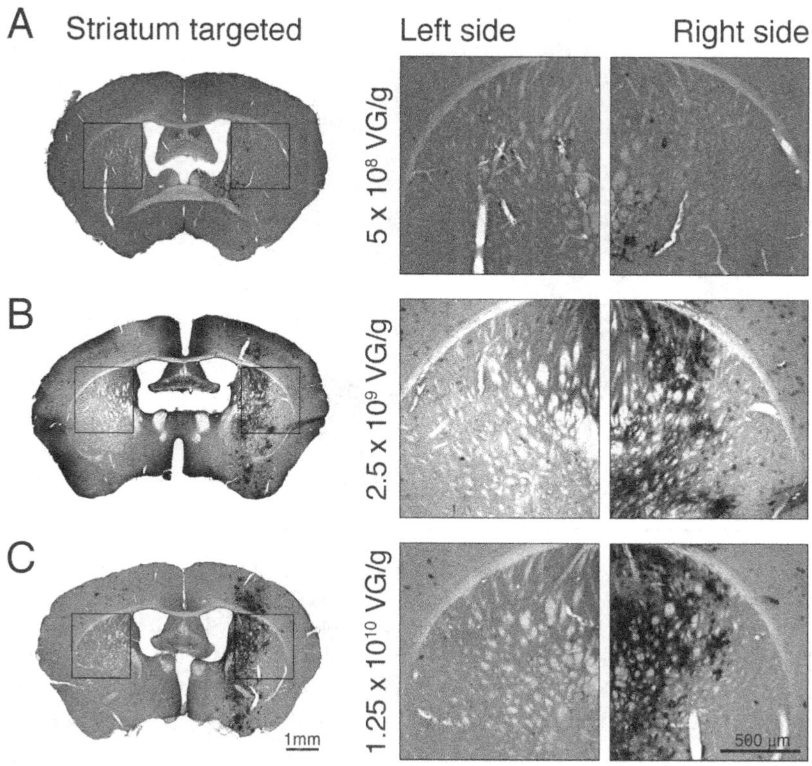

Fig. 3 MRIgFUS for targeted delivery of rAAV9-CMV-GFP to the unilateral mouse striatum. (**a**–**c**) GFP expression was assessed by immunohistochemical DAB staining in the contralateral hemisphere (left) and in FUS-targeted regions (striatum right hemisphere), 2 weeks after treatment. GFP expression was examined after three different doses of intravenous rAAV9-CMV-GFP delivered by MRIgFUS to the right striatum: (**a**) 5×10^8 vg/g, (**B**) 2.5×10^9 vg/g, and (**c**) 1.25×10^{10} vg/g. Strong GFP expression was detected in the FUS-treated striatum (right) compared to the contralateral untreated hemisphere (left) at 2.5×10^9 vg/g and 1.25×10^{10} vg/g (**b**, **c**). The expression of GFP in FUS-treated animals at 5×10^8 vg/g was low-to-undetectable (**a**)

Fig. 3. FUS-facilitated rAAV2 delivery provided detectable GFP expression within the first week, but the optimum expression was achieved 21 days following FUS [29]. As an alternative, luciferase can be used as a transgene, thereby allowing for in vivo imaging of transgene expression levels at multiple time-points in the same animal [47]. Luciferase activity can be measured after tail vein injection of its substrate (luciferin) and measurements can be conducted sequentially in individual animals.

15. In addition to increasing the efficiency of gene expression, MRIgFUS-mediated rAAV delivery into the brain may also alter viral tropism. rAAV tropism after intraparenchymal injection has been well described for a variety of serotypes. However, MRIgFUS delivery exposes astrocytes surrounding the BBB as the primary entry route, thereby creating alternative

accessibility to different cell types and locations in the brain [11, 48]. For example, Foust et al. compared rAAV9 tropism in the brain after intravenous and intracranial injections into the striatum and hippocampus. In adult mice, intravascular injection of rAAV9 produced predominantly glial transgene expression in the brain, whereas intraparenchymal injection resulted in almost exclusively neuronal expression [11]. Similarly, MRIgFUS-mediated delivery of intravenously injected rAAV2 to the brain transduced both astrocytes (41%) and neurons (12%), in contrast to the typical neuronal-biased transduction after direct injection of the rAAV2 serotype [29]. Thus, it is important to take into account the impact of the delivery method on serotype tropism when targeting cell-specific gene expression.

Acknowledgments

We would like to acknowledge Dr. Emmanuel Thévenot for his work, generating animals and images used to create Fig. 3. We thank Hang Yu Lin for her assistance in generating the graphical illustrations, and Sanjana Seerala, Kristina Mikloska, and Dr. Badru Moloo for editing the manuscript. Dr. Melitta Schachner contributed to the initial discussion of the project. This work was supported by the Canadian Institutes of Health Research (KX, Doctoral Scholarship; IA and KH: FRN137064), the Weston Brain Institute (IA, KH), and the Alzheimer Society of Canada (ZN, Doctoral Scholarship).

References

1. Kumar SRP, Markusic DM, Biswas M, High KA, Herzog RW (2016) Clinical development of gene therapy: results and lessons from recent successes. Mol Ther Methods Clin Dev 3:16034. https://doi.org/10.1038/mtm.2016.34
2. Costantini LC, Bakowska JC, Breakefield XO, Isacson O (2000) Gene therapy in the CNS. Gene Ther 7:93. https://doi.org/10.1038/sj.gt.3301119
3. Lim ST, Airavaara M, Harvey BK (2010) Viral vectors for neurotrophic factor delivery: a gene therapy approach for neurodegenerative diseases of the CNS. Pharmacol Res 61 (1):14–26. https://doi.org/10.1016/j.phrs.2009.10.002
4. Hardy J (2018) Neurodegeneration: the first mechanistic therapy and other progress in 2017. Lancet Neurol 17(1):3–5. https://doi.org/10.1016/S1474-4422(17)30414-3
5. Dunbar CE, High KA, Joung JK, Kohn DB, Ozawa K, Sadelain M (2018) Gene therapy comes of age. Science 359(6372):eaan4672. https://doi.org/10.1126/science.aan4672
6. Wilson JM (2017) 2017 is the year we have been waiting for. Hum Gen Ther Clin Dev 28 (4):165–166. https://doi.org/10.1089/humc.2017.29030.wil
7. Chan KY, Jang MJ, Yoo BB, Greenbaum A, Ravi N, Wu WL, Sanchez-Guardado L, Lois C, Mazmanian SK, Deverman BE, Gradinaru V (2017) Engineered AAVs for efficient noninvasive gene delivery to the central and peripheral nervous systems. Nat Neurosci 20 (8):1172–1179. https://doi.org/10.1038/nn.4593
8. Forsayeth JR, Bankiewicz KS (2011) AAV9: over the fence and into the woods. Mol Ther 19(6):1006–1007

9. Manfredsson FP, Rising AC, Mandel RJ (2009) AAV9: a potential blood-brain barrier buster. Mol Ther 17(3):403–405
10. Foust KD, Wang X, McGovern VL, Braun L, Bevan AK, Haidet AM, Le TT, Morales PR, Rich MM, Burghes AH, Kaspar BK (2010) Rescue of the spinal muscular atrophy phenotype in a mouse model by early postnatal delivery of SMN. Nat Biotechnol 28(3):271–274
11. Foust KD, Nurre E, Montgomery CL, Hernandez A, Chan CM, Kaspar BK (2009) Intravascular AAV9 preferentially targets neonatal neurons and adult astrocytes. Nat Biotechnol 27(1):59–65. https://doi.org/10.1038/nbt.1515
12. Mendell JR, Al-Zaidy S, Shell R, Arnold WD, Rodino-Klapac LR, Prior TW, Lowes L, Alfano L, Berry K, Church K, Kissel JT, Nagendran S, L'Italien J, Sproule DM, Wells C, Cardenas JA, Heitzer MD, Kaspar A, Corcoran S, Braun L, Likhite S, Miranda C, Meyer K, Foust KD, Burghes AHM, Kaspar BK (2017) Single-dose gene-replacement therapy for spinal muscular atrophy. N Engl J Med 377(18):1713–1722. https://doi.org/10.1056/NEJMoa1706198
13. Rafii MS, Tuszynski MH, Thomas RG, Barba D, Brewer JB, Rissman RA, Siffert J, Aisen PS, Team ANS (2018) Adeno-associated viral vector (serotype 2)-nerve growth factor for patients with alzheimer disease: a randomized clinical trial. JAMA Neurol 75(7):834–841. https://doi.org/10.1001/jamaneurol.2018.0233
14. Tuszynski MH, Yang JH, Barba D, U HS, Bakay RA, Pay MM, Masliah E, Conner JM, Kobalka P, Roy S, Nagahara AH (2015) Nerve growth factor gene therapy: activation of neuronal responses in Alzheimer disease. JAMA Neurol 72(10):1139–1147. https://doi.org/10.1001/jamaneurol.2015.1807
15. Timbie KF, Mead BP, Price RJ (2015) Drug and gene delivery across the blood-brain barrier with focused ultrasound. J Control Release 219:61–75. https://doi.org/10.1016/j.jconrel.2015.08.059
16. Lipsman N, Meng Y, Bethune AJ, Huang Y, Lam B, Masellis M, Herrmann N, Heyn C, Aubert I, Boutet A, Smith GS, Hynynen K, Black SE (2018) Blood–brain barrier opening in Alzheimer's disease using MR-guided focused ultrasound. Nat Commun 9(1):2336. https://doi.org/10.1038/s41467-018-04529-6
17. Gray SJ, Matagne V, Bachaboina L, Yadav S, Ojeda SR, Samulski RJ (2011) Preclinical differences of intravascular AAV9 delivery to neurons and glia: a comparative study of adult mice and nonhuman primates. Mol Ther 19(6):1058–1069. https://doi.org/10.1038/mt.2011.72
18. Duque S, Joussemet B, Riviere C, Marais T, Dubreil L, Douar AM, Fyfe J, Moullier P, Colle MA, Barkats M (2009) Intravenous administration of self-complementary AAV9 enables transgene delivery to adult motor neurons. Mol Ther 17(7):1187–1196
19. Weber-Adrian D, Thévenot E, O'Reilly MA, Oakden W, Akens MK, Ellens N, Markham-Coultes K, Burgess A, Finkelstein J, Yee AJ, Whyne CM, Foust KD, Kaspar BK, Stanisz GJ, Chopra R, Hynynen K, Aubert I (2015) Gene delivery to the spinal cord using MRI-guided focused ultrasound. Gene Ther 22(7):568–577. https://doi.org/10.1038/gt.2015.25
20. Thévenot E, Jordão JF, O'Reilly MA, Markham K, Weng YQ, Foust KD, Kaspar BK, Hynynen K, Aubert I (2012) Targeted delivery of self-complementary adeno-associated virus serotype 9 to the brain, using magnetic resonance imaging-guided focused ultrasound. Hum Gene Ther 23(11):1144–1155. https://doi.org/10.1089/hum.2012.013
21. Korbelin J, Dogbevia G, Michelfelder S, Ridder DA, Hunger A, Wenzel J, Seismann H, Lampe M, Bannach J, Pasparakis M, Kleinschmidt JA, Schwaninger M, Trepel M (2016) A brain microvasculature endothelial cell-specific viral vector with the potential to treat neurovascular and neurological diseases. EMBO Mol Med 8(6):609–625. https://doi.org/10.15252/emmm.201506078
22. Kügler S (2016) Tissue-specific promoters in the CNS. Methods Mol Biol 1382-(Chapter 6):81–91. https://doi.org/10.1007/978-1-4939-3271-9_6
23. Hynynen K, McDannold N, Vykhodtseva N, Jolesz FA (2001) Noninvasive MR imaging-guided focal opening of the blood-brain barrier in rabbits. Radiology 220(3):640–646. https://doi.org/10.1148/radiol.2202001804
24. Hosseinkhah N, Goertz DE, Hynynen K (2015) Microbubbles and blood-brain barrier opening: a numerical study on acoustic emissions and wall stress predictions. IEEE Trans Biomed Eng 62(5):1293–1304. https://doi.org/10.1109/tbme.2014.2385651
25. Sheikov N, McDannold N, Sharma S, Hynynen K (2008) Effect of focused ultrasound applied with an ultrasound contrast agent on the tight junctional integrity of the brain microvascular endothelium. Ultrasound Med Biol 34

(7):1093–1104. https://doi.org/10.1016/j.ultrasmedbio.2007.12.015

26. Huang Y, Hynynen K (2011) MR-guided focused ultrasound for brain ablation and blood-brain barrier disruption. Methods Mol Biol 711:579–593. https://doi.org/10.1007/978-1-61737-992-5_30
27. Burgess A, Ayala-Grosso CA, Ganguly M, Jordão JF, Aubert I, Hynynen K (2011) Targeted delivery of neural stem cells to the brain using MRI-guided focused ultrasound to disrupt the blood-brain barrier. PLoS One 6(11):e27877. https://doi.org/10.1371/journal.pone.0027877
28. Shen WB (2017) Magnetic enhancement of stem cell–targeted delivery into the brain following MR-guided focused ultrasound for opening the blood–brain barrier. Cell Transplant 26(7):1235–1246. https://doi.org/10.1177/0963689717715824
29. Hsu P-H, Wei K-C, Huang C-Y, Wen C-J, Yen T-C, Liu C-L, Lin Y-T, Chen J-C, Shen C-R, Liu H-L (2013) Noninvasive and targeted gene delivery into the brain using microbubble-facilitated focused ultrasound. PLoS One 8(2):e57682. https://doi.org/10.1371/journal.pone.0057682
30. Wang S, Kugelman T, Buch A, Herman M, Han Y, Karakatsani ME, Hussaini SA, Duff K, Konofagou EE (2017) Non-invasive, focused ultrasound-facilitated gene delivery for optogenetics. Sci Rep 7:39955. https://doi.org/10.1038/srep39955. https://www.nature.com/articles/srep39955#supplementary-information
31. Chopra R, Curiel L, Staruch R, Morrison L, Hynynen K (2009) An MRI-compatible system for focused ultrasound experiments in small animal models. Med Phys 36(5):1867–1874. https://doi.org/10.1118/1.3115680
32. O'Reilly MA, Hynynen K (2012) Blood-brain barrier: real-time feedback-controlled focused ultrasound disruption by using an acoustic emissions-based controller. Radiology 263(1):96–106. https://doi.org/10.1148/radiol.11111417
33. Maguire CA, Crommentuijn MH, Mu D, Hudry E, Serrano-Pozo A, Hyman BT, Tannous BA (2013) Mouse gender influences brain transduction by intravascularly administered AAV9. Mol Ther 21(8):1470–1471. https://doi.org/10.1038/mt.2013.95
34. Jackson KL, Dayton RD, Klein RL (2015) AAV9 supports wide-scale transduction of the CNS and TDP-43 disease modeling in adult rats. Mol Ther Methods Clin Dev 2:15036. https://doi.org/10.1038/mtm.2015.36
35. McDannold N, Zhang Y, Vykhodtseva N (2011) Blood-brain barrier disruption and vascular damage induced by ultrasound bursts combined with microbubbles can be influenced by choice of anesthesia protocol. Ultrasound Med Biol 37(8):1259–1270. https://doi.org/10.1016/j.ultrasmedbio.2011.04.019
36. McDannold N, Zhang Y, Vykhodtseva N (2017) The effects of oxygen on ultrasound-induced blood-brain barrier disruption in mice. Ultrasound Med Biol 43(2):469–475. https://doi.org/10.1016/j.ultrasmedbio.2016.09.019
37. Mullin L, Gessner R, Kwan J, Kaya M, Borden MA, Dayton PA (2011) Effect of anesthesia carrier gas on in vivo circulation times of ultrasound microbubble contrast agents in rats. Contrast Media Mol Imaging 6(3):126–131. https://doi.org/10.1002/cmmi.414
38. Itani M, Mattrey RF (2012) The effect of inhaled gases on ultrasound contrast agent longevity in vivo. Mol Imaging Biol 14(1):40–46. https://doi.org/10.1007/s11307-011-0475-5
39. McDannold N, Vykhodtseva N, Hynynen K (2007) Use of ultrasound pulses combined with Definity for targeted blood-brain barrier disruption: a feasibility study. Ultrasound Med Biol 33(4):584–590. https://doi.org/10.1016/j.ultrasmedbio.2006.10.004
40. DEFINITY® (2017) Lantheus Medical Imaging, Inc., North Billerica, MA
41. O'Reilly MA, Hynynen K (2018) Ultrasound and microbubble-mediated blood-brain barrier disruption for targeted delivery of therapeutics to the brain. In: Sirianni RW, Behkam B (eds) Targeted drug delivery: methods and protocols. Springer, New York, NY, pp 111–119. https://doi.org/10.1007/978-1-4939-8661-3_9
42. McMahon D, Hynynen K (2017) Acute inflammatory response following increased blood-brain barrier permeability induced by focused ultrasound is dependent on microbubble dose. Theranostics 7(16):3989–4000. https://doi.org/10.7150/thno.21630
43. Kovacs ZI, Kim S, Jikaria N, Qureshi F, Milo B, Lewis BK, Bresler M, Burks SR, Frank JA (2017) Disrupting the blood-brain barrier by focused ultrasound induces sterile inflammation. PNAS 114(1):E75–E84. https://doi.org/10.1073/pnas.1614777114
44. Treat LH, McDannold N, Vykhodtseva N, Zhang Y, Tam K, Hynynen K (2007) Targeted delivery of doxorubicin to the rat brain at therapeutic levels using MRI-guided focused ultrasound. Int J Cancer 121(4):901–907. https://doi.org/10.1002/ijc.22732

45. Diehl KH, Hull R, Morton D, Pfister R, Rabemampianina Y, Smith D, Vidal JM, van de Vorstenbosch C (2001) A good practice guide to the administration of substances and removal of blood, including routes and volumes. J Appl Toxicol 21(1):15–23
46. Mason MRJ, Ehlert EME, Eggers R, Pool CW, Hermening S, Huseinovic A, Timmermans E, Blits B, Verhaagen J (2010) Comparison of AAV serotypes for gene delivery to dorsal root ganglion neurons. Mol Ther 18(4):715–724. https://doi.org/10.1038/mt.2010.19
47. Zincarelli C, Soltys S, Rengo G, Rabinowitz JE (2008) Analysis of AAV serotypes 1-9 mediated gene expression and tropism in mice after systemic injection. Mol Ther 16(6):1073–1080
48. Pardridge WM (2012) Drug transport across the blood-brain barrier. J Cereb Blood Flow Metab 32(11):1959–1972. https://doi.org/10.1038/jcbfm.2012.126

Chapter 11

AAV-Mediated Gene Delivery to the Spinal Cord by Intrathecal Injection

Cristina D. Peterson, Alexander G. J. Skorput, Kelley F. Kitto, George L. Wilcox, Lucy Vulchanova, and Carolyn A. Fairbanks

Abstract

Gene therapy targeting the spinal cord is an important tool for analyzing mechanisms of nervous system diseases and the development of gene therapies. Analogous to a lumbar puncture in humans, the rodent spinal cord can be accessed through an efficient, noninvasive injection. Here we describe a method for AAV-mediated gene transfer to cells of the spinal cord by intrathecal injection of small quantities of AAV vector.

Key words Spinal cord, Dorsal root ganglion, Viral vector, Adeno-associated

1 Introduction

The spinal cord represents a key treatment target for both motor and sensory disorders, including chronic pain. The development of chronic neuraxial infusion approaches [1, 2] led to use of a variety of analgesics for the treatment of chronic pain and baclofen for spasticity. The use of the intrathecal route of administration for the treatment of chronic pain is considered for those patients for whom systemic forms of pain control are not effective. These are now well-established therapeutic pain management approaches.

Although the intrathecal route of administration has been used primarily for pain management, there has also been a long-standing interest in accessing the intrathecal route of administration for treating motor dysfunctions, such as amyotrophic lateral sclerosis (ALS), that are refractory to systemic forms of treatment [3]. Early studies of delivery of biotherapeutics released from intrathecally transplanted cells were conducted in ALS patients, albeit with limited success [4]. More positive results have recently been

Cristina D. Peterson and Alexander G. J. Skorput contributed equally to this work.

Michael J. Castle (ed.), *Adeno-Associated Virus Vectors: Design and Delivery*, Methods in Molecular Biology, vol. 1950, https://doi.org/10.1007/978-1-4939-9139-6_11,

reported with intrathecal infusion of small molecule medications via a chronic infusion pump system [5]. Also, recent studies of intrathecal injection of adeno-associated viral (AAV) vectors carrying genes associated with disease-modifying antibodies or microRNAs have shown compelling outcomes in preclinical models of ALS [6, 7]. Clinical trials to assess the safety of intrathecally delivered AAVs to treat motor dysfunctions are presently underway [8]. One Phase I safety and tolerability trial in progress [9] will determine safety and efficacy of intrathecal delivery of an AAV9 viral vector carrying the gene for survival motor neuron protein, the deficiency of which causes spinal muscular atrophy Type I. These developments indicate that the use of the intrathecal route of administration to provide pharmacological or molecular therapy directly to the motor system has been under consideration for many decades and appears to be advancing through clinical translation.

The Institute of Medicine's report on the state of chronic pain in the United States in 2011 [10] and the National Pain Strategy introduced by the Office of the Assistant Secretary of Health and Human Services in 2016 called attention to nation's increasing concern for the prevalence of chronic pain [11]. This report estimated that, in the United States, chronic pain affects 100 million adults, and the annual national economic burden is estimated at up to $635 billion in medical costs and lost productivity. These values highlight the general consensus that chronic pain is an important global problem. The reports feature two key points. First, the present predominant pharmacological approaches for pain management are insufficient. Second, new non-opioid and innovative approaches for pain management are greatly needed. Spinal delivery of gene therapeutics for the treatment of chronic pain may be responsive to this need.

The use of gene therapeutics in the spinal cord for pain management and diseases affecting motor neurons has remained a consistent interest for over 20 years, with a variety of vector delivery approaches in development including, but not limited to, adenovirus vectors [12], adeno-associated virus (AAV) vectors [13], lentivirus vectors [13], and herpes simplex virus (HSV) vectors [14]. HSV has a unique natural infective mechanism at peripheral nerve endings [14] which renders it amenable to simple intradermal injection. In contrast, adenovirus [15], lentivirus [16], and AAV vectors [17] have been developed to treat systemic disorders via intravascular delivery methods, and have also been used to treat central nervous system (CNS) disorders. When considered for CNS therapies, direct CNS injection approaches including intraparenchymal [18] and intraganglionic [19] delivery are frequently employed to ensure accurate delivery to the tissue. However, these approaches are both invasive and inherently tissue-destructive. In contrast, intravenous delivery, although noninvasive, presents the challenges of limited CNS transduction and a requirement for high treatment doses. Compared to these intraparenchymal and systemic routes, intrathecal delivery represents a minimally invasive alternative with a lower dose requirement. Consequently, there has been increasing interest in the use of intrathecal delivery of gene therapeutics for targeting CNS disorders [8].

The spinal cord consists of primary circuitry for sensory and motor relay and processing. The cord is surrounded by the meninges, which include the pia mater, arachnoid mater, and the outermost dura mater. Intrathecal injection by direct lumbar puncture in mice [20] and rats [21] bypasses the dura and arachnoid maters, delivering injectate to the subarachnoid space directly adjacent to the pia mater. An experienced injector can perform the injection in less than 60 s without the need for anesthesia.

Following intrathecal injection of AAV, various pathways distribute the injectate away from the site of delivery. First, the vector can cross the pia mater and transduce cells of the pia as well as selected neurons of the dorsal and ventral horns. We also observe viral vector distribution away from the injection site, presumably along the dorsal root sleeve, to the dorsal root ganglia (DRG). Third, based on the common observation that the liver is transduced following intrathecal delivery, we speculate that a fraction of the AAV exits the arachnoid villi or dura mater to encounter the epidural vasculature and enter systemic circulation. It is possible that viral particles may also exit the CNS from the spinal cord parenchyma via transcellular transport mechanisms on the endothelial cells of microcapillaries. However, studies of the transporter mechanisms by which viral particles cross the blood brain barrier have focused primarily on the blood-to-brain transport pathway [22], rather than the brain-to-blood efflux pathway, and more investigation is needed to understand how viral particles exit the CNS.

The direct lumbar puncture method for intrathecal delivery described here represents an optimal preclinical model for minimally invasive gene transfer to the spinal cord. Gene transfer to the spinal cord may be useful either for the preclinical assessment of potential biotherapeutics for medication development or for the study of molecular mechanisms using cre-recombinase to knock down different genes of interest in the spinal cord. Through the use of cell-specific promoters, specific cell populations can be targeted selectively for gene knockdown. We have previously detailed the method for direct lumbar puncture for intrathecal drug delivery in mice [20] and rats [21], including explicit strategies for training and acquisition of the technique [23]. In this chapter, we present our approach with necessary modifications for injecting adeno-associated viral vectors in the intrathecal space of rodents for gene delivery to the spinal cord. This route of delivery leads to transduction of primary afferent neurons and spinal neurons at all spinal levels, as well as transduction of a number of brain regions

[24–28]. Among the different AAV serotypes, AAV9 and AAVrh10 demonstrate the highest levels of transduction of spinal neurons.

2 Materials

1. Male and female mice or rats. Avoid neonatal mice (younger than 3 weeks), which are challenging to inject due to the positioning and grip required for injection.
2. AAV vector (*see* **Note 1**).
3. Anesthetic such as isoflurane may be used, but is not recommended (*see* **Note 2**).
4. Distilled H_2O (diH_2O).
5. PE20 tubing.
6. Scissors.
7. 27G, ½″ luer hub needle.
8. 27G, 1 ¼″ luer hub needle.
9. 50 μL Hamilton syringe.
10. Soft cloth or paper towel.

3 Methods

3.1 Setup

1. Clip a 250 mm length of PE20 tubing.
2. Extract the 1 ¼″, 27G needle from the plastic connector by using a scissors to slice perpendicularly across the luer hub without severing the needle. Scrape off any residual plastic.
3. Insert the base of the 1 ¼″ needle into one end of the PE20 tubing. This needle will later be inserted into the intrathecal space.
4. Insert the tip of the ½″ needle into the other end of the PE20 tubing.
5. Load the 50 μL Hamilton syringe with diH_2O and connect the ½″ needle hub to the syringe. Depress the plunger to introduce the diH_2O into the PE20 tubing.
6. Introduce the 1 ¼″ needle to the tube containing the injectate and load the AAV vector into the catheter (*see* **Note 3**).

3.2 Injection Procedure

1. Place a soft cloth or towel over the head and back of the rodent, keeping the lumbar region below the iliac crest uncovered (*see* **Note 4**). Firmly hold the rodent by the iliac crest (hip bones) with the thumb and index finger of the nondominant hand. The top of the hand can rest gently over the body and head, framing the subject to keep it calm. With the forefinger of the

dominant hand that holds the syringe, trace the spinal column from lumbar to sacral levels, pushing gently down onto the spinal column, while simultaneously very slightly pulling up on the subject with nondominant hand. The objective here is to open up the spaces in between the vertebral segments to facilitate entry of the needle.

2. Orient the bevel of the 1 ¼″ needle toward the head of the rodent, then insert the needle into the lumbar spinal column at a 70° angle, at the midline of the spinal column. When the needle is felt to encounter bone, lower the angle of the needle to about a 30° angle, and insert the needle in between vertebral segments. This may require several approaches and exits to find an entry point where the needle enters the column without resistance. Indications of proper entry often include a reflexive flick of the tail or of a hind leg.
3. Depress the plunger of the syringe to deliver the injectate into the subarachnoid space (*see* **Note 5**).
4. Following injection, rotate the needle 180° once or twice and withdraw it from the vertebral column. Place the subject back in its home cage to observe its behavior (*see* **Note 6**).

3.3 Representative Results

Figure 1 illustrates the level of transduction achieved in spinal cord with this delivery method using AAV9 vectors that carry the gene for a GFP-Cre fusion protein. Figure 1a, b shows immunohistochemical visualization of GFP-Cre following intrathecal injection of AAV9 vectors in which GFP-Cre expression was controlled by the CMV (Fig. 1a) or the hSyn (Fig. 1b) promoters. The analysis was conducted 8 weeks after AAV delivery. Figure 1e, f shows tdTomato fluorescence 3 weeks after intrathecal injection of AAV9 vectors, carrying GFP-Cre, in mice that express cre-dependent tdTomato (Ai14 mice, Jackson Laboratories #007914). Compared to the broad tdTomato expression achieved in dorsal horn using the hSyn promoter (Fig. 1e), transgene expression under the control of the PKCγ promoter is limited to a subset of neurons (Fig. 1f) (*see* **Note** 7).

4 Notes

1. In our work with AAV5, AAV8, and AAV9, we have observed highest transduction efficiency in spinal cord with AAV9 [25, 27, 28]. rAAVrh10 has also been shown to transduce spinal neurons efficiently [29]. We typically use titers that are approximately 10^{13} vector genomes (vg) per mL.
2. Anesthetic may be used prior to injection, but this is not recommended because it can inhibit the characteristic tail flick

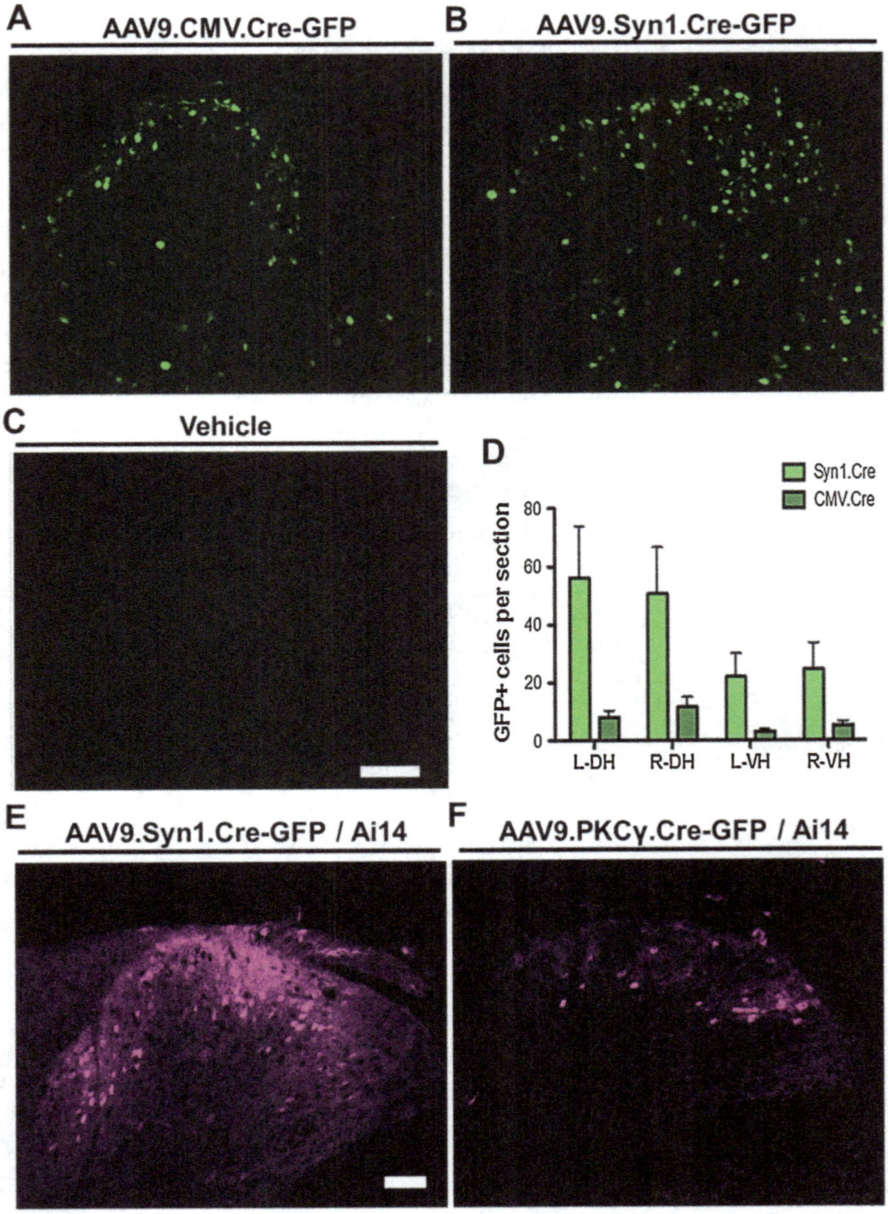

Fig. 1 Representative results after intrathecal delivery of AAV9 by direct lumbar puncture. (**a**–**c**) AAV9.Syn.Cre-GFP (**a**) and AAV9.CMV.Cre-GFP (**b**), at titers of 3.9×10^{13} and 5.5×10^{13} vg/mL, respectively, were injected intrathecally in mice. Transduction was evaluated 8 weeks after treatment by GFP immunolabeling (Abcam ab13970). Labeling was absent in vehicle-treated mice (**c**). Scale bar: 100 μm. (**d**) Comparison of the number of GFP-labeled cells in spinal dorsal and ventral horns of mice treated with AAV9.CMV.Cre-GFP ($n = 7$) or AAV9.Syn.Cre-GFP ($n = 4$). For each animal, counts of cells from three sections were averaged. (**e**, **f**) AAV9.Syn.Cre-GFP (**e**) and AAV9.PKCγ.Cre-GFP (**f**) were injected intrathecally in Ai14 mice (at titers of 5.5×10^{13} and 3.5×10^{13} vg/mL, respectively). Transduction was evaluated 3 weeks after treatment based on the cre-dependent expression of tdTomato. Scale bar: 50 μm. AAV9.Syn.Cre-GFP and AAV9.CMV.Cre-GFP were purchased from the University of Pennsylvania Vector Core. AAV9.PKCγ.Cre-GFP was produced at the University of Minnesota Viral Vector and Cloning Core

which is indicative of dural puncture. This tail flick is a strong indicator of correct placement of the needle prior to injection.

3. The PE20 tubing should still have air separating the diH_2O from the AAV injectate. The tubing can be marked with a marker at the level of the vector and checked following each injection to ensure that product is being distributed into each subject. If the level of vector remains unchanged following an injection, the catheter should be remade and the injection repeated.
4. The cloth is not necessary for injection but may assist in control of the rodent. The cloth should be placed rostrally to the lumbar spinal cord and should leave the iliac crest uncovered to ensure access for holding and injection. Optimizing the injector's hold of the animal is essential for maximizing the likelihood of a successful injection. The finger/thumb hold on the iliac crest should be strong enough to immobilize any movement in the legs of the subject, and should cause the legs to extend outward and downward. New injectors should rehearse the injection hold with the subject prior to attempting the injection to increase the likelihood of injection success. Experienced injectors will frequently readjust the hold prior to performing the injection.
5. The injection should require approximately 10–30 s. A rapid injection can cause temporary hind limb cramping, but such an event typically resolves in 60 s. Following injection and return to the cage, each subject should be observed briefly to confirm return to normal motor function.
6. The average rate of cerebrospinal fluid (CSF) turnover in the mouse is approximately 2 h. This value is based on an estimate of 40 μL of CSF volume in the average mouse and a CSF production rate of 0.325 μL/min [30]. The CSF turnover rate in a rat is about 1.25 h, based on a total CSF volume of 290 μL and a CSF production rate of 3.8 μL/min [30]. The efficiency of transduction following intrathecal injection is impacted by the CSF in each subject, and, therefore, can be enhanced in mice by vertically restricting each subject following injection. We use the expected time of CSF turnover as a reasonable timeframe for such vertical restriction [26].
7. As illustrated in Fig. 1e, f, the use of cell-specific promoters can restrict transgene expression to subpopulations of neurons. When using weaker cell-specific promoters (e.g., PKCγ promoter), we have observed that GFP-Cre expression may be below the limit of detection based on GFP fluorescence or immunolabeling, yet still sufficient for expression of a cre-dependent transgene. Although intrathecal delivery results in robust transduction of spinal neurons at all spinal levels,

intraspinal delivery may be more appropriate for experimental paradigms that require transduction of a maximal number of neurons within restricted spinal segments (*see* Chapter 9) [31].

Acknowledgments

This work was supported by DoD grant W81XWH-15-1-0494 to C.P. and C.F., F32 NS 100438 to A.G.J.S., R01 DE021996 to L.V.

References

1. Yaksh TL, Rudy TA (1976) Chronic catheterization of the spinal subarachnoid space. Physiol Behav 17:1031–1036
2. Wang JK, Nauss LA, Thomas JE (1979) Pain relief by intrathecally applied morphine in man. Anesth 50(2):149–151
3. Krames E, Buchser E, Hassenbusch SJ, Levy R (1999) Future trends in the development of local drug delivery systems: intraspinal, intracerebral, and intraparenchymal therapies. Neuromodulation 2(2):133–148. https://doi.org/10.1046/j.1525-1403.1999.00133.x
4. Munsat TL, Taft J, Jackson IM, Andres PL, Hollander D, Skerry L, Ordman M, Kasdon D, Finison L (1992) Intrathecal thyrotropin-releasing hormone does not alter the progressive course of ALS: experience with an intrathecal drug delivery system. Neurology 42(5):1049–1053
5. Gutierrez J, Federici T, Peterson B, Bartus R, Betourne A, Boulis NM (2016) 321 development of intrathecal riluzole: a new route of administration for the treatment of amyotrophic lateral sclerosis patients. Neurosurgery 63(Suppl 1):193. https://doi.org/10.1227/01.neu.0000489810.52605.80
6. Patel P, Kriz J, Gravel M, Soucy G, Bareil C, Gravel C, Julien JP (2014) Adeno-associated virus-mediated delivery of a recombinant single-chain antibody against misfolded superoxide dismutase for treatment of amyotrophic lateral sclerosis. Mol Ther 22(3):498–510. https://doi.org/10.1038/mt.2013.239
7. Dirren E, Aebischer J, Rochat C, Towne C, Schneider BL, Aebischer P (2015) SOD1 silencing in motoneurons or glia rescues neuromuscular function in ALS mice. Ann Clin Transl Neurol 2(2):167–184. https://doi.org/10.1002/acn3.162
8. Hardcastle N, Boulis NM, Federici T (2018) AAV gene delivery to the spinal cord: serotypes, methods, candidate diseases, and clinical trials. Expert Opin Biol Ther 18(3):293–307. https://doi.org/10.1080/14712598.2018.1416089
9. Avexis (2017) Study of intrathecal administration of AVXS-101 for spinal muscular atrophy. https://www.clinicaltrials.gov/ct2/show/record/NCT03381729. Accessed 1 Aug 2018
10. Institute of Medicine. Committee on Advance Pain Research C, and Education (2011) In: Relieving pain in America: a blueprint for transforming prevention, care, education, and research. The National Academies Collection: Reports funded by National Institutes of Health, Washington (DC). https://doi.org/10.17226/13172
11. Health OotASf (2016) National pain strategy: a comprehensive population health level strategy for pain. US Department of Health and Human Services, Washington, DC
12. Zheng C, Baum BJ, Iadarola MJ, O'Connell BC (2000) Genomic integration and gene expression by a modified adenoviral vector. Nat Biotechnol 18(2):176–180
13. Goss JR, Krisky D, Wechuck J, Wolfe D (2014) Herpes simplex virus-based nerve targeting gene therapy in pain management. J Pain Res 7:71–79. https://doi.org/10.2147/JPR.S36619
14. Guedon JM, Wu S, Zheng X, Churchill CC, Glorioso JC, Liu CH, Liu S, Vulchanova L, Bekker A, Tao YX, Kinchington PR, Goins WF, Fairbanks CA, Hao S (2015) Current gene therapy using viral vectors for chronic pain. Mol Pain 11:27. https://doi.org/10.1186/s12990-015-0018-1
15. Kozarsky KF, Wilson JM (1993) Gene therapy: adenovirus vectors. Curr Opin Genet Dev 3(3):499–503
16. Buchschacher GL Jr, Wong-Staal F (2000) Development of lentiviral vectors for gene therapy for human diseases. Blood 95(8):2499–2504

17. Dunbar CE, Bodine DM, Sorrentino B, Donahue R, McDonagh K, Cottler-Fox M, O'Shaughnessy J, Cowan K, Carter C, Doren S et al (1994) Gene transfer into hematopoietic cells. Implications for cancer therapy. Ann N Y Acad Sci 716:216–224; discussion 224–217
18. Jasmin L, Rabkin SD, Granato A, Boudah A, Ohara PT (2003) Analgesia and hyperalgesia from GABA-mediated modulation of the cerebral cortex. Nature 424(6946):316–320
19. Xu Y, Gu Y, Xu GY, Wu P, Li GW, Huang LY (2003) Adeno-associated viral transfer of opioid receptor gene to primary sensory neurons: a strategy to increase opioid antinociception. Proc Natl Acad Sci U S A 100(10):6204–6209
20. Hylden JL, Wilcox GL (1980) Intrathecal morphine in mice: a new technique. Eur J Pharmacol 67(2–3):313–316
21. Mestre C, Pelissier T, Fialip J, Wilcox G, Eschalier A (1994) A method to perform direct transcutaneous intrathecal injection in rats. J Pharmacol Toxicol Methods 32(4):197–200
22. Merkel SF, Andrews AM, Lutton EM, Mu D, Hudry E, Hyman BT, Maguire CA, Ramirez SH (2017) Trafficking of adeno-associated virus vectors across a model of the blood-brain barrier; a comparative study of transcytosis and transduction using primary human brain endothelial cells. J Neurochem 140 (2):216–230. https://doi.org/10.1111/jnc.13861
23. Fairbanks CA (2003) Spinal delivery of analgesics in experimental models of pain and analgesia. Adv Drug Deliv Rev 55(8):1007–1041
24. Towne C, Pertin M, Beggah AT, Aebischer P, Decosterd I (2009) Recombinant adeno-associated virus serotype 6 (rAAV2/6)-mediated gene transfer to nociceptive neurons through different routes of delivery. Mol Pain 5:52. https://doi.org/10.1186/1744-8069-5-52
25. Vulchanova L, Schuster DJ, Belur LR, Riedl MS, Podetz-Pedersen KM, Kitto KF, Wilcox GL, McIvor RS, Fairbanks CA (2010) Differential adeno-associated virus mediated gene transfer to sensory neurons following intrathecal delivery by direct lumbar puncture. Mol Pain 6:31. https://doi.org/10.1186/1744-8069-6-31
26. Schuster DJ, Dykstra JA, Riedl MS, Kitto KF, Honda CN, McIvor RS, Fairbanks CA, Vulchanova L (2013) Visualization of spinal afferent innervation in the mouse colon by AAV8-mediated GFP expression. Neurogastroenterol Motil 25(2):e89–e100. https://doi.org/10.1111/nmo.12057
27. Schuster DJ, Dykstra JA, Riedl MS, Kitto KF, Belur LR, McIvor RS, Elde RP, Fairbanks CA, Vulchanova L (2014) Biodistribution of adeno-associated virus serotype 9 (AAV9) vector after intrathecal and intravenous delivery in mouse. Front Neuroanat 8:42. https://doi.org/10.3389/fnana.2014.00042
28. Schuster DJ, Belur LR, Riedl MS, Schnell SA, Podetz-Pedersen KM, Kitto KF, McIvor RS, Vulchanova L, Fairbanks CA (2014) Supraspinal gene transfer by intrathecal adeno-associated virus serotype 5. Front Neuroanat 8:66. https://doi.org/10.3389/fnana.2014.00066
29. Bey K, Ciron C, Dubreil L, Deniaud J, Ledevin M, Cristini J, Blouin V, Aubourg P, Colle MA (2017) Efficient CNS targeting in adult mice by intrathecal infusion of single-stranded AAV9-GFP for gene therapy of neurological disorders. Gene Ther 24 (5):325–332. https://doi.org/10.1038/gt.2017.18
30. Artru AA (1999) In: Yaksh TL (ed) Spinal cerebrospinal fluid chemistry and physiology. Spinal Drug Delivery, Amsterdam, pp 177–237, (p. 216)
31. Gutierrez-Mecinas M, Polgar E, Bell AM, Herau M, Todd AJ (2018) Substance P-expressing excitatory interneurons in the mouse superficial dorsal horn provide a propriospinal input to the lateral spinal nucleus. Brain Struct Funct 223(5):2377–2392. https://doi.org/10.1007/s00429-018-1629-x

Chapter 12

Subpial AAV Delivery for Spinal Parenchymal Gene Regulation in Adult Mammals

Mariana Bravo-Hernández, Takahiro Tadokoro, and Martin Marsala

Abstract

The use of adeno-associated virus (AAV) vectors has become an attractive method for treatment of a variety of neurodegenerative disorders by permitting targeted gene upregulation or silencing in the CNS. Systemic and intrathecal infusion, while preferable routes of vector delivery, have shown encouraging but variable efficacy due to the poor permeability of AAV into spinal cord and brain parenchyma in adult mammals. Recently we have developed a novel and relatively noninvasive technique of spinal subpial vector delivery. This technique confers widespread transgene expression throughout the spinal parenchyma, including both white and gray matter. We have demonstrated that this technique can be performed safely, with a high level of accuracy, and is effective in both small (mouse or rat) and large preclinical (adult pig or nonhuman primate) animal models. In this chapter we provide a comprehensive description of the subpial vector delivery technique in adult rodents (mouse and rat) and large preclinical animals (adult pig and nonhuman primates).

Key words Subpial delivery, Gene therapy, Mice, Rats, Pigs, Nonhuman primates, Adeno-associated virus, Spinal cord

1 Introduction

The use of CNS-targeted gene therapy has recently gained significant traction due to the highly effective use of AAV vectors to achieve targeted gene upregulation or gene silencing. Numerous experimental studies which employed well-characterized transgenic or surgically induced models of neurodegenerative disease have successfully validated the potency and safety of CNS-targeted gene therapy.

It has been demonstrated, for example, that brain intraparenchymal injection of various AAV vector serotypes (AAV1, AAV2, AAV5, AAV.rh8) encoding Galc or βgal cDNA in mice or large animal models of neuropathic lysosomal storage disorders (LSDs), such as GM1 and GM2 gangliosidosis and globoid cell leukodystrophy (GCL), induces widespread therapeutic enzyme

Michael J. Castle (ed.), *Adeno-Associated Virus Vectors: Design and Delivery*, Methods in Molecular Biology, vol. 1950, https://doi.org/10.1007/978-1-4939-9139-6_12,

distribution. This effect is associated with biochemical and histological correction in large regions of the mouse brain, resulting in improved behavioral symptoms, motor function, and life span [1, 2]. Several gene therapy strategies have also been evaluated in mice and rats targeting different pathways involved in Alzheimer's and Huntington's diseases with encouraging results [3, 4]. Similarly, protection of dopaminergic neurons from degenerative processes in Parkinson's disease has also been shown following AAV-mediated overexpression of the transcription factors Nurr1 and Foxa2 [5].

Using different routes of AAV delivery (systemic or intrathecal), several other studies have tested the potency and corresponding treatment effect of mutant SOD1 gene silencing in rodent models of ALS. Encouraging but variable outcomes as defined by delayed disease onset, prolonged survival, slowed disease progression, and delayed motor dysfunction and paralysis were seen in those studies [6–8]. In a recent clinical trial, a significant treatment effect in 15 patients with spinal muscular atrophy type 1 (SMA1) was seen after systemic delivery of an AAV9 vector carrying the survival motor neuron 1 (SMN1) gene. The treatment effect was defined by longer survival and a clear improvement in motor ambulatory functions [9].

Despite these advances and early clinical success in treatment of select neurodegenerative diseases using AAV vectors, several limitations persist. These include: (1) poor permeability of AAV vectors through the blood brain barrier after systemic delivery in adult mammals [10, 11], (2) limited deep spinal cord tissue penetration of AAV vectors after intrathecal delivery, and (3) inability to deliver AAV vector into segment-specific regions without direct needle penetration into spinal tissue. In addition, recent studies have demonstrated high toxicity of AAV9 after systemic (intravascular) delivery in piglets and young nonhuman primates (NHPs) [12]. These data and studies suggest that new delivery methods which would permit minimally invasive AAV delivery into the brain and spinal cord, and which would use the lowest AAV titers needed to achieve clinically relevant treatment effects, are urgently needed.

Recently we have developed a novel and relatively noninvasive technique for spinal subpial vector delivery. In contrast to systemic or intrathecal AAV delivery, the subpial delivery approach is highly effective in inducing widespread transgene expression throughout the spinal parenchyma including white and gray matter. We have also demonstrated the comparable effectiveness of subpial AAV9 delivery in small rodents (mouse and rat) and large preclinical animal models (adult pig and NHP) [13, 14]. In addition, by using different volumes of subpially delivered AAV vectors, a multi-segmental or targeted segment-specific delivery can be achieved. In the following chapter we provide a comprehensive description of the subpial vector delivery technique for use in adult mice, rat, pigs, and nonhuman primates.

2 Materials

2.1 Animals (See Note 1)

1. Adult C57BL/6J mice, body weight (BW) of 20–30 g.
2. Adult Sprague-Dawley rats, BW of 250–350 g.
3. Adult minipigs (crossbred from Minnesota and Gottingen strains), BW of 30–40 kg.
4. Adult *cynomolgus macaque* nonhuman primates (NHPs), BW of 3–5 kg.

2.2 Anesthesia and Injectables

2.2.1 Premedication

1. Azaperonum (Minipigs).
2. Prednisolone (Minipigs).
3. Atropine (Minipigs).
4. Ketamine (NHPs).
5. Dexmedetomidine (NHPs).
6. Prednisolone (NHPs).
7. Buprenorphine (NHPs).

2.2.2 Anesthesia

1. Rodents: 5% isoflurane induction and 1–3% isoflurane maintenance in 1 L/min of 50% Oxygen/50% Air mixture.
2. Minipigs: ketamine induction and 1–3% isoflurane maintenance in 1 L/min of 50% Oxygen/50% Air mixture.
3. NHPs: 2–5% isoflurane induction and 1–3% isoflurane maintenance in 1 L/min of 50% Oxygen/50% Air mixture.

2.2.3 Recovery and Pain Medication

1. Buprenorphine (Rodents, Minipigs, NHPs).
2. Cefazolin (Rodents, NHPs).
3. Cetiofur (Minipigs).

2.2.4 Other Drugs

1. Bacitracin, Neomycin and Polimixin topical ointment (BNP).
2. Hydrocortisone topical ointment.
3. 10 mg/mL Prednisolone (Minipigs and NHPs).

2.3 Viral Vector Preparation (See Note 2)

1. 1.2×10^{13} vector genomes (vg) per mL of AAV9-UBI-GFP, AAV9-UBI-RFP, or AAV9-GAG-GFP, or 1.8×10^{13} vg/mL of Anc80-UBI-GFP.
2. 5% dextran (10,000 MW) solution in distilled water.
3. 100–1000 μL mechanical pipettes.
4. 100–1000 μL pipette tips.
5. 0.5 mL and 1.5 mL eppendorf tubes.

2.4 Surgical Procedure

1. 2.5–3.5 F endotracheal tubes (Minipigs and NHPs).
2. Rodent mask.

3. Pulse oximeter (Minipigs and NHPs).
4. Surgical drapes (Minipigs and NHPs).
5. Rigid pillow (NHPs).
6. Anesthesia machine.
7. Heating mat.
8. Spinal immobilization frame or stereotaxic apparatus.
9. Cunningham's spinal clamps (Rodents).
10. Dental and surgical drill.
11. Operating and Iris scissors (straight and curved).
12. Small osteotome.
13. Olsen-Hegar needle holder.
14. Surgical scalpel handle.
15. Surgical scalpel blades.
16. Tissue forceps (serrated and teeth).
17. Hemostatic Forceps (straight and curved).
18. Towel clamps.
19. Wound clips.
20. Laminectomy rongeurs.
21. Laminectomy retractor.
22. Electrocautery pen (Minipigs and NHPs).
23. GelFoam sponges.
24. Appropriate personal protective equipment, including sterile gloves, eyes safety goggles, lab coat or sterile surgical scrubs, and mask covering the mouth and nose.
25. 2% Chlorhexidine.
26. Animal hair clipper.
27. 0.9% NaCl in sterile H_2O.
28. Absorbent sterile pads.
29. Sterile absorbable monofilament sutures (polypropylene 2-0, 4-0, and 6-0).
30. 30G needles.
31. 1 mL syringes.
32. Sterile cotton swabs.
33. 70% ethanol.
34. Povidone-iodine.
35. Sterile povidone-iodine swab sticks.
36. Surgical soap.

2.5 Viral Vector Delivery

1. 36G pia-penetrating and subpial virus delivery needles (Rodents) (*see* **Note 3**).
2. 25G pia-penetrating needles (Minipigs and NHPs) (*see* **Note 3**).
3. 30G subpial virus delivery needles (Minipigs and NHPs) (*see* **Note 3**).
4. Glass capillary beveller.
5. Coarse diamond abrasive plate (5.0–50 μm tip sizes).
6. 20G slave stainless steel tubing (0.91 mm O.D.).
7. Epoxy resin.
8. Stereotaxic apparatus.
9. Two XYZ manipulators (Stoelting Co, IL, USA) (Rodents).
10. Two Modified XYZ manipulators (Minipigs and NHPs, *see* **Note 4**).
11. Two glass capillary holders.
12. Surgical dissecting scope (8–10× magnification).
13. Polyethylene tubing (PE-10).
14. 50 μL Hamilton syringe (Rodents).
15. Manual Hamilton syringe pump (Rodents).
16. 250 μL Hamilton syringe (Minipigs and NHPs).
17. Automatic Hamilton syringe pump (Minipigs and NHPs).
18. 0.9% NaCl in sterile H_2O.
19. Sterile H_2O.

2.6 Tissue Processing

1. Phosphate buffered saline (PBS 1×): 1 mM KH_2PO_4, 155 mM NaCl, 2.966 mM $Na_2HPO_4 \cdot 7H_2O$.
2. 4% Paraformaldehyde (PFA) in PBS 1×.
3. 30% Sucrose in PBS 1×.
4. 1000 IU/mL heparin.
5. 0.9% NaCl with 0.5% heparin.
6. Cryostat.
7. Cryo-embedding medium: Tissue-Tek® O.C.T. compound.
8. Dry ice.
9. Isopentanol.
10. 25 × 75 mm Superfrost® Plus microscope slides.

2.7 Immunohistochemistry (See Note 5)

1. PBS 1×.
2. PBS-TX Wash Buffer: 0.3% Triton X-100 solution in PBS 1×.
3. Blocking and antibody incubation solution: 5% Normal donkey serum in PBS-TX.

4. Primary antibodies.
5. Secondary antibodies.
6. 1.5 mL Eppendorf tubes.
7. Mechanical pipettes.
8. Pipettes tips.
9. Fluorescent and confocal laser scanning microscopes.
10. 12 and 24 well cell culture plates.
11. 25 × 75 mm Superfrost® Plus microscope slides.
12. 24 × 50 mm glass coverslips.
13. Petri dish.
14. Soft brush.
15. ProLong™ Gold Antifade Mountant with DAPI.

2.8 Fluorescence In Situ Hybridization (FISH)

1. RNAscope® Hydrogen Peroxide.
2. RNAscope® 10× Target Retrieval.
3. 100% ethanol.
4. RNAscope® Protease Plus.
5. 3-Plex negative control probe (dapB).
6. 3-Plex positive control probe.
7. Experimental probes of interest.
8. Probe diluent.
9. Multiplex Fluorescent Reagent Kit v2.
10. RNAscope® Multiplex TSA Buffer.
11. RNAscope® 50× Wash Buffer.
12. TSA® Plus fluorophores or Opal™ dyes.
13. Distilled water.
14. HybEZ™ II Oven (110 or 220 V).
15. HybEZ™ Humidity Control Tray (with lid).
16. ACD EZ-Batch™ horizontal 20-slide holder.
17. HybEZ™ Humidifying Paper.
18. Steamer capable of heating to 100 °C.
19. DAPI solution.
20. ProLong™ Gold Antifade Mountant.
21. Vertical 24-slide rack.
22. Solution container.
23. 3 L or larger Carboy.
24. Water bath or incubator capable of holding temperature at 40 ± 1 °C.

25. 0.5 and 1.5 mL Eppendorf tubes.
26. Mechanical pipettes.
27. Pipette tips.
28. Paper towels.
29. 25 × 75 mm Superfrost® Plus microscope slides.
30. 24 × 50 mm glass coverslips.
31. Immedge™ hydrophobic barrier pen.
32. Fluorescent and confocal laser scanning microscopes.

3 Methods

3.1 Virus Preparation

1. Calculate the volume of virus solution to be used for injection:
   (a) Mouse: 0.5, 1.5, 3, or 5 μL per side.
   (b) Rat: 15 μL per side.
   (c) Minipig: 100–200 μL per side.
   (d) NHP: 100–300 μL per side.
2. Prepare the virus solution by mixing 1:1 with 5% dextran solution, for a final dextran concentration of 2.5%.
3. Store on ice until use.

3.2 Subpial Infusion: Rodents

3.2.1 Rodent Anesthesia

1. Use an induction chamber to anesthetize animals with 5% isoflurane in 50% air/50% oxygen at a flow rate of 1 L/min.
2. Maintain deep anesthesia in 2–3% isoflurane by using a nose mask during surgery.
3. Evaluate the anesthetic depth by monitoring the breathing rate and paw pinch response during the procedure and modify the isoflurane percentage accordingly.

3.2.2 Rodent Surgical Procedure (See Note 6)

1. Using an animal clipper, shave the back of the animals, clean the skin 2 or 3 times with 2% chlorhexidine and swab the skin with sterile povidone-iodine swab sticks. Use sterile technique.
2. Perform a skin incision at Th8-L1 vertebrae level with a scalpel. Once exposed, detach the paravertebral muscle surrounding the Th10-Th12 vertebrae using scissors and a small osteotome (Fig. 1a).
3. Mount the animal in a stereotaxic frame and gently fix the lumbar region using the spinal clamps (*see* **Note** 7). Under the dissecting microscope, perform a dorsal laminectomy of the Th10-Th12 vertebrae using a dental drill. Once cracks appear in vertebrae, remove the bone fragments with forceps, allowing visualization of the lumbar spinal cord surface of L1-L2 segments (Fig. 1b, c).

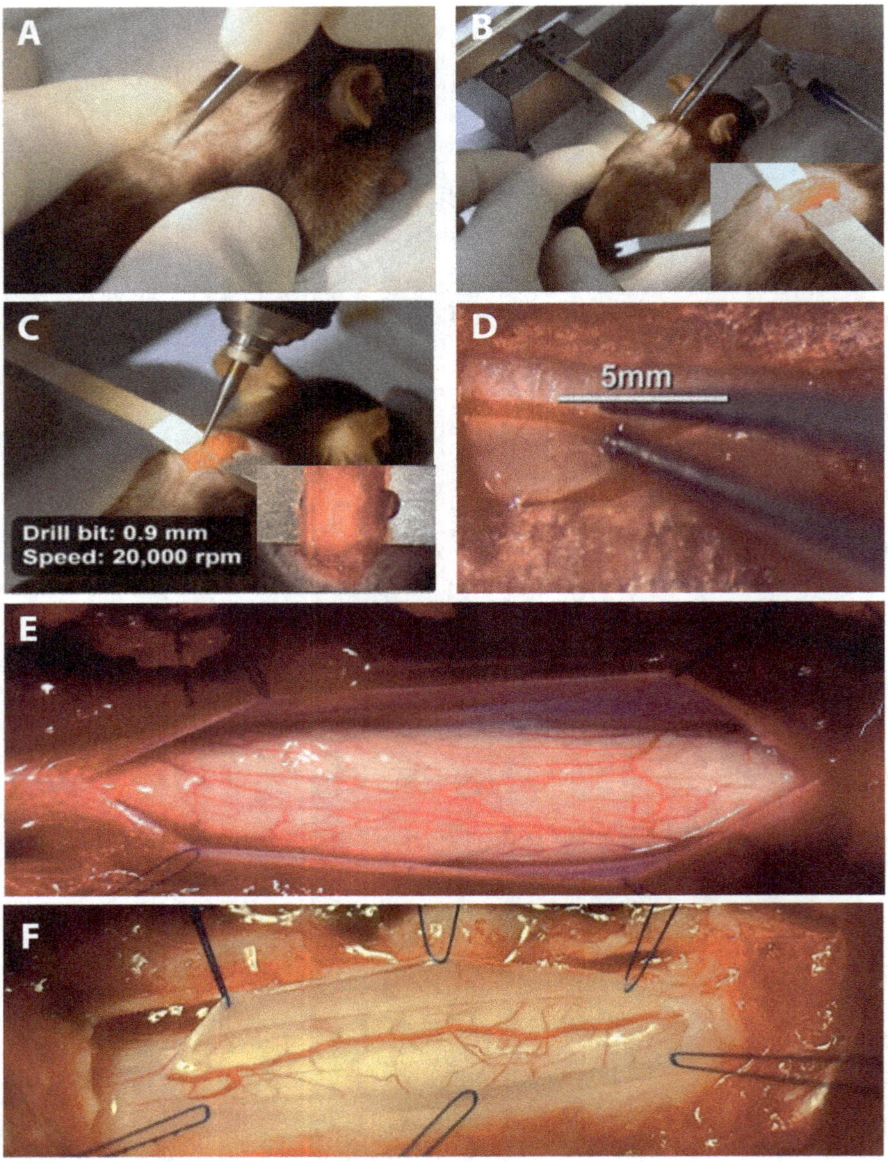

Fig. 1 Surgical approach. (**a–d**) Representative images of the surgical approach to perform lumbar laminectomy and expose the pia membrane in mice. (**e**, **f**) Representative images of the exposed minipig lower thoracic (**e**) or NHP cervical (**f**) spinal cord. The dura was cut open and secured to the surrounding soft tissue. The surface of the pia mater is readily recognizable. Derived from [13] with permission from *Molecular Therapy — Methods & Clinical Development*

4. Cut open and remove 5 mm of dura mater using a 30G sterile stainless steel needle and forceps (*see* **Note 8**) (Fig. 1d).

3.2.3 Rodent Virus Delivery

1. Mount the sterilized 34G pia-penetrating needle onto the Z-arm of an XYZ manipulator using a glass capillary holder (Fig. 2a, b) (*see* **Note 3**).

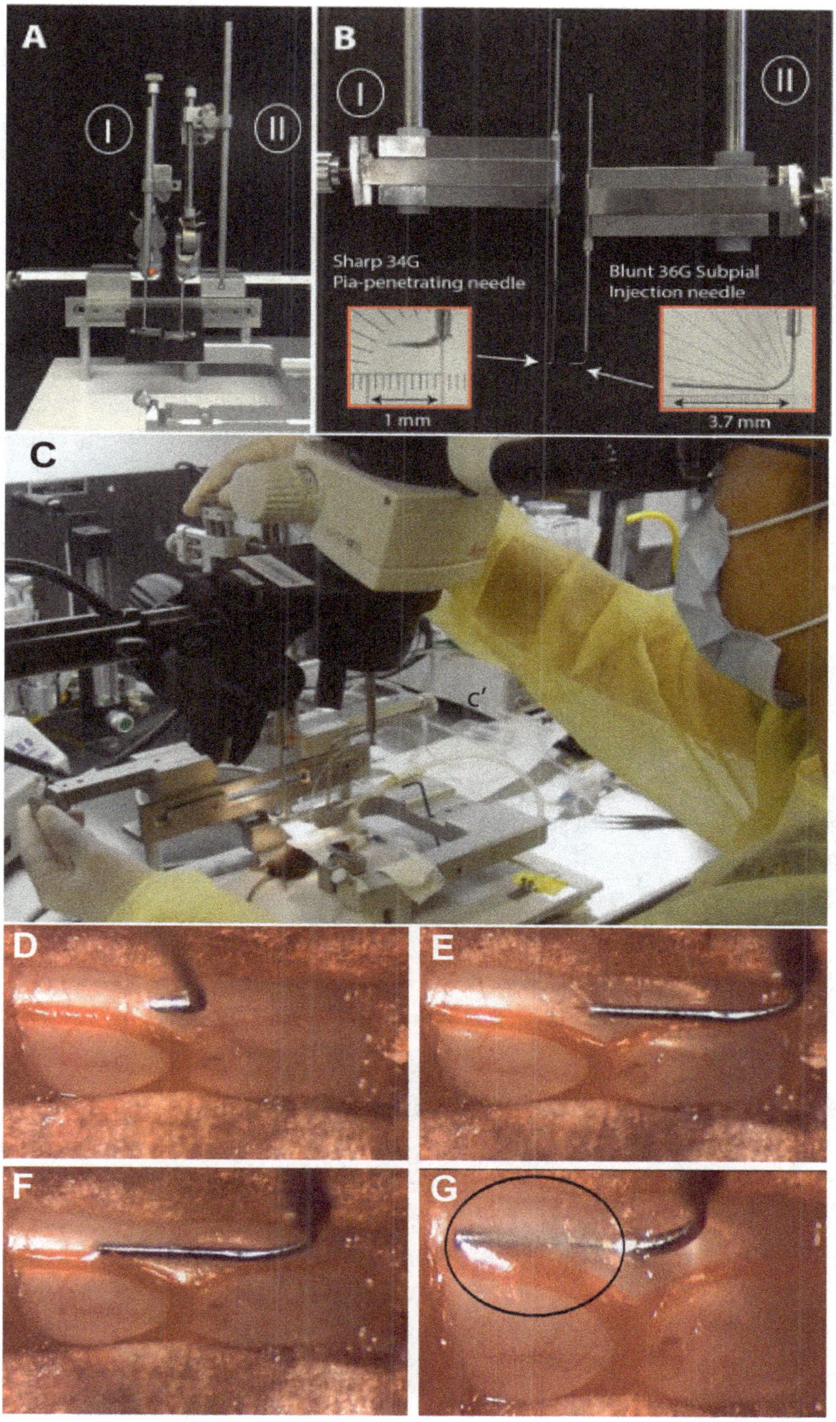

Fig. 2 Stereotaxic setup for subpial injection in adult rodents. (**a**) Two independent XYZ manipulators are mounted on a modified stereotaxic base (I, II). (**b**) The Z arm of each manipulator has a glass capillary holder holding the pia-penetrating needle and the subpial injection needle (I, II). (**c**) Representative image of the complete setup used for subpial injection. (**d–g**) Representative sequence of images showing the penetration of the pia membrane (**d**), placement and advancement of the injection needle in the subpial space (**e**, **f**), and visible expansion of subpial space during virus delivery (**g**) (black circle). Derived from [14] with permission from *JoVE*

2. Under the dissecting microscope (8–10× magnification), use the XYZ manipulator to move the pia-penetrating needle close to the spinal cord and make a 1-mm-long perforation in the pia mater (*see* **Note 9**). Use the X arm to remove the pia-penetrating needle horizontally from the subpial space (*see* **Note 10**) (Fig. 2c, d).
3. Using a manual Hamilton syringe pump and 50 μL Hamilton syringe, load the blunt 36G injection needle with AAV9-UBI-GFP virus (*see* **Notes 11** and **12**).
4. Mount the injection needle onto a second XYZ manipulator using a second glass capillary holder. Using the XYZ manipulator, position the tip of the injection needle into the pial perforation made in **step 2**, and advance 2–3 mm into the subpial space using the X-arm (Fig. 2e, f).
5. Inject the desired volume of AAV9-UBI-GFP into the subpial space using the 50 μL Hamilton syringe and pump. Remove the injection needle from the subpial space after the virus injection is completed (*see* **Notes 11** and **12**) (Fig. 2g).
6. Remove the spinal clamps, loosely close the muscle using a polypropylene 4-0 suture, and close the skin incision with surgical clips. Apply antibiotic ointment (BNP) to the wound.

3.2.4 Rodent Postoperative Care and Pain Management

1. After the surgical procedure, allow the animals to recover on a heating pad.
2. Administer a single subcutaneous (SC) injection of 1 mL (mice) or 3–5 mL (rats) of 0.9% sterile NaCl for rehydration.
3. Administer a single SC antibiotic injection of 100 mg/kg cefazolin.
4. For pain management, administer an SC injection of 0.05 mg/kg buprenorphine every 12 h for 3 days post-surgery.

3.3 Subpial Infusion: Minipigs

3.3.1 Minipig Anesthesia

1. Premedicate the animals with an intramuscular (IM) injection of 2 mg/kg azaperonum and 1 mg/kg atropine. Allow them to relax. Induce anesthesia with an intravenous (IV) injection of 20 mg/kg ketamine.
2. Intubate the animal using a 2.5 F endotracheal tube. Maintain anesthetic depth with 1.5% isoflurane in 50% air/50% oxygen at a constant flow rate of 2 L/min. Monitor anesthetic depth and oxygen saturation throughout the procedure with a pulse oximeter.

3.3.2 Minipig Surgical Procedure (See Note 6)

1. Wash the back of the animals with surgical soap and water at least 3 times.

2. Place the animal into a prone position on the operating table. To immobilize the lumbar spinal cord, use a spinal immobilization apparatus [15] (*see* **Note 7**).
3. Clean the surgical area with povidone-iodine and 70% ethanol at least three times, then position the surgical drapes around the spinal cord.
4. Perform a dorsal laminectomy of L2-L4 vertebrae, corresponding to the L3-L6 spinal segments, and remove the epidural fat using cotton swabs.
5. Cut open the dura mater and secure it to the surrounding tissue using 6.0 polypropylene suture (*see* **Note 8**) (Fig. 1e).

3.3.3 Minipig Virus Delivery

1. Mount the sterilized 25G pia-penetrating needle onto the Z arm of a modified XYZ manipulator (*see* **Notes 3** and **4**).
2. Under the dissecting microscope (8–10× magnification), use the XYZ manipulator to move the pia-penetrating needle close to the spinal cord surface and make a 1-mm-long perforation in the pia mater (*see* **Note 9**). Use the X arm to remove the pia-penetrating needle horizontally from the subpial space (*see* **Note 10**).
3. Using an automatic Hamilton syringe pump and 250 μL Hamilton syringe, load the blunt 30G injection needle with AAV9-UBI-GFP, AAV9-CAG-GFP, AAV9-UBI-RFP, or Anc80-UBI-GFP (*see* **Notes 11** and **13**).
4. Mount the injection needle onto a second modified XYZ manipulator. Using the arms of the manipulator, position the tip of the injection needle into the pial perforation made in **step 2**, and advance 0.4–0.5 cm into the subpial space using the X arm.
5. Inject the desired volume of virus into the subpial space using the 250 μL Hamilton syringe and automatic pump at an infusion rate between 1 and 5 μL per min. Remove the injection needle from the subpial space after the virus injection is complete (*see* **Notes 11** and **13**).
6. Remove the spinal clamps, close the dura matter with 6-0 polypropylene suture, loosely close the muscle using 4-0 polypropylene suture, and close the skin incision with surgical clips or 2.0 nonabsorbable suture.

3.3.4 Minipig Postoperative Care and Pain Management

1. After the surgical procedure, allow the animals to recover on a heated pad.
2. Administer a single IM injection of 1 mg/kg prednisolone.
3. Administer an SC antibiotic injection of 3 mg/kg Ceftiofur.

4. For pain management, administer an SC injection of 0.05 mg/kg buprenorphine every 12 h for 3 days post-surgery.

3.4 Subpial Infusion: Nonhuman Primates

3.4.1 NHP Anesthesia

1. Premedicate the animals with an IM injection of 15–30 μg/kg dexmedetomidine, 0.01–0.03 mg/kg buprenorphine, and 5–20 mg/kg ketamine. Allow them to relax. Induce them with 1–5% isoflurane.
2. Intubate the animal using a 3–3.5 F endotracheal tube. Maintain anesthetic depth with 1–2% isoflurane in 50% air/50% oxygen at a constant flow rate of 2 L/min. Monitor anesthetic depth and oxygen saturation throughout the procedure with a pulse oximeter.

3.4.2 NHP Surgical Procedure (See Note 14)

1. Using an animal clipper, shave the dorsal cervical area, and wash it with surgical soap.
2. Place the animal on the operating table in a prone position.
3. Immobilize the head of animal by placing it into a stereotaxic frame. Place physiologic monitoring equipment on the animal. Adjust the angle of the cervical spine to the desired position by placing a rigid pillow on the ventral aspect of the neck.
4. Clean the surgical area with povidone-iodine and 70% ethanol at least three times, and then position the surgical drapes around the spinal cord.
5. Perform a dorsal skin incision at C1-C4 level with a scalpel. Once exposed, detach the paravertebral muscle surrounding the C1-C4 vertebrae using scissors and osteotome.
6. Immobilize the cervical spinal cord with a spinal clamp secured to C3 or C5 spinal processes (*see* **Note 7**).
7. Perform a dorsal laminectomy of C1-C2 vertebrae, corresponding to the first two spinal segments, and remove the epidural fat using cotton swabs.
8. Cut open the dura mater with a scalpel blade and secure it to the surrounding tissue using 5.0 or 6.0 polypropylene suture (*see* **Note 8**) (Fig. 1f).

3.4.3 NHP Virus Delivery

1. Follow the same injection procedure used in minipigs (Subheading 3.3.3).

3.4.4 NHP Postoperative and Pain Management

1. After the surgical procedure allow the animals to recover on a heated pad.
2. Administer a single IM injection of 2 mg/kg prednisolone.
3. Administer SC antibiotic injections of 20 mg/kg cefazolin every 12 h for 5 days post-surgery.
4. For pain management administer SC injections of 0.05 mg/kg buprenorphine every 12 h for 3 days post-surgery.

3.5 Tissue Processing for Detection of Transgene Expression

Qualitative and quantitative analysis of transgene expression in spinal cord tissue can be performed by: (1) immunohistochemistry (IHC) and western blot to detect AAV-mediated protein expression, (2) fluorescence in situ hybridization (FISH) to detect regional changes AAV-mediated mRNA expression, and (3) quantitative PCR (qPCR) to quantitatively measure transgene mRNA.

The postinjection survival time required to achieve a detectable level of transgene expression depends on several factors, including the serotype of AAV vector used, the promoter, the vector titer, the injected volume, the administration route, and other factors [16]. Here we used AAV9 with a ubiquitin (UBI) or CAG promoter encoding enhanced green fluorescence protein (eGFP) or red fluorescence protein (RFP), and determined the efficacy and distribution of the transgene expression 2–6 weeks after subpial delivery (Fig. 3). The final injected titers were 6×10^{12} vg/mL.

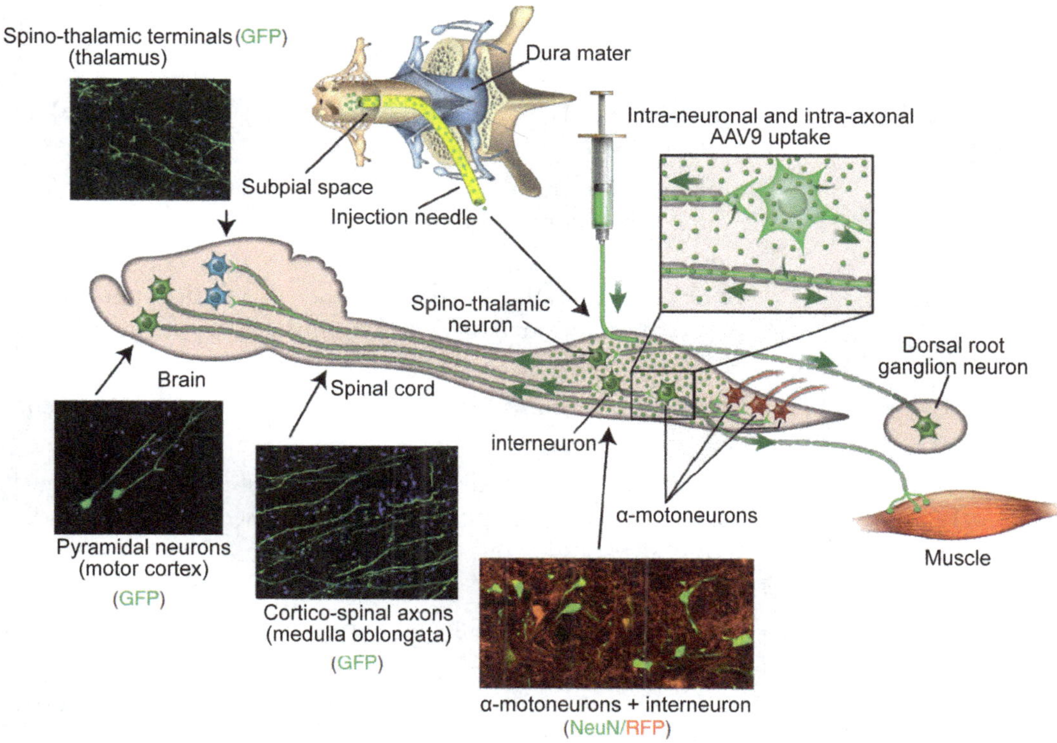

Fig. 3 Schematic drawing of subpial AAV9-GFP delivery to the lumbar spinal cord and resulting transgene expression. We observe AAV9 diffusion and uptake into segmental neurons, resulting in labeling of ascending and descending axon tracts distant from the injection site. Pyramidal neurons in motor cortex and dorsal root ganglia are retrogradely transduced, while motor neuron axons innervating muscle and spinothalamic neurons in the brain are also transduced. Reproduced from [13] with permission from *Molecular Therapy — Methods & Clinical Development*

1. Euthanize the animal and immediately transcardially perfuse cold freshly prepared heparinized saline or PBS 1× (*see* **Note 15**).
2. Dissect the spinal cord and brain, carefully immerse them in cold freshly prepared 4% PFA, and store at 4 °C overnight (*see* **Note 16**).
3. Transfer tissues to 30% cold sucrose solution and store at 4 °C for 72 h.
4. Remove tissues from sucrose and rinse them with PBS 1×. Cut to detach the desired region for analysis (brain or cervical, thoracic, or lumbar spinal cord), embed it in OCT, and flash-freeze using dry ice and isopentanol or liquid nitrogen.
5. Then equilibrate tissue at −20 °C for 10 min. Using a cryostat, cut frozen tissue sections at the desired thickness. For IHC, cut tissue sections at 30 μm and store free-floating in PBS 1× at 4 °C. For FISH, cut tissue sections at 10–15 μm, mount them directly on RNase-free slides, and store at −80 °C (*see* **Note 17**).

3.6 Immunohistochemistry

1. Transfer the sections into a 48 well plate and wash 3× 10 min with PBS-TX and gentle agitation.
2. Incubate the sections in blocking solution for 30–60 min at room temperature (RT) with gentle agitation.
3. Dilute primary antibodies to the desired concentration in blocking solution. Incubate the tissue sections in primary antibody solution at 4 °C overnight with gentle agitation (*see* **Note 18**).
4. Wash the sections 3× 10 min with PBS-TX and gentle agitation.
5. Dilute secondary antibodies to the desired concentration in blocking solution. Incubate the tissue sections in secondary antibody solution at RT for 1 h with gentle agitation in the dark.
6. Wash the sections 3× 10 min with PBS-TX and gentle agitation.
7. Mount the sections on slides using PBS 1× solution in a petri dish and a soft brush. Allow the slides and sections to dry, then add a drop of mounting medium (Prolong® antifade with DAPI). Cover the sections with a coverslip, avoiding bubbles. Store coverslipped slides in the dark at −20 °C and analyze using a fluorescent microscope (Figs. 4 and 5).

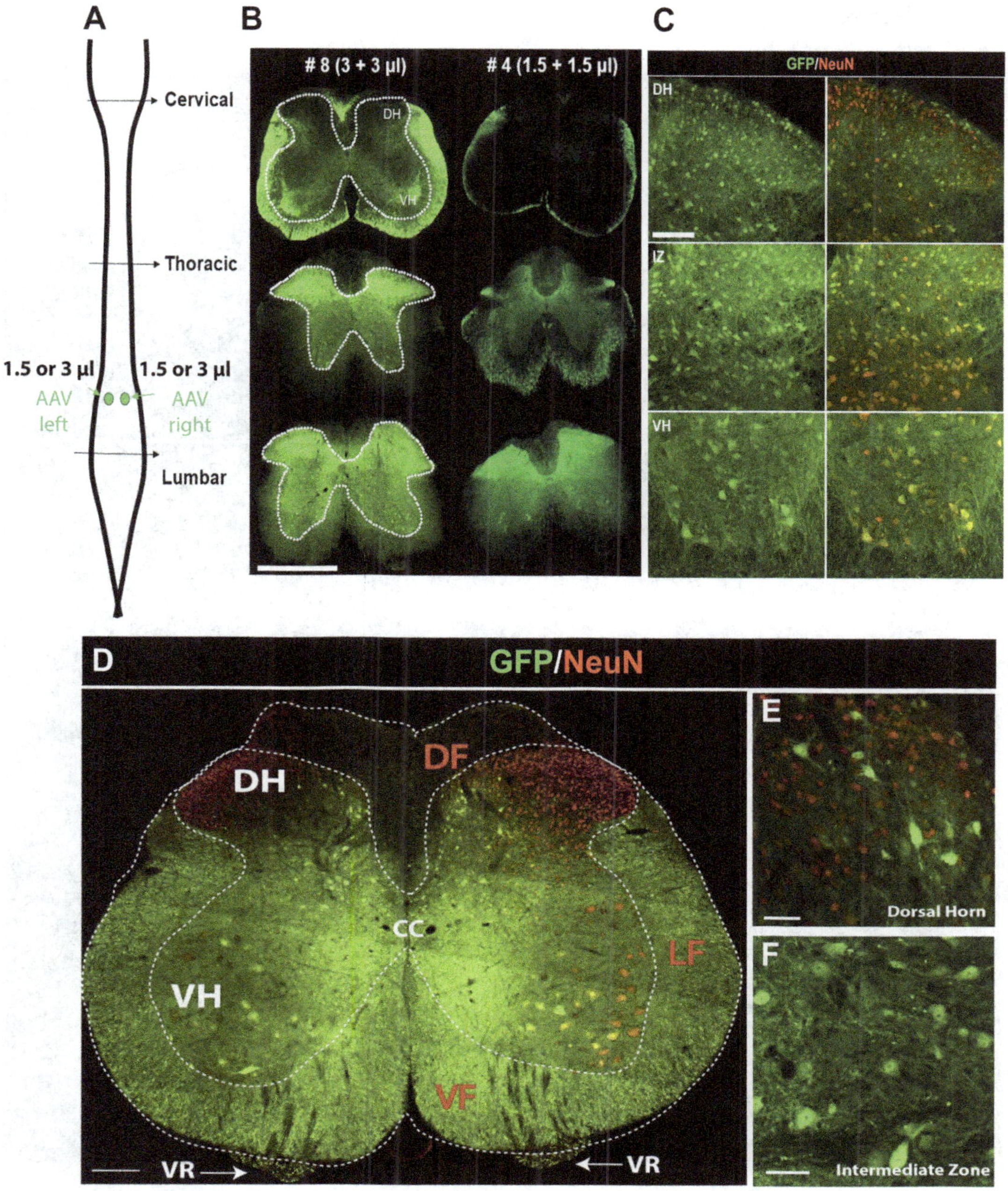

Fig. 4 Spinal parenchymal GFP expression after lumbar subpial AAV9-UBI-GFP delivery in rodents. (**a**) Schematic of bilateral injection in the upper lumbar region delivering 1.5 or 3 μL of virus. (**b**) Volume-dependent GFP expression in gray matter (white dots) and white matter is strongest in lumbar segments but can also be seen in thoracic and cervical segments. (**c**) High-magnification images of lumbar dorsal horn (DH), intermediate zone (IZ), and ventral horn (VH) show colocalization (yellow) of GFP transgene expression (green) with almost all NeuN-positive neurons (red). (**d**) Representative image of lumbar spinal cord in rat showing widespread intense GFP expression after subpial delivery. GFP expression (green) was found to colocalize (yellow) with NeuN-positive neurons (red) throughout the gray matter, including interneurons in the DH (**e**) and the IZ (**f**), as well as α-motoneurons in the VH. Consistent with α-motoneuronal GFP expression, high GFP expression was seen bilaterally in transversely cut motor axons in the ventral roots (VR), lateral funiculi (LF), and ventral funiculi (VF). Scale bars: **b** = 1000 μm, **c** = 100 μm, **d** = 200 μm, **e** = 50 μm, **f** = 20 μm. Derived from [13, 14] with permission from *Molecular Therapy — Methods & Clinical Development* and *JoVE*, respectively

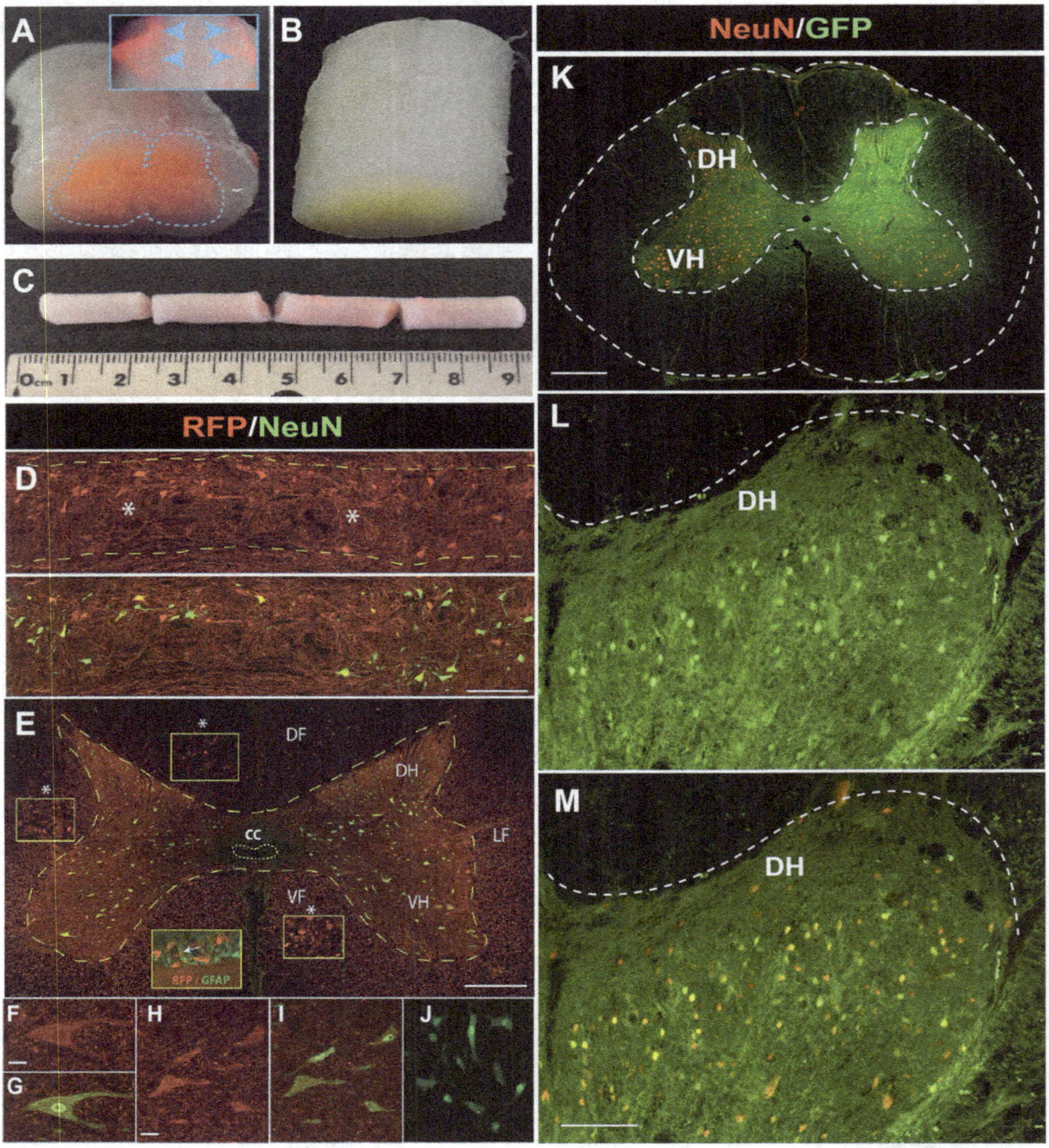

Fig. 5 Spinal parenchymal RFP and GFP expression after thoraco-lumbar subpial AAV9-UBI-RFP or Anc80-UBI-GFP delivery in adult minipigs. (**a**, **c**) Intense RFP fluorescence was detected macroscopically in minipig thoracic spinal cord at 6 weeks after AAV9-UBI-RFP injection, including the ventral roots (**a**, inset). No fluorescence signal in naïve (non-injected) spinal cord can be seen (**b**). (**d**) Horizontal mid thoracic spinal cord sections showing intense RFP expression throughout the whole white and gray matter. Colocalization of RFP-positive neurons (red) with virtually all NeuN-stained positive neurons (green) was observed. (**e**) Transverse spinal cord section showing transversely cut RFP-positive axons in dorsal (DF), lateral (LF) and ventral funiculus (VF) (high-magnification images, yellow box insets). GFAP-positive astrocytes also showed RFP-positive signal (**e**, inset). (**f–i**) High-magnification images in the VH of RFP-expressing α-motoneurons and interneurons. (**j**) Representative image from upper-thoracic spinal cord of a naïve animal showing clear NeuN staining with no detectable RFP signal. (**k–m**) Unilateral dorsal horn GFP expression is observed 4 weeks after unilateral subpial delivery of 50 μL of Anc80-UBI-GFP. Note the highly effective transgene expression in small dorsal horn NeuN-positive neurons. Scale bars: **d**, **e**, **k** = 500 μm; **f–j** = 30 μm; **m** = 200 μm. Derived from [13] with permission from *Molecular Therapy — Methods & Clinical Development*

3.7 Fluorescence In Situ Hybridization (See Note 19)

3.7.1 Pretreatment

1. Defrost the slides for FISH processing. Load slides on a vertical 24-slide rack and wash with PBS 1× in a solution container for 5 min. Move the slide rack up and down to remove OCT.
2. Lay the slides on the bench, let them dry for 5 min, and draw a hydrophobic barrier around the tissue using the Immedge™ hydrophobic pen.
3. Add 5–8 drops of RNAscope® hydrogen peroxide until all tissue is covered, then incubate for 10 min at RT.
4. Remove hydrogen peroxide by tapping the slide on absorbent paper. Immediately load slides on a vertical 24-slide rack and submerge the slide rack in a solution container filled with distilled water. Move the slide rack up and down 3–5 times to wash. Repeat 1 more time with clean distilled water.
5. Prepare for target retrieval by warming a steamer to 99 °C and preparing solution containers (*see* **Note 20**).
6. To perform target retrieval, transfer the slide rack to the prewarmed solution container filled with distilled water and submerge for 10 s, then move the slide rack to the prewarmed solution container filled with RNAscope® 1× target retrieval reagent. Cover the steamer and incubate the slides for 5–10 min (*see* **Note 21**).
7. Transfer the slide rack to a solution container filled with distilled water at RT and rinse them for 15 s.
8. Transfer the slide rack to 100% ethanol and incubate for 3 min.
9. Dry the slides at RT for 1 h or overnight. This is an optional stop point.
10. Perform RNAscope® protease plus treatment. Place dry slides on a horizontal 20-slide holder and add 5 drops of RNAscope® protease plus to cover all tissue sections. Place the horizontal holder inside of the humidity control tray (*see* **Notes 21** and **22**).
11. Place the horizontal 20-slide holder with humidity control tray into the HybEZ II oven and incubate for 30 min at 40 °C (*see* **Note 22**).
12. Remove the slides from the oven and remove excess liquid by tapping and flicking each slide on absorbent paper. Immediately load slides on a vertical 24-slide rack, transfer to a solution container filled with distilled water, and wash slides 3–5 times by moving the slide rack up and down.
13. Proceed immediately to hybridization (*see* **Note 23**).

3.7.2 Hybridization

1. Remove excess liquid from the slides, place the slides in a horizontal 20-slide holder, and add 4–6 drops of probe mix to cover all tissue sections (*see* **Note 23**). Place the slide holder

into the humidity control tray, insert the tray into the oven, and incubate for 2 h at 40 °C.

2. Remove the slides from the oven, place in vertical 24-slide rack, and wash in a solution container filled with 1× wash buffer for 2 min at RT. Repeat 1 more time with clean 1× wash buffer.
3. Remove excess liquid from the slides, place slides on a horizontal 20-slide holder, and add 4–6 drops of RNAscope® Multiplex FL v2 Amp 1 to cover all tissue sections. Place the slide holder into the humidity control tray and insert into the oven and incubate for 30 min at 40 °C.
4. Remove the slides from the oven, place in vertical 24-slide rack, and wash in a solution container filled with 1× wash buffer for 2 min at RT. Repeat 1 more time with clean 1× wash buffer.
5. Remove excess of liquid from the slides, place slides on a horizontal 20-slide holder, and add 4–6 drops of RNAscope® Multiplex FL v2 Amp 2 to cover the sections. Place the slide holder into the humidity control tray and insert into the oven and incubate for 30 min at 40 °C. Remove the slides from the oven, place in vertical 24-slide rack, and wash in a solution container filled with 1× wash buffer for 2 min at RT. Repeat 1 more time with clean 1× wash buffer.
6. Remove the excess of liquid from the slides, place slides on a horizontal 20-slide holder, and add 4–6 drops of RNAscope® Multiplex FL v2 Amp 3 to cover all tissue sections. Place the slide holder into the humidity control tray and insert into the oven and incubate for 15 min at 40 °C.
7. Remove the slides from the oven, place in vertical 24-slide rack, and wash in a solution container filled with 1× wash buffer for 2 min at RT. Repeat 1 more time with clean 1× wash buffer.

3.7.3 Development

1. Prepare TSA® Plus fluorophore dilutions (*see* **Note 24**). Prepare three different fluorophore dilutions: fluorescein, Cy3, and Cy5.
2. Determine the volume of TSA® Plus needed (around 150 μL per slide) and dilute each TSA® stock in TSA buffer (dilutions may vary from 1:100 to 1:3000) (*see* **Note 24**).
3. Remove excess liquid from the slides, place slides on a horizontal 20-slide holder, and add 4–6 drops of the RNAscope® Multiplex FL v2 HRP-C1 to cover all tissue sections. Place the slide holder into the humidity control tray and insert into the oven and incubate for 15 min at 40 °C.
4. Remove the slides from the oven, place in vertical 24-slide rack, and wash in a solution container filled with 1× wash buffer for 2 min at RT. Repeat 1 more time with clean 1× wash buffer.

5. Remove excess liquid from the slides, place slides on a horizontal 20-slide holder, and add 150 μL of diluted TSA® Plus fluorophore. Place the slide holder into the humidity control tray and insert into the oven and incubate for 30 min at 40 °C.
6. Remove the slides from the oven, place in vertical 24-slide rack, and wash in a solution container filled with 1× wash buffer for 2 min at RT. Repeat 1 more time with clean 1× wash buffer.
7. Remove excess liquid from the slides, place slides on a horizontal 20-slide holder, and add 4–6 drops of the RNAscope® Multiplex FL v2 blocker to cover the sections. Place the slide holder into the humidity control tray and insert into the oven incubate for 15 min at 40 °C.
8. Remove the slides from the oven, place in vertical 24-slide rack, and wash in a solution container filled with 1× wash buffer for 2 min at RT. Repeat 1 more time with clean 1× wash buffer.
9. To develop a second fluorophore, repeat **steps 3–8** but use RNAscope® Multiplex FL v2 HRP-C2 instead of HRP-C1 in **step 3**, and use a different TSA plus fluorophore in **step 5**.
10. To develop a third fluorophore, repeat **steps 3–8** but use RNAscope® Multiplex FL v2 HRP-C3 instead of HRP-C1 in **step 3**, and use a third different TSA plus fluorophore in **step 5**.

3.7.4 Counterstain and Mounting

1. Remove excess liquid from the slides and add 4 drops of DAPI to each slide. Incubate for 30 s at RT.
2. Remove DAPI and immediately place 2 drops of mounting medium (Prolong Gold Antifade), then cover with a coverslip, avoiding air bubbles.
3. Dry slides in the dark 30 min to overnight and store at 2–8 °C. Image the slides after 8 h and within 2 weeks (Fig. 6).

3.8 Microscopy

1. Before performing IHC or FISH, we recommend examining the tissue sections under a fluorescence microscope at 10× and 20× magnification to assess the quality of the tissue and cell morphology. After IHC or FISH, we perform conventional fluorescence microscopy, fluorescence microscopy using an apotome accessory on a Zeiss AxioImager M2 Microscope, or confocal microscopy using an Olympus FV1000 microscope. We use the standard combination of 3 microscope filter cubes: GFP/FITC, Cy3, and Cy5 alternating with DAPI to visualize counter-staining of the nuclei.
2. Fluorescence microscopy can be used to capture mosaic images at 10× magnification, producing a reconstructed multichannel image of the entire transverse or longitudinal tissue section from spinal cord or brain.

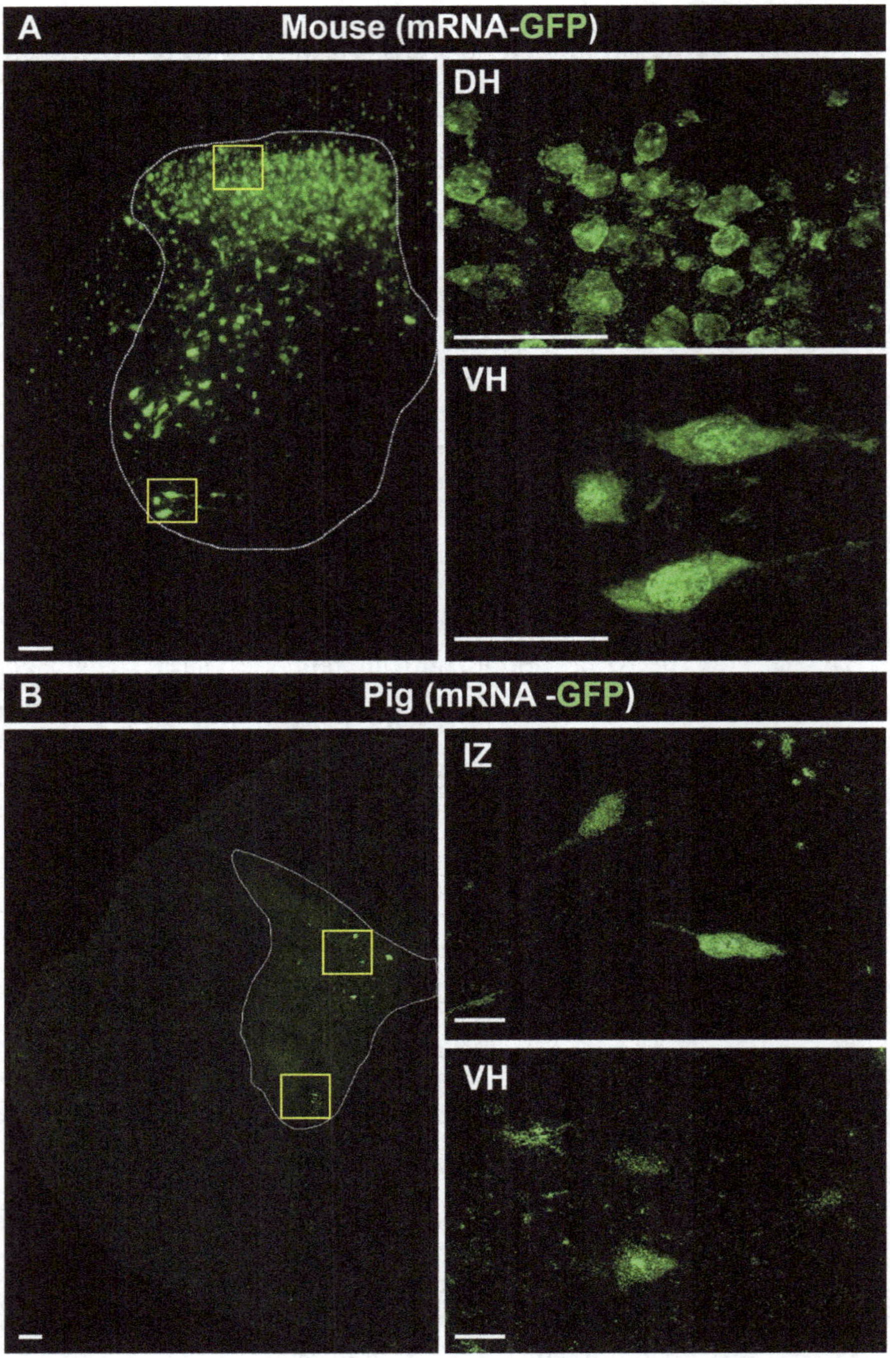

Fig. 6 Fluorescent in situ hybridization (FISH) for GFP mRNA. Animals were sacrificed 14 days or 7 weeks after lumbar subpial AAV9-UBI-GFP or AAV9-CAG-GFP delivery to mouse or minipig, respectively. (**a**) Representative FISH images showing the distribution of mRNA GFP expression in dorsal and ventral horn neurons in mouse L2 spinal cord. (**b**) Representative FISH images in minipig L1 spinal cord depicting positive neurons in the intermediate zone (IZ) and ventral horn (VH). Scale bars: **a** and **b** = 100 μm, insets = 50 μm

3. The apotome-equipped fluorescence microscope and the confocal microscope can be used to control the depth of field and to collect 0.2–0.5 μm serial optical sections at 40× and 63× (oil immersion) magnification, imaging in detail the localization and pattern of the target protein or mRNA. A high-resolution, high-sensitivity cooled CCD camera and image analysis software such as Image J, FV10-ASW version 4, Nuance®, and Photoshop can be used to capture, visualize, and quantify images.

4 Notes

1. The minimum body weight of mice for this procedure is 20 g. Rat pups of similar body weight can also be used. Adjust instrument size and materials accordingly. Screen NHP and pigs for the presence of preexisting neutralizing antibodies (NAbs) against the AAV serotype to be used, and select animals with low neutralizing antibody titers to assure the best possible transgene expression.
2. The scAAV9-GFP plasmids consist of the flanking inverted terminal repeats (ITRs) from AAV2, the 1. 2 kb ubiquitin-C (UBI) or CAG promoter, the eGFP or DsRed (RFP) cDNA, and the SV40 polyadenylation sequence. Vector was produced by transient transfection of HEK293T cells with the vector plasmid, pRep2/Cap9 plasmid, and pAd-helper plasmid [17]. Anc80-UBI-GFP is an in silico-designed synthetic Anc80 AAV vector [18].
3. To manufacture the pia-penetrating needle for rodents, sharpen the beveled tip of the 34G needle using a glass capillary beveller with coarse diamond abrasive plate (5.0–50 μm tip sizes) at an angle of 15–20°. Then bend the tip of the needle (1 mm length) to approximately 90° (Fig. 2b). To manufacture the subpial AAV9 injection needle for rodents, blunt the tip of a 36G needle and polish using a glass capillary beveller with coarse diamond abrasive plate (5.0–50 μm tip sizes). Then gently bend the tip of the needle (2–3 mm length) to approximately 90° (Fig. 2b). Insert the pia-penetrating and subpial injection needles into 1–2 cm-long sections of 20G slave stainless steel tubing (10 mm from the end of the needle) and glue with epoxy. The use of 20G (0.91 mm O.D.) tubing is required for secure attachment to the glass capillary holder.

   To manufacture the pia-penetrating needle for minipigs and NHPs, sharpen the beveled tip of the 25G needle using a glass capillary beveller with coarse diamond abrasive plate (5.0–50 μm tip sizes) at an angle of 15–20°. Then gently bend the tip of the needle (2–3 mm length) to about 30–40°.

To manufacture the subpial AAV9 injection needle for pigs and NHP, blunt the tip of a 30G needle and polish using a glass capillary beveller with coarse diamond abrasive plate (5.0–50 μm tip sizes). Then gently bend the tip of the needle (3–5 mm length) to about 40–50°. Insert the pia-penetrating needle into a 22–23-cm-long section of 20G slave stainless steel tube (10 mm from the end of the needle) and glue with epoxy. Then insert this slave into a 16–24-cm-long external stainless steel tube holder (3 mm in diameter). To assemble the injection apparatus, insert the injection needle into polyethylene tubing (PE 10), glue into a 20G slave stainless steel tube, and insert the slave tube into a stainless steel tube holder, as above. The use of this external tube is required for a secure attachment to the Z arm of the XYZ manipulator.

4. There is a rotating knob attached firmly to the top of the external stainless steel tube holders that allows for rotation of the needles. Attach the pia-penetrating needle and the injection needle to the Z arms of two independent XYZ manipulators at a 15° angle from each other.
5. Primary and secondary antibodies will vary depending on the purposes of the study. In the examples shown in Figs. 4 and 5, we used neuronal marker chicken anti-NeuN (1:1000 dilution, Millipore #ABN91). No anti-GFP or RFP antibodies were used.
6. If cervical subpial injections will be performed in rodents and minipigs, make a 1.5–2 cm incision in the dorsal neck skin using scissors or a scalpel, and expose the C3-C4 vertebrae. Remove the spinous process and lamina on both sides of the vertebrae with forceps (mice) or with the dental drill (rats and minipigs) in order to expose the cervical spinal cord surface. Clean any tissue and bone debris from the incision site using cotton swabs and proceed as described in Subheading 3.2.
7. It is very important to properly place the anesthetized animal in the spinal frame, and to immobilize the spinal column using spinal clamps. Using too much pressure can cause a fracture of the vertebral bone, while using insufficient pressure can lead to instable fixation of the spinal column and risk damaging the spinal cord tissue during laminectomy.
8. After removing the dura, it is critical to keep the surgical site free of blood by electrocauterization (minipigs and NHPs) or by using small pieces of GelFoam sponges or surgical drapes to cover the surrounding tissue. Blood residue may hinder identification of the pia surface.
9. It is important to keep an angle of 5–10° between the pia-penetrating needle and the tissue surface to avoid

damaging the spinal cord parenchyma, and to ensure correct penetration of pia mater.

10. It is helpful to use a landmark, such as a blood vessel, to identify the site of pial perforation.
11. Connect the Hamilton syringe to the injection needle with PE-10 or PE-20 tubing. Mark every 10 μL on the tubing. Use these marks to monitor and control the infusion volume when a manual injector is used.
12. The volume depends on the purpose of the injection, as well as whether it is unilateral or bilateral (based on the spinal central vein). For mice, we recommend 1.5, 3, or 5 μL per side.
13. The volume depends on the purpose of the injection, as well as whether it is unilateral or bilateral (based on the spinal central vein). For minipigs the injection volume can range from 100 μL to 1000 μL, and for NHP from 100 μL to 600 μL.
14. If lumbar injection is performed on NHP, the head does not need to be fixed in the stereotaxic frame. As in minipigs, immobilize the lumbar region using the spinal immobilization apparatus. Perform laminectomy at L2–L4 vertebrae, corresponding to the L3–L6 spinal segments, and proceed as described in Subheading 3.4.
15. We typically perfuse rats, pigs, and NHP with 0.5–1% heparin (10,000 UI/mL) in saline, and perfuse mice with PBS 1× without heparin. However, mice can also be perfused with heparinized saline.
16. The animal may also be perfused with cold freshly prepared 4% PFA before dissection. However, this may increase the autofluorescence background in FISH-processed sections.
17. Tissue for FISH should be mounted on SUPERFROST® Plus slides and can be stored with desiccants at −80 °C for up to 3 months. However, in our experience the use of freshly fixed tissue (1–3 days after fixation) yields the most consistent results.
18. Sometimes GFP transgene expression is very strong and can be visualized under a fluorescent microscope without IHC. *See* **Note 5** for antibodies used in sections with strong endogenous GFP expression.
19. This procedure is based on the RNAscope® Multiplex Fluorescent technique developed by ACDBio. Always asses the quality of the tissue and cell morphology before performing FISH. Assess negative control probes for background and positive control probes for accurate staining before testing the desired target probes.

20. Use a steamer or a bath that can heat to at least 100 °C. Place 2 solution containers in the steamer. Fill one with RNAscope® 1× target retrieval reagent and the other with distilled water. Allow the temperature to reach at least 99 °C.

21. Tissue pretreatment may vary from 5 to 30 min depending on the type of tissue. In our hands, antigen retrieval of brain and spinal cord from mice, rat, minipigs, and NHP requires a 5–10 min incubation for optimal results, while protease plus treatment requires a 30 min incubation.

22. Turn on the oven and maintain it at 40 °C for the remainder of the procedure. Prepare the humidified filter paper and humidity control tray: place the HybEZ™ Humidifying Paper inside of the HybEZ™ Humidity Control Tray and add 50 mL of distilled water to maintain humidity during incubation. Place the horizontal 20-slide holder into the humidity control tray (with humidifying paper below the holder) for all incubation steps. This ensures that the sections do not dry during the procedure.

23. Prepare 6 L of 1× wash buffer in advance. Prewarm 2 bottles (120 mL) of RNAscope 50× wash buffer for 10–20 min at 40 °C, add to 5.88 L of distilled water, and mix well. Store at RT for up to 1 month. To prepare the probes, warm them for 10 min at 40 °C in a water bath, then cool them to RT. Probe C1 (1×) is ready to be used. Dilute probes C2 and C3 (50×) by adding 1 volume of C2 and 1 volume of C3 to 50 volumes of C1 and mixing well. Store combined probes at 2–8 °C for up to 6 months. Equilibrate AMP1-3, HRPC1-3, blocking solution, and DAPI to RT prior to use.

24. Reconstitute the TSA® plus Fluorescein, TSA® Plus Cy3, and TSA® Plus Cy5 stocks with 150 μL of DMSO. TSA® plus Fluorescein is very sensitive to the light, and must be protected from light at all times. The dilution is chosen based on optimization, the recommended starting dilution is 1:1500 and adjust based on signal intensity.

References

1. Lin D, Fantz CR, Levy B, Rafi MA, Vogler C, Wenger DA, Sands MS (2005) AAV2/5 vector expressing galactocerebrosidase ameliorates CNS disease in the murine model of globoid-cell leukodystrophy more efficiently than AAV2. Mol Ther 12(3):422–430. https://doi.org/10.1016/j.ymthe.2005.04.019
2. McCurdy VJ, Johnson AK, Gray-Edwards HL, Randle AN, Brunson BL, Morrison NE, Salibi N, Johnson JA, Hwang M, Beyers RJ, Leroy SG, Maitland S, Denney TS, Cox NR, Baker HJ, Sena-Esteves M, Martin DR (2014) Sustained normalization of neurological disease after intracranial gene therapy in a feline model. Sci Transl Med 6(231):231ra248. https://doi.org/10.1126/scitranslmed.3007733
3. Carty NC, Nash K, Lee D, Mercer M, Gottschall PE, Meyers C, Muzyczka N, Gordon MN, Morgan D (2008) Adeno-associated viral (AAV) serotype 5 vector mediated gene delivery of endothelin-converting enzyme

reduces abeta deposits in APP + PS1 transgenic mice. Mol Ther 16(9):1580–1586. https://doi.org/10.1038/mt.2008.148

4. Connor B, Sun Y, von Hieber D, Tang SK, Jones KS, Maucksch C (2016) AAV1/2-mediated BDNF gene therapy in a transgenic rat model of Huntington's disease. Gene Ther 23(3):283–295. https://doi.org/10.1038/gt.2015.113
5. Oh SM, Chang MY, Song JJ, Rhee YH, Joe EH, Lee HS, Yi SH, Lee SH (2015) Combined Nurr1 and Foxa2 roles in the therapy of Parkinson's disease. EMBO Mol Med 7 (5):510–525. https://doi.org/10.15252/emmm.201404610
6. Foust KD, Salazar DL, Likhite S, Ferraiuolo L, Ditsworth D, Ilieva H, Meyer K, Schmelzer L, Braun L, Cleveland DW, Kaspar BK (2013) Therapeutic AAV9-mediated suppression of mutant SOD1 slows disease progression and extends survival in models of inherited ALS. Mol Ther 21(12):2148–2159. https://doi.org/10.1038/mt.2013.211
7. Stoica L, Todeasa SH, Cabrera GT, Salameh JS, ElMallah MK, Mueller C, Brown RH Jr, Sena-Esteves M (2016) Adeno-associated virus-delivered artificial microRNA extends survival and delays paralysis in an amyotrophic lateral sclerosis mouse model. Ann Neurol 79 (4):687–700. https://doi.org/10.1002/ana.24618
8. Borel F, Gernoux G, Cardozo B, Metterville JP, Toro Cabrera GC, Song L, Su Q, Gao GP, Elmallah MK, Brown RH Jr, Mueller C (2016) Therapeutic rAAVrh10 mediated SOD1 silencing in adult SOD1(G93A) mice and nonhuman primates. Hum Gene Ther 27(1):19–31. https://doi.org/10.1089/hum.2015.122
9. Mendell JR, Al-Zaidy S, Shell R, Arnold WD, Rodino-Klapac LR, Prior TW, Lowes L, Alfano L, Berry K, Church K, Kissel JT, Nagendran S, L'Italien J, Sproule DM, Wells C, Cardenas JA, Heitzer MD, Kaspar A, Corcoran S, Braun L, Likhite S, Miranda C, Meyer K, Foust KD, Burghes AHM, Kaspar BK (2017) Single-dose gene-replacement therapy for spinal muscular atrophy. N Engl J Med 377(18):1713–1722. https://doi.org/10.1056/NEJMoa1706198
10. Foust KD, Poirier A, Pacak CA, Mandel RJ, Flotte TR (2008) Neonatal intraperitoneal or intravenous injections of recombinant adeno-associated virus type 8 transduce dorsal root ganglia and lower motor neurons. Hum Gene Ther 19(1):61–70. https://doi.org/10.1089/hum.2007.093
11. Foust KD, Nurre E, Montgomery CL, Hernandez A, Chan CM, Kaspar BK (2009) Intravascular AAV9 preferentially targets neonatal neurons and adult astrocytes. Nat Biotechnol 27(1):59–65. https://doi.org/10.1038/nbt.1515
12. Hinderer C, Katz N, Buza EL, Dyer C, Goode T, Bell P, Richman LK, Wilson JM (2018) Severe toxicity in nonhuman primates and piglets following high-dose intravenous administration of an adeno-associated virus vector expressing human SMN. Hum Gene Ther 29(3):285–298. https://doi.org/10.1089/hum.2018.015
13. Miyanohara A, Kamizato K, Juhas S, Juhasova J, Navarro M, Marsala S, Lukacova N, Hruska-Plochan M, Curtis E, Gabel B, Ciacci J, Ahrens ET, Kaspar BK, Cleveland D, Marsala M (2016) Potent spinal parenchymal AAV9-mediated gene delivery by subpial injection in adult rats and pigs. Mol Ther Methods Clin Dev 3:16046. https://doi.org/10.1038/mtm.2016.46
14. Tadokoro T, Miyanohara A, Navarro M, Kamizato K, Juhas S, Juhasova J, Marsala S, Platoshyn O, Curtis E, Gabel B, Ciacci J, Lukacova N, Bimbova K, Marsala M (2017) Subpial adeno-associated virus 9 (AAV9) vector delivery in adult mice. J Vis Exp (125). https://doi.org/10.3791/55770
15. Usvald D, Vodicka P, Hlucilova J, Prochazka R, Motlik J, Kuchorova K, Johe K, Marsala S, Scadeng M, Kakinohana O, Navarro R, Santa M, Hefferan MP, Yaksh TL, Marsala M (2010) Analysis of dosing regimen and reproducibility of intraspinal grafting of human spinal stem cells in immunosuppressed minipigs. Cell Transplant 19(9):1103–1122. https://doi.org/10.3727/096368910X503406
16. Naso MF, Tomkowicz B, Perry WL 3rd, Strohl WR (2017) Adeno-associated virus (AAV) as a vector for gene therapy. BioDrugs 31 (4):317–334. https://doi.org/10.1007/s40259-017-0234-5
17. Xiao X, Li J, Samulski RJ (1998) Production of high-titer recombinant adeno-associated virus vectors in the absence of helper adenovirus. J Virol 72(3):2224–2232
18. Carvalho LS, Xiao R, Wassmer SJ, Langsdorf A, Zinn E, Pacouret S, Shah S, Comander JI, Kim LA, Lim L, Vandenberghe LH (2018) Synthetic adeno-associated viral vector efficiently targets mouse and nonhuman primate retina in vivo. Hum Gene Ther 29 (7):771–784. https://doi.org/10.1089/hum.2017.154

Part III

AAV Delivery to Sensory Systems

Chapter 13

Peripheral AAV Injection for Retrograde Transduction of Dorsal Root and Trigeminal Ganglia

David C. Bloom, Zachary L. Watson, and Donna M. Neumann

Abstract

Adeno-associated Virus (AAV) vectors are useful vehicles for delivering transgenes to a number of different tissues and organs in vivo. To date, most of these applications deliver the vectors to their target by either infusion into the bloodstream or direct injection into the target tissue. Recently there has been progress in delivering AAV vectors to neurons of the peripheral nervous system (PNS) following application of vectors to the peripheral epithelium, such as the skin or eye. This delivery only requires treatment of the epithelium to access the underlying nerve termini, and following treatment the vectors are transported retrogradely to the cell bodies of these neurons in the ganglia, such as dorsal root ganglia (DRG) or trigeminal ganglia (TG). Here we describe the methodology for highly efficient transduction of mouse DRG and rabbit TG following application of AAV vectors to the foot, or to the cornea, respectively.

Key words Sensory neurons, Peripheral nerves, Gene transfer, Gene therapy, Axonal transport

1 Introduction

AAV vectors have been widely used for delivery of transgenes to multiple organs and tissues including the central nervous system (CNS) and retina (for reviews *see* [1, 2]). The most common methods of delivery have historically been systemic injection into the bloodstream or direct injection into the target tissue. For delivery of transgenes to the nervous system, these methods have the disadvantages of vector delivery to regions that may not be needed or desired (delivery to bloodstream) or damage to other regions of the nervous system as a result of the injection (injection into the CNS or PNS). It has recently been shown that efficient delivery of transgenes to the PNS is possible using AAV vectors applied to the periphery. Examples include delivery to murine DRG following application to the abraded footpad [3], murine TG following application to the abraded snout [4, 5], and rabbit TG following application to the abraded cornea [3, 6, 7]. Transduction

Michael J. Castle (ed.), *Adeno-Associated Virus Vectors: Design and Delivery*, Methods in Molecular Biology, vol. 1950, https://doi.org/10.1007/978-1-4939-9139-6_13,

efficiencies of >70–90% of neurons can be obtained using these methods.

Here we describe detailed methods for delivery of AAV vectors to the murine DRG and rabbit TG following peripheral inoculation. These methods should be adaptable to different inoculation sites in the periphery for delivery to various peripheral nerve ganglia. Our initial studies suggest that there are caveats to keep in mind, however: *first, not all delivery sites give high levels of transduction, and this varies among animal species.* While applying AAV to the mouse footpad results in very efficient transduction of the DRG [3], we have not been able to transduce more than 10% of TG neurons following ocular inoculation of mice (unpublished data). However, snout inoculation of the mouse efficiently transduces the TG [5], though it remains unclear how neurons of the ophthalmic branch are transduced via this route of inoculation. In contrast to the low transduction efficiencies following ocular inoculation in the mouse, we observe very efficient transduction of rabbit TG following corneal inoculation. *Second, different serotypes of AAV transduce via peripheral inoculation with different efficiencies.* We have only compared AAV2 and AAV8 for mouse footpad/DRG and rabbit eye/TG delivery. AAV8 seems to transduce more efficiently than AAV2, but other serotypes may be more efficient when transducing other sites (and host species). With these caveats in mind, peripheral inoculation of AAV presents itself as a powerful tool for the delivery of therapeutic peptides, performing gene editing, or the knockdown of RNAs within peripheral neurons in vivo. We present detailed methods here for the preparation of the inoculation site, delivery of the AAV vectors, and analysis of the efficiency of nerve ganglia transduction.

2 Materials

Prepare all solutions with distilled water and deionized water.

2.1 Mouse Footpad/DRG Delivery

1. Dulbecco's modified Eagle's medium for dilution of AAV vectors (*see* **Note 1**).
2. 10% Saline: 10 g of NaCl in 100 mL of dH_2O. Filter sterilized.
3. 100 mg/mL Ketamine, sterile solution for injection (USP).
4. 20 mg/mL Xylazine, sterile solution for injection (USP).
5. 10 mg/mL Acepromazine, sterile solution for injection (USP).
6. 5 mg/mL Meloxicam, sterile solution for injection (USP).
7. 1 mL syringes (sterile).
8. 27G × ½″ sterile needles (0.4 mm × 13 mm).
9. 25G × 5/8″ sterile needles (0.5 mm × 16 mm).

10. Emery boards.
11. Isoflurane, liquid for inhalation (USP).
12. Oxygen, compressed gas (USP).
13. Anesthesia vaporizer, with mouse box and gas scavenger (VetEquip).
14. 20–200 μL micropipette.
15. Sterile 200 μL pipet tips.
16. AAV vector: 1×10^{10}–1×10^{12} vector genomes (vg) per mouse.

2.2 Rabbit Ocular/TG Delivery

1. Dulbecco's modified Eagle's medium for dilution of AAV vectors (*see* **Note 1**).
2. 100 mg/mL Ketamine for injection (USP).
3. Xylazine (USP).
4. 0.5% Proparacaine hydrochloride ophthalmic solution (sterile).
5. Sterile 27G beveled edge needles.
6. Sterile 25G beveled edge needles.
7. 1 mL syringes.
8. 2 kg New Zealand White rabbits.
9. 50 μL Micropipette.
10. Sterile pipette tips.
11. AAV vector: 1×10^{12} vg per eye.
12. Slit lamp microscope (to assess corneal topography).

2.3 Immunohistochemistry

1. 10% neutral buffered formalin.
2. 70% ethanol.
3. Paraffin (for tissue embedding).
4. Cryostat.
5. Superfrost Plus Microscope Slides (precleaned).
6. Glass coverslips.
7. Xylenes.
8. Ethanol.
9. Hydrogen peroxide solution (30%).
10. Methanol.
11. Deionized water.
12. Citrate buffer: 10 mM sodium citrate, pH 6.0.
13. TBS-Tween solution: 1× TBS with 1% Tween.
14. Coplin jars.
15. 500 mL beaker.

16. Microwave.
17. Normal goat serum (or normal serum from the species in which the primary antibody was raised).
18. Avidin-Biotin blocking kit (Vector Labs).
19. Zymed antibody diluent (Vector labs).
20. Anti-GFP primary antibody (for detection of AAV vectors with a GFP-label).
21. Biotinylated secondary antibody.
22. ABC Elite kit (vector labs).
23. DAB peroxidase substrate kit (Vector labs).
24. Hematoxylin.
25. VectaMount (Vector Labs).
26. Humidified chamber.
27. Histological microscope for imaging.

2.4 Immunofluorescence

1. 10% neutral buffered formalin.
2. 70% ethanol.
3. Paraffin (for tissue embedding).
4. Cryostat.
5. Superfrost Plus Microscope Slides (precleaned).
6. Glass coverslips.
7. Xylenes.
8. Ethanol.
9. Hydrogen peroxide solution (30%).
10. Methanol.
11. Deionized water.
12. Citrate buffer: 10 mM sodium citrate, pH 6.0.
13. TBS-Tween solution: 1× TBS with 1% Tween.
14. Coplin jars.
15. 500 mL beaker.
16. Microwave.
17. Normal goat serum (or normal serum from the animal in which the secondary antibody was raised).
18. Anti-GFP primary antibody (for detection of AAV vectors with a GFP-label).
19. Fluorophore-conjugated secondary antibody.
20. Prolong gold mounting reagent.
21. Fluorescent microscope for imaging.

3 Methods

All the procedures outlined below for mice and rabbit inoculations have been approved by either the University of Florida (mice) or Louisiana State University (rabbit) Institutional Animal Care and Use Committees and are in accordance with The Guide for Care and Use of Laboratory Animals for Research, eighth Edition.

3.1 Mouse Footpad/ DRG Delivery

1. To facilitate efficient uptake of the vector, and to minimize the amount of abrasion required and enhance vector uptake, saline-treat the feet prior to infection (*see* **Note 2**). First, anesthetize mice with isoflurane by inhalation (vaporizer setting = 4; 1 L O_2 per min). Administer 2–5 mg/kg of Meloxicam subcutaneously (SC) to alleviate any pain associated with the procedure. Then, inject 25–50 μL (no more than 50 μL) of sterile 10% saline subepidermally (subcutaneously) under both rear footpads (Fig. 1). Return the mice to their cages.
2. Four hours later, anesthetize the mice by intraperitoneal (IP) injection of 0.1–0.2 mL of a cocktail of acepromazine (2.5–3.75 mg/kg), xylazine (7.5–11.5 mg/kg), and ketamine (30–45 mg/kg) (*see* **Note 3**). Use the ketamine/xylazine/acepromazine cocktail for the infection step, because it is important to give the vector time to adsorb and enter the cells (30 min for 80–90% efficiency) before the animals recover and start moving around. The anesthesia time for the ketamine/xylazine/acepromazine cocktail is adequate for this requirement.
3. *Lightly* abrade both rear footpads of the anesthetized mice with an emery board to scratch the keratinized layer of the skin and allow the virus to adsorb efficiently. Abrade the surface just until you can pull off the layer (blister) of cornified epithelium. Do not abrade so forcefully that you cause bleeding. Use both feet in order to obtain bilateral infection of both sets of dorsal root ganglia and maximize the amount of material collected from a single mouse. Since ambulation problems are not a

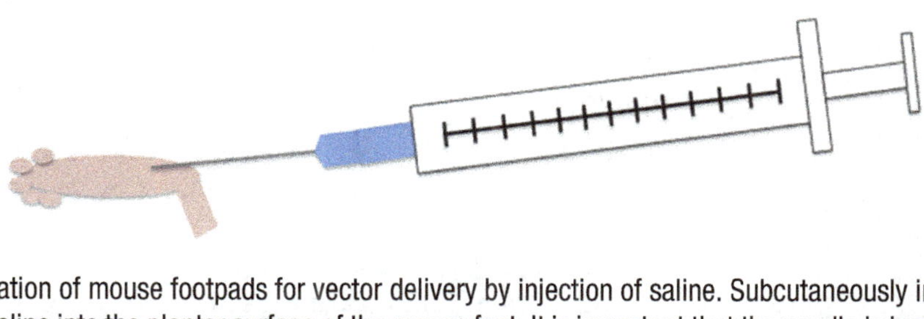

Fig. 1 Preparation of mouse footpads for vector delivery by injection of saline. Subcutaneously inject 50 μL of sterile 10% saline into the plantar surface of the mouse foot. It is important that the needle is inserted near the ankle of the foot, and that the angle of the syringe/needle is nearly parallel with the plane of the foot surface, so as not to inject too deeply into the tissue

common outcome (when using the saline pretreatment), unilateral infection is not necessary, unless one side is to be used as a control.

4. Rest the anesthetized mice on their backs, and place 50 μL of the appropriate dilution of AAV vector on the footpad using a micropipette. Rub the virus into the foot surface using the side of the micropipette tip. Doses of vector range from 1×10^{10} to 1×10^{12} vg per mouse, depending on the transduction efficiency of the particular vector and whether a self-complementary AAV vector or a capsid mutant with higher transduction efficiency is employed.
5. Allow the vector to adsorb until the mice awake (typically 45 min–1 h).
6. Observe the mice continuously until they maintain ventral recumbency to ensure they all awaken from the anesthetic without complications. Administer 2–5 mg/kg of Meloxicam (SC, once per day) after the infection, and subsequently as needed, to alleviate any signs of discomfort associated with the procedure or infection. If the animals appear to be anesthetized too deeply (shallow breathing) and are hypothermic, place a heat lamp 3 ft from them to keep the cages warm during recovery (*see* **Note 4**).
7. Wait at least 2 weeks after surgery, then euthanize the mice and dissect the dorsal root ganglia.

3.2 Rabbit Ocular/TG Delivery

Many species of rabbits can be used for the corneal delivery of AAV vectors, but this chapter will concentrate on New Zealand White rabbits, since they lack pigment in the eye and are commonly used for the study of infectious eye diseases.

1. Anesthetize each rabbit by intramuscular injection of ketamine (30–45 mg/kg) and xylazine (7.5–11.5 mg/kg) using a new sterile syringe and needle for each subject.
2. Place one drop of proparacaine HCl solution on each cornea without touching the bottle to the cornea.
3. After 5 min, use a 27G needle, beveled edge up, to gently abrade the surface of the cornea in a 4 × 4 crosshatch pattern. It is very important to abrade the epithelial layer of the cornea without puncturing the cornea, since this will result in pain.
4. Use the micropipette to pipette the AAV vector suspension onto the surface of the abraded cornea. The vector concentration should be equal to or greater than 1×10^{12} vg per eye to achieve greater than 75% transduction of the trigeminal ganglia. Inoculum volumes greater than 30 μL result in inefficient transduction.

5. After the vector is placed on the cornea, gently rub the vector into the eye by closing the eyelid of the rabbit.
6. Slit lamp examination can be performed to assess corneal topology prior to the start of the experiment or after 3 days following delivery of the vector to the cornea.
7. Wait at least 2 weeks after surgery, then euthanize the rabbit and dissect the trigeminal ganglia.

3.3 Expression Analysis: Immunohistochemistry

At the termination of the experiment, rabbit trigeminal ganglia and mouse dorsal root ganglia can be harvested (*see* **Note 5**) and processed for histology to detect AAV vectors in sensory neurons, so long as the AAV vector carries an appropriate label for detection (Fig. 2). The protocol described below is for the detection of AAV capsids labeled with a green fluorescent protein (GFP) following paraffin embedding and sectioning.

1. Immediately following removal, place harvested ganglia into 10% neutral buffered formalin. These tissues can be stored at room temperature overnight. If the tissues will not be used immediately, transfer the ganglia into a solution of 70% ethanol for long-term storage.
2. Paraffin-embed the ganglia.

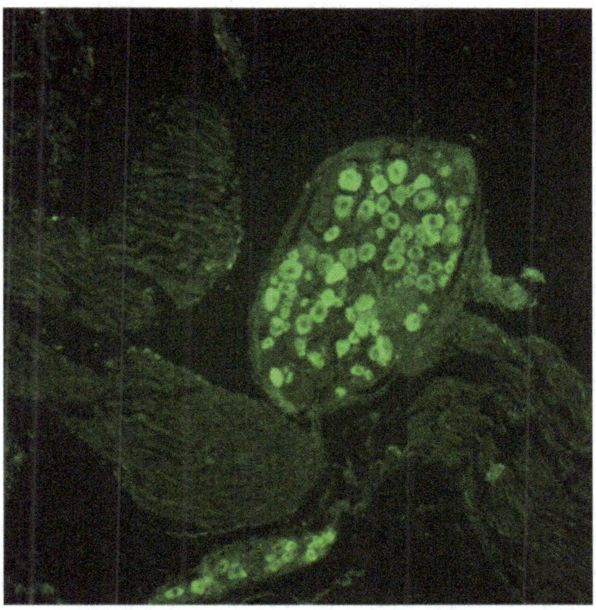

Fig. 2 Detection of AAV vector transduction in mouse DRG neurons following footpad inoculation. Mouse footpads were inoculated with 10^{11} vg of AAV8 expressing GFP driven by the CBA promoter. 14 days post treatment, the DRG (L4, L5, and L6) were dissected. DRG were processed for immunofluorescence using a rabbit anti-GFP antibody (Abcam) followed by a FITC-conjugated goat anti-rabbit antibody (Vector labs)

3. Slice each embedded ganglion using a cryostat and place on the microscope slide. Sections should be 10 μm thick.
4. Deparaffinize the slides using the following series of washes:
   (a) Xylene 1 (5 min).
   (b) Xylene 2 (5 min).
   (c) 100% ethanol 1 (2 min).
   (d) 100% ethanol 2 (2 min).
   (e) 3% hydrogen peroxide: dilute 30% hydrogen peroxide 1:10 in 100% methanol (10 min).
   (f) 95% ethanol (3 min).
   (g) 70% ethanol (1 min).
   (h) Deionized water (1 min).
5. Perform epitope retrieval by microwaving the slides in a Coplin jar with 10 mM citrate buffer at a pH of 6. Prior to microwaving, place the Coplin jar into a 500 mL beaker with enough water to cover the lower portion of the Coplin jar. In the microwave, bring the apparatus to a light boil and then allow the apparatus to remain at room temperature for 25 min.
6. Remove the slides from the citrate buffer and Coplin jar.
7. Wash slides in deionized water, followed by an additional wash in 1× TBS-Tween solution.
8. Incubate for 1 h in 1.5% normal serum (from the species in which the secondary antibody was raised) with four drops of avidin (Vector Labs Avidin-Biotin blocking kit) diluted in Zymed antibody diluent to block the slides.
9. Wash the slides with 1× TBS-Tween for 5 min.
10. Dilute four drops of biotin (Vector Labs Avidin-Biotin blocking kit) and the appropriate anti-GFP primary antibody in Zymed antibody diluent. Incubate the slides overnight at 4 °C in a humidified chamber.
11. Wash the slides in 1× TBS-Tween for 5 min.
12. Incubate the slides in the appropriate biotinylated secondary antibody diluted in Zymed diluent for 2 h at room temperature.
13. Wash the slides in 1× TBS-Tween for 5 min.
14. Incubate the slides in ABC Elite reagent (Vector Labs ABC Elite kit) for 30 min at room temperature.
15. Wash the slides in 1× TBS-Tween for 5 min.
16. Add the DAB chromogen substrate (Vector Labs DAB peroxidase substrate kit) for 1–5 min (watch for color change, then stop the reaction). Exercise care when handling DAB and follow local institutional regulations for disposal of DAB waste.

17. To stop the reaction, wash slides in deionized water for 5 min.
18. Counterstain the slides using hematoxylin (Vector Labs Hematoxylin QS).
19. Dehydrate the slides again using the following series of washes:
    (a) Deionized water (1 min).
    (b) 70% ethanol (1 min).
    (c) 95% ethanol (2 min).
    (d) 100% ethanol (2 min).
    (e) Xylene (1 min).
20. Let the slides dry, then use VectaMount (Vector Labs) to mount the slides and apply the coverslips.
21. Image the slides using an appropriate microscope.

3.4 Expression Analysis: Immunofluorescence

1. Immediately following removal, place harvested ganglia into 10% neutral buffered formalin. These tissues can be stored at room temperature overnight. If the tissues will not be used immediately, transfer the ganglia into a solution of 70% ethanol for long-term storage.
2. Paraffin-embed the ganglia.
3. Slice each embedded ganglion using a cryostat and place on the microscope slide. Sections should be 10 μm thick.
4. Deparaffinize the slides using the following series of washes:
    (a) Xylene 1 (5 min).
    (b) Xylene 2 (5 min).
    (c) 100% ethanol 1 (2 min).
    (d) 100% ethanol 2 (2 min).
    (e) 3% hydrogen peroxide: dilute 30% hydrogen peroxide 1:10 in 100% methanol (10 min).
    (f) 95% ethanol (3 min).
    (g) 70% ethanol (1 min).
    (h) Deionized water (1 min).
5. Perform epitope retrieval by microwaving the slides in a Coplin jar with 10 mM citrate buffer at a pH of 6. Prior to microwaving, place the Coplin jar into a 500 mL beaker with enough water to cover the lower portion of the Coplin jar. In the microwave, bring the apparatus to a light boil and then allow the apparatus to remain at room temperature for 25 min.
6. Remove the slides from the citrate buffer and Coplin jar.
7. Wash slides in deionized water, followed by an additional wash in 1× TBS-Tween solution.

8. Incubate for 1 h in 1.5% normal serum (from the species in which the secondary antibody was raised) diluted in TBS-Tween at room temperature to block the slides.
9. Wash the slides in TBS-Tween for 5 min.
10. Apply primary antibody diluted in TBS-tween to each slide. Incubate at 4 °C overnight in a humidified chamber.
11. Wash the slides in TBS-Tween for 5 min.
12. Apply secondary antibody diluted in TBS-tween to each slide and incubate at room temperature for 2 h. Slides should be kept in the dark from this step onward to avoid degradation of the fluorophore.
13. Wash the slides in TBS-Tween for 5 min.
14. Mount the slides using ProLong gold mounting medium (Invitrogen) and immediately apply the coverslips.
15. Image slides using a fluorescent microscope with the appropriate filters.

4 Notes

1. The AAV vectors used for these studies are based on pseudotyped AAV2 backbones packaged into AAV8 capsids by the UF Powell Gene Therapy Center Vector Core. The packaged vectors are banded on iodixanol gradients and Q sepharose columns and eluted in PBS as previously described [8, 9]. AAV vector stock concentrations were determined by dot blot quantification. Aliquot and store vector stocks at −80 °C. Prior to animal inoculation, dilute stocks in DMEM to the desired working concentration.
2. Saline pretreatment of the mouse feet increases the efficiency of vector uptake by up to 100-fold.
3. To prepare the anesthesia cocktail for 10 mice using typical dose parameters, combine: 0.2 mL of ketamine (100 mg/mL solution) + 0.12 mL of xylazine (20 mg/mL solution) + 0.1 mL of acepromazine (10 mg/mL solution) + 0.58 mL of sterile dH_2O. Administer 0.1 mL of this cocktail per mouse. The addition of acepromazine eliminates twitching of the feet, which may occur when only ketamine and xylazine are used.
4. Occasionally mice (especially the C57BL/6 strain) become hypothermic following anesthesia with ketamine cocktails. This can extend their recovery time from anesthesia. If this occurs, mice can be kept warm by either placing a heat lamp several feet away from their cages, or by placing a microwavable pet warming pad under the cage. Care should be taken to

ensure that temperatures are not excessively elevated if these procedures are used.

5. Detailed procedures for dissecting murine DRG are described in [10].

References

1. Moore NA, Morral N, Ciulla TA, Bracha P (2018) Gene therapy for inherited retinal and optic nerve degenerations. Expert Opin Biol Ther 18(1):37–49. https://doi.org/10.1080/14712598.2018.1389886
2. Blessing D, Deglon N (2016) Adeno-associated virus and lentivirus vectors: a refined toolkit for the central nervous system. Curr Opin Virol 21:61–66. https://doi.org/10.1016/j.coviro.2016.08.004
3. Watson ZL, Ertel MK, Lewin AS, Tuli SS, Schultz GS, Neumann DM, Bloom DC (2016) Adeno-associated virus vectors efficiently transduce mouse and rabbit sensory neurons coinfected with herpes simplex virus 1 following peripheral inoculation. J Virol 90 (17):7894–7901. https://doi.org/10.1128/JVI.01028-16
4. Aubert M, Madden EA, Loprieno M, DeSilva Feelixge HS, Stensland L, Huang ML, Greninger AL, Roychoudhury P, Niyonzima N, Nguyen T, Magaret A, Galleto R, Stone D, Jerome KR (2016) In vivo disruption of latent HSV by designer endonuclease therapy. JCI Insight 1(14). https://doi.org/10.1172/jci.insight.88468
5. Dang CH, Aubert M, De Silva Feelixge HS, Diem K, Loprieno MA, Roychoudhury P, Stone D, Jerome KR (2017) In vivo dynamics of AAV-mediated gene delivery to sensory neurons of the trigeminal ganglia. Sci Rep 7 (1):927. https://doi.org/10.1038/s41598-017-01004-y
6. Watson ZL, Washington SD, Phelan DM, Lewin AS, Tuli SS, Schultz GS, Neumann DM, Bloom DC (2018) In vivo knockdown of the herpes simplex virus 1 latency-associated transcript reduces reactivation from latency. J Virol 92(16). https://doi.org/10.1128/JVI.00812-18
7. Washington SD, Edenfield SI, Lieux C, Watson ZL, Taasan SM, Dhummakupt A, Bloom DC, Neumann DM (2018) Depletion of the insulator protein CTCF results in HSV-1 reactivation in vivo. J Virol. https://doi.org/10.1128/JVI.00173-18
8. Zolotukhin S, Potter M, Zolotukhin I, Sakai Y, Loiler S, Fraites TJ Jr, Chiodo VA, Phillipsberg T, Muzyczka N, Hauswirth WW, Flotte TR, Byrne BJ, Snyder RO (2002) Production and purification of serotype 1, 2, and 5 recombinant adeno-associated viral vectors. Methods 28(2):158–167. https://doi.org/10.1016/S1046-2023(02)00220-7
9. Zeier Z, Kumar A, Bodhinathan K, Feller JA, Foster TC, Bloom DC (2009) Fragile X mental retardation protein replacement restores hippocampal synaptic function in a mouse model of fragile X syndrome. Gene Ther 16 (9):1122–1129. https://doi.org/10.1038/gt.2009.83
10. Bloom DC (1996) HSV vectors for gene therapy. In: Brown SM, MacLean AR (eds) Herpes simplex virus protocols, Methods in molecular medicine, vol 10. Humana Press, Totowa, pp 369–386

Chapter 14

SubILM Injection of AAV for Gene Delivery to the Retina

Paul D. Gamlin, John J. Alexander, Sanford L. Boye,
C. Douglas Witherspoon, and Shannon E. Boye

Abstract

Adeno-associated virus (AAV) has emerged as the vector of choice for delivering genes to the retina. Indeed, the first gene therapy to receive FDA approval in the United States is an AAV-based treatment for the inherited retinal disease, Leber congenital amaurosis-2. Voretigene neparvovec (Luxturna™) is delivered to patients via subretinal (SR) injection, an invasive surgical procedure that requires detachment of photoreceptors (PRs) from the retinal pigment epithelium (RPE). It has been reported that subretinal administration of vector under the cone-exclusive fovea leads to a loss of central retinal structure and visual acuity in some patients. Due to its technical difficulty and potential risks, alternatives to SR injection have been explored in primates. Intravitreally (Ivt) delivered AAV transduces inner retina and foveal cones, but with low efficiency. Novel AAV capsid variants identified via rational design or directed evolution have offered only incremental improvements, and have failed to promote pan-inner retinal transduction or significant outer retinal transduction beyond the fovea. Problems with retinal transduction by Ivt-delivered AAV include dilution in the vitreous, potential antibody-mediated neutralization of capsid in this nonimmune privileged space, and the presence of the inner limiting membrane (ILM), a basement membrane separating the vitreous from the neural retina. We have developed an alternative "subILM" injection method that overcomes all three hurdles. Specifically, vector is placed in a surgically induced, hydrodissected space between the ILM and neural retina. We have shown that subILM injection promotes more efficient retinal transduction by AAV than Ivt injection, and results in uniform and extensive transduction of retinal ganglion cells (RGCs) beneath the subILM bleb. We have also demonstrated transduction of Muller glia, ON bipolar cells, and photoreceptors by subILM injection. Our results confirm that the ILM is a major barrier to transduction by AAV in primate retina and that, when it is circumvented, the efficiency and depth to which AAV2 promotes transduction of multiple retinal cell classes is greatly enhanced. Here we describe in detail methods for vector preparation, vector dilution, and subILM injection as performed in macaque (*Macaca* sp.)

Key words Inner limiting membrane, AAV, Gene delivery, Gene replacement, Retinal ganglion cells, Bipolar cells, Photoreceptors, Novel surgical technique

1 Introduction

Inherited retinal diseases are caused by mutations in genes that encode for proteins expressed in various cells of the retina. In the vast majority of cases, the cells responsible for disease are retinal

Michael J. Castle (ed.), *Adeno-Associated Virus Vectors: Design and Delivery*, Methods in Molecular Biology, vol. 1950,
https://doi.org/10.1007/978-1-4939-9139-6_14,

ganglion cells (RGCs) in the inner retina, and photoreceptors and retinal pigment epithelium (RPE), both of which are in the outer retina. The ideal strategy for treating such diseases is one that ensures efficient delivery of therapy to these target cells. Subretinal (SR) injection of adeno-associated virus (AAV) vectors is the current standard method for delivering therapeutic genes to PRs and RPE, as vector is deposited directly between these two cell layers. Subretinally delivered AAV2-based vectors are being used to treat patients with *RPE65* Leber congenital amaurosis (LCA2), Choroideremia, retinitis pigmentosa (RP), and the *CNGB3* and *CNGA3* forms of Achromatopsia. Alternative AAV capsids that promote more efficient transduction of PRs in nonhuman primate have been identified [1–3]. These capsids, AAV5 and AAV8, are being applied subretinally to treat LCA2, RP, and Achromatopsia. However, targeting of foveal cones by subretinally injected AAV is problematic in patients with retinal degeneration, and may be contraindicated in diseases where underlying pathology complicates delivery or retinal detachment is expected to exacerbate pathology [4–6].

Intravitreal (Ivt) injection has been used to target RGC-mediated diseases such as Leber's hereditary optic neuropathy (LHON), and is now being pursued as an alternative delivery approach to target outer retinal cells for the reasons mentioned above. However, recent studies in nonhuman primates revealed the formidable hurdles associated with this approach. While some transduction of RGCs is accomplished by Ivt injection of AAV2-based vectors in NHPs, the area of transduction is restricted to a "ring" of RGCs around the fovea and scattered foci in the peripheral retina, some foveal cones, and a mixture of retinal cells proximal to large retinal blood vessels [7, 8]. Newer AAV capsid variants identified via rational design or directed evolution are improved in the magnitude of transduction relative to AAV2, but the extent of transduction is essentially limited to the same areas [8, 9]. Transduction of foveal cones is possible, but is efficient only at high vector doses that promote inflammation [8]. Ivt injection is less invasive than SR injection, and may be well suited for treating patients with advanced degenerative disease that are prone to further damage upon surgically induced retinal detachment. However, there has not yet been an AAV capsid identified which promotes efficient transduction of PRs/RPE across the entire retina.

This is due, at least in part, to the following hurdles. First, Ivt-delivered AAV is immediately diluted upon mixing with the vitreous humor (0.05–0.10 mL of vector into ~2.5–4 mL of vitreous). Second, the vitreous is not an immune privileged space, and thus Ivt-delivered AAV can be neutralized by preexisting antibodies [10], and can potentially generate inflammatory responses. Finally, Ivt-delivered AAV must bypass the inner limiting membrane (ILM). The ILM is a typical basement membrane that forms the vitreoretinal junction and poses a biological barrier to Ivt-delivered capsids [11, 12]. Primates have a more impenetrable ILM than rodents, except in and around the fovea, immediately above the optic nerve head, and close to large blood vessels [13]. Enzymatic digestion of the ILM has been shown to improve retinal transduction by Ivt-delivered AAV, but the long-term effects of this approach on retinal structure and function have not been investigated [7, 14].

We recently developed a "subILM" injection method wherein AAV is placed in a surgically induced space between the ILM and neural retina [15]. This overcomes the dilution effect, the potential for immune neutralization and inflammatory responses, and the ILM barrier faced by Ivt-injected AAV. SubILM injection promotes more efficient retinal transduction by AAV than Ivt injection, and may provide a safe and efficient alternative to surgeons administering retinal gene therapies to patients with fragile retinas prone to further damage by subretinal surgery.

2 Materials

2.1 Vector Production

1. AAV plasmids (*see* **Note 1**). Plasmids required for the triple-transfection method of AAV production include:
   - Rep-cap Plasmid containing AAV2 *rep* and AAV *cap* from the serotype of interest without flanking AAV terminal repeats, such as pACG-2, pXX2, and p5E18(2/2), each containing AAV2 *rep* and AAV2 *cap* [16–18].
   - Helper Plasmid containing Adenovirus helper genes required for AAV packaging, such as pXX6 [18]. Alternatively, a helper plasmid that contains AAV2 *rep* and AAV2 *cap* along with the Adenovirus helper genes may be used, such as pDG [19].
   - Vector Plasmid containing AAV2 inverted terminal repeats (ITRs) flanking the transgene of interest. This will typically contain a promoter, splicing signal, reporter gene, and poly adenylation signal, such as pTR-UF11, containing the chimeric CMV-chicken beta actin (CAG) promoter driving "humanized" green fluorescent protein (GFP) [20], or pAAV-GFP, containing the CMV promoter driving GFP [21].
2. Balanced Salt Solution (BSS).
3. Tween (polysorbate) 20.

2.2 Preparation of Vector Solution

1. Micropipettes.
2. Sterile micropipette tips.

3. Wide bore 1 mL pipette tip.
4. Balanced Salt Solution (BSS).
5. BSS/Tween Vector Diluent Buffer: BSS with 0.014% Tween (polysorbate) 20.
6. 10 mg/mL sodium hyaluronate: Healon (McKesson).
7. Sterile individually wrapped microcentrifuge tubes.
8. High-speed microcentrifuge.
9. Vortex mixer.
10. Laminar flow hood.

2.3 Animal Preparation and Postoperative Care

1. 100 mg/mL Ketamine.
2. Glycopyrrolate.
3. 10 mg/mL Cerenia.
4. 1 g Cefazolin.
5. 2 mg/mL Dexamethasone.
6. IV Catheter.
7. Endotracheal tubing.
8. Ventilator.
9. Sustained Release Buprenorphine.
10. Sustained Release Meloxicam.
11. 1% Tropicamide Ophthalmic Solution.
12. 2.5% Phenylephrine Hydrochloride Ophthalmic Solution.
13. 1% Cyclopentolate Ophthalmic Solution.
14. Isoflurane.
15. 0.9% Sodium Chloride.
16. 10% Povidone-Iodine Topical Solution.
17. Benzoin Tincture Swab Stick.
18. 4 mg/mL Dexamethasone.
19. Neomycin/Polymyxin B Sulfates and Dexamethasone Ophthalmic Suspension.

2.4 SubILM Injection

1. BSS Plus Irrigating Solution (Alcon Labs).
2. BSS Solution.
3. Sterile surgical drape (medium).
4. Interlink T-Connector extension set (Baxter).
5. 18G × 1 1/2″ BLUNT fill needles.
6. 1 mL syringe with luer-lock tip (Fisher #309628 is preferred).
7. 0.5 mL syringe.
8. Total Plus 25G or 23G Vitrectomy Pak (Alcon Labs).

9. 25G Valved Trocars (Alcon Labs).
10. Machemer magnifying vitrectomy contact lens (Ocular Instruments).
11. 20G V-Lance Knife (Alcon Labs).
12. 36–42G retinal needle.
13. Syringe pump.
14. 8-0 and 9-0 Vicryl Suture.
15. Bishop-Harmon Irrigating Cannula.
16. 25 mg Indocyanine Green (ICG).
17. Scanning Laser Ophthalmoscope, such as Heidelberg Spectralis®.
18. Fundus Camera, such as Topcon TRC-50EX.
19. Ophthalmic Surgical Microscope with video, such as Zeiss Visu 200.
20. Vitrectomy Surgery System, such as Accurus 800CS with Xenon light source.

3 Methods

Conduct the subILM injection technique in macaque monkeys in a sterile surgical suite with the assistance of an appropriately qualified vitreoretinal surgeon and their staff. An overview of the procedure is provided in Fig. 1.

3.1 Vector Production

1. AAV vectors should be manufactured and purified using methods established to result in highly pure AAV particles free of contamination by helper virus (e.g., Adenovirus, Herpes simplex virus, and baculovirus) and host cell DNA (*see* **Note 2**). Protocols for AAV production can be found in Chapters 3, 7, 19, 21, 22, and 23.
2. Use an AAV storage buffer compatible with intraocular use. Appropriate buffers have physiologic osmolarity (approximately 300 mOsm) and a pH range of 7.0–8.0. Buffers that contain small amounts (2 mM) of divalent cations Mg^{2+} and Ca^{2+} are preferred, as this facilitates AAV capsid stability. Of the commercially available buffers, Balanced Salt Solution (BSS) is the most widely used buffer for storage and dilution of AAV vectors to be injected into the eye.
3. The AAV storage and dilution buffer should also contain a small amount of nonionic surfactant to reduce nonspecific binding of virus to the tubing and syringes. For intraocular delivery, supplement AAV with 0.014% tween (polysorbate) 20 or 0.001% pluronic F68 [22, 23]. BSS with 0.014% tween

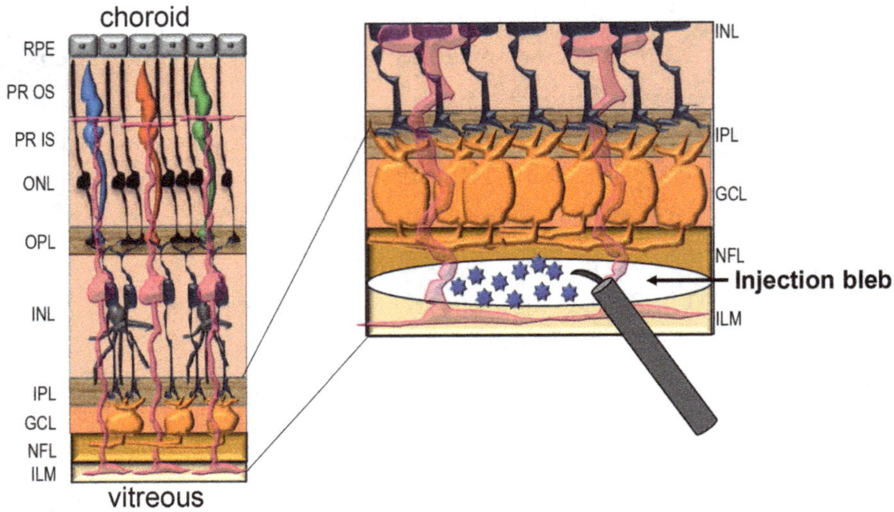

Fig. 1 SubILM injection of AAV and Healon. A schematic of an intact retina (left) is magnified (right) to reveal the location of the subILM injection bleb. A needle is placed posterior to the ILM and anterior to NFL and GCL. Approximately 10 μL of a 1:1 solution of AAV:healon is injected into this space. *RPE* retinal pigment epithelium, *PR OS* photoreceptor outer segments, *PR IS* photoreceptor inner segments, *ONL* outer nuclear layer, *OPL* outer plexiform layer, *IN L* inner nuclear layer, *IPL* inner plexiform layer, *GCL* ganglion cell layer, *NFL* nerve fiber layer, *ILM* inner limiting membrane

has previously been shown to be compatible with the addition of Healon ophthalmic viscoelastic, which may be required for subILM injection [15].

4. Ensure that adequate amounts of the storage buffer are manufactured for later use as diluent buffer.
5. AAV should be titrated relative to a well-characterized standard. Acceptable methods for AAV titration include dot blot, quantitative/real-time polymerase chain reaction (qPCR), and digital PCR (*see* Chapter 4) [24–26].
6. AAV vectors should be confirmed to be free of contaminating protein by visualization via SDS PAGE gel stained with coomassie blue or silver stain. A quantification of the bands conforming to VP1, VP2, and VP3 should represent >95% of the protein visible [24].
7. The final AAV vector should be tested for endotoxin and should contain no more than 5 units of endotoxin per mL.
8. AAV vectors should be stored at −80 °C until shortly before use. Thawing and preparation of dosing solution is detailed in Subheading 2.4.

3.2 Preparation of Vector Solution

1. Calculate vector dilution based on the AAV stock concentration, the number of injections to be performed, the desired dose for each injection, and either 1:1 or 2:1 (v/v) dilution with Healon (1% sodium hyaluronate cohesive solution).

2. A 1:1 AAV:Healon mixture will be more difficult to prepare than a 2:1 AAV:Healon mixture, but may improve bleb formation when surgeons are not experienced in the technique.
3. Depending on the number of injections to be performed, a minimum of 250 μL of vector solution should be prepared prior to mixing with Healon. It is important to note that that the final solution will be very viscous, and it will not be possible to remove all of it from the 1.5 mL centrifuge tube. Prepare a total volume of 500 μL per eye.
4. Follow sterile technique when preparing AAV solutions, buffers, cannulas, and syringes. Prepare vector in a laminar flow hood, if available.
5. Thaw AAV on ice. Once thawed, mix AAV by inverting and flicking the tube. Quickly spin in a microcentrifuge at low speed (≤1000 rpm) to collect AAV solution at the bottom of the tube.
6. Open the AAV vector tube just enough to unseal the cap, and allow the solution to degas for at least 15 min at room temperature. This step is critical to reduce bubble formation in the final injection solution.
7. Thaw BSS/Tween diluent if needed for diluting AAV stock, as described in **steps 5** and **6**.
8. Allow Healon solution to warm to room temperature within the unopened sterile packaging.
9. Pre-dispense Healon into the microcentrifuge tube that will hold the final vector solution. Use an empty sterile 1.5 mL or 2 mL microcentrifuge tube (*see* **Note 3**).
10. Recap the tube containing the AAV stock, briefly vortex for 1–2 s (medium speed), and spin at ~5000 rpm for 2 min in a microcentrifuge. Do the same for the tube containing BSS/Tween diluent (if needed). If the AAV stock does not need to be diluted prior to adding to Healon, proceed to **step 12**.
11. If the AAV stock solution needs to be diluted with BSS/Tween solution before mixing with the Healon, dispense the appropriate amount of BSS/Tween solution into a fresh microcentrifuge tube, then pipette the appropriate volume of AAV stock solution directly into the BSS/Tween. Pipette the mixture up and down ~5 times. Vortex for 1–2 s (medium speed) and pulse spin at 1000 rpm in a microcentrifuge.
12. Pipette the predetermined volume of AAV solution into the microcentrifuge tube containing the Healon (**step 9**). The two solutions will not mix easily; do not attempt to pipette up and down. Use the pipette tip to briefly stir the AAV/Healon mixture together (the solutions will still not appear to mix well together).

13. Vortex the tube on the lowest setting for 30–60 s. Bubbles will form. Spin at max speed in a microcentrifuge for at least 2 min to remove air bubbles. Repeat this step until mixed.
14. The vector solution is now ready to be loaded into the injection device (**step 7** in Subheading 3.4). Use either a wide bore 1 mL pipette tip or a 1 mL syringe with an 18G blunt fill needle. The solution will be very viscous if a 1:1 AAV:Healon mixture is used.

3.3 Animal Preparation

1. Collect serum approximately 6 weeks prior to planned subILM injection.
2. Test serum for the presence of neutralizing antibodies (NAb) to the AAV capsid serotype that will be utilized in the subILM injections, as previously described [1, 27, 28]. Animals with high titers of NAb should be avoided.
3. Obtain reference fundus optical coherence tomography (OCT) and fluorescence images (*see* **Note 4**).
4. A prophylactic steroid and antibiotic regimen is strongly recommended for this procedure. The day before surgery, administer 0.25 mg/kg of intramuscular (IM) Dexamethasone. The day of surgery, administer 0.25 mg/kg of IM Dexamethasone and 25 mg/kg of IM Cefazolin. Additionally, after surgery inject subconjunctivally 0.5 mL of 4 mg/mL Dexamethasone and 0.5 mL of 330 mg/mL Cefazolin. If any postsurgical inflammation is observed without sign of infection, administer 0.25 mg/kg of IM Dexamethasone for 2 days, followed by a 3–5 day taper.
5. Prior to surgery, sedate animals using 20 mg/kg of IM Ketamine, 0.005 mg/kg of subcutaneous Glycopyrrolate, and 1 mg/kg of subcutaneous Cerenia.
6. Place an IV catheter and start a saline drip, then intubate the animal.
7. Once sedated, dilate the eyes using 2.5% phenylephrine, 1% tropicamide, and 1% cyclopentalate.
8. Place the animal on a ventilator and maintain general anesthesia using isoflurane (1.5–2.5%) with continuous monitoring of vital signs.
9. Postsurgically, administer sustained release Buprenorphine and Meloxicam.

3.4 SubILM Injection

1. Prepare eyes with povidone-iodine topical solution plus a Benzoin Tincture swab stick and drape in standard sterile fashion.
2. Carry out the surgical procedure under sterile conditions in a dedicated veterinary ophthalmic surgical suite equipped with anesthesia, ophthalmic surgical microscope with video, and

vitrectomy surgical system. We use an Accurus 800CS surgical system with Xenon light source (*see* **Note 5**), Total Plus gauge 23 or 25 Vitrectomy Pak with valved trocars (*see* **Note 6**), and Zeiss VISU 200 ophthalmic surgical microscope equipped with digital video.

3. Visualize the posterior segment of retina using an irrigating Machemer magnifying vitrectomy contact lens.
4. Perform a standard three-port pars plana vitrectomy using an inferior infusion irrigating cannula and BSS Plus irrigating solution (maintain a pressure of 10–30 mm/Hg).
5. Enlarge the superior-temporal sclerotomy with a 20–25G V-Lance Knife to provide access for the injection cannula.
6. Use a 36–42G retinal needle to deliver vector into the subILM space beneath the macular region (*see* **Note 7**). Couple the needle to a 0.5 mL syringe using a 6″ T-Connector extension set. The injection is made by placing the tip of the cannula almost parallel to the surface of the ILM (*see* **Note 8**), and advancing just enough to penetrate the ILM (Fig. 1). Once subILM access is observed, introduce the vector solution by either manual injection or using a syringe pump (*see* **Note 9**). After a bleb of sufficient size and location is observed, retract the needle slowly (*see* **Note 10**). A typical injection volume is approximately 20 μL. Estimate the amount of introduced vector by manually inspecting the change in volume in the syringe.
7. Close the vitrectomy: remove ports and suture sclera/conjunctiva with 8-0 or 9-0 vicryl.
8. We recommend performing OCT (*see* **Note 4**) immediately after surgery to document the location and extent of the sub-ILM injections. If OCT is not available, document the bleb (s) using the surgical microscope video recording device as well as a surgical dictation describing the bleb locations with respect to anatomical landmarks within the retina (*see* above).
9. Apply 2–3 drops of Neomycin/Polymyxin B Sulfates and Dexamethasone Ophthalmic Suspension to each treated eye.
10. The length of animal survival following this procedure will depend on the experiment. Perform a clinical examination at 3 days after surgery, and at regular intervals thereafter, to ensure that there are no inflammatory responses.
11. In vivo imaging should be performed at regular intervals to confirm that the bleb has resolved without incident, and to monitor expression of the fluorescent reporter (if used).
12. At the end of the experiment, euthanize the animals (*see* **Note 11**) and prepare eyes for histology (*see* **Note 12**).

4 Notes

1. Rep-cap and Helper plasmids can be propagated in a variety of bacterial cells that do not already contain the antibiotic resistance encoded on the plasmids. AAV vector plasmids should be propagated in recombination (recB, recJ) deficient bacteria such as Sure cells to avoid recombination of the AAV ITRs. Bacteria can be grown in terrific broth supplemented with the appropriate antibiotic. Vector plasmid- and Rep-cap plasmid-containing cells are best purified using a Qiagen EndoFree plasmid purification kit of the appropriate scale to meet AAV packaging requirements, typically Maxi or Giga size. Recovery of large Helper plasmids using commercially available plasmid purification kits is inefficient, and requires density gradient purification such as a CsCl gradient. However, this results in higher levels of endotoxin in the purified plasmid DNA, which may carry through to the final AAV prep. Several Contract Research Organizations (CROs) such as Aldevron offer plasmid purification in which most endotoxin is removed, and also sell stock AAV helper plasmids such as pXX6.
2. AAV vectors of suitable purity and sterility for use in NHP studies can be obtained from several academic Vector Cores, such as the Powell Gene Therapy Center (University of Florida), the University of Pennsylvania Vector Core, and the University of North Carolina Gene Therapy Center Vector Core.
3. Full-strength Healon is very difficult to pipette. To avoid pipetting, fill the microcentrifuge tube with an identical volume of sterile BSS. Mark the microcentrifuge at the height of the BSS volume, then remove and discard the BSS. Open the Healon packaging and assemble the dispenser. Slowly dispense Healon into the microcentrifuge tube until it reaches the previously marked volume line (**step 10**). A brief spin in a microcentrifuge may be necessary to collect Healon in the bottom of the tube and evaluate the volume.
4. Normal and fluorescent fundus images should be obtained both prior to surgery and after surgery. In studies using GFP as a reporter, the 488 laser of the Heidelberg Spectralis® confocal scanning laser ophthalmoscope with fluorescence detection capability should be used (*see* below). However, for reporters excited at other wavelengths, a fundus camera (Topcon TRC-50EX) equipped with custom excitation and barrier filters is required. For most in vivo retinal imaging, we recommend a Heidelberg Spectralis® confocal scanning laser ophthalmoscope, with spectral-domain optical coherence tomography (SD-OCT) and fluorescence detection capability. The signal-to-noise ratio of GFP fluorescence is excellent with this system

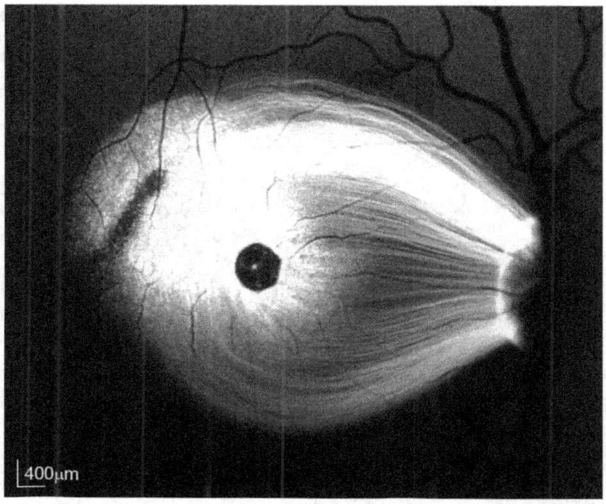

Fig. 2 Fluorescence fundus image of AAV-mediated GFP expression in NHP subject EN-28 taken 4 months postinjection. Healon and AAV (1.7×10^{12} vector genomes per mL) were subILM-injected at a ratio of 1:1. Total bleb volume was 7.5 μL and contained a total of 6.2×10^{9} vector genomes. Scale bars = 400 μm

(Fig. 2). Immediate postsurgical OCT images can confirm that the injection is subILM and does not penetrate the retina. Subsequent imaging can confirm expression of GFP and document that the injection bleb has resolved.

5. The Alcon Accurus 800CS has reached end-of-life and has now been replaced with the Alcon CONSTELLATION® Vision System.
6. The original procedure was performed with 23G ports. We now prefer 25G self-sealing ports.
7. The design of the original cannula was not optimal, and the design of an optimal cannula for subILM injections is ongoing.
8. ICG Indocyanine green may be used to stain and better visualize the ILM prior to subILM injection. Inject a diluted solution of ICG (~0.1–0.3 mL) over the planned injection site and allow it to stain the ILM to effect prior to injection. To prepare the ICG solution, dissolve 25 mg of sterile ICG powder in 0.5–1.0 mL of sterile water. Dilute this in 4.0–4.5 mL of BSS, resulting in a 0.5% ICG solution with an osmolarity of 270 mOsm.
9. We originally used a 500 μL Hamilton syringe operated manually or by a syringe pump to inject the vector. We currently recommend use of a MicroDose™ Injection Kit (Katalyst Surgical) or similar.

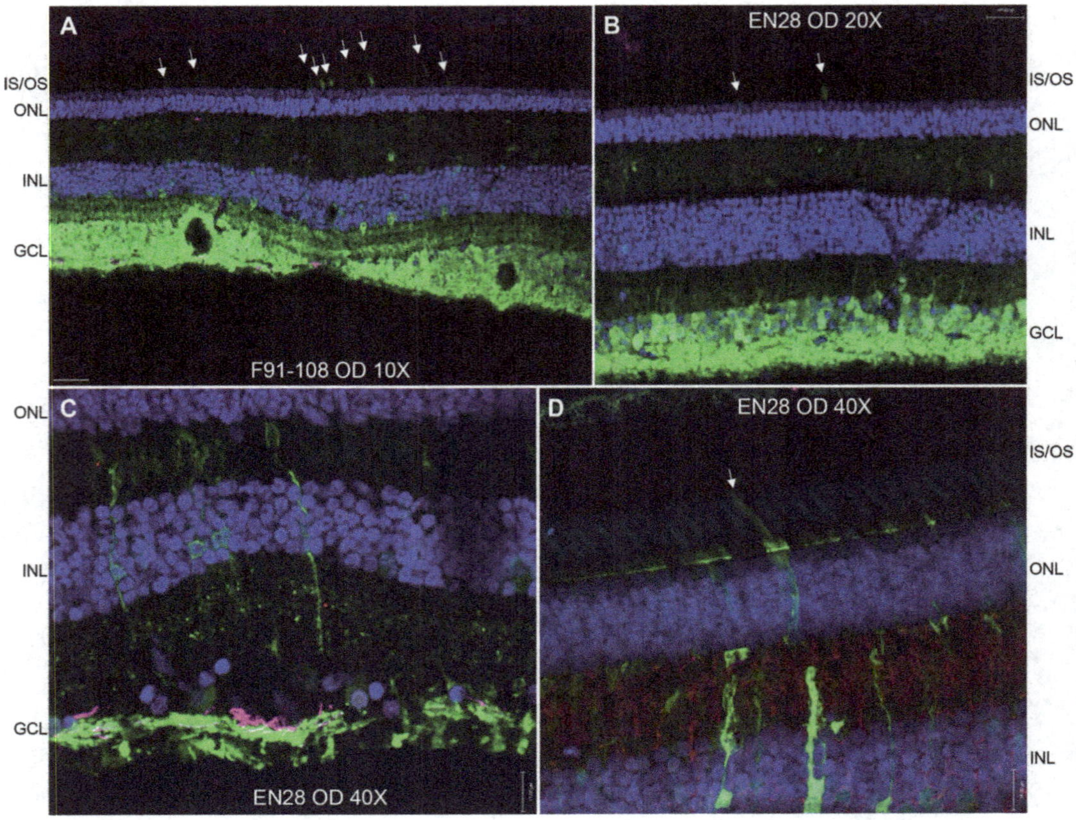

Fig. 3 AAV2-CBA-mediated GFP expression in subjects EN-28 and F91-108. Raw GFP (green), DAPI (blue), and Glial fibrillary acidic protein (purple) are shown in all panels. Glutamine synthetase (red) is shown in panel D. The vast majority of retinal ganglion cells within the subILM injection blebs are GFP positive (**a**, **b**). Transduction of photoreceptors (white arrows in **a**, **b**, **d**) and Muller glia (**c**, **d**) is also observed. *IS/OS* inner segments/outer segments, *ONL* outer nuclear layer, *INL* inner nuclear layer, *GCL* ganglion cell layer. Scale bars = 70 μm (**a**), 35 μm (**b**), 17 μm (**c**, **d**)

10. Clear visualization of the bleb with the surgical camera is strongly recommended to provide video stills and photos for postsurgical analysis.
11. In this and all other cases of euthanasia, the investigator should follow the AVMA Guidelines for the Euthanasia of Animals [29].
12. Expected histology results are shown in Fig. 3. Protocols are outlined in our previously published work [1, 15].

References

1. Boye SE, Alexander JJ, Boye SL, Witherspoon CD, Sandefer KJ, Conlon TJ, Erger K, Sun J, Ryals R, Chiodo VA, Clark ME, Girkin CA, Hauswirth WW, Gamlin PD (2012) The human rhodopsin kinase promoter in an AAV5 vector confers rod- and cone-specific expression in the primate retina. Hum Gene Ther 23(10):1101–1115. https://doi.org/10.1089/hum.2012.125

2. Vandenberghe LH, Bell P, Maguire AM, Cearley CN, Xiao R, Calcedo R, Wang L, Castle MJ, Maguire AC, Grant R, Wolfe JH, Wilson JM, Bennett J (2011) Dosage thresholds for AAV2 and AAV8 photoreceptor gene therapy in monkey. Sci Transl Med 3(88):88ra54. https://doi.org/10.1126/scitranslmed.3002103
3. Vandenberghe LH, Bell P, Maguire AM, Xiao R, Hopkins TB, Grant R, Bennett J, Wilson JM (2013) AAV9 targets cone photoreceptors in the nonhuman primate retina. PLoS One 8(1):e53463. https://doi.org/10.1371/journal.pone.0053463
4. Jacobson SG, Cideciyan AV, Ratnakaram R, Heon E, Schwartz SB, Roman AJ, Peden MC, Aleman TS, Boye SL, Sumaroka A, Conlon TJ, Calcedo R, Pang JJ, Erger KE, Olivares MB, Mullins CL, Swider M, Kaushal S, Feuer WJ, Iannaccone A, Fishman GA, Stone EM, Byrne BJ, Hauswirth WW (2012) Gene therapy for leber congenital amaurosis caused by RPE65 mutations: safety and efficacy in 15 children and adults followed up to 3 years. Arch Ophthalmol 130(1):9–24. https://doi.org/10.1001/archophthalmol.2011.298
5. Yanoff M, Kertesz Rahn E, Zimmerman LE (1968) Histopathology of juvenile retinoschisis. Arch Ophthalmol 79(1):49–53
6. Condon GP, Brownstein S, Wang NS, Kearns JA, Ewing CC (1986) Congenital hereditary (juvenile X-linked) retinoschisis. Histopathologic and ultrastructural findings in three eyes. Arch Ophthalmol 104(4):576–583
7. Yin L, Greenberg K, Hunter JJ, Dalkara D, Kolstad KD, Masella BD, Wolfe R, Visel M, Stone D, Libby RT, Diloreto D Jr, Schaffer D, Flannery J, Williams DR, Merigan WH (2011) Intravitreal injection of AAV2 transduces macaque inner retina. Invest Ophthalmol Vis Sci 52(5):2775–2783. https://doi.org/10.1167/iovs.10-6250
8. Dalkara D, Byrne LC, Klimczak RR, Visel M, Yin L, Merigan WH, Flannery JG, Schaffer DV (2013) In vivo-directed evolution of a new adeno-associated virus for therapeutic outer retinal gene delivery from the vitreous. Sci Transl Med 5(189):189ra176. https://doi.org/10.1126/scitranslmed.3005708
9. Ye GJ, Budzynski E, Sonnentag P, Miller PE, Sharma AK, Ver Hoeve JN, Howard K, Knop DR, Chulay JD (2015) Safety and biodistribution evaluation in cynomolgus macaques of rAAV2tYF-CB-hRS1, a recombinant adeno-associated virus vector expressing retinoschisin. Hum Gene Ther Clin Dev 26(3):165–176. https://doi.org/10.1089/humc.2015.076
10. Kotterman MA, Yin L, Strazzeri JM, Flannery JG, Merigan WH, Schaffer DV (2015) Antibody neutralization poses a barrier to intravitreal adeno-associated viral vector gene delivery to non-human primates. Gene Ther 22(2):116–126. https://doi.org/10.1038/gt.2014.115
11. Dalkara D, Kolstad KD, Caporale N, Visel M, Klimczak RR, Schaffer DV, Flannery JG (2009) Inner limiting membrane barriers to AAV-mediated retinal transduction from the vitreous. Mol Ther 17(12):2096–2102. https://doi.org/10.1038/mt.2009.181
12. Boye SL, Bennett A, Scalabrino ML, McCullough KT, Van Vliet K, Choudhury S, Ruan Q, Peterson J, Agbandje-McKenna M, Boye SE (2016) Impact of heparan sulfate binding on transduction of retina by recombinant adeno-associated virus vectors. J Virol 90(8):4215–4231. https://doi.org/10.1128/JVI.00200-16
13. Matsumoto B, Blanks JC, Ryan SJ (1984) Topographic variations in the rabbit and primate internal limiting membrane. Invest Ophthalmol Vis Sci 25(1):71–82
14. Cehajic-Kapetanovic J, Le Goff MM, Allen A, Lucas RJ, Bishop PN (2011) Glycosidic enzymes enhance retinal transduction following intravitreal delivery of AAV2. Mol Vis 17:1771–1783
15. Boye SE, Alexander JJ, Witherspoon CD, Boye SL, Peterson JJ, Clark ME, Sandefer KJ, Girkin CA, Hauswirth WW, Gamlin PD (2016) Highly efficient delivery of adeno-associated viral vectors to the primate retina. Hum Gene Ther 27(8):580–597. https://doi.org/10.1089/hum.2016.085
16. Li J, Samulski RJ, Xiao X (1997) Role for highly regulated rep gene expression in adeno-associated virus vector production. J Virol 71(7):5236–5243
17. Xiao W, Chirmule N, Berta SC, McCullough B, Gao G, Wilson JM (1999) Gene therapy vectors based on adeno-associated virus type 1. J Virol 73(5):3994–4003
18. Xiao X, Li J, Samulski RJ (1998) Production of high-titer recombinant adeno-associated virus vectors in the absence of helper adenovirus. J Virol 72(3):2224–2232
19. Grimm D, Kern A, Rittner K, Kleinschmidt JA (1998) Novel tools for production and purification of recombinant adenoassociated virus vectors. Hum Gene Ther 9(18):2745–2760. https://doi.org/10.1089/hum.1998.9.18-2745
20. Burger C, Gorbatyuk OS, Velardo MJ, Peden CS, Williams P, Zolotukhin S, Reier PJ, Mandel RJ, Muzyczka N (2004) Recombinant AAV

viral vectors pseudotyped with viral capsids from serotypes 1, 2, and 5 display differential efficiency and cell tropism after delivery to different regions of the central nervous system. Mol Ther 10(2):302–317

21. Gray JT, Zolotukhin S (2011) Design and construction of functional AAV vectors. Methods Mol Biol 807:25–46. https://doi.org/10.1007/978-1-61779-370-7_2

22. Beltran WA, Cideciyan AV, Lewin AS, Iwabe S, Khanna H, Sumaroka A, Chiodo VA, Fajardo DS, Roman AJ, Deng WT, Swider M, Aleman TS, Boye SL, Genini S, Swaroop A, Hauswirth WW, Jacobson SG, Aguirre GD (2012) Gene therapy rescues photoreceptor blindness in dogs and paves the way for treating human X-linked retinitis pigmentosa. Proc Natl Acad Sci U S A 109(6):2132–2137. https://doi.org/10.1073/pnas.1118847109

23. Bennicelli J, Wright JF, Komaromy A, Jacobs JB, Hauck B, Zelenaia O, Mingozzi F, Hui D, Chung D, Rex TS, Wei Z, Qu G, Zhou S, Zeiss C, Arruda VR, Acland GM, Dell'Osso LF, High KA, Maguire AM, Bennett J (2008) Reversal of blindness in animal models of leber congenital amaurosis using optimized AAV2-mediated gene transfer. Mol Ther 16 (3):458–465. https://doi.org/10.1038/sj.mt.6300389

24. Zolotukhin S, Potter M, Zolotukhin I, Sakai Y, Loiler S, Fraites TJ Jr, Chiodo VA, Phillipsberg T, Muzyczka N, Hauswirth WW, Flotte TR, Byrne BJ, Snyder RO (2002) Production and purification of serotype 1, 2, and 5 recombinant adeno-associated viral vectors. Methods 28(2):158–167

25. Veldwijk MR, Topaly J, Laufs S, Hengge UR, Wenz F, Zeller WJ, Fruehauf S (2002) Development and optimization of a real-time quantitative PCR-based method for the titration of AAV-2 vector stocks. Mol Ther 6(2):272–278

26. Lock M, Alvira MR, Chen SJ, Wilson JM (2014) Absolute determination of single-stranded and self-complementary adeno-associated viral vector genome titers by droplet digital PCR. Hum Gene Ther Methods 25 (2):115–125. https://doi.org/10.1089/hgtb.2013.131

27. Day TP, Byrne LC, Flannery JG, Schaffer DV (2018) Screening for neutralizing antibodies against natural and engineered AAV capsids in nonhuman primate retinas. Methods Mol Biol 1715:239–249. https://doi.org/10.1007/978-1-4939-7522-8_17

28. Desrosiers M, Dalkara D (2018) Neutralizing antibodies against adeno-associated virus (aav): measurement and influence on retinal gene delivery. Methods Mol Biol 1715:225–238. https://doi.org/10.1007/978-1-4939-7522-8_16

29. AVMA Guidelines for the Euthanasia of Animals: 2013 Edition

Chapter 15

Intracameral Delivery of AAV to Corneal Endothelium for Expression of Secretory Proteins

Jeffrey O'Callaghan, Matthew Campbell, and Peter Humphries

Abstract

AAV9 drives gene expression in a highly selective manner within the corneal endothelium of mice following intracameral inoculation into the anterior chamber of the eye. In principle, this allows genes encoding protein constituents of the secretome (representing up to 20% of the human proteome) to be delivered directly into the aqueous humor. From here the secreted protein moves with the natural flow of the aqueous humor via a pressure gradient and is directed toward the outflow tissues. Such a delivery can be employed to modulate outflow facility and intraocular pressure through interactions at the trabecular meshwork and Schlemm's canal. We provide a protocol for the delivery of AAV to the corneal endothelium, using a CMV-driven eGFP reporter gene as a marker.

Key words Intracameral injection, Anterior chamber, Protein secretion, Corneal endothelium, Outflow facility, Intraocular pressure modulation, AAV transduction

1 Introduction

The human secretome consists of a very large portfolio of proteins including cytokines, chemokines, proteolytic enzymes, extracellular matrix components, hormones, and antimicrobial peptides synthesized in the endoplasmic reticulum [1]. These proteins could, in principle, be used in experimental therapies for diseases of the anterior chamber of the eye, including ophthalmitis, uveitis, and glaucoma. As a prerequisite, such proteins must be expressed in such a manner that they are able to gain immediate access to the aqueous humor. In this regard, the corneal endothelium, the single layer of cells lining the inner part of the cornea, represents an ideal target for such expression, as these cells secrete proteins directly into the aqueous humor. The corneal endothelium retains proliferative capacity in humans but is maintained in a non-replicative state, rendering it an attractive cell type for long-term adeno-associated virus (AAV) vector-mediated gene expression [2]. These cells, with an average cell density of 3200 cells/mm^2 in subjects

Michael J. Castle (ed.), *Adeno-Associated Virus Vectors: Design and Delivery*, Methods in Molecular Biology, vol. 1950, https://doi.org/10.1007/978-1-4939-9139-6_15,

aged 50 and older, are entirely encompassed by the aqueous humor on the apical side, making them a readily accessible target for intracamerally injected AAV [3].

Disease pathology in primary open-angle glaucoma begins within the tissues of the conventional outflow pathway, the trabecular meshwork (TM) and Schlemm's canal. Here, through mechanisms that are incompletely understood, the natural flow of aqueous humor is impeded, resulting in elevated intraocular pressure and subsequent degeneration of the optic nerve head and retinal ganglion cells [4]. Interestingly, the pressure-reducing medications routinely used in disease management largely either slow down aqueous humor production by the ciliary body or increase aqueous humor drainage though the bundles of the ciliary muscles (the unconventional pathway for aqueous removal) [5]. Hence, pressure-reducing medications targeting conventional outflow are under active development.

A recent gene therapy approach targeting the TM involves delivering AAV9 carrying the matrix metalloproteinase-3 (MMP-3) gene to the corneal endothelium by intracameral injection of the vector [6]. The AAV drives gene expression selectively in corneal endothelium, resulting in secretion and activation of MMP-3 within the aqueous humor. This results in remodeling of the extracellular matrix of the TM, increasing aqueous outflow through the Schlemm's canal and reducing intraocular pressure. This method of delivery to the anterior chamber can also be used to routinely administer compounds including siRNAs or antibodies to the outflow tissues, as well as AAV serotypes capable of transducing cells of the TM other than (or in addition to) corneal endothelium [7–10]. A detailed and optimized protocol for AAV-mediated gene delivery to corneal endothelium in mice using a CMV-driven eGFP reporter gene is described, utilizing a simple, single intracameral injection (Fig. 1). This method involves a puncture into the anterior chamber to remove aqueous humor, followed by the infusion of viral solution from a guided syringe.

2 Materials

2.1 Intracameral Injection

1. AAV9 carrying eGFP: use this vector to assess transduction and reporter gene expression in the anterior chamber of wild-type mice (C57BL/6J). EGFP is expressed using the constitutive CMV promoter (*see* **Note 1**).
2. AAV9 null vector: use a vector of the same capsid serotype carrying a null cassette containing the CMV promoter and no transgene as a contralateral control (*see* **Note 1**).
3. 45° angled forceps.
4. Blunt forceps.

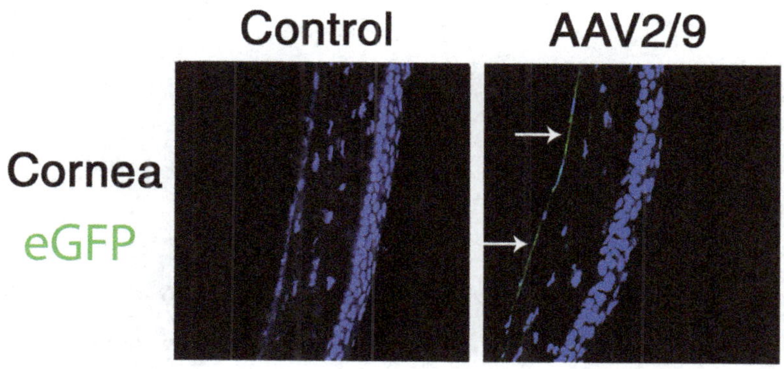

Fig. 1 AAV-mediated GFP expression from the corneal endothelium. GFP expression is observed in the murine eye 4 weeks after intracameral inoculation of AAV9 constitutively expressing eGFP. Expression is found to be exclusive to the corneal endothelium after immunostaining and confocal microscopy

5. Micromanipulator.
6. Hamilton syringes.
7. Hot bead sterilizer.
8. U.V. sterilizer cabinet.
9. Surgical microscope.
10. Vertical micropipette puller.
11. Murine head holder.
12. Anesthesia system including vaporizer, scavenger, induction chamber, and oxygen tank with regulator.
13. 1 mm diameter glass capillaries.
14. Sterile water.
15. Sterile drapes.
16. Parafilm.
17. Isoflurane.
18. Vidisic gel.
19. Fucithalmic antibacterial eyedrops.
20. Phenylephrine eye drops.
21. Tropicamide eye drops.
22. Tissue paper.

2.2 Detection of eGFP by Immunohistochemistry

1. Curved forceps.
2. Microdissection scissors.
3. 25G needle.
4. Stereo microscope.
5. Cryostat.

6. PAP pen.
7. Humid chamber.
8. Polylysine glass slides.
9. Coverslips.
10. Hydromount mounting medium.
11. Confocal microscope.
12. Imaging software.
13. 4% paraformaldehyde in PBS.
14. 10, 20, and 30% sucrose solutions in PBS.
15. OCT compound.
16. Small cryomolds.
17. Isopropanol.
18. Liquid nitrogen.
19. Blocking buffer: 1% BSA, 5% normal goat serum, and 0.1% Triton-X 100 in TBS.
20. Anti-eGFP rabbit primary antibody (Cell Signaling).
21. Anti-rabbit cy-3 conjugated secondary antibody (Abcam).
22. Counterstaining solution: 1:5000 dilution of 1 mg/mL DAPI in TBS.
23. 1× TBS: 50 nm Tris–Cl (pH 7.5), 150 nM NaCl.

3 Methods

3.1 Needle Preparation (See Note 2)

1. Insert glass capillaries into the micropipette puller and set heater to 90 °C. Secure the needle by tightening the lock screw and turn on the heater. One the heating coil getts to the set temperature, the weight will pull the capillary into two needles.
2. Place newly pulled needles into a clear petri dish and leave in a U.V. cabinet for 2–3 h to sterilize. Place forceps in the hot bead sterilizer for 2–3 min, then let cool.
3. Place a sterile drape on workspace. Prior to injection, use large, blunt forceps to break 1 mm off the tip of the needle to create an "injection needle." On a separate needle, break approximately 3 mm off the tip of the needle to create the slightly larger diameter "puncture needle."
4. Insert the injection needle into the ferrule and then into the barrel of a Hamilton syringe and secure the compression fitting by tightening the locknut.
5. Place a few drops of sterile water onto a small sheet of parafilm. Withdraw this water into the needle, loosening and tightening

the compression fittings as necessary, until the needle is filled and no air is present in the needle or syringe barrel (*see* **Note 3**).

6. Withdraw 2 μL of air. Withdraw 1.5 μL of vector. Withdraw a further 0.5 μL of air (*see* **Note 3**).

3.2 Animal Preparation

1. Expose the adult mouse to 2.5% isoflurane in oxygen at a flow rate of 1 L/min in an induction chamber. After 2 min, transfer the animal to a head holder with a vaporizer/scavenger attachment. Animals should be placed on a heating pad atop a surgical drape.
2. Dilate the pupils by applying one drop of phenylephrine and tropicamide eye drops to each eye (*see* **Note 4**).
3. Adjust the micromanipulator so that it is approximately 10° to the horizontal plane in a downward direction, and secure the Hamilton syringe containing the virus.
4. Adjust the head holder so that the animal is on its side and the entire circumference of the eye can be seen through the surgical microscope. Dab any residual eye drops using tissue paper to remove them.

3.3 Intracameral Injection (See Notes 5 and 6)

1. Using the angled forceps, secure the eye by gripping the far and near sides of the sclera. Take the puncture needle and hold it at approximately 10° to match the angle of the micromanipulator.
2. Gently press the puncture needle through the cornea just above the limbus, taking care not to scratch the corneal surface or the lens. Due to the intraocular pressure, aqueous humor will enter the needle by capillary action. After 5–10 s, 2–3 μL will have entered the needle. Slowly remove the needle and release the forceps (*see* **Note 7**).
3. Align the micromanipulator to the puncture site on the corneal surface, adjusting to attain the same insertion angle as the puncture needle. Using the micromanipulator, insert the injection needle into the puncture site 2–3 mm so that the puncture site is sealed around the needle.
4. Slowly press on the syringe plunger to release the 0.5 μL air bubble and 1.5 μL vector solution into the eye (*see* **Note 8**). Wait for 1 min to allow any increased intraocular pressure to disperse and equilibrate. Ensure that the air bubble lies near the puncture site (*see* **Notes 9** and **10**).
5. Use the micromanipulator to slowly remove the needle from the eye. Apply a drop of fucithalmic antibacterial eye drop solution to the eye. Vidisic gel may also be applied during recovery to hydrate the eye if necessary.
6. Repeat the procedure for the other eye using a control null vector, or remove the animal from the head holder and allow it to recover.

3.4 Tissue Preparation

1. Wait 4 weeks to allow vector-mediated eGFP expression to peak. After 4 weeks, enucleate eyes using curved forceps and place them into fixative for 24 h at 4 °C.
2. Remove eyes from fixative and dissect under a stereo microscope. Using a 25G needle or similar, pierce the sclera just below the limbus. Insert one blade of the microdissection scissors into this opening and cut around the circumference of the eye. This dissects an anterior segment containing all of the cornea, iris, and outflow tissues. Remove the lens before proceeding.
3. Place the anterior segment into a sucrose gradient (subsequent solutions of 10, 20, and 30% sucrose) until the tissue sinks to the bottom (*see* **Note 11**).
4. Embed the tissue in OCT compound in a cryomold, and place the mold into a container filled with isopropanol. Flash-freeze the tissue by placing this isopropanol bath into a trough of liquid nitrogen.

3.5 Detection of eGFP by Immunohistochemistry

1. Attach the OCT block to the specimen disk with OCT compound. Place in the cryostat's object head and trim excess OCT from the block.
2. Section the tissue at a thickness of 12 μm and collect sections on glass slides.
3. Draw around sections with a PAP pen and let dry.
4. Perform one wash by immersing the sections in 1× TBS for 5 min, then drain the TBS off the slide. Add blocking buffer and leave for 1 h at room temperature.
5. Apply primary antibody (1:250 dilution in TBS) and place into a humid chamber at 4 °C overnight.
6. Wash slides three times for 5 min each in TBS. Drain and add secondary antibody (1:500 dilution in TBS) to the slides. Add TBS without secondary antibody to one slide as a negative control. Incubate protected from light for 2 h at room temperature.
7. Wash slides two times for 5 min each in TBS, then add counterstaining solution for 30–60 s. Perform one final wash for 5 min in TBS and remove all liquid from the slides. Apply hydromount mounting medium and gently attach a coverslip.
8. Once mounting medium has dried, use a light microscope or confocal microscope to determine the intensity of fluorescence. Imaging software can be employed to perform stitching, stacking, or quantitative analysis across sections.

4 Notes

1. Dilute viral solutions to a working titer of 2×10^{12} vector genomes (vg)/mL using nuclease-free water. Vectors were purchased from Vector Biolabs.
2. Glass needles are extremely brittle and care must be taken to preserve the tips. Blue tack can be used to secure needles in parallel along the diameter of a petri dish for storage and transport.
3. Filling the injection needle with water while in the syringe improves responsiveness when withdrawing small volumes, as it removes air compression and expansion. Mineral oil can also be used instead of the 2 μL of air to prevent dilution or mixing of the virus and the water in the needle.
4. The length of time that phenylephrine and tropicamide eye drops should stay on the eye may vary between animals, but will take on average 3–5 min for the pupil to fully dilate. When these drops are removed from the first eye by gentle dapping of tissue paper, the drops on the other eye can be left until the first eye is injected so as to prevent dehydration.
5. This is an effective and repeatable method that takes less than 15 min for the inoculation of both eyes in the mouse from the onset of anesthesia. Variations of this technique have also been described [9, 10].
6. Care should be taken to avoid any scratches on any ocular surfaces. Aseptic technique should be employed to reduce the risk of infection or cataract. Needles should minimally penetrate the eye as to reduce the likelihood of infection or hemorrhage. Puncturing too close to the limbus may result in piercing of the iris or an iris retraction into the wound site when healing.
7. When puncturing the cornea, having the puncture needle at the same angle as the micromanipulator makes insertion of the injection needle easier. More than 2–3 μL can be removed from the anterior chamber if greater volumes will be injected. However, deflation of the eye can result in a depressed puncture site, making insertion of the injection needle difficult.
8. A syringe pump may be used in place of a Hamilton syringe. An infusion rate of 1–2 μL per min can be used to minimize any sudden transient increases in intraocular pressure.
9. It is preferable for the air bubble to reside at the puncture site after injection to prevent reflux. The air bubble can be guided through the cornea using forceps, but it is recommended that the eye be tilted so that the bubble will naturally reside at the

puncture site. A small amount (0.5–1 μL) of reflux is tolerated with this method.

10. Fucithalmic eye drops may be applied before removal of the injection needle to further reduce reflux.
11. Enucleated eyes can be stored in 0.01% sodium azide in PBS for long-term storage. Sections on slides may also be stored at −20 °C prior to staining.

References

1. Sage H, Pritzl P, Bornstein P (1981) Secretory phenotypes of endothelial cells in culture: comparison of aortic, venous, capillary, and corneal endothelium. Arteriosclerosis 1(6):427–442
2. Joyce NC (2003) Proliferative capacity of the corneal endothelium. Prog Retin Eye Res 22 (3):359–389
3. Wilson RS, Roper-Hall MJ (1982) Effect of age on the endothelial cell count in the normal eye. Br J Ophthalmol 66(8):513–515
4. Campbell M, Cassidy PS, O'Callaghan J, Crosbie DE, Humphries P (2018) Manipulating ocular endothelial tight junctions: applications in treatment of retinal disease pathology and ocular hypertension. Prog Retin Eye Res 62:120–133. https://doi.org/10.1016/j.preteyeres.2017.09.003
5. O'Callaghan J, Cassidy PS, Humphries P (2017) Open-angle glaucoma: therapeutically targeting the extracellular matrix of the conventional outflow pathway. Expert Opin Ther Targets 21(11):1037–1050. https://doi.org/10.1080/14728222.2017.1386174
6. O'Callaghan J, Crosbie DE, Cassidy PS, Sherwood JM, Flugel-Koch C, Lutjen-Drecoll E, Humphries MM, Reina-Torres E, Wallace D, Kiang AS, Campbell M, Stamer WD, Overby DR, O'Brien C, Tam LCS, Humphries P (2017) Therapeutic potential of AAV-mediated MMP-3 secretion from corneal endothelium in treating glaucoma. Hum Mol Genet 26(7):1230–1246. https://doi.org/10.1093/hmg/ddx028
7. Tam LC, Reina-Torres E, Sherwood JM, Cassidy PS, Crosbie DE, Lutjen-Drecoll E, Flugel-Koch C, Perkumas K, Humphries MM, Kiang AS, O'Callaghan J, Callanan JJ, Read AT, Ethier CR, O'Brien C, Lawrence M, Campbell M, Stamer WD, Overby DR, Humphries P (2017) Enhancement of outflow facility in the murine eye by targeting selected tight-junctions of Schlemm's canal endothelia. Sci Rep 7:40717. https://doi.org/10.1038/srep40717
8. Raghuram A, Saravanan VR, Narendran V (2007) Intracameral injection of bevacizumab (Avastin) to treat anterior chamber neovascular membrane in a painful blind eye. Indian J Ophthalmol 55(6):460–462
9. Wang L, Xiao R, Andres-Mateos E, Vandenberghe LH (2017) Single stranded adeno-associated virus achieves efficient gene transfer to anterior segment in the mouse eye. PLoS One 12(8):e0182473. https://doi.org/10.1371/journal.pone.0182473
10. Bogner B, Boye SL, Min SH, Peterson JJ, Ruan Q, Zhang Z, Reitsamer HA, Hauswirth WW, Boye SE (2015) Capsid mutated adeno-associated virus delivered to the anterior chamber results in efficient transduction of trabecular meshwork in mouse and rat. PLoS One 10 (6):e0128759. https://doi.org/10.1371/journal.pone.0128759

Chapter 16

AAV-Mediated Gene Delivery to the Inner Ear

Omar Akil and Lawrence Lustig

Abstract

Cochlear gene therapy has made tremendous strides over the past 5 years. The first study documenting successful restoration of congenital hearing loss using AAV-mediated gene therapy occurred in a mouse model of deafness lacking vesicular glutamate transporter 3 (VGLUT 3). This study utilized a trans-bulla round window membrane (RWM) delivery approach. Since this study, these methodologies have been applied to a number of other mouse models of genetic deafness with varying degrees of success, lending promise for future clinical application of this burgeoning technology. Here we describe a method of virally mediated gene delivery into the cochlear scala tympani through the RWM. This method involves negligible damage to essential structures of the middle and inner ear while preserving hearing. The efficacy of this surgical technique will be demonstrated by the restoration of hearing to the VGLUT3 knockout mice (a mouse model of congenital deafness) after delivery of VGLUT3 gene to the inner ear using an adeno-associated virus as a vector.

Key words Gene therapy, Cochlea, Hearing, Adeno-associated virus, Vesicle glutamate transporter 3, Inner hair cells, Mouse, Transfection

1 Introduction

Hearing impairment is the most common sensory deficit in humans [1]. About 1.5 in 1000 newborns suffer from congenital hearing loss (HL) [2], and genetic forms of sensorineural deafness account for almost half of all patients with HL [3]. Treatments such as amplification and cochlear implantation [1, 4] have substantial limitations and do not restore normal hearing [1]. As a result, cochlear gene therapy has been suggested as an alternative route for treatment for both genetic and acquired forms of HL. Gene therapy for inner ear disorders has developed along two major lines of investigation: hair cell or sensory regeneration and gene therapy for genetic deafness. However, it is important to note that breakthroughs in cellular regeneration may not benefit those with genetic deafness, since the underlying genetic background remains unchanged [5, 6] and the regenerated hair cells may be subject to the same genetic mutation that led to their loss in the first place.

Michael J. Castle (ed.), *Adeno-Associated Virus Vectors: Design and Delivery*, Methods in Molecular Biology, vol. 1950, https://doi.org/10.1007/978-1-4939-9139-6_16,

Thus, for genetic disorders leading to congenital or early-onset hearing loss, the optimal solution may be defective gene replacement. Significant recent advances in gene therapy in mouse models of deafness [7, 8] support the promise of hearing restoration in humans.

In order for gene therapy to be effective, the required gene must be appropriately delivered both at the right time and to the right target(s). A key to this process is utilizing a vector that achieves these goals. To date both adenovirus (AV) and adeno-associated virus (AAV) have been used for cochlear gene delivery. While AV achieves transfection of a wide variety of cell types within the cochlea, its chief limitation relates to the limited transgene expression time, typically on the order of months [9], in addition to its potential immunogenicity (though reduced with modern vector design). While this is advantageous for hair cell regeneration where limited expression is all that is required, this is inadequate for the long-term gene expression needed for genetic forms of hearing loss.

In contrast, AAV has shown promise for gene therapy for genetic forms of hearing loss due to its proven safety profile and high transduction efficiency. AAVs are replication-deficient viruses and can efficiently transfer transgenic molecules to different cochlear cell types including hair cells [10] and neurons with cellular specificity for each AAV serotype. Furthermore, AAV can incorporate into the host genome, resulting in stable, long-term transgene expression for over a year [11], an important advantage over other viruses. Lastly, because AAV is well tolerated (no demonstrated immune response) [12, 13], is not associated with any human disease or infection, and has demonstrated no ototoxicity [14–16], it remains an ideal candidate for use in gene therapy for inherited forms of hearing loss [17].

To date there have been a number of studies that have documented partial or complete rescue of hearing loss in mouse models of genetic deafness. The first to achieve this utilized a mouse lacking the vesicular glutamate transporter 3 (VGLUT3), a mouse model of congenital deafness DFNA25, which lost hearing due to loss of glutamate transmission at the inner hair cells (IHCs) afferent synapse [18]. It was shown in this model that local monogenetic therapy using an AAV carrying the VGLUT3 gene, delivered through the cochlear round window membrane (RWM), restored hearing to near normal levels for an extended period of time [7]. Following this publication, a number of other groups utilized similar techniques in other models of genetic deafness with varying degrees of success and have employed delivery through other access points within the labyrinth, including the oval window and the semicircular canals [19–24].

The following review describes the original technique that delivers AAV1 containing a VGLUT3 insert to the inner ear of the VGLUT-3 knockout mice using a round window membrane approach. The methodology allows for a high degree of transfection (up to 100%) with appropriately concentrated viral subtype and sufficient titer.

2 Materials

2.1 Surgical Procedures and Vector Injection

1. Ketamine-Xylazine solution: add 2 mL of ketamine (100 mg/mL) to 1 mL of xylazine (20 mg/mL) and adjust the total volume to 5 mL using sterile bi-distilled water, mix well, place in a rubber sealed vial, and store at 4 °C until used (*see* **Notes 1** and **2**).
2. Acepromazine.
3. Carprofen analgesia.
4. 25G disposable needle.
5. 100 mm borosilicate capillary pipette (1 mm/0.58 mm OD/ID).
6. Antibacterial suture.
7. Tissue glue.
8. Shaving clipper.
9. 70% ethanol.
10. Dissecting microscope.
11. 5% Povidone-iodine (Betadine) antiseptic-microbicide solution.
12. Dexamethasone ophthalmic ointment.
13. Nonelectric heating pad.
14. Rectal Probe for body temperature monitoring.
15. Clean empty cages.
16. Tweezers style 5.

2.2 Assessment of Cochlear Transgene Protein Expression Using Immunofluorescence

1. Ketamine-Xylazine solution.
2. Rabbit Anti-GFP antibody (Invitrogen #A11122).
3. Guinea pig Anti-VGLUT3 antibody.
4. Goat anti-Rabbit Cy2-conjugated secondary antibody.
5. 1.5 μg/mL DAPI in PBS.
6. Glass slides.
7. Mounting medium.
8. Confocal microscope.

2.3 Assessment of Cochlear Function Following Viral Delivery Using Auditory Brainstem Response (ABR) Recordings

1. Ketamine-Xylazine solution.
2. 27G × 25 mm disposable monopolar needle electrodes for electromyography.
3. TDT BioSig III system (Tucker Davis technologies) for ABR recording.
4. Moist electric heating pad.
5. Rectal Probe for body temperature monitoring.
6. Clean empty cages.

3 Methods

All procedures and animal handling described in this protocol should be done according to approved national ethical guidelines and comply with all protocol requirements of the Institutional Animal Care and Use Committee (*see* **Note 3**)

3.1 Surgical Procedure and Vector Injection

1. Perform surgical procedures in a clean, dedicated space. Autoclave all surgical instruments and sterilize with a glass-bead sterilizer prior to surgery (*see* **Notes 4–6**).
2. Sedate the mice with intraperitoneal injection of a mixture of ketamine hydrochloride (100 mg/kg), xylazine hydrochloride (10 mg/kg), and acepromazine (2 mg/kg). Begin surgical preparation only after the animal no longer responds to painful stimuli, such as toe pinch. If necessary, administer a booster dose (one-fifth the original dose) of the anesthetic cocktail to restore the original anesthetic plane (*see* **Notes 7** and **8**).
3. Apply a protective ophthalmic ointment to the animal's eyes to keep them moist during anesthesia and suppress the animal's blink reflex.
4. Place the mouse with neck extended on a heating pad throughout the procedure until the mouse is totally awake, to prevent post-anesthesia hypothermia (*see* **Note 9**).
5. Shave the post-auricular region with a clipper and disinfect with 70% ethanol and povidone-iodine before surgical manipulation.
6. Use a post-auricular approach to expose the tympanic bulla. Incise the subcutaneous tissue with small scissors to expose the post-auricular muscle. After retracting the adipose tissue to the posterior side of the incision, separate the muscles using fine tweezers to the right and left side perpendicular to the incision to expose the temporal bone. Ensure that this incision is long enough to create a sufficient visual field.

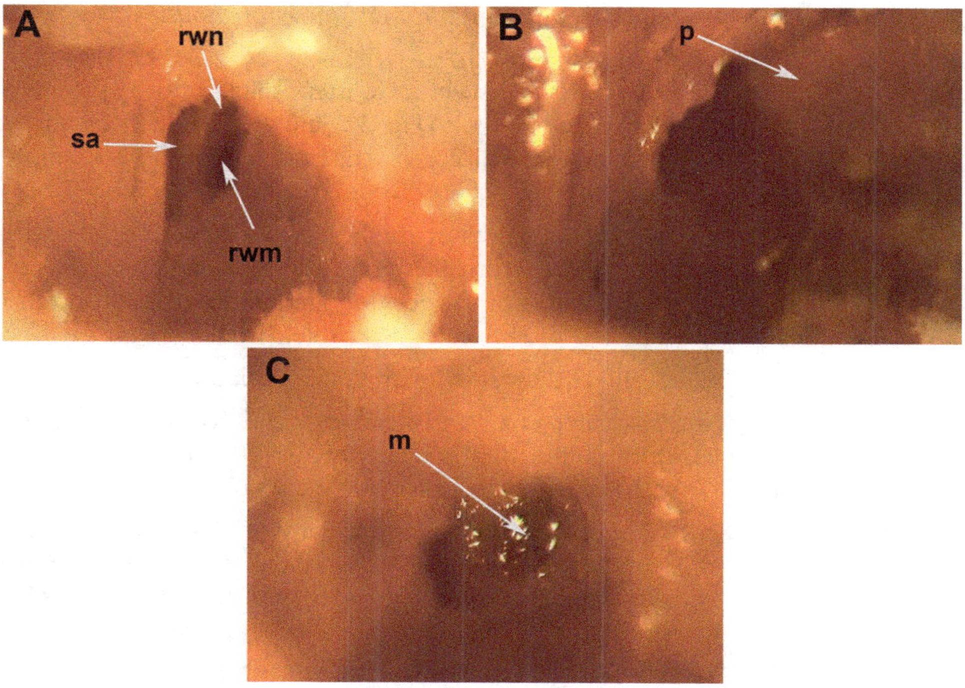

Fig. 1 Photomicrographs illustrating the most important steps of the surgical approach to the mouse ear. The bony bulla is opened laterally to expose the round window niche. The round window niche (rwn) and the stapedial artery (sa) are visualized in (**a**). The rwn is usually oval, and gives a direct and free visualization to the round window membrane (rwm). The stapedial artery (sa) runs over the round window niche and traverses between the crura of stapes into the middle cranial fossa. (**b**) shows the microinjection of the virus using a micropipette (p) into the scala tympani through the rwm. (**c**) shows rwn sealed after injection with a small plug of muscle (m)

7. Perforate the tympanic bulla with a 25G needle and expand the hole as necessary with forceps by peeling the bone back to allow access to the basal turn of the cochlea, and then widen sufficiently to visualize the stapedial artery and the round window membrane (RWM) (Fig. 1a).
8. Puncture the RWM gently in the center with a borosilicate capillary pipette. Observe fluid efflux through the RWM at this point; this is normal. Wait until the efflux has stabilized (the effluxed fluid from the cochlea is dried with sterile filter paper) (*see* **Note 10**).
9. Prepare the pipettes for injection using a pipette puller the same way patch clamp pipettes are prepared. Ensure that the tip diameter is large (about 15 μm in diameter) so that pipetting the virus in and out is done easily.
10. Draw up 0.6–2 μL of AAV1-VGLUT3, AAV2-GFP, or AAV5-GFP into an injection pipette. After the efflux is stabilized (5–10 min), microinject this fixed volume into the scala

tympani through the same hole previously made in RWM (Fig. 1b).

11. Seal the RW niche quickly after pulling out the pipette with a small plug of muscle (Fig. 1c) and secure it with a small drop of tissue adhesive placed on the muscle to avoid perilymph leakage from the cochlea through the RWM (*see* **Note 11**).
12. After injection, cover the hole in the auditory bulla with adipose tissue, return the post-auricular muscles and adipose tissue to their normal position, and suture the wound in layers with a 6–0 or smaller absorbable chromic suture.
13. Disinfect the wound with povidone-iodine.
14. Place mice in a warm clean cage and do not leave them unattended until they are fully recovered, then move them back with the mother (*see* **Notes 12** and **13**).
15. Administer subcutaneous Carprofen (2 mg/kg) for analgesia postoperatively and every 24 h thereafter for 3 days, to manage inflammation and pain. Monitor the animals daily for signs of distress, abnormal weight loss, pain, or infection. In general all mice should be acting normally by the third day after the surgery. If any signs of distress or disease appear in a mouse after the third day, consider euthanizing the mouse as per institutional guidelines.

3.2 Assessment of Cochlear Transgene Protein Expression Using Immunofluorescence

1. Anesthetize mice by an overdose of intraperitoneal injection of a mixture of ketamine hydrochloride and xylazine hydrochloride as described above (*see* **Note 14**). Decapitate the head only after the animal no longer responds to painful stimuli, such as a toe pinch.
2. Dissect and remove the cochleae and process for whole-mount immunofluorescence as described [7]. A detailed protocol of cochlear whole-mount immunofluorescence using anti-VGLUT3 and anti-Myosin 7a antibodies is also described in Akil and Lustig [25].
3. For GFP labeling, incubate the cochlear whole-mounts overnight at 4 °C with a rabbit anti-GFP antibody at 1:250 dilution.
4. Rinse the cochleae twice for 10 min with PBS and then incubate for 2 h in goat anti-rabbit IgG conjugated to Cy2, diluted 1:4000 in PBS.
5. Rinse the cochleae with PBS twice for 10 min and incubate them with DAPI for 15 min.
6. Mount the cochleae on glass slides and observe under a microscope with confocal immunofluorescence (Figs. 2 and 3).

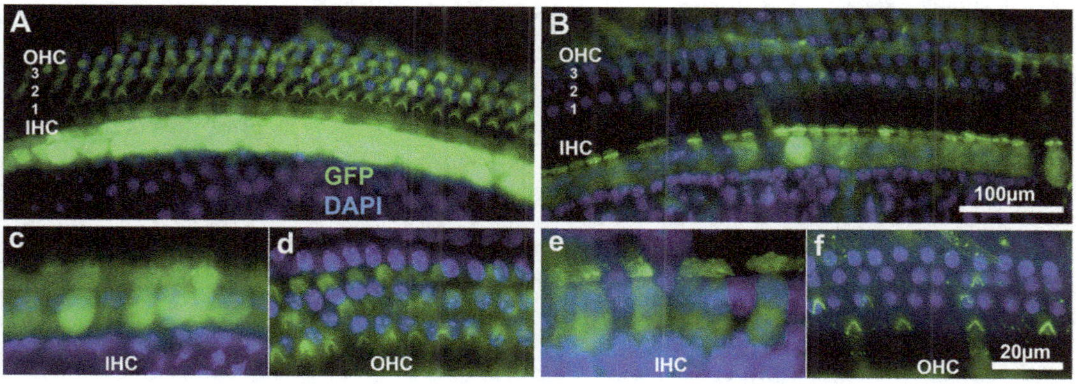

Fig. 2 Mouse cochlear whole-mount immunofluorescence after RWM AAV2-GFP and AAV5-GFP delivery to wild-type P1012 mice. To determine the optimal AAV serotype to be used for a study, AAV2-GFP and AAV5-GFP were delivered into P10-12 mice cochlea via the RWM using the post-auricular approach. After AAV2-GFP delivery to the cochlea expression of GFP is seen in inner hair cells (IHC), outer hair cells (OHCs), and some supporting cells (SC) ((**a**) overview and (**c**, **d**) magnified view), whereas AAV5-GFP delivery to the cochlea induces expression of GFP mostly in IHCs with some OHC expression ((**b**) overview and (**e**, **f**) magnified view). The observation of different cell types expressing GFP seen in these images demonstrates that the choice of AAV serotype must be compatible with the cell types desired to be transduced. For example, AAV2 serotype is a better choice if the targeted hair cells are both IHCs and OHCs, whereas AAV5 serotype is superior for only IHC transfection. In this study AAV1 serotype was chosen as a vector for the delivery of VGLUT3 to VGLUT3 KO mice ears based on prior studies that documented robust IHC transduction [29]

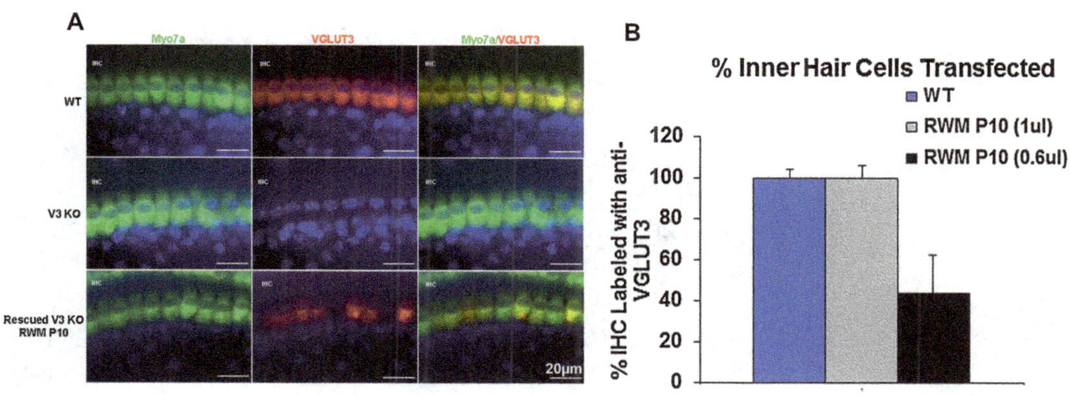

Fig. 3 VGLUT3 IHC transfection after RWM viral delivery to P10–P12 KO mice. (**a**) The transfected cochleae were examined at P30 by immunofluorescence using anti-Myo7a antibody, a hair-cell marker, and an anti-VGLUT3 antibody. Myo7a and VGLUT3 were expressed in all IHCs only in the WT (row 1), while IHCs from KO mice expressed only Myo7a (row 2). The rescued mouse cochleae showed some IHCs expressing VGLUT3, and all IHCs expressed Myo7a (row 3). (**b**) The number of IHCs expressing VGLUT3 in the VGLUT3 KO mice cochlea after AAV1-VGLUT3 delivery depends on the amount of virus delivered to the inner ear; for example, 40% of IHCs expressed VGLUT3 when 0.6 μL of the virus was injected, whereas 100% of IHCs expressed VGLUT3 after 1 μL of the virus was delivered. There was no difference in VGLUT3 expression between the apex, mid-turn, or base when 0.6 μL of virus was injected. The distribution of IHCs expressing VGLUT3 after AAV1-VGLUT3 transfection (Fig. 3b) [7] suggests that transfection via the RWM is more effective and uniform through the whole cochlea than other methods that have reported higher transfection rates closer to the site of viral delivery. *IHC* inner hair cells. Reproduced from Akil et al. [7] with permission from *Neuron*

3.3 Assessment of Cochlear Function Following Viral Delivery Using Auditory Brainstem Response (ABR) Recordings

1. Anesthetize mice by intraperitoneal injection of a mixture of ketamine hydrochloride and xylazine hydrochloride as described above (*see* **Note 15**).
2. Place the mouse on a heating pad throughout the hearing test until the mouse is totally awake, to prevent post-anesthesia hypothermia (*see* **Note 16**).
3. Place subdermal needle electrodes at the vertex, below the pinna of the left ear (reference), and below the contralateral ear (ground) and start recording ABRs as previously described in a soundproof chamber [26, 27]. Record ABR waveforms in 5 dB sound pressure level intervals down from maximum amplitude (*see* **Notes 16** and **17**). A detailed protocol of auditory brainstem recording is also described in Akil et al. [28].
4. Determine ABR thresholds postoperatively as early as 7 days after viral delivery (*see* **Note 18**).
5. Measure Wave I to analyze the activity from the cochlear nerve. The lowest stimulus level that yields a detectable ABR waveform is defined as the threshold (Fig. 4).

4 Notes

1. This solution of ketamine and xylazine can be used as long as the ketamine and xylazine are not expired and are sterile.
2. The solution of ketamine and xylazine should be stored at 4 °C when it is not in use and at room temperature when it is in use.
3. Appropriate site-specific IACUC approvals are required to perform the following procedures.
4. The *JoVe* video titled "Virally Mediated Gene Delivery to the Mouse Inner Ear through the Round Window Membrane" visualizes all steps of this method [28].
5. In this protocol, we used postnatal day 10–12 (P10–12) FVB mice. Different ages and strains of mice can be used to meet the needs of a specific project.
6. Mice older than P40 can be challenging because the bulla bone gets harder.
7. Take care at this step when injecting the anesthetic because most of the mortality observed at this age is caused by anesthetic overdose.
8. Due to the duration of the surgery (~30–40 min), one fifth of the original dose of the anesthetic should be injected at around 20 min to avoid the likelihood of the mouse waking up during the surgery.

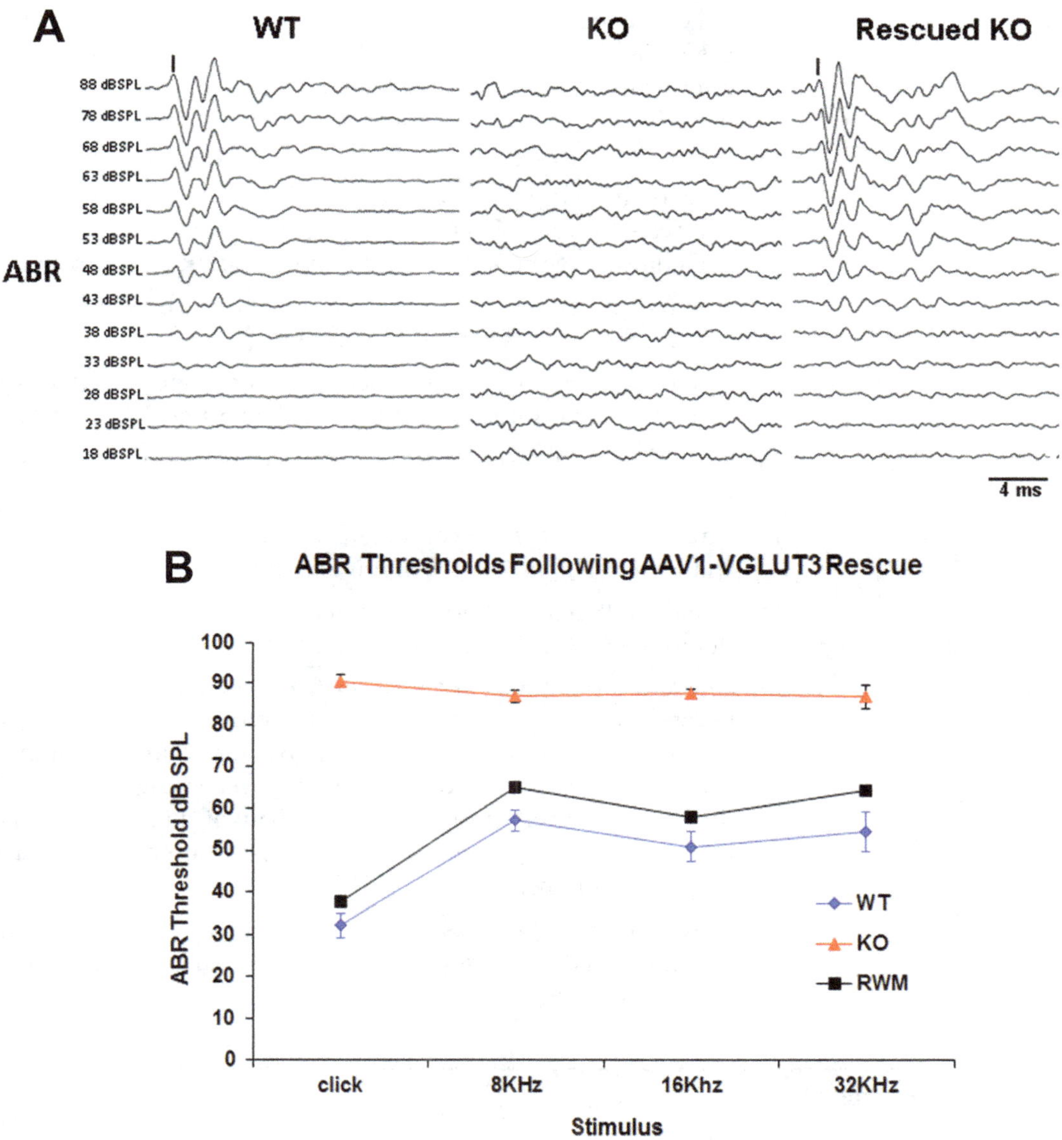

Fig. 4 Auditory brainstem response (ABR) assessment. Representative ABR waveforms after AAV1-VGLUT3 delivery were similar between rescued and WT mice; in contrast, no ABR waveforms were seen in VGLUT3 KO mice (**a**). Rescued KO mice also showed measurable ABR thresholds. These ABR thresholds were comparable to those seen in WT mice for all frequencies measured (**b**). I: ABR wave I. These results document the ability of post-auricular RWM delivery of AAV1-VGLUT3 into the inner ear of the VGLUT3 KO mice to produce hearing rescue in this mouse model of congenital hearing loss. Reproduced from Akil et al. [7] with permission from *Neuron*

9. Animal body temperature can be monitored with a rectal probe throughout surgery and the ABR recordings and the mouse's body temperature must be maintained warm using a heating pad.

10. Take care when holding and advancing the glass needle so that the RWM perforation is as small as possible. Depending on the microscope used, hold the pipettes using a micromanipulator or by hand using a pipette holder.
11. Seal the hole very well after the pipette is removed with a small plug of muscle to prevent any perilymph leakage from the cochlea—this step is critical to preserve hearing. Failure to seal completely the RWM will result in hearing loss over time.
12. It is recommended to remove the male from the cage before putting the pups back with the mother.
13. If mice are weaned multiple mice of the same group/litter can be placed in the same cage (no more than 5 mice).
14. There is no need for Acepromazine at this step because it is a terminal procedure.
15. There is no need for Acepromazine at this step because there will be no pain in this step and it is not a surgical procedure.
16. Use the following sound stimuli in this experiment: click (5 ms duration; 31 Hz presentation rate).
17. The auditory brainstem response (ABR) is an auditory evoked potential extracted from ongoing electrical activity in the brain and recorded via electrodes placed under the scalp. The animal is stimulated with sound. The resulting recording consists of five waves that reflect the electrical activity of successive points in the auditory pathway in the first 10 ms after onset of an auditory stimulus.
18. Threshold is defined as the lowest stimulus level at which response peaks for waves I–V were clearly and repeatedly present on visual inspection.

Acknowledgments

The authors would like to acknowledge the financial support provided by Hearing Research Inc. (HRI). We thank *Neuron* for some illustrating images we used in this chapter.

References

1. Kral A, O'Donoghue GM (2010) Profound deafness in childhood. N Engl J Med 363 (15):1438–1450
2. Di Domenico M, Ricciardi C, Martone T, Mazzarella N, Cassandro C, Chiarella G, D'Angelo L, Cassandro E (2011) Towards gene therapy for deafness. J Cell Physiol 226 (10):2494–2499
3. Shearer AE, Hildebrand MS, Sloan CM, Smith RJ (2011) Deafness in the genomics era. Hear Res 282(1–2):1–9
4. Petersen MB, Willems PJ (2006) Non-syndromic, autosomal-recessive deafness. Clin Genet 69(5):371–392
5. Izumikawa M, Minoda R, Kawamoto K, Abrashkin KA, Swiderski DL, Dolan DF,

Brough DE, Raphael Y (2005) Auditory hair cell replacement and hearing improvement by Atoh1 gene therapy in deaf mammals. Nat Med 11:271–276

6. Praetorius M, Hsu C, Baker K, Brough DE, Plinkert P, Staecker H (2009) Adenovector-mediated hair cell regeneration is affected by promoter type. Acta Otolaryngol 130 (2):215–222
7. Akil O, Seal RP, Burke K, Wang C, Alemi A, During M, Edwards RH, Lustig LR (2012) Restoration of hearing in the VGLUT3 knockout mouse using virally mediated gene therapy. Neuron 75(2):283–293
8. Lentz JJ, Jodelka FM, Hinrich AJ, McCaffrey KE, Farris HE, Spalitta MJ, Bazan NG, Duelli DM, Rigo F, Hastings ML (2013) Rescue of hearing and vestibular function by antisense oligonucleotides in a mouse model of human deafness. Nat Med 19(3):345–350
9. Staecker H, Schlecker C, Kraft S, Praetorius M, Hsu C, Brough DE (2014) Optimizing atoh1-induced vestibular hair cell regeneration. Laryngoscope 124:S1–S12
10. Ryan AF, Mullen LM, Doherty JK (2009) Cellular targeting for cochlear gene therapy. Adv Otorhinolaryngol 66:99–115
11. Xia L, Yin S, Wang J (2012) Inner ear gene transfection in neonatal mice using adeno-associated viral vector: a comparison of two approaches. PLoS One 7(8):e43218
12. Bennett J, Ashtari M, Wellman J, Marshall KA, Cyckowski LL, Chung DC, McCague S, Pierce EA, Chen Y, Bennicelli JL, Zhu X, Ying GS, Sun J, Wright JF, Auricchio A, Simonelli F, Shindler KS, Mingozzi F, High KA, Maguire AM (2012) AAV2 gene therapy readministration in three adults with congenital blindness. Sci Transl Med 4(120):120ra15
13. Simonelli F, Maguire AM, Testa F, Pierce EA, Mingozzi F, Bennicelli JL, Rossi S, Marshall K, Banfi S, Surace EM, Sun J, Redmond TM, Zhu X, Shindler KS, Ying GS, Ziviello C, Acerra C, Wright JF, McDonnell JW, High KA, Bennett J, Auricchio A (2010) Gene therapy for Leber's congenital amaurosis is safe and effective through 1.5 years after vector administration. Mol Ther 18(3):643–650
14. Husseman J, Raphael Y (2009) Gene therapy in the inner ear using adenovirus vectors. Adv Otorhinolaryngol 66:37–51
15. Ballana E, Wang J, Venail F, Estivill X, Puel JL, Arbonès ML, Bosch A (2008) Efficient and specific transduction of cochlear supporting cells by adeno-associated virus serotype 5. Neurosci Lett 442(2):134–139
16. Praetorius M, Brough DE, Hsu C, Plinkert PK, Pfannenstiel SC, Staecker H (2009) Adenoviral vectors for improved gene delivery to the inner ear. Hear Res 248(1–2):31–38
17. Kay MA, Glorioso CG, Naldini L (2001) Viral vectors for gene therapy: the art of turning infectious agents into vehicles of therapeutics. Nat Med 7(1):33–40
18. Seal RP, Akil O, Yi E, Weber CM, Grant L, Yoo J, Clause A, Kandler K, Noebels JL, Glowatzki E, Lustig LR, Edwards RH (2008) Sensorineural deafness and seizures in mice lacking vesicular glutamate transporter 3. Neuron 57:263–275
19. Askew C, Rochat C, Pan B, Asai Y, Ahmed H, Child E, Schneider BL, Aebischer P, Holt JR (2015) Tmc gene therapy restores auditory function in deaf mice. Sci Transl Med 7 (295):295ra108
20. Emptoz A, Michel V, Lelli A, Akil O, Boutet de Monvel J, Lahlou G, Meyer A, Dupont T, Nouaille S, Ey E, Franca de Barros F, Beraneck M, Dulon D, Hardelin JP, Lustig L, Avan P, Petit C, Safieddine S (2017) Local gene therapy durably restores vestibular function in a mouse model of Usher syndrome type 1G. Proc Natl Acad Sci U S A 114(36):9695–9700
21. Geng R, Akil O, Gopal SR, Chen DH, Stepanyan R, Basch ML, Dinculescu A, Furness DN, Saperstein D, Hauswirth W, Lustig LR, Alagramam KN (2017) Modeling and preventing progressive hearing loss in Usher syndrome III. Sci Rep 7:13480
22. Isgrig K, Shteamer JW, Belyantseva IA, Drummond MC, Fitzgerald TS, Vijayakumar S, Jones SM, Griffith AJ, Friedman TB, Cunningham LL, Chien WW (2017) Gene therapy restores balance and auditory functions in a mouse model of Usher syndrome. Mol Ther 25(3):780–791
23. Pan B, Askew C, Galvin A, Heman-Ackah S, Asai Y, Indzhykulian AA, Jodelka FM, Hastings ML, Lentz JJ, Vandenberghe LH, Holt JR, Géléoc GS (2017) Gene therapy restores auditory and vestibular function in a mouse model of Usher syndrome type 1c. Nat Biotechnol 35 (3):264–272
24. Suzuki J, Hashimoto K, Xiao R, Vandenberghe LH, Liberman MC (2017) Cochlear gene therapy with ancestral AAV in adult mice: complete transduction of inner hair cells without cochlear dysfunction. Sci Rep 7:45524
25. Akil O, Lustig LR (2013) Mouse cochlear whole mount immunofluorescence. Bio Protoc 3(5):e332
26. Akil O, Chang J, Hiel H, Kong JH, Yi E, Glowatzki E, Lustig LR (2006) Progressive

deafness and altered cochlear innervation in knock-out mice lacking prosaposin. J Neurosci 26(5):13076–13088

27. Fremeau RT, Kam K, Qureshi T, Johnson J, Copenhagen DR, Storm-Mathisen J, Chaudhry FA, Nicoll RA, Edwards RH (2004) Vesicular glutamate transporters 1 and 2 target to functionally distinct synaptic release sites. Science 304(5678):1815–1819

28. Akil O, Oursler AE, Fan K, Lustig LR (2016) Mouse auditory brainstem response testing. Bio Protoc 6(6):e1768

29. Liu Y, Okada T, Sheykholeslami K, Shimazaki K, Nomoto T, Muramatsu S, Kanazawa T, Takeuchi K, Ajalli R, Mizukami H, Kume A, Ichimura K, Ozawa K (2005) Specific and efficient transduction of cochlear inner hair cells with recombinant adeno-associated virus type 3 vector. Mol Ther 12 (4:725–733

Chapter 17

Intranasal Delivery of Adenoviral and AAV Vectors for Transduction of the Mammalian Peripheral Olfactory System

Cedric R. Uytingco and Jeffrey R. Martens

Abstract

Intranasal delivery of solutions is a straightforward methodology for viral vector transduction and gene transfer to the epithelia within the nasal cavity. Beyond the simplicity of the technique, intranasal delivery has demonstrated restricted transduction of the olfactory and respiratory epithelial tissues. Here we outline the procedure of viral vector intranasal delivery in early postnatal and adult mice, as well as adult rats. The procedure allows for robust transduction and ectopic gene delivery that can be used for the visualization of cellular structures, protein distribution, and assessment of viral vector-mediated therapies.

Key words Olfactory epithelium, Nasal epithelium, Intranasal delivery, Adenovirus, Adeno-associated virus, Gene therapy

1 Introduction

Viral vector-mediated therapies have demonstrated significant promise toward the treatment of several inherited and rare diseases. While proven successful, the efficiency and efficacy of in vivo vector-mediated therapies is hampered significantly by the accessibility of the target tissue, administration of the vectors, and the restriction of treatment within the target tissue. Over the last two decades, multiple studies have demonstrated the nasal cavity to be a highly privileged area that is readily accessible and amenable to adenoviral and adeno-associated viral (AAV) vectors [1–8], as well as to herpes simplex viral (HSV) and lentiviral vector transduction [9–13].

The mammalian nasal cavity is a bilateral space that is divided medially by the bony and cartilaginous nasal septum. Laterally, other bony protrusions called nasal turbinates or conchae extend toward the midline facilitating airflow, air humidification, and increasing surface area for sensory detection. Lining the surface of the nasal cavities are the respiratory and olfactory epithelia (also

Michael J. Castle (ed.), *Adeno-Associated Virus Vectors: Design and Delivery*, Methods in Molecular Biology, vol. 1950, https://doi.org/10.1007/978-1-4939-9139-6_17,

known as the neuroepithelium). Both epithelial regions are pseudostratified tissues that differ in their cellular composition and function. The respiratory epithelium is largely composed of columnar respiratory cells that project multiple motile cilia and are responsible for mucociliary clearance of debris and mucus from the nasal cavity. The olfactory epithelium (OE) predominantly houses the olfactory sensory neurons (OSNs), ciliated neurons that are involved in odorant detection and the relay of odorant information toward the olfactory bulb for processing. The OE also contains sustentacular and microvillar cells, which play supporting roles in the maintenance of the OE, and the Bowman's glands, which are responsible for the production of mucus.

The mechanisms of viral vector transduction within the OE are understood to follow the same processes as within other cell types. OSNs and sustentacular cells were shown to be enriched in heparan sulfate proteoglycans (HSPG) at their apical surface [14]. Cell surface HSPG play a role in the attachment and mechanism of cellular entry for HSV and some AAVs [15–17]. Recent studies have demonstrated the capacity of AAV1, AAV2, AAV7, AAV8, AAV9, and AAV12 to transduce OSNs and sustentacular cells, with AAV8 and AAV9 exhibiting the highest preference toward OSNs [5, 8, 18]. Interestingly, the distribution of HSPG in respiratory cells is restricted to their basolateral surface and absent at the apical surface, which may limit the cells' capacity to be infected by certain viruses and viral serotypes [14]. However, AAV5 and AAV6 have demonstrated efficient transduction of respiratory cells, potentially through an HSPG-independent mechanism of binding and entry [19–21]. In addition to HSPG, OSNs and other cells of the OE also express the coxsackievirus and adenovirus receptor (CAR) [22, 23]. CAR is a cell adhesion protein that is predominantly utilized for adenoviral vector surface binding, specifically adenovirus serotype 5 [16, 24, 25]. The robust transduction by adenovirus and its ease of preparation make it a flexible tool to understand fundamental olfactory cellular signaling, in vivo protein trafficking, and OSN axon pathfinding [1, 2, 18, 26, 27]. Although usage of adenoviral vectors for gene therapy is hampered by their strong immunogenic responses, the peripheral olfactory system has demonstrated resilience toward this response [8, 28]. This could pave the way for developing a gene therapy for congenital loss of olfactory function.

Here, we demonstrate the methodology by which viral vectors are administered intranasally to early postnatal and adult mice, as well as adult rats. Although similar methods were performed in the past for the delivery of small molecules and viral vectors, our modified procedure minimizes damage to the tissues while restricting transduction to the nasal cavity. We also present a modified version for rat intranasal delivery, which combines anesthesia and an elongated cannula for efficient transduction of the olfactory epithelium. Together, these approaches have yielded consistent and compelling results for examination of the nasal cavity tissues and the capacity of viral vector-mediated therapies for the peripheral olfactory system.

2 Materials

2.1 Animals and Anesthesia

1. House laboratory mouse and rat strains in conditions according to the American Association for Accreditation of Laboratory Animal Care (AAALAC) and the Institutional Animal Care and Use Committee (IACUC), or equivalent.
2. Prior, during, and after intranasal delivery of viral vectors, house animals in a biosafety level 2 (BSL2) facility or higher, depending on the administered chemical or viral vector, in order to prevent contamination and personal exposure.
3. Various inbred wildtype mouse strains of both sexes and across multiple ages have demonstrated amenability to adenoviral and AAV transduction.
4. For adult mice, use isoflurane as the primary anesthetic.
5. 20 mL scintillation vials or equivalent, to prevent direct contact of the animal with isoflurane.
6. Filter paper or small cotton ball, trimmed to fit inside the scintillation vial.
7. Bell jar or another equivalent small container with a secure lid. Container must be nonporous, translucent, and easily cleaned and sanitized. The container must also be large enough to accommodate the animal, while not being so large to require excessive anesthetic.
8. Use adult rats at 8 weeks of age for both males and females (approximately 300 g and 200 g, respectively).
9. For adult rat anesthesia, use a cocktail of 80–100 mg/kg ketamine and 5–10 mg/kg xylazine prepared fresh and injected intraperitoneally. Prepare an appropriate volume for delivery of 0.1 mL of ketamine/xylazine cocktail per 100 g rat weight.

2.2 Adenoviral and AAV Vectors

1. The respiratory and olfactory epithelia are susceptible to both adenoviral and AAV-mediated transduction (*see* **Note 1**).
2. Adenoviral vector solutions with titers ranging between 1×10^{10} and 1×10^{11} plaque forming units (PFU), or around 1×10^{10} viral particles (VP).
3. AAV vector solutions with titers at least 1×10^{11} PFU or around 2×10^{12} viral genomes (vg)/mL.
4. Microcentrifuge tubes.
5. Micropipettes and micropipette tips.

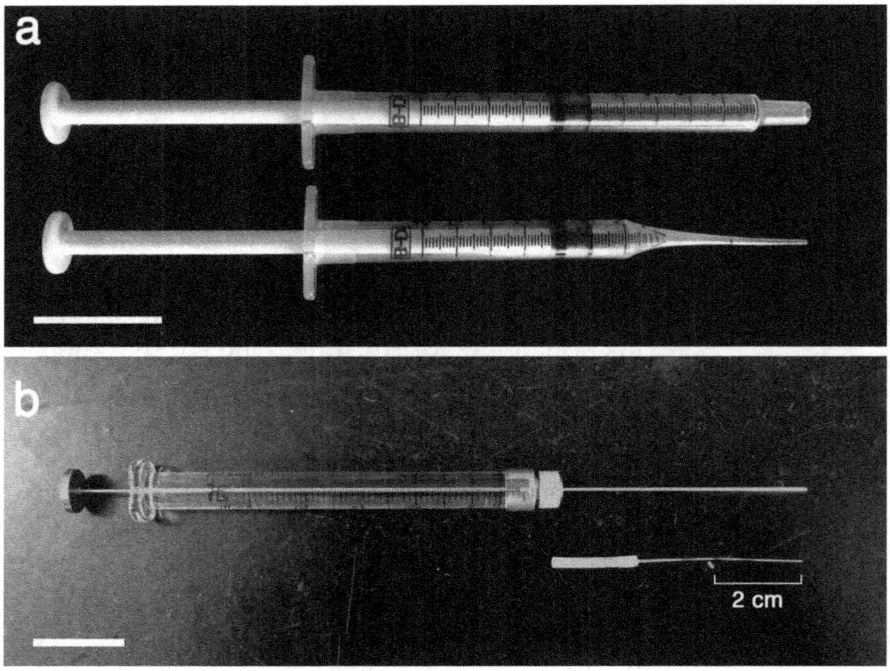

Fig. 1 Custom syringes for intranasal delivery of viral vectors

2.3 Delivery Syringes and Other Materials

1. For mouse intranasal delivery of viral vectors, prepare custom microsyringes by heating standard 1 mL plastic syringes or equivalent (Fig. 1a) (*see* **Note 2**).
2. Heating element such as a Bunsen burner.
3. Tabletop scissors.
4. For rat intranasal delivery of viral vectors, modify a 25 μL Hamilton 1702 Gastight syringe or equivalent. Fit the syringe with a long narrow silicone cannula with an opening approximately 0.25 μm in diameter. The tip must be at least 2 cm in length for insertion deep into the rat nasal cavity (Figs. 1b and 2c).
5. 500 mL disposable container that accommodates the size of the syringe.
6. Disinfectant solution: bleach mixed with water at a 1:9 ratio (10% bleach).
7. Bleach is susceptible to degradation, and disinfectant solution must be made fresh daily. For procedures requiring multiple days, disinfectant solutions can be made with bleach mixed with water at a 1:4 ratio (20% bleach).
8. BSL2 Biosafety Cabinet.
9. BSL2 personal protective equipment.
10. 70% ethanol.

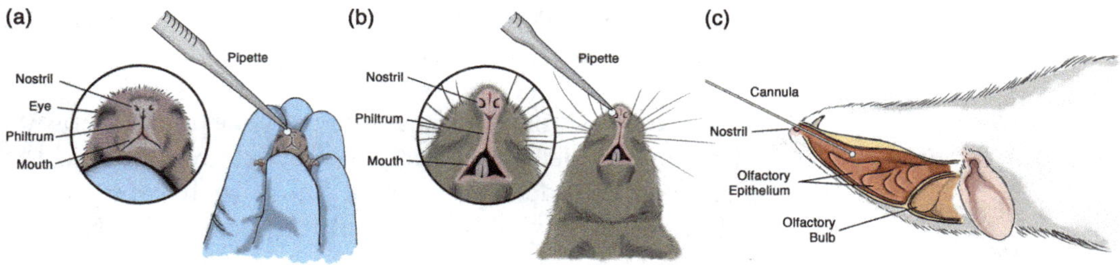

Fig. 2 Intranasal delivery of viral vectors to early postnatal mice, adult mice, and adult rats

3 Methods

3.1 Preparation of Mouse Vector Delivery Syringe

1. Prepare custom viral vector delivery syringes from standard 1 mL plastic syringes (Fig. 1a) (*see* **Note 2**).
2. Prior to heating, pull the plunger and gasket to its back as far as possible, to prevent damage and melting.
3. Using a Bunsen burner with an open yellow flame, slowly and carefully heat the tip and distal segment of the syringe barrel until it melts.
4. Over a waste container, quickly hold the syringe vertically and allow gravity to pull the tip to approximately 0.5 mm in diameter.
5. Using standard tabletop scissors, trim the pulled syringe tip to the desired length and external diameter. If necessary, excess material can be cut off.
6. Push the plunger to force air from the syringe and ensure proper flow through the pulled tip. If the tip is obstructed, continue to trim until the plunger can easily be depressed, indicating proper flow (*see* **Note 3**).
7. Ensure that you prepare a sufficient number of syringes for each experiment. Each individual treatment should have its own disposable syringe.
8. Store the pulled syringes until use.

3.2 Mouse Intranasal Delivery

Carry out all vector delivery at room temperature within a BSL2 room and cabinet. Individuals must wear proper BSL2 personal protective equipment, including a gown, gloves, sleeve covers, and face mask. Before cabinet use and intranasal deliveries, all surfaces of the biosafety cabinet and hands must be wiped with disinfectant and 70% ethanol in order to remove any particulate or possible contaminants. The disinfection procedure must be performed in between mice receiving different viral constructs or mixtures to prevent cross contamination. For sufficient transduction and therapeutic gene expression, administer all intranasal deliveries in three doses across three consecutive days [7, 8, 27].

3.2.1 Early Postnatal Mouse Intranasal Delivery

1. Perform early postnatal intranasal delivery starting at P7 or P8, and administer daily on three consecutive days. Prior to the experiment, consider experimental factors such as target area and the amount of viral vector solution to be used (*see* **Notes 1** and **4**).
2. Each daily dosage consists of approximately 10–15 μL of recombinant adenoviral or AAV vector solution (*see* **Note 5**).
3. Using a disposable pulled syringe with a 0.5 mm tip (Fig. 1), aspirate one dose of the viral vector solution and set aside until the animal is ready for delivery.
4. Hold a single early postnatal mouse vertically with all five fingers (*see* **Note 6**). Ensure that the mouse head is fully restrained to avoid injury to the mouse and prevent potential flicking of the viral solution (Fig. 2a). Anesthesia is not required for treatment of early postnatal mice.
5. Holding the animal in an upright position, deliver the viral solution as 1.0 mm diameter small droplets (equivalent to 3.0–4.0 μL) by closely placing a droplet onto the opening of one nostril, avoiding the philtrum, and allowing the animal to inhale the solution (*see* **Note 7**).
6. Once the viral solution is no longer blocking the nostril, repeat the process on the opposite nostril. Repeat with alternating nostrils until the entire dose of the vector solution has been used (*see* **Notes 8** and **9**).
7. Following each daily dosage, monitor the mice for 5–10 min to verify consciousness and movement, and ensure they are properly cared for by the dam (*see* **Note 10**).
8. Discard used syringes and empty tubes in a container filled with a 10% bleach disinfectant solution. Aspirate the disinfectant solution into the syringe and ensure sufficient surface contact.
9. Upon completion, wipe all surfaces and hands with disinfectant and 70% ethanol.
10. For the second and third doses on subsequent days, repeat **steps 2** through **9**.
11. Following treatment, wait 10 days for adenoviral-mediated gene expression or 3–6 weeks for AAV-mediated gene expression (*see* **Note 11**).
12. After waiting for gene expression to peak, treated mice can be examined by following standard protocols for euthanasia, cardiac perfusion, cryosectioning, and immunostaining.

3.2.2 Adult Mouse Intranasal Delivery.

1. For adult mice (at least 5 weeks of age), perform intranasal delivery on three consecutive days. Prior to the experiment, consider experimental factors such as the target area and the amount of viral vector solution to be used. Unlike juvenile mice, adult mice should be lightly anesthetized during viral vector administration (*see* **Notes 1**, **4**, and **12**).
2. Prior to intranasal delivery, aliquot a single dose of 15–20 μL of recombinant adenovirus or AAV (*see* **Note 5**).
3. Using a disposable pulled syringe with a 0.5 mm tip (Fig. 1), aspirate one dose of the viral vector solution and set aside until delivery.
4. Under a hood, transfer 500 μL of isoflurane to a 20 mL scintillation vial containing either a strip of filter paper or a cotton ball, then place the vial within a bell jar or other small container (*see* **Note 13**).
5. Once the container is sufficiently saturated, place a single mouse into the container with the lid closed tightly and monitor the mouse closely (*see* **Note 14**).
6. After approximately 1 min, the animal should demonstrate loss of consciousness, decreased respiration, lack of a righting reflex, and exhibit hyporeflexia.
7. Using standard manual mouse restraint technique, firmly grasp the scruff of the neck close to the back of the skull [29]. It is essential that you have complete control of the head, in order to avoid injury to the mouse and prevent potential flicking of the viral solution (Fig. 2b).
8. Holding the animal in an upright position, deliver the viral solution as 1.0 mm diameter small droplets (equivalent to 3.0–4.0 μL) by closely placing a droplet onto the opening of one nostril, avoiding the philtrum, and allowing the animal to inhale the solution (*see* **Note 7**).
9. Once the viral solution is no longer blocking the nostril, repeat the process on the opposite nostril. Repeat with alternating nostrils until the entire dose of the vector solution has been used (*see* **Notes 8** and **9**).
10. Upon completion, return mice to their cage and monitor closely for at least 10 min to verify for consciousness, normal behavior, and breathing (*see* **Note 10**).
11. Discard used syringes and empty tubes in a container filled with a 10% bleach solution. Aspirate the disinfectant solution into the syringe and ensure sufficient surface contact.
12. Upon completion, wipe all surfaces and hands with disinfectant and 70% ethanol.

13. For the second and third doses on subsequent days, repeat **steps 2** through **12**.
14. Following treatment, wait 10 days for adenoviral-mediated gene expression or 3–6 weeks for AAV-mediated gene expression (*see* **Note 11**).
15. After waiting for gene expression to peak, treated mice can be examined by following standard protocols for euthanasia, cardiac perfusion, cryosectioning, and immunostaining.

3.3 Adult Rat Intranasal Delivery

Carry out all vector delivery at room temperature within a BSL2 room and cabinet. Individuals must wear proper BSL2 personal protective equipment, including a gown, gloves, sleeve covers, and face mask. Before cabinet use and intranasal deliveries, all surfaces of the biosafety cabinet and hands must be wiped with disinfectant and 70% ethanol in order to remove any particulate or possible contaminants. The disinfection procedure must be performed in between rats receiving different viral constructs or mixtures to prevent cross contamination.

1. Due to the size and sensitivity of rats, additional care must be taken to ensure sufficient transduction of the olfactory epithelial tissue. For adult rats, perform intranasal delivery on three consecutive days. Prior to the experiment, consider experimental factors such as the target area and the amount of viral vector solution to be used (*see* **Notes 1**, **4**, and **12**).
2. Prior to intranasal delivery, aliquot two single doses of 25 μL of recombinant adenoviral or AAV vector solution, one for each nostril (*see* **Note 5**).
3. Anesthetize adult rats by intraperitoneal injection of a ketamine/xylazine cocktail (0.1 mL of cocktail per 100 g rat weight), which should provide anesthesia for 60–80 min. Allow sufficient time for the anesthesia to take effect. The rat will demonstrate loss of consciousness, decreased respiration, lack of a righting reflex, and exhibit hyporeflexia.
4. Once fully unconscious, place the rat in a supine position on top of a disposable cloth.
5. Fill a 25 μL Hamilton syringe fitted with a 0.25 mm diameter tip with 25 μL of viral vector solution (Fig. 1b).
6. To efficiently deliver vector solution to the OE, thread the syringe tip through the nostril and approximately 2 cm into the nasal cavity (Fig. 2c).
7. Slowly and carefully, manually deliver the solution into the nasal cavity at a rate of <5 μL per min, while visually monitoring the respiratory rate and skin coloration.

8. Following delivery into the first nasal cavity, repeat **steps 5–7** for the second nostril and nasal cavity.
9. Upon completion, wipe all surfaces and hands with disinfectant and 70% ethanol.
10. Thoroughly wipe the Hamilton syringe and flush the interior surfaces with disinfectant and 70% ethanol.
11. Upon completion of the procedure, monitor treated rats for at least 3 h or until conscious, with periodic monitoring across subsequent days.
12. For the second and third doses on subsequent days, repeat **steps 2** through **11**.
13. Following treatment, wait 10 days for adenoviral-mediated gene expression or 3–6 weeks for AAV-mediated gene expression (*see* **Note 11**).
14. After waiting for gene expression to peak, treated mice can be examined by following standard protocols for euthanasia, cardiac perfusion, cryosectioning, and immunostaining.

4 Notes

1. The selection of viral vector for nasal tissue delivery will depend highly on the questions being asked, resources available to the study, and the need for cell specificity. Although both adenoviral and AAV vectors have demonstrated the capacity to transduce the olfactory system, the adenoviral vector system has been used for the majority of peripheral olfactory system studies due in part to its rapid and robust expression, ability to transduce multiple cell types, generate high titer preparations, and accommodate an 3.0–8.0 kb gene insert in its ~36 kb linear dsDNA genome. In contrast, AAV has an ~4.7 kb ssDNA genome with an ~4.3 kb genetic insert. On the other hand, there are 13 different serotypes of AAV and several AAV pseudotypes, each having distinct cell specificity. Within the nasal epithelia, AAV8 and AAV9 have been shown to have higher specificity toward OSNs [8], while AAV5 and AAV6 have exhibited preference toward respiratory cells [21, 30, 31]. In addition to cell specificity, certain AAV serotypes demonstrate reduced immunogenic response and cellular toxicity when compared to adenoviral-mediated transduction at equivalent doses [32, 33]. However, the nasal epithelia appears to be an exception, with recent studies demonstrating no significant changes in macrophage infiltration and apoptotic cells [8, 34].
2. While a standard micropipette and pipette tips are sufficient to deliver small volumes of virus to the nostrils of mice, we found droplet formation and control to be inconsistent. In addition,

the repeated use of the same micropipette or microsyringe may increase the likelihood of cross contamination. Therefore, we pull 1 mL plastic syringes in order to deliver more consistent and controlled droplets, while adding disposability and limiting cross contamination.

3. When making the disposable plastic syringes, it is critical that the opening allows for continuous and unobstructed airflow. Too small of an opening may lead to clogging or added resistance during delivery, which may result in loss of control and solution loss. This may be the case when delivering crude viral vectors. Syringes can be further trimmed in order to reduce resistance, or suboptimal syringes can simply be disposed.

4. Determination of viral vector dosage volume and frequency depends on the intended application, and also on the targeted tissue and available vector solution. The adult mouse nasal cavity volume remains largely consistent across adulthood and between sexes, between 31.5 and 32.5 mm^3 (or μL) [35]. Early postnatal and juvenile mice have nasal cavities that are significantly smaller with postnatal day 7 mice demonstrating 60% and 40% smaller nasal cavity height and length, respectively [36]. These numbers equate to P7 mice having an approximately three-fold smaller nasal cavity volume (~10 mm^3) than adults. Compared to mice, the adult rat nasal cavity is significantly larger, with an estimated volume of 156 mm^3 at 7 weeks of age and 257 mm^3 at 16 weeks of age [35]. Due to this large volume, rats can tolerate large intranasal doses of viral solutions. However, too large of a solution volume increases the likelihood of regional spread of transduction for all conditions. Studies have demonstrated that intranasal delivery of 10 μL of radiolabeled dye is largely restricted to the head, with increased volumes expanding the labeled areas to include the lungs [37]. The impact of volume and spread was also observed following the intranasal delivery of AAV and bacteria [38, 39]. In addition to the impact of nasal cavity volume, transduction can also be impaired by continual mucociliary clearance occurring within the nasal cavity. Studies have estimated that at least 35% of particles can be cleared in as little as 60 min following exposure [40]. Therefore, repetitive treatments are performed in order to ensure sufficient viral vector-mediated transduction.

5. Prior to intranasal delivery, aliquot the necessary amount of virus or viral mixture into disposable 0.5 mL microcentrifuge tubes. The total number of microcentrifuge tubes should be sufficient for the entire treatment regimen (3 tubes for each animal). The aliquoted tubes can be stored at −80 °C or 4 °C until use. However, repeated freeze-thaw cycles will lead to

reduction in virulence and must be avoided as much as possible [41].

6. Other studies have examined the impact of position on intranasal delivery. Variations of the intranasal delivery where adult mice were held close to vertical position demonstrated greater AAV transduction of the lungs compared to a supine position (or dorsal recumbency) [42]. However, the total volume may have impacted the areas of transduction. Contrary to this observation, orientation of the animal had no impact on the percentage of radiolabeled cells following administration of 25 μL of ^{99m}Tc-SC [37]. Thus, differences in AAV spread based on the delivery position may be attributed to the use of 80 μL of AAV6 viral solution at a titer of 1×10^{11} vg/mL in this study [42].
7. During viral solution delivery, special care must be taken to avoid contact with the philtrum of both early postnatal and adult mice (Fig. 2a, b). The philtrum is the narrow vertical groove that extends from the upper lip toward the nose in mice and other vertebrates. Contact with the philtrum often results in siphoning of solution away from the nostrils and into the mouth, resulting in solution loss and reduced transduction efficiency. For early postnatal intranasal deliveries, this will lead to increased risk of airway obstruction and drowning. If this occurs, either postpone the delivery until the airway is clear or wait until the following day to resume administration.
8. As mice and other rodents are largely obligate nose breathers, it is critical that viral solution delivery is performed slowly and in small consecutive volumes. Delivery of excessive viral solution may lead to obstruction of the airway and labored breathing, especially when treating early postnatal mice. Therefore, sufficient time must be allowed for the solution to be inhaled thoroughly through one nostril before delivering solution to the second nostril.
9. When performing intranasal delivery, experimenters must be cognizant of the airflow in animals and its potential impact on transduction. Naturally occurring infections, such as rhinitis, and other forms of nasal blockages will impact viral vector transduction and access to different areas of the nasal cavity. Intranasal deliveries have proven difficult in models of primary ciliary dyskinesia and cystic fibrosis, where mucosal accumulation prevents viral vector binding to the surface of targeted cells [43]. Under these conditions, air-assisted methods could increase nasal transduction, but may also increase the risk of transducing other parts of the respiratory tract, such as the trachea and the lungs [34].

10. If the animal loses consciousness or appears under duress at any point, as characterized by labored breathing and/or absence of movement, immediately euthanize according to IACUC or University guidelines. If the animal has been in contact with the viral vector, the carcass must be properly disposed of according to standard biohazard safety methods.
11. The timing for viral vector gene expression depends on the viral vector and serotype being used. Adenoviral-mediated gene expression can occur very quickly, with in vitro expression occurring as early as 24 h, and maximum expression in vivo observed between 3 and 10 days post administration [2]. Waiting 10 days post-treatment is sufficient for ectopic gene expression, incorporation of protein into various organelles [27], and cilia restoration in ciliopathy mouse models [7]. When combined with the relative ease of preparation, adenoviral-mediated gene expression serves as a rapid and practical tool for expressing tagged proteins and markers within the olfactory system. Longevity studies have observed the persistence of gene products up to 21 days post-delivery and as long as 90 days post-delivery; however, there is a dramatic reduction in expression levels 30 days post-delivery due to loss of adenoviral DNA [1, 2]. AAV-mediated transduction within the olfactory epithelium has been demonstrated to be stable from anywhere from 2 to 4 weeks following intranasal delivery, which is sufficient for cell labeling and OSN cilia restoration [5, 8, 21].
12. Intranasal viral delivery is highly dependent on the state of the animal, with deeply anesthetized mice demonstrating decreased nasal transduction and increased exposure of the lungs [9, 37, 39]. On the other hand, intranasal administration to fully awake adult animals will prove stressful for the animal, and significant loss of viral solutions due to flicking or excessive exhalation may occur. As a compromise, light anesthesia using isoflurane is adequate for handling and administration, and restricts delivery of viral vectors and radiolabeled solutions to the nasal cavity [8, 37, 39].
13. Isoflurane anesthesia can be performed using a drop jar or equivalent container, with 0.26–0.30 mL of isoflurane per 1000 mL container volume (resulting in ~5% vaporized isoflurane concentration). Place the filter strip or cotton ball in a 20 mL scintillation vial to prevent the animal from coming into direct contact with isoflurane, which is an irritant.
14. Although anesthesia by isoflurane occurs quickly, it also results in rapid recovery. On average, the drop jar method will leave the mice anesthetized for approximately 30–45 s. Due to this time constraint, it is critical to firmly grasp the mouse and perform intranasal deliveries quickly.

Acknowledgments

Intellectual support was provided by members of the University of Florida Center for Smell and Taste. This work was supported by National Institutes of Health R01DC009606 (J.R.M.).

Competing Interests: *The authors declare no competing or financial interests.*

References

1. Doi K, Nibu K, Ishida H et al (2005) Adenovirus-mediated gene transfer in olfactory epithelium and olfactory bulb: a long-term study. Ann Otol Rhinol Laryngol 114:629–633
2. Zhao H, Otaki JM, Firestein S (1996) Adenovirus-mediated gene transfer in olfactory neurons in vivo. J Neurobiol 30:521–530. https://doi.org/10.1002/(SICI)1097-4695
3. Holtmaat AJGD, Hermens WTJMC, Beate Oestreicher A et al (1996) Efficient adenoviral vector-directed expression of a foreign gene to neurons and sustentacular cells in the mouse olfactory neuroepithelium. Mol Brain Res 41:148–156. https://doi.org/10.1016/0169-328X(96)00085-X
4. Mitchell AL, Dwyer A, Pitteloud N, Quinton R (2011) Genetic basis and variable phenotypic expression of Kallmann syndrome: towards a unifying theory. Trends Endocrinol Metab 22:249–258. https://doi.org/10.1016/j.tem.2011.03.002
5. Quinn K, Quirion MR, Lo CY et al (2011) Intranasal administration of adeno-associated virus type 12 (AAV12) leads to transduction of the nasal epithelia and can initiate transgene-specific immune response. Mol Ther 19:1990–1998. https://doi.org/10.1038/mt.2011.146
6. Arimoto Y, Nagata H, Isegawa N et al (2002) In vivo expression of adenovirus-mediated lac Z gene in murine nasal mucosa. Acta Otolaryngol 122:627–633. https://doi.org/10.1080/000164802320396303
7. McIntyre JC, Davis EE, Joiner A et al (2012) Gene therapy rescues cilia defects and restores olfactory function in a mammalian ciliopathy model. Nat Med 18:1423–1428. https://doi.org/10.1038/nm.2860
8. Williams CL, Uytingco CR, Green WW et al (2017) Gene therapeutic reversal of peripheral olfactory impairment in Bardet-Biedl syndrome. Mol Ther 25. https://doi.org/10.1016/j.ymthe.2017.02.006
9. Shivkumar M, Milho R, May JS et al (2013) Herpes simplex virus 1 targets the murine olfactory neuroepithelium for host entry. J Virol 87:10477–10488. https://doi.org/10.1128/JVI.01748-13
10. Cmielewski P, Donnelley M, Parsons DW (2014) Long-term therapeutic and reporter gene expression in lentiviral vector treated cystic fibrosis mice. J Gene Med 16:291–299. https://doi.org/10.1002/jgm.2778
11. Sadrian B, Chen H, Gong Q (2011) Lentivirus-mediated genetic manipulation and visualization of olfactory sensory neurons in vivo. J Vis Exp (51). https://doi.org/10.3791/2951
12. Patel M, Giddings AM, Sechelski J, Olsen JC (2013) High efficiency gene transfer to airways of mice using influenza hemagglutinin pseudotyped lentiviral vectors. J Gene Med 15:51–62. https://doi.org/10.1002/jgm.2695
13. Sinn PL, Arias AC, Brogden KA, PB MC (2008) Lentivirus vector can be readministered to nasal epithelia without blocking immune responses. J Virol 82:10684–10692. https://doi.org/10.1128/JVI.00227-08
14. Milho R, Frederico B, Efstathiou S, Stevenson PG (2012) A heparan-dependent herpesvirus targets the olfactory neuroepithelium for host entry. PLoS Pathog 8. https://doi.org/10.1371/journal.ppat.1002986
15. Summerford C, Samulski RJ (1998) Membrane-associated heparan sulfate proteoglycan is a receptor for adeno-associated virus type 2 virions. J Virol 72:1438–1445
16. Zaiss AK, Foley EM, Lawrence R et al (2016) Hepatocyte heparan sulfate is required for adeno-associated virus 2 but dispensable for adenovirus 5 liver transduction in vivo. J Virol 90:412–420. https://doi.org/10.1128/JVI.01939-15
17. Kern A, Schmidt K, Leder C et al (2003) Identification of a heparin-binding motif on adeno-associated virus type 2 capsids. J Virol

77:11072–11081. https://doi.org/10.1128/JVI.77.20.11072-11081.2003

18. Cao L, Schrank BR, Rodriguez S et al (2012) Aβ alters the connectivity of olfactory neurons in the absence of amyloid plaques in vivo. Nat Commun 3:1009. https://doi.org/10.1038/ncomms2013
19. Seiler MP, Miller AD, Zabner J, Halbert CL (2006) Adeno-associated virus types 5 and 6 use distinct receptors for cell entry. Hum Gene Ther 17:10–19. https://doi.org/10.1089/hum.2006.17.10
20. Kurosaki F, Uchibori R, Mato N et al (2017) Optimization of adeno-associated virus vector-mediated gene transfer to the respiratory tract. Gene Ther 24:290–297. https://doi.org/10.1038/gt.2017.19
21. Limberis MP, Vandenberghe LH, Zhang L et al (2009) Transduction efficiencies of novel AAV vectors in mouse airway epithelium in vivo and human ciliated airway epithelium in vitro. Mol Ther 17:294–301. https://doi.org/10.1038/mt.2008.261
22. Venkatraman G, Behrens M, Pyrski M, Margolis FL (2005) Expression of Coxsackie-Adenovirus receptor (CAR) in the developing mouse olfactory system. J Neurocytol 34:295–305. https://doi.org/10.1007/s11068-005-8359-8
23. Soudais C, Laplace-Builhe C, Kissa K, Kremer EJ (2001) Preferential transduction of neurons by canine adenovirus vectors and their efficient retrograde transport in vivo. FASEB J 15:2283–2285. https://doi.org/10.1096/fj.01-0321fje
24. Bergelson JM, Cunningham JA, Droguett G et al (1997) Isolation of a common receptor for coxsackie B viruses and adenoviruses 2 and 5. Science 275:1320–1323. https://doi.org/10.1126/science.275.5304.1320
25. Tuve S, Wang H, Jacobs JD et al (2008) Role of cellular heparan sulfate proteoglycans in infection of human adenovirus serotype 3 and 35. PLoS Pathog 4. https://doi.org/10.1371/journal.ppat.1000189
26. Zhao H, Ivic L, Otaki JM et al (1998) Functional expression of a mammalian odorant receptor. Science 279:237–242. https://doi.org/10.1126/science.279.5348.237
27. Williams CL, McIntyre JC, Norris SR et al (2014) Direct evidence for BBSome-associated intraflagellar transport reveals distinct properties of native mammalian cilia. Nat Commun 5:5813. https://doi.org/10.1038/ncomms6813
28. Thacker EE, Timares L, Matthews QL (2009) Strategies to overcome host immunity to adenovirus vectors in vaccine development. Expert Rev Vaccines 8:761–777
29. Machholz E, Mulder G, Ruiz C et al (2012) Manual restraint and common compound administration routes in mice and rats. J Vis Exp. https://doi.org/10.3791/2771
30. Zabner J, Seiler M, Walters R et al (2000) Adeno-associated virus type 5 (AAV5) but not AAV2 binds to the apical surfaces of airway epithelia and facilitates gene transfer. J Virol 74:3852–3858. https://doi.org/10.1128/JVI.74.8.3852-3858.2000
31. Halbert CL, Allen JM, Miller AD (2001) Adeno-associated virus type 6 (AAV6) vectors mediate efficient transduction of airway epithelial cells in mouse lungs compared to that of AAV2 vectors. J Virol 75:6615–6624. https://doi.org/10.1128/JVI.75.14.6615-6624.2001
32. Muruve DA (2004) The innate immune response to adenovirus vectors. Hum Gene Ther 15:1157–1166. https://doi.org/10.1089/hum.2004.15.1157
33. Mingozzi F, High KA (2013) Immune responses to AAV vectors: overcoming barriers to successful gene therapy. Blood 122:23–36
34. Gau P, Rodriguez S, De Leonardis C et al (2011) Air-assisted intranasal instillation enhances adenoviral delivery to the olfactory epithelium and respiratory tract. Gene Ther 18:432–436. https://doi.org/10.1038/gt.2010.153
35. Gross EA, Swenberg JA, Fields S, Popp JA (1982) Comparative morphometry of the nasal cavity in rats and mice. J Anat 135:83–88
36. Vora SR, Camci ED, Cox TC (2016) Postnatal ontogeny of the cranial base and craniofacial skeleton in male C57BL/6J mice: a reference standard for quantitative analysis. Front Physiol 6. https://doi.org/10.3389/fphys.2015.00417
37. Southam DS, Dolovich M, O'Byrne PM, Inman MD (2002) Distribution of intranasal instillation in mice: effects of volume, time, body position and anesthesia. Am J Phys Lung Cell Mol Phys 282:L833–L839. https://doi.org/10.1152/ajplung.00173.2001
38. Gruntman AM, Mueller C, Flotte TR, Gao G (2012) Gene transfer in the lung using recombinant adeno-associated virus. Curr Protoc Microbiol Chapter 14:Unit 14D.2. https://doi.org/10.1002/9780471729259.mc14d02s26
39. Miller MA, Stabenow JM, Parvathareddy J et al (2012) Visualization of murine intranasal dosing efficiency using luminescent Francisella

tularensis: effect of instillation volume and form of anesthesia. PLoS One 7. https://doi.org/10.1371/journal.pone.0031359

40. Hua X, Zeman KL, Zhou B et al (2010) Non-invasive real-time measurement of nasal muco-ciliary clearance in mice by pinhole gamma scintigraphy. J Appl Physiol 108:189–196. https://doi.org/10.1152/japplphysiol.00669.2009
41. Ugai H, Watanabe S, Suzuki E et al (2002) Stability of a recombinant adenoviral vector: optimization of conditions for storage, transport and delivery. Jpn J Cancer Res 93:598–603. https://doi.org/10.1111/j.1349-7006.2002.tb01296.x
42. Santry LA, Ingrao JC, Yu DL et al (2017) AAV vector distribution in the mouse respiratory tract following four different methods of administration. BMC Biotechnol 17. https://doi.org/10.1186/s12896-017-0365-2
43. Ostrowski LE, Yin W, Patel M et al (2014) Restoring ciliary function to differentiated primary ciliary dyskinesia cells with a lentiviral vector. Gene Ther 21:253–261. https://doi.org/10.1038/gt.2013.79

Chapter 18

AAV-Mediated Gene Delivery to Taste Cells of the Tongue

Akiyuki Taruno and Makiko Kashio

Abstract

Taste sensation is initiated in sensory cells within the taste buds (taste cells), in which the cooperation of many signaling molecules leads to the coding and transmission of information on the quality and intensity of taste to the afferent gustatory nerves. Here, we describe our method for inducing foreign gene expression in taste cells of fungiform papillae in a living mouse using a recombinant adeno-associated virus (AAV) vector, enabling us to study and control the function of a gene product in vivo. Among the serotypes tested to date, only AAV-DJ, a synthetic serotype, can transduce taste cells in vivo. We also describe how to validate intragemmal foreign gene expression in fungiform taste buds using an immunohistochemical approach.

Key words Taste, Taste bud, Tongue, AAV-DJ, Adeno-associated virus

1 Introduction

The sense of taste plays a critical role in the survival of animals by providing information on the nutritional values of foodstuffs. Five basic taste qualities are generally recognized: sweetness, bitterness, sourness, saltiness, and umami. As sweetness and umami respectively signal the presence of sugar and amino acids, animals show hedonic responses to these qualities. In contrast, bitterness and sourness elicit avoidance responses as they signal the potential presence of toxic ingredients. Saltiness is preferred and avoided at low and high concentrations, respectively. This finely tuned taste recognition is facilitated by specialized cellular and molecular arrangements in taste buds. The majority of taste buds are embedded in the tongue epithelium and typically composed of ~100 elongated taste cells. Although each of the five basic taste qualities is generally sensed by dedicated taste cells [1], there are also data to suggest that one taste quality may be perceived by the activation of one or two dedicated taste cell subsets [2–4]. Accordingly, there are distinct functional taste cell types that can be distinguished by their gene expression profiles. Transferring foreign genetic materials into

Michael J. Castle (ed.), *Adeno-Associated Virus Vectors: Design and Delivery*, Methods in Molecular Biology, vol. 1950, https://doi.org/10.1007/978-1-4939-9139-6_18,

taste cells in vivo would be a useful technique for elucidating the roles of a specific gene product in gustation and the development of taste cells.

In the gustatory system, earlier studies demonstrated that recombinant adenoviral and herpes simplex virus type I vectors can successfully induce foreign gene expression in primary rat taste cell culture [5] and freshly isolated rat taste buds [6]. Recently, we established a method for delivering genetic materials into taste cells in fungiform taste buds, which are located in the anterior part of the tongue, in a living mouse. To achieve long-term transgene expression, we tested various serotypes of recombinant adeno-associated virus (AAV) vectors, including AAV1, 2, 5, 6, DJ, and DJ/8, as well as a lentivirus vector, and found that only the AAV-DJ vector, a synthetic serotype, can effectively transduce taste cells in vivo [7]. Of note, the AAV-DJ vector may not directly transduce mature taste cells, but instead transduce progenitor cells located in the basal layer of the surrounding epithelium, giving rise to all functional taste cell types through migration and differentiation [7]. As the difficulty of this method is mostly due to the complicated procedure of virus injection into the submucosal space of the tongue of a living, anesthetized mouse, here we provide a step-by-step protocol for our virus injection technique as well as the protocol for immunofluorescence staining of fungiform taste buds as a method to validate successful foreign gene expression.

While many genes involved in taste function have been identified over the past two decades [8], over 2000 taste bud-associated genes with no known function in gustation have also been identified [9, 10]. Overexpression or RNA interference via the expression of shRNA in taste cells may help define the roles of those genes in taste transduction or taste cell fate. Cell lineage tracing in the development of taste cells may also be possible using the AAV-DJ vector [7]. The use of recently developed optogenetic tools may facilitate the measurement and control of activities of taste cells in live animals. Thus, the AAV-DJ-mediated in vivo gene transfer into taste cells will facilitate the application of an expanding repertoire of genetic technology and has the potential to promote future research on the biology of taste. Limitations of this technique include the lack of targeting specificity. The AAV-DJ vector transduces all functional taste cell types as well as nongustatory epithelial cells and underlying mesenchymal and muscle cells. Cell-type-specific transgene expression may be achieved by injecting a viral vector containing the loxP-Stop-loxP cassette upstream of a transgene into tissue-specific Cre-driver mouse lines.

2 Materials

2.1 Virus Injection

1. Mice.
2. Recombinant AAV-DJ vector containing a gene of interest: we prepare the AAV-DJ vector containing enhanced green fluorescent protein using AAV Helper Free Packaging System (Cell Biolabs, San Diego, CA, USA). The resulting viral titer should be 1–2 × 10^{10} vector genomes (vg)/μL.
3. Borosilicate glass capillary with a filament inside (*see* **Note 1**).
4. Micropipette puller (*see* **Note 2**).
5. Micropipette grinder (*see* **Note 3**).
6. 25 μL Hamilton® syringe and PTFE tubing.
7. Microinjection oil: 3M™ Fluorinert™ electronic liquid.
8. Anesthetic reagents: ketamine hydrochloride and xylazine hydrochloride.
9. Custom-made platform (Fig. 1).
10. 27G needles.
11. Tissue adhesive (*see* **Note 4**).
12. Stereo microscope.

2.2 Immunohistochemistry

Prepare all solutions using ultrapure water (prepared by purifying deionized water, to attain a sensitivity of 18 MΩ cm at 25 °C) and analytical-grade reagents.

1. Sodium pentobarbital.
2. Phosphate-buffered saline (PBS): 0.137 M NaCl, 0.0027 M KCl, 0.01 M Na_2HPO_4, 0.0018 M KH_2PO_4, pH 7.4 (*see* **Note 5**).

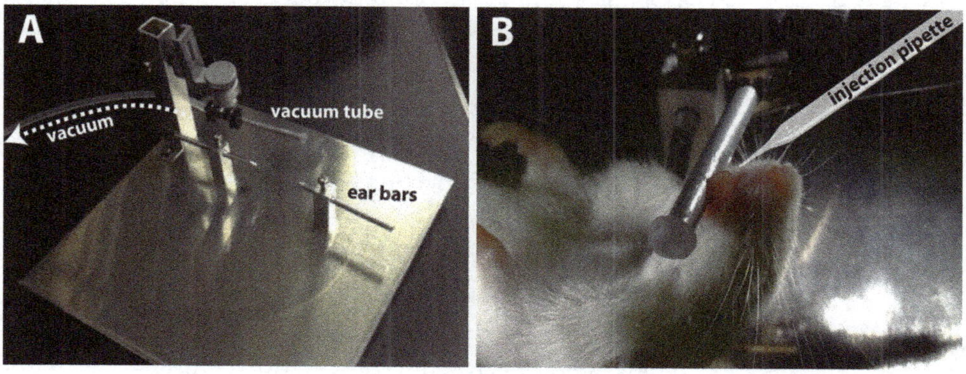

Fig. 1 A platform for holding the head and tongue of a mouse. (**a**) Our custom-made platform is equipped with ear bars and a vacuum tube for holding the head and tongue, respectively. The vacuum tube is connected to a vacuum pump with silicon tubing. (**b**) An anesthetized mouse is set on the platform. The angle of the injection pipette needs to be adjusted for each preparation

3. 4% PFA-PBS fixative solution: PBS containing 4% (w/v) paraformaldehyde (PFA) (*see* **Note 6**).
4. 30% sucrose-PBS: PBS containing 30% (w/v) sucrose (*see* **Note 7**).
5. Optimal cutting temperature (O.C.T.) compound.
6. Slide glasses with coated surfaces for bonding tissue sections and cover glasses.
7. PBS-XD blocking solution: PBS containing 5% (v/v) normal donkey serum and 0.1% (v/v) Triton™ X-100 (*see* **Note 8**).
8. A taste cell marker primary antibody and a primary antibody against a foreign gene product.
9. Alexa Fluor®-conjugated secondary antibodies.
10. Mounting medium.
11. Confocal scanning microscope.

3 Methods

3.1 Virus Injection into Submucosa of the Tongue

1. Fabricate a glass injection pipette from a borosilicate capillary with a filament inside. Pull the capillary using a micropipette puller (*see* **Note 9**), and grind the tip of the pulled pipette to produce a smooth, sharp, 100 μm tip using a microgrinder (*see* **Note 10**).
2. Build the microinjection assembly by connecting the fabricated injection pipette and a 25 μL Hamilton syringe with PTFE tubing.
3. Fill the inside of the microinjection system with microinjection oil.
4. Aspirate 10 μL of the AAV-DJ vector suspension from the tip of the injection pipette (*see* **Note 11**). The interface between the viral suspension and oil is visible.
5. Anesthetize a mouse by intraperitoneal injection of ketamine hydrochloride (100 mg/kg) and xylazine hydrochloride (5 mg/kg) (*see* **Note 12**). Wait approximately 10 min for the induction of deep anesthesia.
6. Clamp the head of the anesthetized mouse with the face up using ear bars. Then, pull the tongue out of its mouth with tweezers and fix it in a stuck-out position on a vacuum tube (Fig. 1) (*see* **Note 13**).
7. Under stereomicroscopic observation, while gently holding the tongue with tweezers, pierce the tongue epithelium with a 27G needle at the tip of the tongue to create a hole in the epithelium. Caution is required not to wound the underlying muscle layer.

8. Insert the injection pipette filled with the AAV-DJ suspension through the small incision in the epithelium, and advance the tip further (~3 mm) into the submucosal space between the epithelial and muscle layers under the medial sulcus. Carefully perform this under stereomicroscopic observation to avoid causing any additional damage to the epithelium except for the injection hole (*see* **Note 14**).
9. Slowly inject 10 μL of the AAV-DJ suspension into the submucosal space while holding the injection hole and pipette together with tweezers to block leakage of the injected solution through the injection hole (*see* **Notes 15** and **16**). Liquid injection inflates the tongue and no solution leakage should be observed (*see* **Note 17**).
10. Carefully remove the injection pipette while pinching the epithelial injection hole with tweezers to avoid solution leakage. While pinching the injection wound, apply tissue adhesive to close it (*see* **Note 18**). Wait 3 min for it to completely harden.
11. Maintain the mouse on a normal diet until analyses (*see* **Note 19**).

3.2 Immunofluorescence Detection of Foreign Gene Expression

1. Seven days after viral injection (*see* **Note 20**), sacrifice the mouse with an overdose of sodium pentobarbital (70 mg/kg, intraperitoneal).
2. Open the thoracic cavity using the standard surgical procedure, and transcardially perfuse the vascular system with ice-cold PBS followed by ice-cold 4% PFA-PBS fixative solution.
3. Dissect out the anterior half of the tongue and immerse it in 4% PFA-PBS at 4 °C for 5 h, and then in 30% sucrose-PBS at 4 °C overnight to 2 days (until the tissue sinks to the bottom).
4. Embed the fixed, cryoprotected tissue block in O.C.T. compound at −80 °C, and make sagittal sections using a cryostat at 9 μm thickness. Mount the tissue sections on s glasses. Store the slides at −80 °C until analyses.
5. Wash the slides 2× with PBS for 5 min each at room temperature with gentle agitation (*see* **Note 21**).
6. Incubate the tissue sections with the PBS-XD blocking solution for 1 h at room temperature in a humidified container.
7. Incubate the sections with primary antibodies diluted in PBS-XD overnight at 4 °C in a humidified container. Use a taste cell marker antibody and an antibody against the product of the foreign gene delivered by the AAV-DJ vector to identify taste cells and transduced cells, respectively (*see* **Note 22**).
8. Wash the slides 3× with PBS at room temperature for 10 min each with gentle agitation and incubate with Alexa Fluor®-conjugated secondary antibodies diluted in PBS-XD in the

dark for 1 h at room temperature in a humidified container (*see* **Note 23**).

9. Wash the slides 3× with PBS in the dark for 10 min each at room temperature with gentle agitation and mount them in a mounting medium with a thin cover glass (*see* **Note 24**).
10. Observe and image the stained sections with a confocal scanning microscope (*see* **Note 25**).

4 Notes

1. The filament inside the capillary helps when aspirating the viral suspension from the tip of the injection pipette. We use GD-1 (Narishige, Tokyo, Japan) with outer and inner diameters of 1 and 0.6 mm, respectively, and a length of 90 mm.
2. We use P-1000 (Sutter Instruments, Novato, CA, USA).
3. We use EG-45 (Narishige, Tokyo, Japan).
4. We use Aron Alpha 30434 (Toagosei, Tokyo, Japan).
5. Make 10× PBS (1.37 M NaCl, 0.027 M KCl, 0.1 M Na_2HPO_4, 0.018 M KH_2PO_4, pH 7.4 adjusted with HCl), sterilize the solution by autoclaving, and store it at room temperature. Mix 1 part 10× PBS and 9 parts water to make 1× PBS.
6. In a ventilated hood, add 4 g PFA to 100 mL of PBS. Heat while gently stirring to 60-65 °C. Do not boil. When dissolved, cool down and filter the solution through filter paper to remove any remaining particles. Make fresh prior to use.
7. Dissolve 30 g of sucrose in 60 mL of PBS. Bring to 100 mL with PBS. Store at 4 °C.
8. Mix 1 mL of 1× PBS, 0.5 mL of normal donkey serum, and 10 μL of Triton™ X-100, and bring to 10 mL with water. Make fresh prior to use. The species from which the normal serum is obtained should be chosen appropriately depending on the host species of secondary antibodies.
9. A long and fine taper is desired because the pipette tip must be inserted deep into the submucosal space through a small incision in the lingual epithelium. Cut the tip of the pulled pipette with a scalpel under stereomicroscopic observation to produce an orifice diameter of approximately 100 μm.
10. Wet-polishing produces a smooth and sharp tip, and the resulting tip angle is approximately 45°.
11. It is very important to avoid introducing air bubbles inside the injection system, as this will cause problems during the subsequent injection step. Aspirate the virus suspension slowly

and ensure that the inside of the injection system is kept free of air bubbles.

12. We normally use 8-week-old male C57BL/6 mice (Japan SLC, Hamamatsu, Japan). Female, younger, or older mice can also be used.
13. To fix the head and tongue of the anesthetized mouse, we use a custom-made platform with ear bars and a vacuum tube (Fig. 1). Here, it is important to ensure that the head and tongue will not move during the subsequent steps. There may be other ways to achieve this.
14. Additional epithelial holes will allow leakage of the injected virus suspension, thereby markedly reducing the infection efficiency.
15. This step is carried out by two people: one person inserts the injection pipette into the tongue, and the other person holds the Hamilton syringe and depresses the plunger as requested.
16. We determined the largest possible injection volume per tongue (10 μL) for 8-week-old male C57BL/6 mice. The injection volume needs to be optimized if female, younger, or older mice, or other strains are used.
17. It is important to confine as much viral vector as possible around the taste buds. Care should be taken to avoid leakage of the injected viral suspension, which results in lower infectivity. We also inject the prepared viral suspensions without further dilution, at the highest possible concentrations, to achieve the highest transduction efficiency, but a lower titer may be used upon optimization.
18. It is best to apply the minimum amount of tissue adhesive necessary to the wound. We place a drop of the adhesive on a plastic sheet and dip the tip of a 27G needle into it in order to take up and apply a tiny amount of the adhesive.
19. As the tongue of the injected mouse is transiently swollen, it may experience difficulty in eating standard food pellets. Therefore, we crush the pellets into small pieces and place them on the floor of the home cage.
20. As we previously reported [7], it is considered that the AAV-DJ vector transduces $K5^+/K14^+$ progenitor cells located in the basal epithelium, which migrate slowly into taste buds and differentiate into all functional taste cell types [6, 11, 12], but not directly into mature intragemmal taste cells. Thus, there is a delay between virus injection and the emergence of transduced taste cells, and the size of the population of transduced taste cells peaks at seven days after injection. Thereafter, the population gradually decreases over the next few weeks due to cell turnover.

21. Antigen retrieval may be necessary depending on the primary antibodies used in the subsequent steps. We incubate the slides in a preheated target retrieval solution (S1700, Dako, Glostrup, Denmark) at 80 °C for 20 min. After cooling down to room temperature in the target retrieval solution, wash them 3× with PBS for 10 min each.
22. We use goat anti-KCNQ1 (1500 dilution, sc-10,646, Santa Cruz Biotechnology, Santa Cruz, CA, USA) as the taste cell marker antibody [13] and mouse anti-GFP (1500 dilution, 632,381, Clontech Laboratories, Mountain View, CA, USA).
23. We use Alexa Fluor® 488 donkey anti-mouse IgG and Alexa Fluor® 546 donkey anti-goat IgG at 1:500 dilution (Thermo Fisher Scientific, Waltham, USA).
24. We use VectaShield with 4′,6-diamidino-2-phenylindole (DAPI) as a mounting medium to counterstain the nuclei (H-1500, Vector Laboratories, Burlingame, CA, USA) [13, 14].
25. With our antibody combination, taste cells and transduced cells are respectively indicated by immunoreactivities to KCNQ1 (red) and GFP (green). The transduction efficiency can be calculated by dividing the number of GFP^+ cells by that of $KCNQ1^+$ cells.

Acknowledgments

This work was supported by Grants-in-Aid from Japan Society of the Promotion of Science [26713008 and 16K15181 to A.T.; 15K18974 to M.K.]; Salt Science [1429 and 1542 to A.T.; 1634 to M.K.].

References

1. Yarmolinsky DA, Zuker CS, Ryba NJ (2009) Common sense about taste: from mammals to insects. Cell 139(2):234–244
2. Yoshida R, Ninomiya Y (2016) Taste information derived from T1R-expressing taste cells in mice. Biochem J 473(5):525–536
3. Oka Y, Butnaru M, von Buchholtz L, Ryba NJ, Zuker CS (2013) High salt recruits aversive taste pathways. Nature 494(7438):472–475
4. Kusuhara Y, Yoshida R, Ohkuri T, Yasumatsu K, Voigt A, Hubner S, Maeda K, Boehm U, Meyerhof W, Ninomiya Y (2013) Taste responses in mice lacking taste receptor subunit T1R1. J Physiol 591(7):1967–1985
5. Kishi M, Emori Y, Tsukamoto Y, Abe K (2001) Primary culture of rat taste bud cells that retain molecular markers for taste buds and permit functional expression of foreign genes. Neuroscience 106(1):217–225
6. Stone LM, Tan SS, Tam PP, Finger TE (2002) Analysis of cell lineage relationships in taste buds. J Neurosci 22(11):4522–4529
7. Taruno A, Kashio M, Sun H, Kobayashi K, Sano H, Nambu A, Marunaka Y (2017) Adeno-associated virus-mediated gene transfer into taste cells in vivo. Chem Senses 42 (1):69–78
8. Liman ER, Zhang YV, Montell C (2014) Peripheral coding of taste. Neuron 81 (5):984–1000
9. Moyer BD, Hevezi P, Gao N, Lu M, Kalabat D, Soto H, Echeverri F, Laita B, Yeh SA, Zoller M,

Zlotnik A (2009) Expression of genes encoding multi-transmembrane proteins in specific primate taste cell populations. PLoS One 4 (12):e7682

10. Hevezi P, Moyer BD, Lu M, Gao N, White E, Echeverri F, Kalabat D, Soto H, Laita B, Li C, Yeh SA, Zoller M, Zlotnik A (2009) Genome-wide analysis of gene expression in primate taste buds reveals links to diverse processes. PLoS One 4(7):e6395
11. Stone LM, Finger TE, Tam PP, Tan SS (1995) Taste receptor cells arise from local epithelium, not neurogenic ectoderm. Proc Natl Acad Sci U S A 92(6):1916–1920
12. Barlow LA (2015) Progress and renewal in gustation: new insights into taste bud development. Development 142(21):3620–3629
13. Taruno A, Vingtdeux V, Ohmoto M, Ma Z, Dvoryanchikov G, Li A, Adrien L, Zhao H, Leung S, Abernethy M, Koppel J, Davies P, Civan MM, Chaudhari N, Matsumoto I, Hellekant G, Tordoff MG, Marambaud P, Foskett JK (2013) CALHM1 ion channel mediates purinergic neurotransmission of sweet, bitter and umami tastes. Nature 495(7440):223–226
14. Taruno A, Sun H, Nakajo K, Murakami T, Ohsaki Y, Kido MA, Ono F, Marunaka Y (2017) Post-translational palmitoylation controls the voltage gating and lipid raft association of CALHM1 channel. J Physiol 595 (18):6121–6145

Part IV

AAV Delivery to Other Tissues

Chapter 19

AAV Vectors for Efficient Gene Delivery to Rodent Hearts

Estrella Lopez-Gordo, Erik Kohlbrenner, Michael G. Katz, and Thomas Weber

Abstract

Currently, gene therapy is one of the most promising fields in biomedicine, with great therapeutic potential for an array of inherited and acquired diseases. Adeno-associated viral (AAV) vectors have emerged as promising tools to deliver selectively a therapeutic payload to target organs, including the heart. In this chapter, we describe the production and quality control of recombinant AAV (rAAV) vectors of the serotype 9, the most cardiotropic AAV serotype when delivered systemically in rodents. We also describe the systemic administration of rAAV vectors and the local delivery of rAAV vectors by direct intramyocardial injection. Taken together, the methods described in this chapter will allow the reader to deliver efficiently therapeutic genes to the rodent heart, both globally and regionally.

Key words Adeno-associated virus, AAV production, AAV characterization, AAV9, AAV delivery, Rat, Mouse

1 Introduction

AAV is a small, nonenveloped virus of the family *parvoviridae* and has an approximately 4.7 kb-long, single-stranded DNA genome, which is flanked by two inverted terminal repeats (ITRs). The AAV genome encodes only two genes: the *rep* gene encoding the REP proteins, which are responsible for DNA replication and packaging, and the *cap* gene, which encodes three capsid proteins (VP1, VP2, and VP3) with overlapping reading frames. On an alternative reading frame, *cap* also encodes the assembly-activating protein (AAP), which aids in capsid assembly [1].

AAV can infect dividing and nondividing cells, is nonpathogenic, and displays limited immunogenicity. Importantly for cardiac gene delivery, AAV vectors mediate long-term transgene expression in postmitotic cells such as cardiomyocytes, even in the absence of genome integration. Because of their beneficial characteristics, AAV-based vectors are arguably the top choice of vectors for in vivo gene therapy and have been used in a large number of

Michael J. Castle (ed.), *Adeno-Associated Virus Vectors: Design and Delivery*, Methods in Molecular Biology, vol. 1950, https://doi.org/10.1007/978-1-4939-9139-6_19,

clinical trials. Importantly, the treatment of Leber's congenital amaurosis type 2 by an AAV2 vector delivering the defective gene, RPE65, has recently been approved by the FDA for clinical use [2]. This represents the first AAV gene therapeutic treatment approved in the United States.

In addition to the most commonly used serotypes AAV1–6, AAV8, and AAV9, more than a hundred AAV variants have been described and, together, they display a diverse and broad tissue and cell tropism. However, for the efficient delivery of genes to the entire rodent myocardium via systemic injection, AAV9 has clearly emerged as the most promising serotype ([3, 4] and reviewed in [5]).

In the first part of this chapter, we describe the production and characterization of rAAV serotype 9 vectors and the importance of quality control, especially for studies in animal models. The generation of rAAV vectors begins with the co-transfection of HEK293T (or HEK293) cells with a plasmid that contains the experimental or therapeutic expression cassette flanked by two AAV2 ITRs, along with one or two plasmids that together express the AAV2 *rep* gene, the *cap* gene of the desired serotype, and the helper functions from adenovirus, including the virus-associated (VA) RNA, *E2A* and *E4* genes [6]. HEK293T cells provide adenovirus *E1A* and *E1B* genes, which are also necessary for AAV production [6]. Triple co-transfection, where the helper functions are distributed within two plasmids (a *rep/cap* plasmid and an adenoviral helper plasmid), is currently the most widely used methodology in research laboratories and allows AAV production of up to 10^5 viral genomes (vg) per cell ([7, 8] and reviewed in [9]). In our laboratory, we use a two-plasmid approach to produce rAAV vectors [10]. This method combines the AAV2 *rep* gene, the AAV *cap* gene (for the purpose of this chapter AAV9 *cap*), and all the adenoviral helper functions (except E1A and E1B) in a single plasmid. We use polyethylenimine (PEI) transfection because of its simplicity and its high level of reproducibility. Nevertheless, transfection can also be performed with calcium phosphate or other transfection approaches (reviewed in [11]). Seventy-two hours following transfection, rAAV virions are harvested from the cells or from both the cells and the cell culture supernatant. While we routinely harvest AAV9 virions from both the cells and the cell culture supernatant, this is not necessary for certain serotypes such as AAV2 ([12] and unpublished results). We then purify the rAAV virions via iodixanol gradient centrifugation [13], a method that can be used for any AAV serotype or variant. Next, purified rAAV virions undergo a series of rigorous quality control tests such as qPCR titration of viral genomes, Coomassie staining of virions run on an SDS-PAGE gel, and alkaline agarose gel electrophoresis, to quantify rAAV vector genome titers, to determine rAAV vector capsid titers and vector purity, and to ensure that the majority of packaged viral genomes are full-length (and to confirm qPCR titers), respectively. In rodents, AAV is usually delivered either systemically by tail vein injection or retro-orbital injection (for global transduction of the myocardium), or locally by open chest, intramyocardial injection (for transduction of specific areas of the heart). These approaches are described below. A third delivery method that can be used instead of tail vein injection for global transduction of the myocardium is intraventricular injection with cross-clamping of the pulmonary artery and the aorta. Because of its invasiveness, this method is rarely used; *see* [14] for a detailed description. For reviews describing delivery methods that are suitable for large animal models and humans, *see* [15–19].

2 Materials

2.1 Plasmid Production (See Note 1)

1. Cis-plasmid containing the rAAV genome to be packaged.
2. pDG9 helper plasmid: AAV9 capsid sequence cloned into the *Swa*I/*Cla*I digested pDG backbone (PlasmidFactory).
3. SURE2 (Agilent Technologies) or Stbl3 (Life Technologies) competent bacteria for transformation with cis-plasmid.
4. Plasmid Maxiprep Kit.
5. Lysogeny broth (LB) medium (also commonly, but incorrectly, referred to as Luria-Bertani medium).
6. *Sma*I (or the isoschizomer XmaI) restriction enzyme.

2.2 Cell Culture

1. HEK 293T/17 cells (ATCC).
2. Cell culture medium: DMEM with 4.5 g/L glucose, L-glutamate, and sodium pyruvate, supplemented with 10% fetal bovine serum and 1× penicillin/streptomycin.
3. 0.025% Trypsin.
4. Calcium and magnesium free PBS.
5. 175 cm^2 tissue culture flask.
6. BSL2 laminar flow tissue culture hood.

2.3 Transfection

1. Vented tissue culture TripleFlask(s) (Thermo Fisher Scientific).
2. 50 μg of cis-plasmid and 150 μg of helper plasmid (pDG9) per triple flask to be transfected.
3. Transfection media: DMEM with 4.5 g/L glucose, L-glutamate, and sodium pyruvate; and DMEM with 4.5 g/L glucose, L-glutamate, and sodium pyruvate, supplemented with 2% fetal bovine serum (no penicillin/streptomycin).

4. Post-transfection medium: DMEM with 4.5 g/L glucose, L-glutamate, and sodium pyruvate, supplemented with 2% fetal bovine serum and 1× penicillin/streptomycin.
5. Linear 40 kDa polyethyleneimine (PEI)-Max (Polysciences Inc.).
6. 12 N hydrochloric acid.
7. 250 mL sterile filter (0.22 μm) bottles.

2.4 Harvesting and Processing

1. 200 mL polypropylene conical centrifuge tubes.
2. 50 mL polypropylene conical tubes.
3. Lysis buffer: 2 mM $MgCl_2$, 150 mM NaCl, 50 mM Tris–HCl, pH 8.5.
4. Sorvall RC-6+ centrifuge.
5. SH-3000BK, F14-6x-250y, and F13-14x50cy rotors.
6. Pierce Universal Nuclease for Cell Lysis (Thermo Fisher Scientific) (*see* **Note 2**).
7. Ammonium sulfate.
8. 18G needle.

2.5 Preparation of Ultracentrifugation Gradients

1. Optiprep 60% iodixanol (Thermo Fisher Scientific).
2. 5× Optiprep dilution buffer (ODB): 5× PBS, 5 mM magnesium chloride, and 12.5 mM potassium chloride.
3. Iodixanol Gradient Layer Solutions (*see* Table 1).
4. Ti70 rotor (Beckman Coulter).
5. OptiSeal polypropylene 37.4 mL (26 × 77 mm) ultracentrifuge tubes (Beckman Coulter).
6. 5 and 10 mL syringes.
7. 18G needles.
8. 18G blunt-end Hamilton needles.

Table 1
Components to prepare iodixanol gradient layer solutions sufficient for one complete (8 tubes) Ti70 Rotor

Iodixanol gradient layer (%)	Optiprep (mL)	1 M $MgCl_2$ (μL)	5 M NaCl (mL)	10× PBS (mL)	Phenol Red (μL)	Water (mL)	Total for 1 full rotor (mL)
15	14.6	58.4	11.68	5.84	–	26.02	58.2
25	16.33	39.2	–	3.92	78.4	18.83	39.2
40	21.33	32	–	3.2	–	7.44	32
60	31.95	32	–	–	16	–	32

2.6 Dialysis

1. 10 mm flat width, 12–14 kDa MWCO, regenerated cellulose dialysis tubing (Spectrum Labs).
2. Dialysis tubing closures.
3. Lactated Ringer's solution.
4. 0.22 μm sterile syringe filter.
5. Large sterile bottle or container.
6. Orbital shaker.

2.7 Quantitative Real-Time PCR

1. Real-time PCR machine.
2. Real-time PCR plate or tubes and caps compatible with the specific machine.
3. iTaq Universal SYBR Green Supermix (Bio-Rad).
4. Forward and reverse primers that anneal perfectly to both the reference standard and the AAV sample to be quantified (*see also* **Note 3**).
5. AAV2 Reference Standard Material (ATCC VR1616) or calibrated in-house AAV reference standard (*see* **Note 4**).

2.8 Alkaline Gel

1. Horizontal ("submarine style") electrophoresis apparatus and power supply.
2. Electrophoresis grade agarose.
3. 50× Alkaline Gel Buffer: 2.5 M sodium hydroxide, 50 mM EDTA.
4. Alkaline Gel loading dye: 4× Alkaline Gel Buffer, 1.2% SDS, 20% Glycerol, 0.01% Xylene Cyanol.
5. 1 kb DNA Ladder (New England Biolabs).
6. 0.1 M Tris–HCl pH 8.5.
7. Gel Red (Biotium), SYBR Gold (Thermo Fisher Scientific), or equivalent dye that allows for the sensitive detection of single-stranded DNA.
8. UV transilluminator (Image capturing system, e.g. ChemiDoc Touch (Bio-Rad)).

2.9 Anesthesia

1. Research anesthesia machine with veterinary anesthesia vaporizer, nose cone device, and small animals' inhalational isoflurane anesthetic chamber.
2. Inhaled anesthetic: isoflurane.
3. Injectable anesthetic: 100 mg/mL ketamine HCl and 20 mg/mL xylazine.
4. Animal scale.
5. 26G needles (for mice) and 21G needles (for rats).
6. 1 mL syringes.

2.10 Intravascular Injection Via the Tail Vein

1. 27-30G 1 mL insulin syringes, 5/8″ needle (for mice).
2. 28-30G catheter (for rats).
3. 1 mL syringes (for rats).
4. 25G needles (for rats).
5. Restraining device, such as Tailveiner restrainer (Braintree Scientific) or anesthetic.

2.11 Intravascular Injection Via the Venous Sinus (Retro-orbital Injection)

1. Isoflurane and research anesthesia machine.
2. Isothermal pad, such as PhysioSuite with RightTemp module (Kent Scientific).
3. 27.5G 0.5 mL insulin syringes, 1/2″ needle (for mice).
4. 1 mL syringes (for rats).
5. 23-28G needles (for rats).

2.12 Local Injection into the Ventricular Wall of the Heart

1. Rodent ventilator, such as VentElite small animal ventilator (Harvard Apparatus) for rats or VP3 veterinary anesthesia vaporizer (DRE Veterinary) for mice.
2. Isothermal pad, such as PhysioSuite with RightTemp module (Kent Scientific).
3. Rectal temperature probe.
4. High intensity fiber optic gooseneck light (Cole-Parmer).
5. Soft catheter for endotracheal intubation: 16G for rats above 200 g, 18G for rats below 200 g, and 20G for mice.
6. Soft catheter for the thoracic cavity: 18G catheter (for rats) and 22G catheter (for mice).
7. Electric hair shaver.
8. 70% ethanol.
9. Povidone-iodine 10%.
10. 1 mL syringes (for rats) and 0.3-0.5 mL syringes (for mice).
11. 25 and 30G needles (for rats) and 26 and 31G needles (for mice).
12. 0.3 mg/mL Buprenex HCl.
13. Standard small animal surgical instruments, including rib spreader, scalpel handle and blades, surgical scissors, and forceps.
14. 5-0 (for rats) and 6-0 (for mice) Vicryl sutures with cutting needle.
15. 4-0 (for rats) and 5-0 (for mice) Prolene sutures with non-cutting needle.
16. Battery-operated electrocautery device.
17. Stereo microscope.

3 Methods

3.1 Plasmid Cloning and Production

1. To prevent the potential loss of intact ITRs, all cis-plasmids are maintained in either SURE2 or Stbl3 cells and grown at 30–32 °C (*see* **Note 5**). pDG9 helper plasmid can be maintained in a regular *E. coli* strain such as DH5α and grown at 37 °C. Purify plasmids using a Maxiprep kit. For the production of the cis-plasmid, one 2 L Erlenmeyer flask with 600 mL of LB medium should produce approximately 500 μg of plasmid. Since the amount of pDG9 plasmid needed for the same number of triple flasks is three times the amount of the cis-plasmid, grow pDG9 bacterial cultures in three 2 L flasks with 600 mL of LB medium each.
2. To confirm that the majority of the purified cis-plasmids contain two intact ITRs, digest the cis-plasmid with *Sma*I (or the SmaI isoschizomer XmaI) restriction enzyme and analyze the digested DNA on an agarose gel. Each ITR contains two SmaI/XmaI sites.

3.2 Cell Maintenance (See Note 6)

1. Maintain HEK293T cells in DMEM supplemented with 10% FBS and 1× Penicillin/Streptomycin for all growth and passaging steps unless otherwise stated. For continuous culture, grow the cells to approximately 70–80% confluency and then split 1:10, so that cells will be ready to be split again 3 or 4 days later, respectively.
2. To split cells from a 175 cm^2 flask, gently remove the culture medium, add 20 mL of PBS without Mg/Ca to cover the surface, and gently rock the flask to completely cover the cells. Remove the PBS and add 5 mL of 0.025% Trypsin to the side of the flask, not directly onto the cells. Spread over the cells by gently "rocking" the flask several times. Incubate at 37 °C for 1–2 min. Then rock the flask to completely dislodge the cells. Pipet 10 mL of medium onto the cells and pipet up and down to prepare a homogenous cell suspension. Add 25 mL of a 1:10 dilution of cell suspension in fresh medium to a new flask (*see* **Note 7**).

3.3 Transfection

1. To prepare the PEI solution for transfection, add 200 mg of PEI-Max powder to 200 mL of sterile, deionized water and stir with magnetic stir bar. Adjust pH to 4.5 by adding 12 N hydrochloric acid. Stir overnight to assure complete dissolution of the PEI-Max. The next day, sterile-filter the solution with a 0.22 μm filter bottle, and store aliquots at −80 °C. Once thawed, the solutions can be stored at 4 °C for up to 2 months.
2. The day before transfection, detach cells with Trypsin and use the cells to seed a triple flask (*see* **Note 8**). Prior to seeding the

triple flask, remove a sufficient number of cells to seed a 175 cm^2 flask for the continuation of cell passaging.

3. The morning after seeding the triple flasks, check the flasks for cell confluency. If ~70% confluent, proceed to transfection immediately. If the confluency is less than ~70%, perform the transfection later in the day.
4. For each triple flask to be transfected, heat 90 mL of DMEM (2% FBS, no Pen/Strep) to 37 °C.
5. Prepare the transfection mix by adding in the following order to 20 mL of room temperature DMEM (no FBS, no Pen/-Strep): 50 μg of cis-plasmid, 150 μg of pDG9 plasmid, and 250 μL of PEI-Max solution (pH 4.5). Vortex for 10–20 s and incubate at room temperature for 15 min.
6. Add the transfection mix to the prewarmed 2% FBS medium and mix by swirling.
7. Gently remove the medium from the cells in the triple flasks and replace with transfection mix medium.
8. Following incubation of cells overnight at 37 °C, replace medium with DMEM (2% FBS, 1× Pen/Strep) and further incubate the transfected cells at 37 °C.
9. Three days from the day of transfection, harvest rAAV virions from cells.

3.4 Harvesting and Processing

3.4.1 Preparation of Cell Lysate

1. Detach the cells by tapping the flask vigorously and transfer the solution to a sterile 200 mL conical tube. Pellet the cells by centrifugation at 1000 × *g* for 15 min in a Sorvall RC-6+ centrifuge and SH-3000BK swinging bucket rotor.
2. Collect the cell culture supernatant in a sterile bottle for later processing.
3. Gently add 10–15 mL of PBS to the side of the conical tube. Swirl gently to dislodge, but not break apart, the pellet. Transfer the pellet to a 50 mL conical tube. Multiple pellets can be combined into one 50 mL tube for further processing. If cells remain in the large conical tube, use the PBS to rinse the cells off the large tube and transfer to the small tube.
4. Pellet the cells by centrifugation in a Sorvall RC-6+ centrifuge and SH-3000BK rotor at 1000 × *g* for 10 min. Pour the supernatant into the bottle containing the cell culture supernatant (*see* **Note 9**). This cell culture supernatant will be used for Subheading 3.4.2, and can be frozen at −80 °C to be processed later.
5. Resuspend the cell pellet in 5 mL of lysis buffer (*see* **Note 10**).

6. To lyse the cells, repeatedly freeze and thaw the cell suspension. The cell suspension can either be frozen in a −80 °C freezer or on dry ice. Thaw the cell suspension in a 37 °C water bath followed by brief vortexing. Repeat this freeze/thaw cycle twice. Aggregation of cellular debris during the freeze/thaw cycles is normal. Any of the freezing steps can be used as a stopping point.
7. To digest genomic and plasmid DNA, non-encapsidated viral DNA, and cellular RNA, add 1 μL (250 U) of Pierce Universal Nuclease to the thawed, crude lysate and incubate for 30 min at 37 °C (*see* **Note 2**).
8. Centrifuge the crude lysate in a Sorvall RC-6+ centrifuge and F13-14x50cy fixed angle rotor at 5000 × *g* for 20 min and transfer the supernatant to a new tube (discard the pellet). The new tube is ready for iodixanol gradient ultracentrifugation. It can be stored at −80 °C at this point.

3.4.2 Processing of the Cell Culture Supernatant

1. If the cell culture supernatant was frozen (**step 4** of Subheading 3.4.1), thaw it at 37 °C, but do not incubate longer than necessary. Add 31.3 g ammonium sulfate per 100 mL of supernatant, and shake for 1–2 min to dissolve the ammonium sulfate completely. Incubate the mixture on ice for at least 30 min (*see* **Note 11**).
2. To pellet the virus precipitate, centrifuge in a Sorvall RC-6+ centrifuge and an F14-6x250y fixed angle rotor at 8300 × *g* for 30 min. Because the pellet may be loose, pour off the supernatant as soon as possible after centrifugation. DO NOT STERILIZE THE SUPERNATANT WITH BLEACH. INSTEAD, THE SUPERNATANT MUST BE AUTOCLAVED AND THEN DISCARDED. Ammonium Sulfate is incompatible with bleach, and treatment of the supernatant with bleach would result in the production of TOXIC GASES AND EXPLOSIVE COMPOUNDS.
3. The pellet will accumulate on the side of the centrifuge tube. Resuspend the pellet with a pipette in 4 mL of lysis buffer (*see* **Note 10**). The resuspension is complete when there is no remaining pellet on the wall of the tube and the solution is homogeneous.
4. Add 1 μL (250 U) of Pierce Universal Nuclease for digestion of possible cellular and plasmid DNA or RNA (*see* **Note 2**) and incubate 30 min at 37 °C (or according to manufacturer's instructions if using an alternative nuclease). The virus suspension is now ready for iodixanol gradient purification of the virus.

3.5 Preparation of Ultracentrifugation Gradients

3.5.1 First Gradient

1. To form the gradient, slowly underlay the following solutions: (a) 7.3 mL of 15% iodixanol, (b) 4.9 mL of 25% iodixanol, (c) 4 mL of 40% iodixanol, and (d) 4 mL of 60% iodixanol. To add each layer, use a 10 mL syringe fitted with an 18G blunt-end Hamilton needle, and lower it into the tube until the tip of the needle reaches the bottom and center of the tube (*see* **Note 12**).
2. Once the gradient is ready, add the viral solution on top of the gradient by placing the Hamilton needle on the side of the tube and dispensing the viral solution dropwise onto the top layer of the gradient.
3. If at this point the tube is not completely filled, add lysis buffer to the top of the tube to fill the tube up to the neck. Any bubbles must be removed prior to centrifugation to prevent the collapse of the tubes during centrifugation. To remove potential bubbles, slightly squeeze the tube until the bubbles rise above the neck of the tube and remove them with a pipet. Seal the tube by firmly inserting the plastic stopper.
4. Use a permanent marker to draw a line at the 40–60% interface. The interfaces are easy to see before centrifugation but more difficult to visualize after centrifugation.
5. Centrifuge at 350,333 × g in a Ti70 rotor for 1 h (*see* **Note 13**).
6. The majority of AAV particles will be located at the 60–40% iodixanol interface. To collect the virus, remove the stopper of the tube, insert an 18G needle connected to a 5 mL syringe just below the interface and slowly withdraw 3 mL (these 3 mL will consist of approximately the top 1 mL of the 60% layer and the bottom 2 mL of the 40% layer). Alternatively, the virus can be collected in fractions from the bottom of the gradient as described in **steps 5** and **6** of Subheading 3.5.2. If a second gradient is not performed, collecting the virus in fractions will yield virus preparations with fewer empty particles.
7. Determine the AAV titer by qPCR (Subheading 3.7). If the viral titer is too low (*see* **Note 14**), combine the virus from up to four tubes and perform a second gradient (Subheading 3.5.2). If the viral titer is sufficiently high, proceed to Subheading 3.6.

3.5.2 Second Gradient

1. For a second round of iodixanol gradient purification, dilute the harvested AAV/iodixanol solution in 1× ODB such that the percentage of iodixanol is lower than 25%. For this calculation, the AAV/iodixanol harvested from the first gradient is assumed to be 47% iodixanol (1 mL of 60% + 2 mL of 40%). If the virus was collected in fractions, calculate the approximate

percentage of iodixanol for the corresponding fractions to be purified.

2. Add 22 mL of the AAV/iodixanol dilution into an OptiSeal ultracentrifuge tube.
3. Underlay this solution sequentially with 4 mL of 40% iodixanol and 4 mL of 60% iodixanol as described in **step 1** of Subheading 3.5.1.
4. Centrifuge the tubes at 350,333 × *g* in a Ti70 rotor for 2 h.
5. Collect 1.25 mL fractions from the bottom of the tube. To do this, remove the stopper and cover the opening with a finger (this will form a vacuum that prevents flow once a hole is punched in the bottom of the tube). With the finger still placed on the opening, make hole with an 18G needle at the bottom of the tube. Now, place a microcentrifuge tube (pre-marked at the 1.25 ml level) under the hole and release the finger to let the content drip into the microcentrifuge tube to collect the first 1.25 ml fraction. Next, place the finger back on the top opening to stop the flow, place the next microcentrifuge tube under the hole, release the finger and collect the second 1.25 ml fraction. Repeat this process until you have collected 8 fractions. The remaining content can be collected in a large conical tube to be discarded.
6. Run an alkaline gel of the fractions to identify the fractions containing the majority of full-length viral genomes and then combine those fractions.

3.6 Dialysis

1. Soak dialysis tubing in lactated Ringer's solution overnight.
2. The next morning, squeeze out any lactated Ringer's solution left in the tubing, clamp one side of the tubing with a dialysis clip, pipet the AAV/iodixanol solution into the tubing, and clamp the other end with a dialysis clip. To allow for easy recovery of the virus, leave at least 2 in. of tubing hanging outside one of the clamped ends.
3. Place the closed dialysis tubing in a large sterile bottle with at least 100 dialyzate volumes of lactated Ringer's solution.
4. Agitate the dialysis solution gently for 1–2 h at 4 °C on an orbital shaker platform and replace the lactated Ringer's solution with an equal volume of fresh solution.
5. Agitate the dialysis solution for an additional 4–5 h at 4 °C and replace with fresh lactated Ringer's solution with an equal volume of fresh solution.
6. Agitate overnight at 4 °C. Remove the clamp on the side with additional tubing and collect the contents of the tubing in a 50 mL tube. Filter through a 0.22 μm sterile syringe filter. Aliquot and store at −80 °C (*see* **Note 15**).

3.7 Viral Titer Determination by Quantitative Real-Time PCR (qPCR)

1. Prepare a master mix of reagents (5 μL of 2X iTaq Universal SYBR Green Supermix, 0.5 μL of 10 μM forward and reverse primers, and 2 μL of water) that will be sufficient for duplicates of the following samples: four dilutions of standards, two dilutions of AAV sample, and one no-template control.
2. Thaw an aliquot of the ATCC AAV reference virus (or a well-characterized in-house AAV standard) (*see* **Notes 3, 4** and **16**), as well as an aliquot of the AAV to be quantified.
3. Prepare a standard curve of the reference virus by diluting the virus stock: 1:1,000; 1:10,000; 1:100,000 and 1:1,000,000.
4. Dilute the AAV sample 1:10,000 and 1:100,000.
5. Add 8 μL of master mix and 2 μL of virus dilutions or water to each tube/plate well for a final reaction volume of 10 μL, and perform the qPCR according to the PCR reagents and instrument manufacturer's instructions.
6. Calculate the viral titer based on the standard curve and dilution of virus sample using the qPCR machine software (*see* **Note 4**).

3.8 Alkaline Gel Electrophoresis

1. Perform alkaline gel electrophoresis to determine vector genome integrity and titer. Prepare the alkaline agarose gel by adding 1 g of agarose to 98 mL of water and microwave until all the agarose is dissolved. Let the solution cool until it can be handled without gloves, then add 2 mL of 50× Alkaline Electrophoresis Buffer, swirl to mix, and pour the solution into the tray to cast the gel.
2. Place the gel into an electrophoresis apparatus, fill with cold 1× Alkaline Running Buffer, and place everything in a cold room.
3. Dilute 500 ng of DNA ladder in 25 μL of water.
4. Mix 25 μL of AAV sample or diluted DNA ladder with 8.5 μL of freshly prepared 4× alkaline sample loading buffer, heat to 95 °C for 3 min, and cool on ice prior to loading on the gel.
5. Load the gel and run overnight at 20 V in the cold room using a dedicated power supply (*see* **Note 17**).
6. Remove the gel from the electrophoresis apparatus, place the gel in a container, and cover it with 0.1 M Tris–HCl pH 8.5 and rock gently for 1 h.
7. Discard the buffer and replace it with 4× GelRed or SYBR Gold in 0.1 M NaCl and rock protected from light for 2 h.
8. Rinse the gel briefly in tap water.
9. Visualize the gel with a UV transilluminator and capture a digital picture (*see* Fig. 1 and **Note 18**).

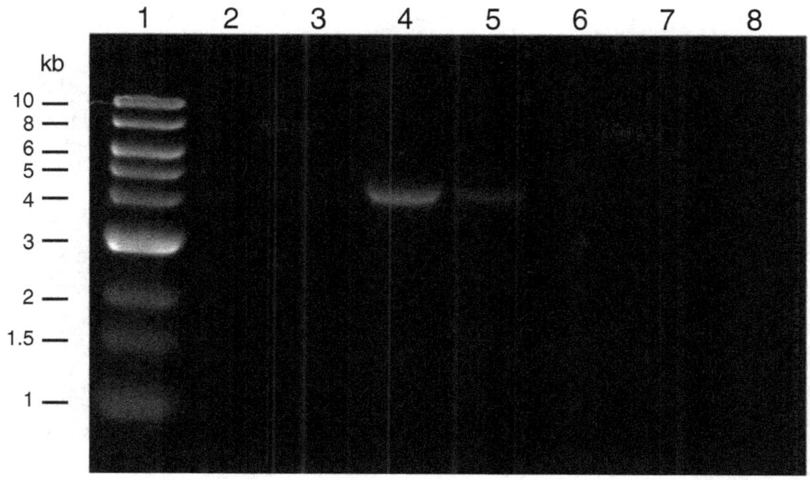

Fig. 1 Alkaline gel of fractions collected from an iodixanol gradient. *Lane 1*: 1 kb DNA Ladder (500 ng). *Lanes 2–8*, 20 μL of fractions 2–8 of a first iodixanol gradient of AAV9-LMNA. The capsid protein content in fractions 5 is higher than in fraction 4, indicating that fraction 5 contains empty capsids (SDS-PAGE, not shown)

3.9 Vector Delivery Approaches for Cardiac Transduction

In the following sections, we describe the intravascular injection of rAAV vectors to rodents via the tail vein or the venous sinus, as well as local injection into the ventricular wall of the heart. Vector delivery by intraventricular injection combined with cross-clamping of the aorta and pulmonary artery and local injection in the pericardium are described elsewhere [20].

1. Approval of the study by the local Institutional Animal Care and Use Committee is required before the initiation of the study.
2. The animals should be free of all endoparasites, ectoparasites, mycoplasma species, and common rodent viruses.
3. House rodents in a room at 22 ± 2 °C with a relative humidity of 50 ± 10%, and feed rodents a commercial laboratory diet and water *ad libitum* [21]. Ventilation in the room should be between 12 and 15 air changes per hour of 100% prefiltered outside air. The light cycle period should be controlled on a 12/12 h light/dark cycle with no twilight transition.
4. Males and females must be housed separately, as well as males from different litters, and overcrowding of cages must be avoided (*see* ref. 22 for further information on housing recommendations).

3.10 Anesthesia

Different methods of anesthesia can be employed, depending on the AAV vector administration procedure to be performed, the animal model used, and the available methods in the laboratory facilities. Here, animals are anesthetized with inhaled anesthetic for retro-orbital injections and with injectable anesthetic for local intra-myocardial injection into the ventricular wall of the heart.

3.10.1 Injectable Anesthesia

1. Prepare a mixture of 100 mg/mL ketamine and 20 mg/mL xylazine.
2. Weigh the animal and inject intraperitoneally a dose equivalent to 50 mg/kg ketamine and 10 mg/kg xylazine with a 26G needle (for mice) or 21G needle (for rats) fitted with a 1 mL syringe.

3.10.2 Inhaled Anesthesia

1. Place the animal into an appropriately scavenged inhalational isoflurane anesthetic chamber. Introduce isoflurane into the chamber at a rate of 1.5–3 L/min at a 5% (v/v isoflurane/ oxygen) concentration.
2. After 3–5 min, assess the level of anesthetic depth by observing the loss of the righting and palpebral reflexes such as pedal reflex (firm toe pinch), and by muscular tone, response to painful stimulation, and rate and depth of respiration.
3. Once the animal is completely anesthetized, switch the system to flow through a nose cone, and quickly remove the animal from the chamber and position it in the nose cone (*see* **Note 19**). Adjust the anesthetic delivery to 2.5% in 50% oxygen/50% air concentration to maintain surgical plane.
4. Proceed with AAV vector delivery, monitoring anesthetic depth throughout the procedure and adjusting the vaporizer as needed.

3.11 Intravascular Injection Via the Tail Vein

1. Prepare the AAV vector solution by diluting the dialyzed virions to the desired concentration in either PBS or lactated Ringer's solution (*see* **Note 20**).
2. Place the rodent under a heat lamp for a few minutes to promote peripheral vasodilation.
3. Immobilize the animal with a restraining device or lightly anesthetize the animal.
4. Hold the needle parallel to the posterior tail vein of the animal with the bevel side up at the lower portion of the tail (*see* **Note 21**). Use a 27–30G 1 mL insulin syringe for mice, or a 28–30G catheter or 1 mL syringe fitted with a 25G needle for rats (reviewed in [23]).
5. Puncture the vein and inject the vector solution (*see* **Note 22**).
6. Return the animal to the cage for recovery.

3.12 Intravascular Injection Via the Venous Sinus (Retro-Orbital Injection) (See Note 23)

1. Prepare the AAV vector solution by diluting the dialyzed virions in a volume of PBS or lactated Ringer's solution not exceeding 150 μL [24].
2. Anesthetize the animal with a volatile anesthetic drug such as isoflurane (*see* Subheading 3.10.2) and place it in left lateral recumbency with the head facing to the right on an isothermal

heating pad (*see* **Note 24**). Apply gentle pressure to the skin dorsal and ventral to the eye to partially protrude the mouse right eyeball from the eye socket (*see* **Note 25**).

3. Using a 27.5G insulin syringe (for mice) or a 1 mL syringe fitted with a 23–28G needle (for rats), introduce vector solution carefully into the medial canthus with the bevel down at an angle of approximately 30 degrees (*see* **Note 25**). When the needle reaches the base of the eye, slowly inject the vector solution (do not aspirate before the injection) and, when completed, withdraw the needle carefully and slowly to prevent the solution to leak out.
4. Return the animal to the cage for recovery.

3.13 Local Intramyocardial Injection into the Ventricular Wall of the Heart

1. Prepare the AAV vector solution by diluting the dialyzed virions in a volume of PBS or lactated Ringer's solution not exceeding 30 μL (for mice) or 400 μL (for rat). The temperature of the solution should be approximately body temperature.
2. Thirty minutes prior to anesthetization, inject all animals intraperitoneally with 0.1 mg/kg of the analgesic buprenorphine with a 26G needle (for mice) or 25G needle (for rats) fitted with a 1 mL syringe.
3. Anesthetize the animal with an intraperitoneal injection of the anesthetic ketamine/xylazine mixture (*see* Subheading 3.10.1). This injection will sedate the animal within 15–20 min.
4. When the animal is fully sedated, perform an endotracheal intubation with a 16–18G catheter (for rats) or a 20G catheter (for mice) using a fiber optic gooseneck light and connect the endotracheal tube to the ventilator. Test if the animal is breathing either by observing the heaving of the chest or by using a cold mirror to observe condensation of humidity from the animal's exhaled breath.
5. Set the ventilator to the correct tidal volume and respiratory rate (*see* **Note 26**). Maintain the surgical plane with inhaled anesthesia (*see* Subheading 3.10.2).
6. Place the anesthetized animal in a right lateral semi-decubitus position on an isothermal heating pad to maintain body temperature, and place a rectal temperature probe in the animal for temperature monitoring.
7. Remove the hair on the chest and disinfect the surgical area by washing with 10% povidone-iodine, followed by 70% ethanol.
8. Perform a left antero-lateral thoracotomy incision in the fourth intercostal space. Insert a small self-retaining rib spreader (*see* **Note 27**). Move the lung towards the posterior to allow access to the heart, and incise the pericardium. Once the myocardium

is exposed, inject the virus directly into the left ventricle with a 1 mL syringe fitted with a 30G needle (for rats) or a 0.3–0.5 mL syringe fitted with a 31G needle (for mice) at an angle of 45 degrees in relation to the left ventricle wall. The puncture should go only a few millimeters deep as the ventricular wall is very thin (*see* **Note 28**).

9. Once hemostasis is achieved, begin closing the ribs and intercostal muscles with 4-0 Prolene suture (for rats) or 5-0 Prolene suture (for mice). Before tying the sutures, insert an 18G catheter (for rats) or a 22G catheter (for mice) into the thoracic cavity, then finish tying the sutures. Next, close the skin and subcutaneous space with running 5-0 Vicryl (absorbable) suture (for rats) or 6-0 Vicryl suture (for mice).
10. After tying the skin sutures, use a 3 mL syringe (for rats) or a 0.5 mL syringe (for mice) to evacuate air and fluid from the pleural cavity through the catheter that was inserted into the thoracic cavity, in order to regain the normal negative physiologic pressure.
11. Administer sterile lactated Ringer's solution subcutaneously to the animals (2 mL for rats and 0.3 mL for mice) and cease administration of isoflurane.
12. When the animal starts making spontaneous breathing motions and neck movements, disconnect the ventilator from the endotracheal tube and remove the endotracheal tube from the animal.
13. Return the animal to the cage for recovery.
14. For pain control, inject the animals subcutaneously with 0.3 mg/kg of Buprenex (for mice) or 0.1 mg/kg of Buprenex (for rats) immediately after surgery and again every 12 h for 3 days.

4 Notes

1. Subheadings 2.1–2.8 and 3.1–3.8, as well as associated Notes and display items, are reproduced with permission from [25].
2. Alternative nucleases that digest double-stranded DNA, single-stranded DNA, and RNA, such as Benzonase (Sigma), can also be used to achieve the same result. If Benzonase is used, add 2 μL of Benzonase (25 units/μL) to every 1 mL of crude lysate and incubate for 30 min at 37 °C.
3. Extensively characterized AAV2 and AAV8 reference standard material can be obtained from ATCC (#VR-1616 and #VR-1816, respectively). The broad use of these reference standards should facilitate the reproducibility of titer determinations

among different laboratories. For economic reasons, it might be preferable to employ the ATCC standard viruses to prepare a thoroughly characterized in-house reference virus preparation. However, unless ITR-specific primers are used [26], the reference standards can only be used for the titration of AAVs that share other common sequence elements with the AAV standards. A detailed protocol for in-house production of reference standard can be found in [27].

4. If a single-stranded in-house reference standard is used and the virus to be titrated is self-complementary, the final titer should be divided by 2. In this context, the accurate titer determination of double-stranded AAVs by qPCR has additional pitfalls, as previously described [28].
5. For further information on designing and constructing cis-plasmids, we refer the reader to [29].
6. Cells can be grown in the presence or absence of antibiotics. HEK293T cells are a human cell line, and all work must be performed in accordance with Biosafety Level 2 (BSL2) regulations. This includes the use of a BSL2 laminar flow tissue culture hood. Adeno-associated viruses are BSL1, but all work with AAV should be performed under sterile conditions.
7. HEK293T cells can be passaged for about 1–2 months after thawing the initial vial, as long as they are regularly split at least twice a week.
8. To achieve approximately 70% cell confluency seed the cells of 2 confluent 175 cm^2 flasks per triple flask.
9. We routinely combine this PBS pellet wash solution with the cell culture supernatant from the previous step, rather than discarding it, because it can contain AAV virions.
10. The maximum volume of crude lysate that can be loaded in a single ultracentrifugation tube is 10 mL. However, the cell pellets from up to three triple flasks can be combined and resuspended in 10 mL of lysis buffer. After freeze/thaw and clearing of the lysate, the combined lysate can be loaded into a single tube. Similarly, it is possible to resuspend the pellets of precipitated AAV from cell culture supernatant of up to three triple flasks.
11. Although incubation on ice for as little as 30 min is sufficient for virus precipitation, keeping the ammonium sulfate/cell culture supernatant mixture for several days at 4 °C will not affect the quality of the virus preparation. However, once the virus precipitate has been spun down, the pellet should be resuspended and loaded onto a gradient the same day. Resuspending the pellet and freezing for a later gradient is not recommended as aggregation appears upon thawing.

12. The creation of multiple iodixanol gradients can be facilitated by using a multichannel peristaltic pump (Watson-Marlow) to assemble several gradients simultaneously. With such a system the different iodixanol solutions can be delivered to multiple tubes at the same time, reducing the number of necessary pipetting steps. Using this method, up to four centrifugations (with Ti70 rotors, holding 8 tubes each) can be performed in one day.
13. If a different rotor is used, the speed and run times must be adjusted using the k-factors of the Ti70 rotor at 350,333 × *g* ($k = 44.9$) and the respective *k*-factor of the alternate rotor at a given speed. A convenient tool for this calculation can be found at: https://www.beckmancoulter.com.
14. Calculate whether the AAV titer is sufficiently high for the planned experiments. Especially for in vivo experiments, when small injection volumes are required, it might be necessary to concentrate the virus further. In addition to concentrating the virus, if the virus was collected as a single fraction during the first gradient purification, this second ultracentrifugation gradient also allows the removal of empty capsids that may have been collected together with the genome-containing AAV particles from the first gradient.
15. For long-term storage, AAV is preferably stored at −80 °C. However, AAVs are extraordinarily stable; recent reports have shown no loss of activity after one week of storage at room temperature [30]. This can have important implications for the shipping of AAVs. In particular, the quality of the virus is unlikely to be affected if all dry ice has evaporated due to unanticipated delays in the delivery of the virus. Because of the danger of damaging the AAV genome, never expose AAV to UV light (such as that from a biosafety cabinet).
16. The quality of AAVs purchased from academic vector cores or especially commercial vendors can vary significantly. This is especially true for double-stranded AAVs or AAVs with a genome size approaching the maximum packaging capacity of the AAV capsid. Therefore, we recommend a thorough in-house characterization before performing any experiments. Capsid titers and virus purity can be assessed by Coomassie staining of virions run on an SDS-PAGE gel [25]. Although this is not routine for in vitro or small animal experiments, we recommend that researchers that are not familiar with the procedures described in this chapter use SDS-PAGE and Coomassie staining to confirm the purity of their viral preparations. Once the routine production of high-quality AAV vector preparations has been firmly established in a laboratory, SDS-PAGE followed by Coomassie Blue staining may be

omitted. However, we strongly recommend the use of alkaline gel electrophoresis to confirm vector genome integrity.

17. Moving the power supply frequently from the cold room to room temperature will damage the power supply due to water condensation.

18. Alkaline gel electrophoresis can also be used to determine AAV titers. For this, load a dilution series of a mass DNA ladder, or dilutions of a linear DNA fragment of known concentration and, preferably, of similar size to the AAV genome (e.g., the cis-plasmid digested with *Sma*I if the only *Sma*I sites are in the viral ITRs) onto the same gel as the AAV samples to be quantified. Quantification can then be performed with the freely available ImageJ (http://imagej.nih.gov/ij/) or similar software. This method is especially useful if the AAV samples to be quantified do not share sequence elements with the reference standard and for double-stranded AAVs.

19. During the system change, the animal might start to awaken. In this case, wait until it is fully anesthetized again before starting with the AAV vector administration procedure. The use of a nose cone can cause eye dryness. Use an eye lubricant compatible with the vector delivery procedure while using the nose cone, if necessary.

20. For systemic delivery, the dose of AAV vector usually ranges from 2×10^{10} vg to 5×10^{11} vg for adult mice and 5×10^{11} vg to 2×10^{12} vg for rats. The injection volume should be around 5 mL of solution per kg (reviewed in [31]). For reference, commonly used injection volumes are 100–200 μL for mice and should not exceed 500 μL for rats.

21. For administration of the vector solution with a needle into small rodents such as mice, slightly bending the needle 30 degrees up prevents having to excessively bend the rodent tail and, thus, helps with the injection procedure.

22. When injecting the vector solution intravenously, the vein loses its purple color and acquires a whiter color due to the vector solution displacing the blood contained in the vein. This is a positive sign that the vector solution is being properly injected into the vein. Swelling of the tail is indicative of the vector solution being injected subcutaneously instead of intravenously. In the latter case, remove the needle and try a second administration attempt puncturing the vein closer to the tail base than the initial attempt, as the vein might have been ruptured at the location of the initial attempt.

23. Despite the fact that this technique might seem aesthetically distasteful, it has been argued to be a more humane alternative compared to the more commonly used tail vein injection because retro-orbital injection has a lower failure rate, and it

can easily be applied to adult and neonatal mice ([32] and reviewed in [24]). It is important to not perform more than one injection per eye per day, and not more than two injections in total per eye (with one day interval between injections).

24. A drop of ophthalmic anesthetic can also be applied on the eye to minimize pain to the animal upon completion of the procedure. Excess ophthalmic anesthetic can be removed carefully from the medial canthus with an absorbent gauze pad.

25. Special care must be taken while performing this technique because excessive pressure to the ventral cervical vessels (jugular veins and carotid arteries) will obstruct blood flow and impede administration of the vector solution, and pressure to the trachea may cause it to collapse and restrict air flow. Also, the needle must have minimal movement once inserted into the retro-orbital plexus to prevent vessel rupture and subsequent bleeding and loss of vector solution into the tissue behind the eye. There should be little-to-no bleeding following withdrawal of the needle, and the whole procedure should not take more than 1 min.

26. Use a convenient calculator for appropriate tidal volumes and respiratory rates at https://www.kentscientific.com/Products/Specs/SpecPopUp.asp?Mode=SpecBody&SpecTypeId=3&SpecId=216, or use a formula found in [33]. Keep the peak pressure at or below 10 cmH_2O.

27. For easier evaluation, perform the surgical procedure with a stereo microscope at ×10 magnification.

28. The viral solution should be injected in the apex and posterolateral wall of the left ventricle.

Acknowledgments

We would like to thank Elena Chepurko, Malik Bisserier, and Anthony Fargnoli for helpful discussions. This work is supported by NIH P50 HL112324, R01 HL119046, R01 HL117505, R01 HL128099, R01 HL129814, R01 HL131404 and Trans-Atlantic Network of Excellence grants 13CVD01 and 14CVD03 from the Leducq Foundation.

References

1. Sonntag F, Schmidt K, Kleinschmidt JA (2010) A viral assembly factor promotes AAV2 capsid formation in the nucleolus. Proc Natl Acad Sci U S A 107(22):10220–10225. https://doi.org/10.1073/pnas.1001673107

2. Kumaran N, Michaelides M, Smith AJ, Ali RR, Bainbridge JWB (2018) Retinal gene therapy. Br Med Bull. https://doi.org/10.1093/bmb/ldy005

3. Zincarelli C, Soltys S, Rengo G, Rabinowitz JE (2008) Analysis of AAV serotypes 1-9 mediated

gene expression and tropism in mice after systemic injection. Mol Ther 16(6):1073–1080. https://doi.org/10.1038/mt.2008.76

4. Moulay G, Ohtani T, Ogut O, Guenzel A, Behfar A, Zakeri R, Haines P, Storlie J, Bowen L, Pham L, Kaye D, Sandhu G, O'Connor M, Russell S, Redfield M (2015) Cardiac AAV9 gene delivery strategies in adult canines: assessment by long-term serial SPECT imaging of sodium iodide symporter expression. Mol Ther 23(7):1211–1221. https://doi.org/10.1038/mt.2015.78
5. Chamberlain K, Riyad JM, Weber T (2017) Cardiac gene therapy with adeno-associated virus-based vectors. 1Curr Opin Cardiol. https://doi.org/10.1097/HCO.0000000000000386
6. Samulski RJ, Muzyczka N (2014) AAV-mediated gene therapy for research and therapeutic purposes. Annu Rev Virol 1 (1):427–451. https://doi.org/10.1146/annurev-virology-031413-085355
7. Snyder RO, Francis J (2005) Adeno-associated viral vectors for clinical gene transfer studies. Curr Gene Ther 5(3):311–321
8. Wright JF (2008) Manufacturing and characterizing AAV-based vectors for use in clinical studies. Gene Ther 15(11):840–848. https://doi.org/10.1038/gt.2008.65
9. Ayuso E, Mingozzi F, Bosch F (2010) Production, purification and characterization of adeno-associated vectors. Curr Gene Ther 10 (6):423–436
10. Grimm D, Kern A, Rittner K, Kleinschmidt JA (1998) Novel tools for production and purification of recombinant adenoassociated virus vectors. Hum Gene Ther 9(18):2745–2760
11. Wright JF (2009) Transient transfection methods for clinical adeno-associated viral vector production. Hum Gene Ther 20(7):698–706. https://doi.org/10.1089/hum.2009.064
12. Vandenberghe LH, Xiao R, Lock M, Lin J, Korn M, Wilson JM (2010) Efficient serotype-dependent release of functional vector into the culture medium during adeno-associated virus manufacturing. Hum Gene Ther 21(10):1251–1257. https://doi.org/10.1089/hum.2010.107
13. Zolotukhin S, Byrne BJ, Mason E, Zolotukhin I, Potter M, Chesnut K, Summerford C, Samulski RJ, Muzyczka N (1999) Recombinant adeno-associated virus purification using novel methods improves infectious titer and yield. Gene Ther 6 (6):973–985. https://doi.org/10.1038/sj.gt.3300938
14. Zincarelli C, Soltys S, Rengo G, Koch WJ, Rabinowitz JE (2010) Comparative cardiac gene delivery of adeno-associated virus serotypes 1-9 reveals that AAV6 mediates the most efficient transduction in mouse heart. Clin Transl Sci 3(3):81–89. https://doi.org/10.1111/j.1752-8062.2010.00190.x
15. Watanabe S, Leonardson L, Hajjar RJ, Ishikawa K (2017) Cardiac gene delivery in large animal models: antegrade techniques. Methods Mol Biol 1521:227–235. https://doi.org/10.1007/978-1-4939-6588-5_16
16. Bonnet G, Ishikawa K, Hajjar RJ, Kawase Y (2017) Direct myocardial injection of vectors. Methods Mol Biol 1521:237–248. https://doi.org/10.1007/978-1-4939-6588-5_17
17. Hinkel R, Kupatt C (2017) Selective pressure-regulated retroinfusion for gene therapy application in ischemic heart disease. Methods Mol Biol 1521:249–260. https://doi.org/10.1007/978-1-4939-6588-5_18
18. Byrne MJ, Kaye DM (2017) Cardiac gene delivery using recirculating devices. Methods Mol Biol 1521:261–269. https://doi.org/10.1007/978-1-4939-6588-5_19
19. Katz MG, Fargnoli AS, Kendle AP, Bridges CR (2017) Molecular cardiac surgery with recirculating delivery (MCARD): procedure and vector transfer. Methods Mol Biol 1521:271–289. https://doi.org/10.1007/978-1-4939-6588-5_20
20. Federica del Monte KI, Roger J. Hajjar (2017) Gene transfer to rodent hearts in vivo. In: Ishikawa K (ed) Cardiac gene therapy: methods and protocols. Methods Mol Biol, vol 1521. Springer Science + Business Media, New York, pp 195–204. https://doi.org/10.1007/978-1-4939-6588-5_13
21. Speakman JR, Keijer J (2012) Not so hot: optimal housing temperatures for mice to mimic the thermal environment of humans. Mol Metab 2(1):5–9. https://doi.org/10.1016/j.molmet.2012.10.002
22. Garber JC, Wayne Barbee R, Bielitzki JT, Clayton LA, Donovan JC, Hendriksen CFM, Kohn DF, Lipman NS, Locke PA, Melcher J, Quimby FW, Turner PV, Wood GA, Wurbel H (2011) Environment, housing and management. In: Fletcher CH (ed) Guide for the care and use of laboratory animals, 8th edn. National Academy of Sciences, Washington, DC, pp 45–112
23. Turner PV, Pekow C, Vasbinder MA, Brabb T (2011) Administration of substances to laboratory animals: equipment considerations, vehicle selection, and solute preparation. J Am Assoc Lab Anim Sci 50(5):614–627

24. Yardeni T, Eckhaus M, Morris HD, Huizing M, Hoogstraten-Miller S (2011) Retro-orbital injections in mice. Lab Anim (NY) 40(5):155–160. https://doi.org/10.1038/laban0511-155
25. Kohlbrenner E, Weber T (2017) Production and characterization of vectors based on the cardiotropic AAV serotype 9. In: Ishikawa K (ed) Cardiac gene therapy: methods and protocols, vol 1521. Methods in molecular biology. Springer Science + Business Media, New York, pp 91–107
26. Aurnhammer C, Haase M, Muether N, Hausl M, Rauschhuber C, Huber I, Nitschko H, Busch U, Sing A, Ehrhardt A, Baiker A (2012) Universal real-time PCR for the detection and quantification of adeno-associated virus serotype 2-derived inverted terminal repeat sequences. Hum Gene Ther Methods 23(1):18–28. https://doi.org/10.1089/hgtb.2011.034
27. Kohlbrenner E, Henckaerts E, Rapti K, Gordon RE, Linden RM, Hajjar RJ, Weber T (2012) Quantification of AAV particle titers by infrared fluorescence scanning of coomassie-stained sodium dodecyl sulfate-polyacrylamide gels. Hum Gene Ther Methods 23(3):198–203
28. Fagone P, Wright JF, Nathwani AC, Nienhuis AW, Davidoff AM, Gray JT (2012) Systemic errors in quantitative polymerase chain reaction titration of self-complementary adeno-associated viral vectors and improved alternative methods. Hum Gene Ther Methods 23 (1):1–7. https://doi.org/10.1089/hgtb.2011.104
29. Gray JT, Zolotukhin S (2011) Design and construction of functional AAV vectors. Methods Mol Biol 807:25–46. https://doi.org/10.1007/978-1-61779-370-7_2
30. Gruntman AM, Su L, Su Q, Gao G, Mueller C, Flotte TR (2015) Stability and compatibility of recombinant adeno-associated virus under conditions commonly encountered in human gene therapy trials. Hum Gene Ther Methods 26(2):71–76. https://doi.org/10.1089/hgtb.2015.040
31. Turner PV, Brabb T, Pekow C, Vasbinder MA (2011) Administration of substances to laboratory animals: routes of administration and factors to consider. J Am Assoc Lab Anim Sci 50 (5):600–613
32. Steel CD, Stephens AL, Hahto SM, Singletary SJ, Ciavarra RP (2008) Comparison of the lateral tail vein and the retro-orbital venous sinus as routes of intravenous drug delivery in a transgenic mouse model. Lab Anim (NY) 37 (1):26–32. https://doi.org/10.1038/laban0108-26
33. Pacher P, Nagayama T, Mukhopadhyay P, Batkai S, Kass DA (2008) Measurement of cardiac function using pressure-volume conductance catheter technique in mice and rats. Nat Protoc 3(9):1422–1434. https://doi.org/10.1038/nprot.2008.138

Chapter 20

AAV-Mediated Gene Delivery to the Liver: Overview of Current Technologies and Methods

Brett Palaschak, Roland W. Herzog, and David M. Markusic

Abstract

Adeno-associated virus (AAV) vectors to treat liver-specific genetic diseases are the focus of several ongoing clinical trials. The ability to give a peripheral injection of virus that will successfully target the liver is one of many attractive features of this technology. Although initial studies of AAV liver gene transfer revealed some limitations, extensive animal modeling and further clinical development have helped solve some of these issues, resulting in several successful clinical trials that have reached curative levels of clotting factor expression in hemophilia. Looking beyond gene replacement, recent technologies offer the possibility for AAV liver gene transfer to directly repair deficient genes and potentially treat autoimmune disease.

Key words Liver gene therapy, Hemophilia, Adeno-associated virus, Immune tolerance, Hepatocyte-specific promoter, Gene delivery, Hepatocytes, Inherited metabolic disease

1 Introduction

The past few years have seen incredible advances for gene therapy [1]. The adeno-associated virus (AAV) biologic Luxturna received landmark FDA approval as the first in vivo gene therapy product in the United States for inherited retinal disease [2]. In the past year, ongoing phase I/II clinical trials of AAV liver gene therapy for the coagulation disorders hemophilia A [3] and B [4] have published data demonstrating stable factor VIII (FVIII) and factor IX (FIX) protein expression, respectively, in treated patients, and have recently initiated phase III studies. Both vector-based products are anticipated to receive FDA approval following completion of their respective studies. Thus, with the ongoing success of AAV gene therapy for hemophilia and expected FDA approval, there is renewed interest in developing AAV gene therapies for inherited liver disorders for clinical use. Yet, these successes have not come without challenges [5]. Adaptive immune responses against viral capsid antigen presented on transduced hepatocytes led to a vector dose-dependent loss of FIX protein expression in some hemophilia

Michael J. Castle (ed.), *Adeno-Associated Virus Vectors: Design and Delivery*, Methods in Molecular Biology, vol. 1950, https://doi.org/10.1007/978-1-4939-9139-6_20,

B patients receiving different AAV vectors [6–8]. In some instances, loss of FIX expression occurred without evidence of T cell activation [8], suggesting other factors may be involved in modulating long-term transgene expression in the human liver.

The liver plays an important role in regulating metabolism, detoxification, and synthesis of plasma proteins, including albumin and coagulation factors such as FVIII in liver sinusoidal endothelial cells (LSEC) and FIX in hepatocytes [9, 10]. Hepatocytes account for approximately 70% of liver cell mass and are responsible for the majority of metabolic and biosynthetic functions of the liver. Thus, disruption caused by monogenic disorders of metabolic and detoxification pathways and coagulation factor function or expression can have far reaching effects. The liver has many properties that make it an ideal target organ for AAV gene therapy. The biosynthetic function of the liver and access to the circulatory system provide an efficient factory and distribution system for secreted hepatic or extrahepatic proteins. In healthy liver, hepatocytes have a slow turnover that prevents dilution of transgenes delivered by AAV vectors, which rarely integrate into the host genome. A single administration of an AAV vector has provided greater than 6 years of stable FIX protein expression in humans [11] and greater than 10 years of expression in dogs [12]. Preclinical and clinical studies for hemophilia have demonstrated that peripheral vein administration of an AAV vector can transduce a sufficient number of hepatocytes to express therapeutic levels of FVIII and FIX proteins [13]. And finally, preclinical studies using AAV vectors have shown that hepatocyte-restricted transgene expression, above a threshold level, results in the induction of transgene-specific immunological tolerance [14–19].

2 Adeno-associated Virus (AAV) Vectors

2.1 AAV Vectors

AAV is a small nonpathogenic parvovirus that naturally infects humans, nonhuman primates (NHPs), and other mammals. Viral replication is dependent on a helper virus such as adenovirus or herpes virus. AAV was identified as a contaminant in adenoviral vector preparations [20] and the first AAV viral genome was cloned [21] and sequenced [22] nearly 20 years later. AAV is a single-stranded DNA virus with a genome of approximately 4.7 kb that carries the replication (*rep*) and viral capsid (*cap*) genes, as well as two inverted terminal repeats (ITRs) that are required for packaging and replication [23]. Viral genomes are packaged into an icosahedral viral capsid which defines the infectious host range through interactions with cell surface receptors [24].

The AAV2 vector in use today was first produced by cloning the wild-type AAV genome into a plasmid backbone [25] and replacing the *rep* and *cap* genes with an enhancer/promoter and marker gene

[23]. The flanking ITRs are the only wild-type sequence retained in AAV vectors. Several groups identified and cloned the essential adenoviral proteins for AAV replication into a plasmid, forming a three [26, 27] or two plasmid [28] system for production of AAV vectors by transient transfection of HEK293 cells. To scale up vector production, several groups have developed a baculovirus and Sf9 insect-based production system for preclinical [29–31] and clinical use [3]. Detailed protocols for AAV vector production and purification have been previously described [32] and can be found in Chapters 3, 7, 19, 21–23.

Several features of AAV vectors make them ideal for in vivo liver gene therapy. AAV vectors do not strongly activate innate immunity compared to other viral vectors [33] and are considered weakly immunogenic. AAV vectors can efficiently transduce quiescent cells such as hepatocytes. The majority of AAV transgene expression cassettes within a cell do not integrate [34], reducing the risk for insertional mutagenesis.

Early attempts at in vivo AAV2 vector delivery to the liver were met with an equal measure of successes and challenges. Three reports in 1997 described the in vivo transduction of murine liver with AAV2 vectors delivered through a variety of routes including tail vein, portal vein, and direct injection of a liver lobe [35–37]. Independent investigators noted that intravenous delivery of AAV2 vectors in mice resulted in a limited fraction of hepatocytes that expressed the viral transgene product, despite evidence that the majority of hepatocytes were transduced [36, 38, 39].

The following sections aim to provide an overview of the major determinants of efficient AAV transduction and transgene expression in the liver: (1) the capsid, (2) the AAV genome, and (3) the delivery route of vector. Next, we discuss optimal AAV vectors and delivery routes for liver gene transfer in commonly used small and large animal models and humans, and present an example of gene correction of hemophilia and applications of transgene immune tolerance to treat disease. Finally, we provide an overview of some of the limitations of AAV liver gene therapy and discuss approaches to overcome these limitations.

2.2 AAV Capsids

In addition to determining which cells are susceptible to transduction by AAV vectors, the viral capsid plays an important role in intracellular processing and trafficking upon receptor-mediated uptake of viral particles.

2.2.1 Natural Capsids

The majority of naturally occurring capsid serotypes used in the lab and clinic have been isolated from humans and NHPs, and are divided into clades based on shared sequence homology and seroreactivity [40, 41]. AAV serotypes identified from humans and NHPs were initially numbered AAV1 to AAV9. Of these nine serotypes, AAV4 and AAV5 have the highest sequence divergence

[40]. Conservation between different serotypes allows for the packaging of an AAV2 ITR expression cassette into a capsid from another serotype ("pseudotyping") [42, 43]. This allowed investigators to generate AAV vectors using capsids from AAV serotypes with natural tropism for liver [24]. A biodistribution study performed on AAV1 through AAV9 vectors revealed several serotypes with a natural affinity for murine hepatocytes following tail vein injection: AAV7, AAV8, and AAV9 [44]. AAV3B vectors, long thought to be of little use for liver gene transfer as they poorly transduced mouse liver, were shown to have an affinity for NHP and human hepatocytes [45, 46]. Vectors are often referred to as "AAVx/y," where x is a number designating the source of the ITR (usually 2) and y is a number referring to the capsid serotype. Recently identified serotypes now include a two-letter abbreviation to identify the species from which they were derived, such as "AAVrh10," in which "rh" indicates the rhesus macaque. New AAV serotypes now number in the hundreds and have been isolated from a variety of mammalian and nonmammalian sources. Serotypes from nonmammalian sources with specificity for human hepatocytes may be of interest, as they may avoid neutralization by antibodies that cross-react with many of the commonly used human and primate derived capsids.

2.2.2 Engineered Capsids

Capsid engineering for AAV liver gene transfer using a structure-based rational design has successfully enhanced the in vivo murine hepatocyte transduction efficiency of the AAV2 capsid. Study of AAV2 biology revealed that phosphorylation of tyrosine residues on the AAV2 capsid signals polyubiquitination and shunting of viral particles to proteasomal degradation [47]. Using structural data, investigators identified and substituted seven surface-exposed tyrosine residues on the AAV2 capsid individually or combinatorially with phenylalanine, then screened viable vectors in vivo for murine hepatocyte transduction efficiency [48, 49]. Of the combinations tested, a triple AAV2 (Y444/500/730F) variant was identified with significantly enhanced murine hepatocyte transduction and reduced capsid antigen presentation on MHC class I molecules, which protected transduced hepatocytes from killing by cytotoxic $CD8^+$ T cells [50]. Though effective, successful rationally designed capsid engineering is labor intensive and requires foreknowledge of AAV capsid biology, structure, and function.

2.2.3 Capsid Library Screening

The main advantage of generating a capsid library for functional screening is that there is no requirement for prior knowledge of capsid structure and function. There are several well described approaches for generating a capsid library. The first approach used error prone PCR to introduce random mutations in the capsid coding sequence, followed by iterative screening. This approach was applied to the AAV2 capsid to identify variants with altered affinity to heparin and the ability to escape antibody neutralization [51]. While this process can generate a high complexity library, random mutagenesis will result in the generation of nonviable capsid variants. However, these nonfunctional variants are rapidly lost during repeated iterative screenings.

A second approach used DNA family shuffling for capsid library generation, taking cDNA from eight representative capsids from different clades with greater than 50% homology. These cDNAs were fragmented and reassembled to generate a library of hybrid capsids consisting of fragments of various sizes from different donors. This library was screened for in vitro transduction of human hepatocytes in the presence of purified human in vitro immunoglobulin (IVIG), which contains neutralizing antibodies to AAV capsids [52]. The screening identified a hybrid capsid containing fragments of AAV2, AAV8, and AAV9, named AAV-DJ, with comparable in vivo murine hepatocyte transduction efficiency to AAV8. A similar DNA family shuffling approach selected capsids for in vivo human hepatocyte transduction, using a recently described chimeric human hepatocyte mouse model [53, 54]. Four rounds of in vivo selection resulted in the selection of a hybrid capsid named LK-03, consisting primarily of the parental AAV3B with some fragments from AAV2, AAV4, AAV8, and AAV9. An extended study further refined the screen to identify neutralizing antibody-resistant capsids [55].

A third method for generating a capsid library combines both rational design and DNA shuffling to generate a highly diverse capsid library for directed evolution selection on target tissues and cells. In this approach, an AAV2-based capsid library was generated using shuffled variants of the surface-exposed loops, restricted to only those amino acids found on 150 aligned naturally occurring AAV serotypes [56]. Additionally, four amino acids were modified: Y444F and Y500F, shown to reduce capsid ubiquitination and enhance murine hepatocyte transduction [49], and R585 and R588, shown to prevent binding with the primary AAV2 receptor, heparan sulfate proteoglycan [57]. After three rounds of in vivo selection on murine liver, two variants, AAV2-LiA and AAV2-LiC, were found to possess comparable murine hepatocyte transduction efficiency to AAV8.

2.3 AAV Genomic Elements

AAV vectors have a single-stranded DNA genome and must undergo rate-limiting second-strand synthesis for mRNA transcription and transgene expression [58]. This process, however, is capsid dependent [59, 60]. To bypass this rate-limiting step, a self-complementary AAV (scAAV) genome was generated by mutating one of the viral ITRs [61, 62]. These scAAV vectors provide improved murine hepatocyte transduction efficiency and expression kinetics compared to ssAAV vectors [63]. A major limitation of scAAV vectors is that their configuration reduces the size of the transgene expression cassette by half, which excludes large transgenes and limits the size of nontranscribed *cis* transcriptional regulatory elements, such as enhancer/promoters, introns, and polyadenylation signals. Nevertheless, scAAV2/8 and scAAV2/5 vectors have been used in several clinical trials targeting human hepatocytes for FIX expression in adult hemophilia B patients [7, 8, 64].

The inclusion of a hepatic-specific enhancer/promoter is often necessary to prevent off-target expression of the transgene, as no capsid is truly liver specific. Importantly, in murine studies inclusion of a hepatocyte-specific promoter in an AAV2-*F9* vector enhanced the induction of immune tolerance to the FIX protein [14, 65]. There are several commonly used hepatocyte-specific enhancer/promoters that function well within the limited packaging capacity of AAV vectors. A hybrid of the enhancer element from human *ApoE* and the promoter from human *A1AT*, ApoE-hAAT (710 bp), is a strong hepatocyte-specific promoter that has been used in several preclinical studies and clinical trials for FIX expression in hemophilia B patients [4, 66, 67]. A reduced size ApoE-hAAT promoter, LP1 (448 bp) [68], was used to drive FIX protein expression in two clinical trials for hemophilia B [7, 64]. At 260 bp, an even smaller version of LP1, named the hybrid liver-specific promoter (HLP), was created for use in AAV vectors expressing FVIII protein [69]. Several groups have applied in silico prediction and computational analysis to generate small and strong hepatocyte-specific promoters. The first group identified a 71 bp fragment of the *A1AT* promoter, termed a hepatocyte-specific cis-regulatory module (HSCRM8), and fused this to the small liver-specific transthyretin (TTR) promoter [70], resulting in an ~ten-fold increase in transgene expression compared to TTR alone [71, 72]. In a more recent study, a small and strong 146 bp hepatocyte-specific synthetic promoter called the hepatic combinatorial bundle was described to accommodate the larger B-domain deleted FVIII (BDD-FVIII) cDNA within an AAV expression cassette [73]. All of the above enhancer/promoter sequences are published and can be easily synthesized.

Codon optimization is frequently used to increase transgene expression in preclinical and clinical AAV vectors used in liver gene transfer. In addition to codon replacement, proprietary algorithms are used to remove cryptic splice sites, avoid high GC content, and minimize CpG methylated sequences. As an example, codon optimization was used to enhance expression of the BDD-FVIII protein, a clotting factor deficient in patients with the coagulation disorder hemophilia A, expressed from an AAV2/8 vector [74]. More recently, a new tissue-specific codon optimization approach was described, which utilized a codon usage table based on genes highly expressed in the mouse liver to further enhance BDD-FVIII protein expression [73].

Even though AAV vectors are described as weakly immunogenic compared to other viral vectors, they are sensed by the innate immune system [33] and provoke adaptive immune responses in animal models and humans [75]. In murine studies, investigators have identified the innate receptor toll-like receptor 9 (TLR9) as a critical sensor of both ssAAV and scAAV vectors [76, 77]. TLR9 is located within endosomes and signals via the adaptor protein MyD88 following activation by its substrate, DNA containing CpG motifs. To underscore the role of TLR9 activation in adaptive immune responses against the AAV capsid, elimination of CpG motifs in a vector with a naturally immunogenic capsid, AAV2/rh32.33, resulted in stable expression and evaded elimination by the immune system [78]. Minimization of CpG motifs is commonly included in codon optimization algorithms and has been used in several clinical trials for hemophilia B [7, 64].

3 Routes of Vector Delivery

For liver-directed AAV gene transfer, the route of vector administration can impact the degree of hepatocyte transduction, but this is also dependent on other factors such as the capsid serotype and genome configuration. Here we present the most commonly used delivery routes in small (Fig. 1) and large animal models and humans, and also provide guidance for selecting the optimal delivery route to achieve the desired experimental results (Fig. 2).

3.1 Direct Liver Delivery

Direct liver delivery routes in animal models were investigated to overcome poor transduction with AAV2 vectors following intravenous (IV) delivery into the tail vein of mice. The most common routes include direct injection into the liver lobes [79, 80], portal vein injection [81], splenic capsule injection [82, 83], and hepatic artery delivery in large animal models [84] and humans [67]. However, all of these routes require minor to major surgery, anesthesia, and a level of proficiency to ensure survival and effective gene delivery [81]. Further, such invasive procedures conducted in coagulation-deficient animal models put the subject at risk for operative and postoperative complications. In hemophilia mouse models, injection into the splenic capsule was adopted as an alternative less invasive delivery route than portal vein injection [82, 83]. Given the availability of natural and engineered AAV capsids that demonstrate highly efficient liver transduction following IV delivery into peripheral veins, there has been reduced interest in pursuing these invasive delivery routes.

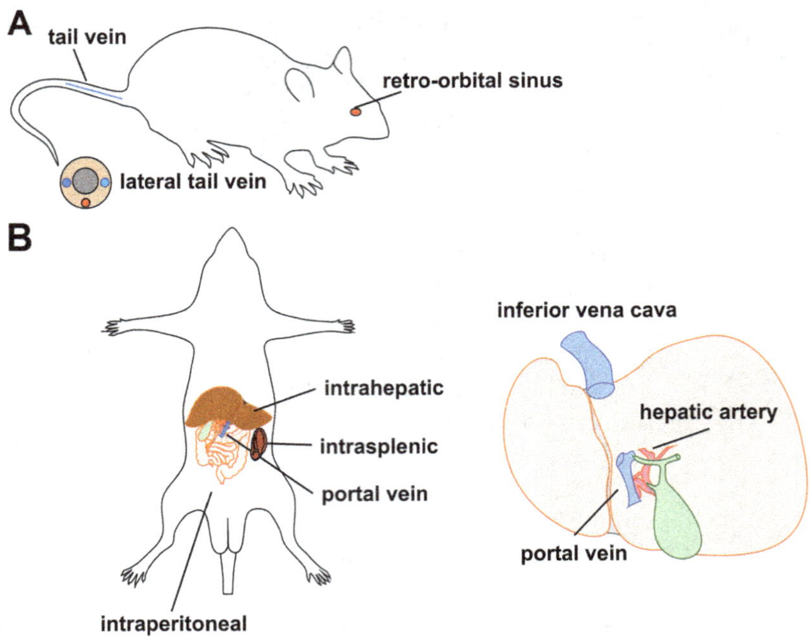

Fig. 1 Common vector delivery routes for AAV liver gene transfer in mouse. There are a variety of vector delivery routes for AAV liver gene transfer in the mouse. (**a**) Side view of mouse depicting IV injection routes into lateral tail vein or retro-orbital sinus. (**b**) Supine view of mouse depicting the nonsurgical intraperitoneal and surgical intrahepatic, intrasplenic, and portal vein delivery routes

Species:	mouse	rat	NHP	dog	human
Weight (kg):	0.025	0.250	6-8	10-40	60-90
Dose ($1x10^{12}$ vg/kg):	$2.5x10^{10}$ vg	$2.5x10^{11}$ vg	$6\text{-}8x10^{12}$ vg	$1\text{-}4x10^{13}$ vg	$6\text{-}9x10^{13}$ vg
Route:	IV[a] and IP[b]	PV	PV, HA, and IV	IV and PV	IV
Serotype:	AAV8	AAV1	AAV3 or AAV8	AAV8	AAV3 (LK03), AAV5, AAV8

Fig. 2 Inter-species comparison of recommended delivery routes and serotypes. This chart summarizes the most effective capsid and delivery route for different animal models and humans. Assuming a vector dose of 1×10^{12} vg/kg, approximate vector dose requirements are listed based on average species weight. [a]IV injection via tail vein is recommended in mice older than 4 weeks due to difficulties accessing the small tail veins. If IV delivery is needed in younger mice, retro-orbital sinus injection is suggested. [b]IP injections are suggested for AAV vector delivery in utero or in neonatal mice. *Abbreviations*: *HA* hepatic artery, *IP* intraperitoneal, *IV* intravenous, *NHP* nonhuman primate, *PV* portal vein

3.2 Peripheral Vector Delivery

In contrast to direct liver delivery, both intraperitoneal (IP) and IV delivery routes do not require surgery or anesthesia and can be conducted by an individual with minimal training. IP delivery of AAV vectors has been reported in mice over a range of ages from in utero to adult [85, 86]. Comparable murine hepatocyte transduction to portal vein injection was reported following IP administration of the hepatotropic capsids AAV7, AAV8, and AAV9 [87]. IP injection in utero and into neonatal mice results in a significant reduction in transgene expression due to the rapid doubling of the mouse liver within the first four to six weeks of life [85]. IP delivery offers several advantages including the relative ease of injecting pups in utero and neonatal mice compared to IV delivery. Further, IP injections can generally be safely administered in higher volumes compared to IV, allowing for administration of higher AAV vector doses or use of vector preparations with lower titers. However, there is little experience with IP AAV vector delivery beyond rodents, and it will likely be less effective in larger animal models and humans due to the increased volume of the peritoneal cavity.

IV delivery can be used for AAV vector administration to juvenile and adult rodents, as well as to neonatal and adult large animal models and humans. In mice, the two most common injection sites are the lateral tail vein and the retro-orbital sinus [88, 89] (Fig. 1a). Both techniques require practice to obtain proficiency. Tail vein injections in adult mice are generally performed without anesthesia using a restraining chamber. Retro-orbital injections should be initially practiced on cadavers and terminal experimental mice under anesthesia, to avoid movement of the animal during injection, prior to delivery to experimental animals [89]. Injections into the lateral tail vein in mice younger than four weeks may be challenging due to the smaller size of the lateral tail vein. However, it is possible to administer vector to the retro-orbital sinus in younger mice, including neonatal animals, but care must be taken to avoid damaging the eye and this requires more extensive training [89]. IV injections in larger animals and humans are generally performed in the veins found in the limbs.

4 AAV Liver Gene Transfer in Animal Models and Humans

4.1 Mice

Of the natural AAV capsids, AAV8 [43, 59, 90] is considered one of the best capsids for murine hepatocyte transduction. Importantly, peripheral administration of an AAV2/8 vector results in equivalent hepatocyte transduction compared to portal vein or splenic injection, and at higher doses can transduce more than 95% of all mouse hepatocytes [91, 92]. While liver-targeted delivery routes such as intrahepatic, portal vein, and splenic capsule injection are possible to perform in adult mice [81, 93], they require specialized training and Institutional Animal Care and Use Committee (IACUC)-approved survival surgery, and they offer little improvement of hepatocyte transduction efficiency compared to noninvasive tail vein or retro-orbital sinus vector delivery using AAV2/8 vectors.

Additionally, AAV2 capsid variants (AAV2-LiA and AAV2-LiC) isolated from an iterative in vivo murine liver AAV2 capsid library screen confer similar expression levels to AAV8 following IV tail vein delivery [56]. However, immune staining revealed that livers transduced by AAV2-LiA and AAV2-LiC showed zonal hepatocyte transduction similar to their parental AAV2 capsid [94]. Therefore, if the therapeutic protein is secreted, these AAV2 variants are indistinguishable from AAV8. But, if your disease model requires transduction of a large fraction of mouse hepatocytes, AAV8 is the better choice. Another highly active capsid for murine hepatocyte gene transfer, AAV-DJ, was identified through in vitro screening on primary human hepatocytes of a shuffled DNA capsid library generated from eight different capsids [52].

When working with mouse models that require early intervention with gene therapy in utero or neonatally, IP and intrahepatic injections [80, 86, 95] are suggested to avoid difficulties with finding accessible veins. Alternatively, retro-orbital sinus delivery may be attempted [89]. Detailed protocols for AAV vector administration to mice have been previously published [88, 93, 96], and can be found in Chapters 8 and 19.

As a concluding note, our experience and that of others show that hepatocyte transduction efficiency differs both between mouse strains [14] and between male and female mice [87, 97–99]. Therefore, it is important to empirically determine therapeutic vector doses for the specific strain and gender to be used. Pretreating female mice with the proteasome inhibitor bortezomib [99] or utilizing liver-targeted delivery routes may reduce the gender difference [100]. However, due to the potential side effects of bortezomib and the challenges of portal vein or intrahepatic injections, it may be simpler to adjust the vector dose used in female mice.

4.2 Rats

Given their increased size, direct liver delivery of AAV vectors to rat hepatocytes is relatively easier than in mice. Similar to mice, the delivery route of an AAV2 vector has a strong impact on rat hepatocyte transduction, with hepatic arterial infusion and increased dwell time of vector within the liver providing the best outcome [101]. In contrast to mice, AAV8 vectors are not the most efficient serotype for transducing rat hepatocytes. A study in Gunn rats, a rat model of Crigler-Najjar syndrome (UGT1A1 deficiency), compared portal vein delivery of AAV2/1, AAV2, AAV2/6, and AAV2/8 vectors expressing UGT1A1 and found that AAV1 provided the highest level of transgene expression and disease correction [102]. In a follow-up study, an AAV2/5 UGT1A1 vector failed to correct disease in the Gunn rat [103]. Different liver-directed AAV1 delivery routes were tested in healthy rats and rats with cirrhotic livers, and the authors concluded that intrahepatic injection and hepatic artery injection were equally effective [104].

4.3 Large Animal Models: Cats, Dogs, and Nonhuman Primates

Large animals, such as dogs, have a sensitivity to endotoxin contamination in AAV vector preparations. Therefore, we strongly recommend that all plasmids used for AAV vector production be purified using an endotoxin-free kit or protocol.

There are limited published studies of AAV liver gene transfer in cats. In an MPS VI feline model, IV administration by catheter into the jugular vein (postnatal day 5) or cephalic vein (postnatal day 50) of an AAV2/8 vector expressing feline arylsulfatase B (ARSB) resulted in long-term expression [105, 106]. In an MPS I feline model, cephalic vein administration of an AAV2/8 vector expressing feline IUDA was also reported [107].

Due to the existence of several naturally occurring canine disease models including hemophilia A, hemophilia B, and inherited metabolic disease, there have been substantial investigational studies into canine liver gene transfer with AAV vectors. IV injection of an AAV2 vector expressing glucose-6-phosphatase was reported in a glycogen storage disease type Ia dog model [108]. Long-term expression from an AAV2 vector expressing the canine FIX protein was reported using liver-directed mesenteric and portal vein administration in a hemophilia B dog model [12, 109]. Out of the different preclinical hemophilia B animal models tested, canine FIX expression levels from an AAV2 vector best predicted expression of human FIX in hemophilia B patients [67, 110]. Two AAV2/8 vectors expressing light chain and heavy chain canine FVIII were injected into the saphenous vein of hemophilia A dogs and provided stable FVIII protein levels [111, 112]. A single dosing of the proteasome inhibitor bortezomib was reported to enhance AAV2/8 canine FVIII expression following portal vein injection of a limiting dose of vector [113]. Long-term expression of canine FVIII was reported when a single or split AAV2/8 vector expressing FVIII or light chain and heavy chain FVIII, respectively, was injected into the portal vein or saphenous vein [114]. Finally, a comparative capsid study (AAV2, AAV6.2, AAV8, AAVrh8, AAVrh64R1, AAVhu37) reported that AAV8 provided the highest level of gene expression in the liver of young dogs injected in the cephalic vein [115].

NHPs are an important resource for determining gene expression levels, safety, and biodistribution prior to clinical study of an AAV gene therapy vector. However, it is often difficult to measure efficacy, as natural and genetically engineered NHP disease models are lacking. A study compared the liver gene transfer efficiency of AAV capsid serotypes AAV2, AAV5, and AAV8 expressing human FIX protein following mesenteric vein administration [59]. Although FIX levels in AAV2/8-treated NHPs were initially higher, levels declined and stabilized over time, comparable to levels in AAV5/5-treated NHPs. Pre-existing neutralizing antibodies to the AAV capsid effectively prevented efficient liver gene transfer. Importantly, NHPs with pre-existing neutralizing antibodies to AAV2 could be treated with an AAV5/5 vector, avoiding neutralization by using the divergent AAV5 capsid. A recent report extended this study by demonstrating successful sequential gene delivery to NHP liver by administration of AAV vectors to the saphenous vein, using an AAV2/5(CH) vector followed by a second AAV2/1 vector one month later [116]. An scAAV2/7 vector provided superior liver gene transfer efficiency following portal vein injection in NHPs compared to a scAAV2/8 vector [117]. In a separate study, enhanced transgene expression was reported in NHPs by using an scAAV2/8 vector infused into the mesenteric vein [68]. Importantly, peripheral vein infusion of scAAV2/5 and scAAV2/8 vectors was shown to provide equivalent transgene expression [118]. Intrauterine gene transfer into late gestation NHP fetuses was studied by delivery of scAAV2/5 and scAAV2/8 vectors into the fetal hepatic vein, and it was found that the scAAV2/8 vector provided higher levels of transgene expression both in the short term [119] and the long term [120]. Finally, the impact of the vector infusion rate was studied using saphenous vein vector delivery and no significant impact on transduction efficiency was observed, although NHPs injected at the fastest infusion rate had a higher magnitude and frequency of innate immune activation [121].

4.4 Human Hepatocytes: Models and Clinical Experience

To date, large animal models such as dogs and NHPs have been the best predictors of outcomes following AAV vector liver gene transfer in humans. However, emerging mouse models with a chimeric liver containing both human and murine hepatocytes have been used to evaluate AAV tropism for human hepatocytes [53]. One such model is the FRG mouse, which is a triple knockout of the *Fah* gene involved in tyrosine catabolism as well as two immune genes, *Rag2* and *IL2Rg*, to prevent immunological rejection of human hepatocytes. Mice and humans with a mutant *Fah* gene develop severe liver disease due to the accumulation of a toxic tyrosine metabolite in hepatocytes. Pharmaceutical treatment with an upstream inhibitor of the tyrosine catabolism pathway, NTBC, prevents hepatotoxicity in the FRG mouse. Withdrawal of NTBC following transplantation of *Fah*-expressing human hepatocytes promotes their engraftment and expansion in the mouse liver. The hybrid AAV capsid LK03 is currently being evaluated in a clinical trial for hemophilia A (NCT03003533). LK03 was identified by sequential screening of a capsid library in the above FRG mouse system, selecting for variants with enhanced human hepatocyte tropism and resistance to capsid neutralizing antibodies [55]. Using the same human hepatocyte FRG mouse, the specificity for mouse and human hepatocytes of the capsids AAV3B-ST (S663V and T492V), AAV5, AAV8, and AAV9 was compared, with AAV3B-ST displaying the highest specificity for human hepatocytes [45].

The majority of AAV liver gene transfer clinical trials have been hybrid phase I/II trials focused on hemophilia A and hemophilia B. With the exception of the first hemophilia B trial, which used hepatic artery infusion of an AAV2 vector [67], all subsequent trials have used peripheral vein delivery. Hemophilia B clinical studies have used scAAV2/8, scAAV2/5, ssAAV8, and a ssAAV8 variant (SPK-9001) to express wild-type or a hyperactive variant of human FIX protein [4, 7, 64]. There are currently two ongoing clinical trials for hemophilia A using ssAAV2/5 and ssAAV2/LK-03 vectors to express BDD-FVIII protein [3]. One non-hemophilia phase I clinical dose escalation trial evaluated AAV2/5 liver gene transfer for acute intermittent porphyria, which was safe but failed to achieve therapeutic correction [122].

4.5 Summary

Due to easy access and low technical difficulty, peripheral vein injection should be used to deliver AAV vectors to the liver in research animals, with the exception of young mice. Figure 2 summarizes the recommended AAV capsid serotypes and delivery routes for AAV liver gene transfer in various research animals and humans. It is clear that the liver tropism of a specific AAV capsid is not necessarily conserved across different species. For example, inverse zonation of hepatocyte transduction was reported between mice and NHPs transduced with an AAV2/8 vector [123]. Two different AAV2 variants with improved murine hepatocyte gene transfer efficiency failed to demonstrate improved canine hepatocyte transduction when injected into hemophilia B dogs [124]. AAV2/3B vectors poorly transduce mouse liver, but are effective in NHPs and humans. In large animal models, pre-existing cross-reactive capsid neutralizing antibodies can block liver gene transfer, and it is important to screen subject plasma prior to vector infusion [125, 126]. When developing a gene therapy treatment protocol based on a mouse disease model, it is important to consider that reduced hepatocyte gene transfer efficiency with current AAV vectors in humans may result in 10 to 100-fold less transgene expression as compared to mice.

5 Applications of Liver Gene Transfer

5.1 Gene Replacement Therapy: Hemophilia

Hemophilia A and B are coagulation disorders resulting from the loss of functional FVIII and FIX protein expression. Both of these clotting factors are naturally synthesized in the liver, FVIII in LSECs and FIX in hepatocytes. Hemophilia patients are currently treated with plasma-derived or recombinant clotting factor proteins, which require infusion two to three times per week throughout life. Even though prophylactic clotting factor protein therapy effectively prevents bleeding episodes, this treatment is expensive and burdensome for the patient. Since the half-life of both clotting factors is short, patients often fail to maintain steady state levels of clotting factor between infusions. The formation of inhibitory antibodies directed against the clotting factor protein is a major complication of therapy and often puts patients at risk for serious bleeding episodes. Hemophilia patients with such inhibitors must be managed with less effective bypassing agents and may respond to immune tolerance therapy, which requires frequent high dose clotting factor infusions. Thus, AAV liver gene therapy represents a potentially life altering therapy for hemophilia, in which a single infusion of a vector could replace life-long infusion of clotting factor protein.

Over the course of decades, AAV liver gene therapy has progressed from an effective therapy for murine and canine hemophilia models to a method for providing stable FVIII and FIX expression in adult hemophilia patients [13]. Indeed, initial experiments investigating in vivo AAV liver gene transfer expressed the FIX protein in mouse hepatocytes [35, 37]. Due to the large size of the gene coding for FVIII, AAV vectors were initially not capable of packaging the coding sequence. Removal of the B domain of the FVIII protein, which was found to be dispensable for function, reduced the coding sequence to a size that could fit into an AAV vector. Thus, historically there are fewer studies conducted on hemophilia A animal models, and this section will focus principally on highlighting advances in AAV-based hemophilia B gene therapy. However, despite the delay, several clinical trials are now ongoing for hemophilia A, and it is likely that AAV-based therapies for hemophilia A and B will soon receive FDA approval.

Similar to hemophilia B patients, hemophilia B mouse and dog models develop inhibitors to FIX protein therapy. Depending on the mouse strain background, AAV capsid, and promoter, hemophilia B mice may also form inhibitors to FIX protein derived from AAV-transduced hepatocytes. The formation of inhibitors following AAV gene therapy initially represented a serious translational challenge. However, when FIX protein expression was restricted to hepatocytes mice developed immune tolerance and did not form inhibitors [14, 65]. Follow-up studies found that AAV liver gene transfer induced a subset of transgene-specific $CD4^+$ suppressor T cells, regulatory T cells, with a hallmark phenotype of $CD4^+$ $CD25^+$ $FoxP3^+$ [127]. Using different mouse models, tolerance induction was found to require a threshold level of antigen expression [18], as well as the immune suppressive cytokines IL-10 and TGF-beta [19, 128]. AAV-mediated liver tolerance induction to both FVIII and FIX proteins has been documented in murine and canine hemophilia models [13]. Importantly, tolerance induction has also been reported for nonsecreted transgenes [17]. It is not known how robust AAV-mediated liver immune tolerance is in humans, as all treated hemophilia patients are adults with long-term exposure to clotting factor protein therapy and no history of inhibitor formation. However, the absence of inhibitors in hemophilia patients receiving AAV liver gene transfer thus far is promising. The answer to this question will be forthcoming when clinical trials are conducted on younger patients with less exposure to clotting factor, or on patients with inhibitor or a previous history of inhibitor formation.

5.2 Gene Editing for Hemophilia B

Gene editing has emerged as an alternative to gene replacement therapy and can facilitate site-specific DNA integration for repair or replacement of a mutated gene through the homologous recombination or nonhomologous end joining DNA repair pathways. The low natural frequency of homologous recombination with a donor template is significantly increased when a double strand DNA break is induced within the site of homology. Site-specific double strand DNA breaks can be generated by a synthetic DNA binding protein fused to an endonuclease, zinc finger nuclease (ZFN), or by an adapted RNA-guided nuclease from the bacterial adaptive immune system (Cas9/CRISPR). Several different gene editing systems using AAV vector liver gene transfer have now been described, including ZFNs for FIX expression [129, 130] and Cas9/CRISPR for correction of metabolic liver disease [131, 132]. Due to the limited packaging capacity of AAV, both systems require two separate vectors to express both the nuclease/guide RNA and the donor template with homology arms, as well as high vector doses to achieve a sufficient level of recombination. Interestingly, it appears that AAV can facilitate nuclease-free site-specific integration into the albumin locus using a promoterless *F9* cassette, although the reported frequency is likely too low for disease correction outside of hemophilia B [133].

When designing an optimal AAV gene therapy vector for a liver-related disease, such as inherited disorders of metabolism and lysosomal storage disorders, several important questions must be addressed. What is the therapeutic expression window? Can correction be achieved with a fraction of transduced hepatocytes, or is a majority required? Is hepatic zonal expression needed for disease correction? If so, zonal expression can be controlled by the choice of AAV capsid, delivery route, and promoter. If the disease is lethal at an early age, what is the best capsid and delivery route for therapeutic gene expression and, if necessary, how will vector be successfully re-administered? Answering basic questions such as these will help choose the most appropriate capsid, delivery route, and vector dose.

5.3 AAV Gene Immunotherapy

The formation of inhibitory antibodies is a major complication of clotting factor protein therapy in hemophilia, and also of enzyme replacement therapies (ERT) in other inherited disorders, such as acid alpha glucosidase (GAA) protein ERT for the lysosomal storage disorder Pompe disease. Because AAV liver gene therapy induced immunological tolerance to FIX protein, investigators asked whether immunological tolerance could eliminate pre-existing inhibitors. Using a hemophilia B mouse model that develops fatal pathogenic antibodies to FIX protein therapy, investigators demonstrated that an AAV2/8 vector expressing FIX protein in hepatocytes could rapidly and durably eliminate inhibitors [134]. Similar studies were effective in a mouse model of Pompe disease [135, 136]. Further, AAV2/8 vectors expressing FVIII and a hyperactive FIX variant protein eliminated low titer inhibitors in hemophilia A [112] and B [137] dogs, respectively, suggesting that this mechanism is not restricted to mice. In all of these instances the gene therapy not only eliminated pre-existing immunity to the transgene product, but also corrected disease.

Given the success of AAV liver gene therapy for eradication of inhibitors in hemophilia and Pompe disease animal models, mediated by induction of transgene-specific regulatory T cells, it was hypothesized that AAV liver gene therapy could be used to treat autoimmune disease [138]. In a landmark study, investigators demonstrated that AAV liver gene transfer of a neuroantigen could both prevent and reverse paralysis in a mouse model of multiple sclerosis, experimental autoimmune encephalomyelitis [139]. This first proof of concept application of AAV liver gene transfer to treat autoimmune disease is unique, transcending traditional gene replacement therapies.

6 Limitations of AAV for Liver Gene Transfer

6.1 Dilution of Vector Genomes with Hepatocyte Turnover

Nonintegrating AAV vector genomes are diluted when gene delivery occurs at an early age, or if there is a high rate of hepatocyte turnover. Although there is evidence that genomic integration occurs at a low frequency, most animal models treated at an early age show a drastic drop in transgene expression. Vector re-administration may be considered, using a different capsid, or using transient immune suppression such as a nondepleting CD4 antibody [140] or a combination of the B cell-depleting anti-CD20 antibody and rapamycin [141] during first exposure. Another approach is to use gene editing for targeted integration into the deficient gene or a safe harbor locus, such as AAVS1 or albumin, to ensure continued transgene expression.

6.2 Neutralizing Antibodies to the AAV Capsid

Animals that are naturally infected by wild-type AAV, including humans, may have pre-existing neutralizing antibodies to a particular AAV serotype that can effectively block liver gene transfer directly or by cross-reactivity [125, 126, 142–144]. Therefore, it is important to prescreen for neutralizing antibodies in subjects that are known to be naturally infected by AAV and related parvoviruses. A different capsid serotype can be utilized, although it may be less optimal for liver transduction [116]. An alternative approach is to use capsid decoys by varying the ratio of empty AAV capsids to genome-containing capsids [145]. A more recent approach describes exosome-enveloped AAV vectors that are resistant to antibody neutralization and more strongly transduce hepatocytes [146]. Alternatively, it may be possible to temporarily remove antibodies locally or from circulation prior to vector administration through flushing of tissue, plasmapheresis, or the use of immune suppressive drugs [147].

6.3 Limited Packaging Capacity

One of the major limitations of AAV vectors is a maximum packaging capacity of approximately 4.5–5.0 kb. Exceeding 5.0 kb often results in the incomplete packaging of vector genomes and a reduction in transgene expression. This is exacerbated when using scAAV, which has half the packaging capacity. Depending on the transgene size and protein configuration there are several approaches that can be used. The first approach was initially described for expression of FVIII protein, and used two different vectors to express the light chain and heavy chain portions of the FVIII protein which self-assemble in the cytoplasm [148]. A second approach splits a large transgene between two or more AAV vectors, and the large cassette is recombined after infection by trans-splicing or homologous recombination [149]. Techniques for AAV-mediated delivery of large transgenes are described in detail in Chapter 2.

6.4 Risks of Genotoxicity

While the majority of AAV genomes remain episomal following liver gene transfer, there is growing evidence that low frequency integration events occur preferentially during cell cycling and within active genes [34, 150, 151]. Conflicting reports have claimed that AAV genomic integration results in the induction of hepatocellular carcinoma (HCC) in mice receiving AAV liver gene transfer [152–156]. However, there are no reports of HCC in AAV-treated dogs and NHPs, which are followed comparably longer than mice [157, 158]. This concern was again raised by a report claiming that AAV2 integrations were present in human HCC [159]. However, there was no direct evidence that the integrations caused the HCC or were acquired after cell transformation [160]. Finally, there have been no reports of HCC developing in human hemophilia patients treated with AAV vectors in multiple clinical trials, as well as in a recent phase I study for acute intermittent porphyria [158], although the follow-up time and patient numbers are not large enough to draw definitive conclusions. More treated patients and longer follow-up times will be needed in order to fully describe this risk.

6.5 Vector-Related Toxicity

Asymptomatic transient transaminitis in human hemophilia patients receiving AAV liver gene transfer, related in part to $CD8^+$

T cell targeting of AAV-transduced hepatocytes, is the primary concern involving AAV vector-related toxicity [8, 75]. To date, no preclinical study or model of this $CD8^+$ T cell response against AAV capsid antigen presented on hepatocytes has completely replicated the conditions and timing observed in humans [8, 50, 94]. The main approach to reduce this immune toxicity has been to modify the vector for increased transgene expression, in order to achieve therapeutic expression at lower vector doses and thereby avoid the killing of transduced hepatocytes by activated $CD8^+$ T cells. This has been accomplished in hemophilia B by using a hyperactive FIX variant with an eight-fold specific activity over wild-type FIX, novel capsid variants, and optimized expression cassettes [4]. Alternatively, in a hemophilia A trial, subjects treated with an AAV2/5 vector expressing BDD-FVIII protein received prophylactic steroids [3]. Others are working to understand how the immune system senses and responds to AAV vector transduction, to identify key pathways to target for inhibition [77]. While these approaches may be sufficient for treating hemophilia, better models are needed to predict outcomes for AAV liver gene therapy in other liver-related diseases, such as inherited metabolic disorders and lysosomal storage disorders.

7 Conclusions

AAV technology as it applies to the liver has lately seen considerable progress. Several liver-specific promoters and novel capsid variants can achieve high levels of transduction and secretion following a simple peripheral injection of virus. With the first FDA-approved AAV gene therapy treatment now available, and several more on the horizon in various stages of clinical trials, the future for many people affected by genetic disorders is brighter than ever.

References

1. Dunbar CE, High KA, Joung JK, Kohn DB, Ozawa K, Sadelain M (2018) Gene therapy comes of age. Science 359(6372). https://doi.org/10.1126/science.aan4672
2. FDA approves hereditary blindness gene therapy (2018) Nat Biotechnol 36(1):6. https://doi.org/10.1038/nbt0118-6a
3. Rangarajan S, Walsh L, Lester W, Perry D, Madan B, Laffan M, Yu H, Vettermann C, Pierce GF, Wong WY, Pasi KJ (2017) AAV5-factor VIII gene transfer in severe hemophilia A. N Engl J Med 377(26):2519–2530. https://doi.org/10.1056/NEJMoa1708483
4. George LA, Sullivan SK, Giermasz A, Rasko JEJ, Samelson-Jones BJ, Ducore J, Cuker A, Sullivan LM, Majumdar S, Teitel J, McGuinn CE, Ragni MV, Luk AY, Hui D, Wright JF, Chen Y, Liu Y, Wachtel K, Winters A, Tiefenbacher S, Arruda VR, van der Loo JCM, Zelenaia O, Takefman D, Carr ME, Couto LB, Anguela XM, High KA (2017) Hemophilia B gene therapy with a high-specific-activity factor IX variant. N Engl J Med 377(23):2215–2227. https://doi.org/10.1056/NEJMoa1708538
5. Colella P, Ronzitti G, Mingozzi F (2018) Emerging issues in AAV-mediated in vivo gene therapy. Mol Ther Methods Clin Dev 8:87–104. https://doi.org/10.1016/j.omtm.2017.11.007

6. Mingozzi F, Maus MV, Hui DJ, Sabatino DE, Murphy SL, Rasko JE, Ragni MV, Manno CS, Sommer J, Jiang H, Pierce GF, Ertl HC, High KA (2007) CD8(+) T-cell responses to adeno-associated virus capsid in humans. Nat Med 13(4):419–422. https://doi.org/10.1038/nm1549
7. Nathwani AC, Tuddenham EG, Rangarajan S, Rosales C, McIntosh J, Linch DC, Chowdary P, Riddell A, Pie AJ, Harrington C, O'Beirne J, Smith K, Pasi J, Glader B, Rustagi P, Ng CY, Kay MA, Zhou J, Spence Y, Morton CL, Allay J, Coleman J, Sleep S, Cunningham JM, Srivastava D, Basner-Tschakarjan E, Mingozzi F, High KA, Gray JT, Reiss UM, Nienhuis AW, Davidoff AM (2011) Adenovirus-associated virus vector-mediated gene transfer in hemophilia B. N Engl J Med 365(25):2357–2365. https://doi.org/10.1056/NEJMoa1108046
8. Ertl HCJ, High KA (2017) Impact of AAV capsid-specific T-cell responses on design and outcome of clinical gene transfer trials with recombinant adeno-associated viral vectors: an evolving controversy. Hum Gene Ther 28 (4):328–337. https://doi.org/10.1089/hum.2016.172
9. Gordillo M, Evans T, Gouon-Evans V (2015) Orchestrating liver development. Development 142(12):2094–2108. https://doi.org/10.1242/dev.114215
10. Kattenhorn LM, Tipper CH, Stoica L, Geraghty DS, Wright TL, Clark KR, Wadsworth SC (2016) Adeno-associated virus gene therapy for liver disease. Hum Gene Ther 27 (12):947–961. https://doi.org/10.1089/hum.2016.160
11. Nathwani AC, Reiss UM, Tuddenham EG, Rosales C, Chowdary P, McIntosh J, Della Peruta M, Lheriteau E, Patel N, Raj D, Riddell A, Pie J, Rangarajan S, Bevan D, Recht M, Shen YM, Halka KG, Basner-Tschakarjan E, Mingozzi F, High KA, Allay J, Kay MA, Ng CY, Zhou J, Cancio M, Morton CL, Gray JT, Srivastava D, Nienhuis AW, Davidoff AM (2014) Long-term safety and efficacy of factor IX gene therapy in hemophilia B. N Engl J Med 371 (21):1994–2004. https://doi.org/10.1056/NEJMoa1407309
12. Niemeyer GP, Herzog RW, Mount J, Arruda VR, Tillson DM, Hathcock J, van Ginkel FW, High KA, Lothrop CD Jr (2009) Long-term correction of inhibitor-prone hemophilia B dogs treated with liver-directed AAV2-mediated factor IX gene therapy. Blood 113 (4):797–806. https://doi.org/10.1182/blood-2008-10-181479
13. Rogers GL, Herzog RW (2015) Gene therapy for hemophilia. Front Biosci (Landmark Ed) 20:556–603
14. Mingozzi F, Liu YL, Dobrzynski E, Kaufhold A, Liu JH, Wang Y, Arruda VR, High KA, Herzog RW (2003) Induction of immune tolerance to coagulation factor IX antigen by in vivo hepatic gene transfer. J Clin Invest 111(9):1347–1356. https://doi.org/10.1172/JCI16887
15. Dobrzynski E, Fitzgerald JC, Cao O, Mingozzi F, Wang L, Herzog RW (2006) Prevention of cytotoxic T lymphocyte responses to factor IX-expressing hepatocytes by gene transfer-induced regulatory T cells. Proc Natl Acad Sci U S A 103 (12):4592–4597. https://doi.org/10.1073/pnas.0508685103
16. Kumar SRP, Hoffman BE, Terhorst C, de Jong YP, Herzog RW (2017) The balance between CD8(+) T cell-mediated clearance of AAV-encoded antigen in the liver and tolerance is dependent on the vector dose. Mol Ther 25(4):880–891. https://doi.org/10.1016/j.ymthe.2017.02.014
17. Martino AT, Nayak S, Hoffman BE, Cooper M, Liao G, Markusic DM, Byrne BJ, Terhorst C, Herzog RW (2009) Tolerance induction to cytoplasmic beta-galactosidase by hepatic AAV gene transfer: implications for antigen presentation and immunotoxicity. PLoS One 4(8):e6376. https://doi.org/10.1371/journal.pone.0006376
18. LoDuca PA, Hoffman BE, Herzog RW (2009) Hepatic gene transfer as a means of tolerance induction to transgene products. Curr Gene Ther 9(2):104–114
19. Keeler GD, Markusic DM, Hoffman BE (2017) Liver induced transgene tolerance with AAV vectors. Cell Immunol. https://doi.org/10.1016/j.cellimm.2017.12.002
20. Atchison RW, Casto BC, Hammon WM (1965) Adenovirus-associated defective virus particles. Science 149(3685):754–756
21. Samulski RJ, Berns KI, Tan M, Muzyczka N (1982) Cloning of adeno-associated virus into pBR322: rescue of intact virus from the recombinant plasmid in human cells. Proc Natl Acad Sci U S A 79(6):2077–2081
22. Srivastava A, Lusby EW, Berns KI (1983) Nucleotide sequence and organization of the adeno-associated virus 2 genome. J Virol 45 (2):555–564
23. McLaughlin SK, Collis P, Hermonat PL, Muzyczka N (1988) Adeno-associated virus general transduction vectors: analysis of proviral structures. J Virol 62(6):1963–1973

24. Grimm D, Pandey K, Nakai H, Storm TA, Kay MA (2006) Liver transduction with recombinant adeno-associated virus is primarily restricted by capsid serotype not vector genotype. J Virol 80(1):426–439. https://doi.org/10.1128/JVI.80.1.426-439.2006
25. Samulski RJ, Srivastava A, Berns KI, Muzyczka N (1983) Rescue of adeno-associated virus from recombinant plasmids: gene correction within the terminal repeats of AAV. Cell 33(1):135–143
26. Xiao X, Li J, Samulski RJ (1998) Production of high-titer recombinant adeno-associated virus vectors in the absence of helper adenovirus. J Virol 72(3):2224–2232
27. Matsushita T, Elliger S, Elliger C, Podsakoff G, Villarreal L, Kurtzman GJ, Iwaki Y, Colosi P (1998) Adeno-associated virus vectors can be efficiently produced without helper virus. Gene Ther 5(7):938–945. https://doi.org/10.1038/sj.gt.3300680
28. Grimm D, Kern A, Rittner K, Kleinschmidt JA (1998) Novel tools for production and purification of recombinant adenoassociated virus vectors. Hum Gene Ther 9 (18):2745–2760. https://doi.org/10.1089/hum.1998.9.18-2745
29. Kohlbrenner E, Aslanidi G, Nash K, Shklyaev S, Campbell-Thompson M, Byrne BJ, Snyder RO, Muzyczka N, Warrington KH Jr, Zolotukhin S (2005) Successful production of pseudotyped rAAV vectors using a modified baculovirus expression system. Mol Ther 12(6):1217–1225. https://doi.org/10.1016/j.ymthe.2005.08.018
30. Gray JT, Zolotukhin S (2011) Design and construction of functional AAV vectors. Methods Mol Biol 807:25–46. https://doi.org/10.1007/978-1-61779-370-7_2
31. Mietzsch M, Grasse S, Zurawski C, Weger S, Bennett A, Agbandje-McKenna M, Muzyczka N, Zolotukhin S, Heilbronn R (2014) OneBac: platform for scalable and high-titer production of adeno-associated virus serotype 1-12 vectors for gene therapy. Hum Gene Ther 25(3):212–222. https://doi.org/10.1089/hum.2013.184
32. Wang L, Blouin V, Brument N, Bello-Roufai-M, Francois A (2011) Production and purification of recombinant adeno-associated vectors. Methods Mol Biol 807:361–404. https://doi.org/10.1007/978-1-61779-370-7_16
33. Zaiss AK, Liu Q, Bowen GP, Wong NC, Bartlett JS, Muruve DA (2002) Differential activation of innate immune responses by adenovirus and adeno-associated virus vectors. J Virol 76(9):4580–4590
34. Nakai H, Yant SR, Storm TA, Fuess S, Meuse L, Kay MA (2001) Extrachromosomal recombinant adeno-associated virus vector genomes are primarily responsible for stable liver transduction in vivo. J Virol 75 (15):6969–6976. https://doi.org/10.1128/JVI.75.15.6969-6976.2001
35. Koeberl DD, Alexander IE, Halbert CL, Russell DW, Miller AD (1997) Persistent expression of human clotting factor IX from mouse liver after intravenous injection of adeno-associated virus vectors. Proc Natl Acad Sci U S A 94(4):1426–1431
36. Ponnazhagan S, Mukherjee P, Yoder MC, Wang XS, Zhou SZ, Kaplan J, Wadsworth S, Srivastava A (1997) Adeno-associated virus 2-mediated gene transfer in vivo: organ-tropism and expression of transduced sequences in mice. Gene 190(1):203–210
37. Snyder RO, Miao CH, Patijn GA, Spratt SK, Danos O, Nagy D, Gown AM, Winther B, Meuse L, Cohen LK, Thompson AR, Kay MA (1997) Persistent and therapeutic concentrations of human factor IX in mice after hepatic gene transfer of recombinant AAV vectors. Nat Genet 16(3):270–276. https://doi.org/10.1038/ng0797-270
38. Miao CH, Nakai H, Thompson AR, Storm TA, Chiu W, Snyder RO, Kay MA (2000) Nonrandom transduction of recombinant adeno-associated virus vectors in mouse hepatocytes in vivo: cell cycling does not influence hepatocyte transduction. J Virol 74 (8):3793–3803
39. Nakai H, Thomas CE, Storm TA, Fuess S, Powell S, Wright JF, Kay MA (2002) A limited number of transducible hepatocytes restricts a wide-range linear vector dose response in recombinant adeno-associated virus-mediated liver transduction. J Virol 76 (22):11343–11349. https://doi.org/10.1128/JVI.76.22.11343-11349.2002
40. Gao G, Vandenberghe LH, Alvira MR, Lu Y, Calcedo R, Zhou X, Wilson JM (2004) Clades of adeno-associated viruses are widely disseminated in human tissues. J Virol 78 (12):6381–6388. https://doi.org/10.1128/JVI.78.12.6381-6388.2004
41. Agbandje-McKenna M, Kleinschmidt J (2011) AAV capsid structure and cell interactions. Methods Mol Biol 807:47–92. https://doi.org/10.1007/978-1-61779-370-7_3
42. Rabinowitz JE, Rolling F, Li C, Conrath H, Xiao W, Xiao X, Samulski RJ (2002) Cross-packaging of a single adeno-associated virus (AAV) type 2 vector genome into multiple AAV serotypes enables transduction with broad specificity. J Virol 76(2):791–801

43. Gao GP, Alvira MR, Wang L, Calcedo R, Johnston J, Wilson JM (2002) Novel adeno-associated viruses from rhesus monkeys as vectors for human gene therapy. Proc Natl Acad Sci U S A 99(18):11854–11859. https://doi.org/10.1073/pnas.182412299

44. Zincarelli C, Soltys S, Rengo G, Rabinowitz JE (2008) Analysis of AAV serotypes 1-9 mediated gene expression and tropism in mice after systemic injection. Mol Ther 16 (6):1073–1080. https://doi.org/10.1038/mt.2008.76

45. Vercauteren K, Hoffman BE, Zolotukhin I, Keeler GD, Xiao JW, Basner-Tschakarjan E, High KA, Ertl HC, Rice CM, Srivastava A, de Jong YP, Herzog RW (2016) Superior in vivo transduction of human hepatocytes using engineered AAV3 capsid. Mol Ther 24 (6):1042–1049. https://doi.org/10.1038/mt.2016.61

46. Li S, Ling C, Zhong L, Li M, Su Q, He R, Tang Q, Greiner DL, Shultz LD, Brehm MA, Flotte TR, Mueller C, Srivastava A, Gao G (2015) Efficient and targeted transduction of nonhuman primate liver with systemically delivered optimized AAV3B vectors. Mol Ther 23(12):1867–1876. https://doi.org/10.1038/mt.2015.174

47. Zhao W, Zhong L, Wu J, Chen L, Qing K, Weigel-Kelley KA, Larsen SH, Shou W, Warrington KH Jr, Srivastava A (2006) Role of cellular FKBP52 protein in intracellular trafficking of recombinant adeno-associated virus 2 vectors. Virology 353(2):283–293. https://doi.org/10.1016/j.virol.2006.04.042

48. Markusic DM, Herzog RW, Aslanidi GV, Hoffman BE, Li B, Li M, Jayandharan GR, Ling C, Zolotukhin I, Ma W, Zolotukhin S, Srivastava A, Zhong L (2010) High-efficiency transduction and correction of murine hemophilia B using AAV2 vectors devoid of multiple surface-exposed tyrosines. Mol Ther 18 (12):2048–2056. https://doi.org/10.1038/mt.2010.172

49. Zhong L, Li B, Mah CS, Govindasamy L, Agbandje-McKenna M, Cooper M, Herzog RW, Zolotukhin I, Warrington KH Jr, Weigel-Van Aken KA, Hobbs JA, Zolotukhin S, Muzyczka N, Srivastava A (2008) Next generation of adeno-associated virus 2 vectors: point mutations in tyrosines lead to high-efficiency transduction at lower doses. Proc Natl Acad Sci U S A 105 (22):7827–7832. https://doi.org/10.1073/pnas.0802866105

50. Martino AT, Basner-Tschakarjan E, Markusic DM, Finn JD, Hinderer C, Zhou S, Ostrov DA, Srivastava A, Ertl HC, Terhorst C, High KA, Mingozzi F, Herzog RW (2013) Engineered AAV vector minimizes in vivo targeting of transduced hepatocytes by capsid-specific CD8+ T cells. Blood 121 (12):2224–2233. https://doi.org/10.1182/blood-2012-10-460733

51. Maheshri N, Koerber JT, Kaspar BK, Schaffer DV (2006) Directed evolution of adeno-associated virus yields enhanced gene delivery vectors. Nat Biotechnol 24(2):198–204. https://doi.org/10.1038/nbt1182

52. Grimm D, Lee JS, Wang L, Desai T, Akache B, Storm TA, Kay MA (2008) In vitro and in vivo gene therapy vector evolution via multispecies interbreeding and retargeting of adeno-associated viruses. J Virol 82 (12):5887–5911. https://doi.org/10.1128/JVI.00254-08

53. Lisowski L, Dane AP, Chu K, Zhang Y, Cunningham SC, Wilson EM, Nygaard S, Grompe M, Alexander IE, Kay MA (2014) Selection and evaluation of clinically relevant AAV variants in a xenograft liver model. Nature 506(7488):382–386. https://doi.org/10.1038/nature12875

54. Azuma H, Paulk N, Ranade A, Dorrell C, Al-Dhalimy M, Ellis E, Strom S, Kay MA, Finegold M, Grompe M (2007) Robust expansion of human hepatocytes in Fah−/−/Rag2−/−/Il2rg−/− mice. Nat Biotechnol 25(8):903–910. https://doi.org/10.1038/nbt1326

55. Paulk NK, Pekrun K, Zhu E, Nygaard S, Li B, Xu J, Chu K, Leborgne C, Dane AP, Haft A, Zhang Y, Zhang F, Morton C, Valentine MB, Davidoff AM, Nathwani AC, Mingozzi F, Grompe M, Alexander IE, Lisowski L, Kay MA (2018) Bioengineered AAV capsids with combined high human liver transduction in vivo and unique humoral seroreactivity. Mol Ther 26(1):289–303. https://doi.org/10.1016/j.ymthe.2017.09.021

56. Marsic D, Govindasamy L, Currlin S, Markusic DM, Tseng YS, Herzog RW, Agbandje-McKenna M, Zolotukhin S (2014) Vector design Tour de Force: integrating combinatorial and rational approaches to derive novel adeno-associated virus variants. Mol Ther 22 (11):1900–1909. https://doi.org/10.1038/mt.2014.139

57. Summerford C, Samulski RJ (1998) Membrane-associated heparan sulfate proteoglycan is a receptor for adeno-associated virus type 2 virions. J Virol 72(2):1438–1445

58. Ferrari FK, Samulski T, Shenk T, Samulski RJ (1996) Second-strand synthesis is a rate-limiting step for efficient transduction by

recombinant adeno-associated virus vectors. J Virol 70(5):3227–3234

59. Davidoff AM, Gray JT, Ng CY, Zhang Y, Zhou J, Spence Y, Bakar Y, Nathwani AC (2005) Comparison of the ability of adeno-associated viral vectors pseudotyped with serotype 2, 5, and 8 capsid proteins to mediate efficient transduction of the liver in murine and nonhuman primate models. Mol Ther 11(6):875–888. https://doi.org/10.1016/j.ymthe.2004.12.022
60. Wang J, Xie J, Lu H, Chen L, Hauck B, Samulski RJ, Xiao W (2007) Existence of transient functional double-stranded DNA intermediates during recombinant AAV transduction. Proc Natl Acad Sci U S A 104 (32):13104–13109. https://doi.org/10.1073/pnas.0702778104
61. McCarty DM, Monahan PE, Samulski RJ (2001) Self-complementary recombinant adeno-associated virus (scAAV) vectors promote efficient transduction independently of DNA synthesis. Gene Ther 8 (16):1248–1254. https://doi.org/10.1038/sj.gt.3301514
62. McCarty DM, Fu H, Monahan PE, Toulson CE, Naik P, Samulski RJ (2003) Adeno-associated virus terminal repeat (TR) mutant generates self-complementary vectors to overcome the rate-limiting step to transduction in vivo. Gene Ther 10(26):2112–2118. https://doi.org/10.1038/sj.gt.3302134
63. Wang Z, Ma HI, Li J, Sun L, Zhang J, Xiao X (2003) Rapid and highly efficient transduction by double-stranded adeno-associated virus vectors in vitro and in vivo. Gene Ther 10(26):2105–2111. https://doi.org/10.1038/sj.gt.3302133
64. Miesbach W, Meijer K, Coppens M, Kampmann P, Klamroth R, Schutgens R, Tangelder M, Castaman G, Schwable J, Bonig H, Seifried E, Cattaneo F, Meyer C, Leebeek FWG (2018) Gene therapy with adeno-associated virus vector 5-human factor IX in adults with hemophilia B. Blood 131 (9):1022–1031. https://doi.org/10.1182/blood-2017-09-804419
65. Herzog RW (2017) Complexity of immune responses to AAV transgene products—example of factor IX. Cell Immunol. https://doi.org/10.1016/j.cellimm.2017.05.006
66. Miao CH, Ohashi K, Patijn GA, Meuse L, Ye X, Thompson AR, Kay MA (2000) Inclusion of the hepatic locus control region, an intron, and untranslated region increases and stabilizes hepatic factor IX gene expression in vivo but not in vitro. Mol Ther 1 (6):522–532. https://doi.org/10.1006/mthe.2000.0075
67. Manno CS, Pierce GF, Arruda VR, Glader B, Ragni M, Rasko JJ, Ozelo MC, Hoots K, Blatt P, Konkle B, Dake M, Kaye R, Razavi M, Zajko A, Zehnder J, Rustagi PK, Nakai H, Chew A, Leonard D, Wright JF, Lessard RR, Sommer JM, Tigges M, Sabatino D, Luk A, Jiang H, Mingozzi F, Couto L, Ertl HC, High KA, Kay MA (2006) Successful transduction of liver in hemophilia by AAV-Factor IX and limitations imposed by the host immune response. Nat Med 12(3):342–347. https://doi.org/10.1038/nm1358
68. Nathwani AC, Gray JT, Ng CY, Zhou J, Spence Y, Waddington SN, Tuddenham EG, Kemball-Cook G, McIntosh J, Boon-Spijker-M, Mertens K, Davidoff AM (2006) Self-complementary adeno-associated virus vectors containing a novel liver-specific human factor IX expression cassette enable highly efficient transduction of murine and nonhuman primate liver. Blood 107(7):2653–2661. https://doi.org/10.1182/blood-2005-10-4035
69. McIntosh J, Lenting PJ, Rosales C, Lee D, Rabbanian S, Raj D, Patel N, Tuddenham EG, Christophe OD, McVey JH, Waddington S, Nienhuis AW, Gray JT, Fagone P, Mingozzi F, Zhou SZ, High KA, Cancio M, Ng CY, Zhou J, Morton CL, Davidoff AM, Nathwani AC (2013) Therapeutic levels of FVIII following a single peripheral vein administration of rAAV vector encoding a novel human factor VIII variant. Blood 121(17):3335–3344. https://doi.org/10.1182/blood-2012-10-462200
70. Costa RH, Grayson DR, Darnell JEJ (1989) Multiple hepatocyte-enriched nuclear factors function in the regulation of transthyretin and alpha 1-antitrypsin genes. Mol Cell Biol 9 (4):1415–1425
71. Chuah MK, Petrus I, De Bleser P, Le Guiner C, Gernoux G, Adjali O, Nair N, Willems J, Evens H, Rincon MY, Matrai J, Di Matteo M, Samara-Kuko E, Yan B, Acosta-Sanchez A, Meliani A, Cherel G, Blouin V, Christophe O, Moullier P, Mingozzi F, VandenDriessche T (2014) Liver-specific transcriptional modules identified by genome-wide in silico analysis enable efficient gene therapy in mice and non-human primates. Mol Ther 22(9):1605–1613. https://doi.org/10.1038/mt.2014.114
72. Nair N, Rincon MY, Evens H, Sarcar S, Dastidar S, Samara-Kuko E, Ghandeharian O, Man Viecelli H, Thony B,

De Bleser P, VandenDriessche T, Chuah MK (2014) Computationally designed liver-specific transcriptional modules and hyperactive factor IX improve hepatic gene therapy. Blood 123(20):3195–3199. https://doi.org/10.1182/blood-2013-10-534032

73. Brown HC, Zakas PM, George SN, Parker ET, Spencer HT, Doering CB (2018) Target-cell-directed bioengineering approaches for gene therapy of hemophilia A. Mol Ther Methods Clin Dev 9:57–69. https://doi.org/10.1016/j.omtm.2018.01.004

74. Sack BK, Merchant S, Markusic DM, Nathwani AC, Davidoff AM, Byrne BJ, Herzog RW (2012) Transient B cell depletion or improved transgene expression by codon optimization promote tolerance to factor VIII in gene therapy. PLoS One 7(5): e37671. https://doi.org/10.1371/journal.pone.0037671

75. Vandamme C, Adjali O, Mingozzi F (2017) Unraveling the complex story of immune responses to AAV vectors trial after trial. Hum Gene Ther 28(11):1061–1074. https://doi.org/10.1089/hum.2017.150

76. Martino AT, Suzuki M, Markusic DM, Zolotukhin I, Ryals RC, Moghimi B, Ertl HC, Muruve DA, Lee B, Herzog RW (2011) The genome of self-complementary adeno-associated viral vectors increases Toll-like receptor 9-dependent innate immune responses in the liver. Blood 117 (24):6459–6468. https://doi.org/10.1182/blood-2010-10-314518

77. Rogers GL, Shirley JL, Zolotukhin I, Kumar SRP, Sherman A, Perrin GQ, Hoffman BE, Srivastava A, Basner-Tschakarjan E, Wallet MA, Terhorst C, Biswas M, Herzog RW (2017) Plasmacytoid and conventional dendritic cells cooperate in crosspriming AAV capsid-specific CD8(+) T cells. Blood 129 (24):3184–3195. https://doi.org/10.1182/blood-2016-11-751040

78. Faust SM, Bell P, Cutler BJ, Ashley SN, Zhu Y, Rabinowitz JE, Wilson JM (2013) CpG-depleted adeno-associated virus vectors evade immune detection. J Clin Invest 123 (7):2994–3001. https://doi.org/10.1172/JCI68205

79. Sferra TJ, Backstrom K, Wang C, Rennard R, Miller M, Hu Y (2004) Widespread correction of lysosomal storage following intrahepatic injection of a recombinant adeno-associated virus in the adult MPS VII mouse. Mol Ther 10(3):478–491. https://doi.org/10.1016/j.ymthe.2004.05.029

80. Chandler RJ, Chandrasekaran S, Carrillo-Carrasco N, Senac JS, Hofherr SE, Barry MA, Venditti CP (2011) Adeno-associated virus serotype 8 gene transfer rescues a neonatal lethal murine model of propionic acidemia. Hum Gene Ther 22(4):477–481. https://doi.org/10.1089/hum.2010.164

81. Sherman A, Schlachterman A, Cooper M, Merricks EP, Raymer RA, Bellinger DA, Herzog RW, Nichols TC (2014) Portal vein delivery of viral vectors for gene therapy for hemophilia, 1114. Humana Press, Totowa, pp 413–426. https://doi.org/10.1007/978-1-62703-761-7_27

82. Nakai H, Herzog RW, Hagstrom JN, Walter J, Kung SH, Yang EY, Tai SJ, Iwaki Y, Kurtzman GJ, Fisher KJ, Colosi P, Couto LB, High KA (1998) Adeno-associated viral vector-mediated gene transfer of human blood coagulation factor IX into mouse liver. Blood 91(12):4600–4607

83. Fisher KJ, Gao GP, Weitzman MD, DeMatteo R, Burda JF, Wilson JM (1996) Transduction with recombinant adeno-associated virus for gene therapy is limited by leading-strand synthesis. J Virol 70 (1):520–532

84. Zabaleta N, Salas D, Paramo M, Hommel M, Sier-Ferreira V, Hernandez-Alcoceba R, Prieto J, Bilbao JI, Gonzalez-Aseguinolaza G (2017) Improvement of adeno-associated virus-mediated liver transduction efficacy by regional administration in Macaca fascicularis. Hum Gene Ther Clin Dev 28(2):68–73. https://doi.org/10.1089/humc.2016.183

85. Cunningham SC, Dane AP, Spinoulas A, Alexander IE (2008) Gene delivery to the juvenile mouse liver using AAV2/8 vectors. Mol Ther 16(6):1081–1088. https://doi.org/10.1038/mt.2008.72

86. Lipshutz GS, Gruber CA, Cao Y, Hardy J, Contag CH, Gaensler KM (2001) In utero delivery of adeno-associated viral vectors: intraperitoneal gene transfer produces long-term expression. Mol Ther 3(3):284–292. https://doi.org/10.1006/mthe.2001.0267

87. Dane AP, Wowro SJ, Cunningham SC, Alexander IE (2013) Comparison of gene transfer to the murine liver following intraperitoneal and intraportal delivery of hepatotropic AAV pseudo-serotypes. Gene Ther 20 (4):460–464. https://doi.org/10.1038/gt.2012.67

88. Ahmed SS, Li J, Godwin J, Gao G, Zhong L (2013) Gene transfer in the liver using recombinant adeno-associated virus. Curr Protoc Microbiol Chapter 14:Unit14D 16. https://

doi.org/10.1002/9780471729259.mc14d06s29

89. Yardeni T, Eckhaus M, Morris HD, Huizing M, Hoogstraten-Miller S (2011) Retro-orbital injections in mice. Lab Anim (NY) 40(5):155–160. https://doi.org/10.1038/laban0511-155
90. Wang L, Wang H, Bell P, McCarter RJ, He J, Calcedo R, Vandenberghe LH, Morizono H, Batshaw ML, Wilson JM (2010) Systematic evaluation of AAV vectors for liver directed gene transfer in murine models. Mol Ther 18(1):118–125. https://doi.org/10.1038/mt.2009.246
91. Nakai H, Fuess S, Storm TA, Muramatsu S, Nara Y, Kay MA (2005) Unrestricted hepatocyte transduction with adeno-associated virus serotype 8 vectors in mice. J Virol 79 (1):214–224. https://doi.org/10.1128/JVI.79.1.214-224.2005
92. Cooper M, Nayak S, Hoffman BE, Terhorst C, Cao O, Herzog RW (2009) Improved induction of immune tolerance to factor IX by hepatic AAV-8 gene transfer. Hum Gene Ther 20(7):767–776. https://doi.org/10.1089/hum.2008.161
93. Sands MS (2011) AAV-mediated liver-directed gene therapy. Methods Mol Biol 807:141–157. https://doi.org/10.1007/978-1-61779-370-7_6
94. Palaschak B, Marsic D, Herzog RW, Zolotukhin S, Markusic DM (2017) An immune-competent murine model to study elimination of AAV-transduced hepatocytes by capsid-specific CD8(+) T cells. Mol Ther Methods Clin Dev 5:142–152. https://doi.org/10.1016/j.omtm.2017.04.004
95. Daly TM, Vogler C, Levy B, Haskins ME, Sands MS (1999) Neonatal gene transfer leads to widespread correction of pathology in a murine model of lysosomal storage disease. Proc Natl Acad Sci U S A 96 (5):2296–2300
96. McKay TR, Rahim AA, Buckley SM, Ward NJ, Chan JK, Howe SJ, Waddington SN (2011) Perinatal gene transfer to the liver. Curr Pharm Des 17(24):2528–2541. https://doi.org/10.2174/138161211797247541
97. Davidoff AM, Ng CYC, Zhou JF, Spence Y, Nathwani AC (2003) Sex significantly influences transduction of murine liver by recombinant adeno-associated viral vectors through an androgen-dependent pathway. Blood 102 (2):480–488. https://doi.org/10.1182/blood-2002-09-2889
98. Paneda A, Vanrell L, Mauleon I, Crettaz JS, Berraondo P, Timmermans EJ, Beattie SG, Twisk J, van Deventer S, Prieto J, Fontanellas A, Rodriguez-Pena MS, Gonzalez-Aseguinolaza G (2009) Effect of adeno-associated virus serotype and genomic structure on liver transduction and biodistribution in mice of both genders. Hum Gene Ther 20(8):908–917. https://doi.org/10.1089/hum.2009.031
99. Nathwani AC, Cochrane M, McIntosh J, Ng CY, Zhou J, Gray JT, Davidoff AM (2009) Enhancing transduction of the liver by adeno-associated viral vectors. Gene Ther 16 (1):60–69. https://doi.org/10.1038/gt.2008.137
100. Berraondo P, Crettaz J, Ochoa L, Paneda A, Prieto J, Troconiz IF, Gonzalez-Aseguinolaza G (2006) Intrahepatic injection of recombinant adeno-associated virus serotype 2 overcomes gender-related differences in liver transduction. Hum Gene Ther 17 (6):601–610. https://doi.org/10.1089/hum.2006.17.601
101. Ohashi K, Nakai H, Couto LB, Kay MA (2005) Modified infusion procedures affect recombinant adeno-associated virus vector type 2 transduction in the liver. Hum Gene Ther 16(3):299–306. https://doi.org/10.1089/hum.2005.16.299
102. Seppen J, Bakker C, de Jong B, Kunne C, van den Oever K, Vandenberghe K, de Waart R, Twisk J, Bosma P (2006) Adeno-associated virus vector serotypes mediate sustained correction of bilirubin UDP glucuronosyltransferase deficiency in rats. Mol Ther 13 (6):1085–1092. https://doi.org/10.1016/j.ymthe.2006.01.014
103. Montenegro-Miranda PS, Paneda A, ten Bloemendaal L, Duijst S, de Waart DR, Gonzalez-Aseguinolaza G, Bosma PJ (2013) Adeno-associated viral vector serotype 5 poorly transduces liver in rat models. PLoS One 8(12):e82597. https://doi.org/10.1371/journal.pone.0082597
104. Sobrevals L, Enguita M, Rodriguez C, Gonzalez-Rojas J, Alzaguren P, Razquin N, Prieto J, Fortes P (2012) AAV vectors transduce hepatocytes in vivo as efficiently in cirrhotic as in healthy rat livers. Gene Ther 19 (4):411–417. https://doi.org/10.1038/gt.2011.119
105. Ferla R, O'Malley T, Calcedo R, O'Donnell P, Wang P, Cotugno G, Claudiani P, Wilson JM, Haskins M, Auricchio A (2013) Gene therapy for mucopolysaccharidosis type VI is effective in cats without pre-existing immunity to

AAV8. Hum Gene Ther 24(2):163–169. https://doi.org/10.1089/hum.2012.179

106. Cotugno G, Annunziata P, Tessitore A, O'Malley T, Capalbo A, Faella A, Bartolomeo R, O'Donnell P, Wang P, Russo F, Sleeper MM, Knox VW, Fernandez S, Levanduski L, Hopwood J, De Leonibus E, Haskins M, Auricchio A (2011) Long-term amelioration of feline Mucopolysaccharidosis VI after AAV-mediated liver gene transfer. Mol Ther 19(3):461–469. https://doi.org/10.1038/mt.2010.257
107. Hinderer C, Bell P, Gurda BL, Wang Q, Louboutin JP, Zhu Y, Bagel J, O'Donnell P, Sikora T, Ruane T, Wang P, Haskins ME, Wilson JM (2014) Liver-directed gene therapy corrects cardiovascular lesions in feline mucopolysaccharidosis type I. Proc Natl Acad Sci U S A 111(41):14894–14899. https://doi.org/10.1073/pnas.1413645111
108. Beaty RM, Jackson M, Peterson D, Bird A, Brown T, Benjamin DK Jr, Juopperi T, Kishnani P, Boney A, Chen YT, Koeberl DD (2002) Delivery of glucose-6-phosphatase in a canine model for glycogen storage disease, type Ia, with adeno-associated virus (AAV) vectors. Gene Ther 9(15):1015–1022. https://doi.org/10.1038/sj.gt.3301728
109. Mount JD, Herzog RW, Tillson DM, Goodman SA, Robinson N, McCleland ML, Bellinger D, Nichols TC, Arruda VR, Lothrop CD Jr, High KA (2002) Sustained phenotypic correction of hemophilia B dogs with a factor IX null mutation by liver-directed gene therapy. Blood 99 (8):2670–2676
110. Nichols TC, Whitford MH, Arruda VR, Stedman HH, Kay MA, High KA (2015) Translational data from adeno-associated virus-mediated gene therapy of hemophilia B in dogs. Hum Gene Ther Clin Dev 26 (1):5–14. https://doi.org/10.1089/humc.2014.153
111. Callan MB, Haskins ME, Wang P, Zhou S, High KA, Arruda VR (2016) Successful phenotype improvement following gene therapy for severe hemophilia A in privately owned dogs. PLoS One 11(3):e0151800. https://doi.org/10.1371/journal.pone.0151800
112. Finn JD, Ozelo MC, Sabatino DE, Franck HW, Merricks EP, Crudele JM, Zhou S, Kazazian HH, Lillicrap D, Nichols TC, Arruda VR (2010) Eradication of neutralizing antibodies to factor VIII in canine hemophilia A after liver gene therapy. Blood 116 (26):5842–5848. https://doi.org/10.1182/blood-2010-06-288001
113. Monahan PE, Lothrop CD, Sun J, Hirsch ML, Kafri T, Kantor B, Sarkar R, Tillson DM, Elia JR, Samulski RJ (2010) Proteasome inhibitors enhance gene delivery by AAV virus vectors expressing large genomes in hemophilia mouse and dog models: a strategy for broad clinical application. Mol Ther 18 (11):1907–1916. https://doi.org/10.1038/mt.2010.170
114. Sabatino DE, Lange AM, Altynova ES, Sarkar R, Zhou S, Merricks EP, Franck HG, Nichols TC, Arruda VR, Kazazian HH Jr (2011) Efficacy and safety of long-term prophylaxis in severe hemophilia A dogs following liver gene therapy using AAV vectors. Mol Ther 19(3):442–449. https://doi.org/10.1038/mt.2010.240
115. Bell P, Gao G, Haskins ME, Wang L, Sleeper M, Wang H, Calcedo R, Vandenberghe LH, Chen SJ, Weisse C, Withnall E, Wilson JM (2011) Evaluation of adeno-associated viral vectors for liver-directed gene transfer in dogs. Hum Gene Ther 22 (8):985–997. https://doi.org/10.1089/hum.2010.194
116. Majowicz A, Salas D, Zabaleta N, Rodriguez-Garcia E, Gonzalez-Aseguinolaza G, Petry H, Ferreira V (2017) Successful repeated hepatic gene delivery in mice and non-human primates achieved by sequential administration of AAV5(ch) and AAV1. Mol Ther 25 (8):1831–1842. https://doi.org/10.1016/j.ymthe.2017.05.003
117. Gao GP, Lu Y, Sun X, Johnston J, Calcedo R, Grant R, Wilson JM (2006) High-level transgene expression in nonhuman primate liver with novel adeno-associated virus serotypes containing self-complementary genomes. J Virol 80(12):6192–6194. https://doi.org/10.1128/JVI.00526-06
118. Nathwani AC, Gray JT, McIntosh J, Ng CY, Zhou J, Spence Y, Cochrane M, Gray E, Tuddenham EG, Davidoff AM (2007) Safe and efficient transduction of the liver after peripheral vein infusion of self-complementary AAV vector results in stable therapeutic expression of human FIX in nonhuman primates. Blood 109(4):1414–1421. https://doi.org/10.1182/blood-2006-03-010181
119. Mattar CN, Nathwani AC, Waddington SN, Dighe N, Kaeppel C, Nowrouzi A, McIntosh J, Johana NB, Ogden B, Fisk NM, Davidoff AM, David A, Peebles D, Valentine MB, Appelt JU, von Kalle C, Schmidt M, Biswas A, Choolani M, Chan JK (2011) Stable human FIX expression after 0.9G intrauterine gene transfer of self-complementary adeno-associated viral vector 5 and 8 in

macaques. Mol Ther 19(11):1950–1960. https://doi.org/10.1038/mt.2011.107

120. Mattar CNZ, Gil-Farina I, Rosales C, Johana N, Tan YYW, McIntosh J, Kaeppel C, Waddington SN, Biswas A, Choolani M, Schmidt M, Nathwani AC, Chan JKY (2017) In utero transfer of adeno-associated viral vectors produces long-term factor IX levels in a cynomolgus macaque model. Mol Ther 25(8):1843–1853. https://doi.org/10.1016/j.ymthe.2017.04.003

121. Greig JA, Nordin JM, Bote E, Makaron L, Garnett ME, Kattenhorn LM, Bell P, Goode T, Wilson JM (2016) Impact of intravenous infusion time on AAV8 vector pharmacokinetics, safety, and liver transduction in cynomolgus macaques. Mol Ther Methods Clin Dev 3(Suppl C):16079. https://doi.org/10.1038/mtm.2016.79

122. D'Avola D, Lopez-Franco E, Sangro B, Paneda A, Grossios N, Gil-Farina I, Benito A, Twisk J, Paz M, Ruiz J, Schmidt M, Petry H, Harper P, de Salamanca RE, Fontanellas A, Prieto J, Gonzalez-Aseguinolaza G (2016) Phase I open label liver-directed gene therapy clinical trial for acute intermittent porphyria. J Hepatol 65 (4):776–783. https://doi.org/10.1016/j.jhep.2016.05.012

123. Bell P, Wang L, Gao G, Haskins ME, Tarantal AF, McCarter RJ, Zhu Y, Yu H, Wilson JM (2011) Inverse zonation of hepatocyte transduction with AAV vectors between mice and non-human primates. Mol Genet Metab 104 (3):395–403. https://doi.org/10.1016/j.ymgme.2011.06.002

124. Markusic DM, Nichols TC, Merricks EP, Palaschak B, Zolotukhin I, Marsic D, Zolotukhin S, Srivastava A, Herzog RW (2017) Evaluation of engineered AAV capsids for hepatic factor IX gene transfer in murine and canine models. J Transl Med 15(1):94. https://doi.org/10.1186/s12967-017-1200-1

125. Rapti K, Louis-Jeune V, Kohlbrenner E, Ishikawa K, Ladage D, Hajjar RJ, Weber T (2012) Neutralizing antibodies against AAV serotypes 1, 2, 6 and 9 in sera of commonly used animal models and their significance for in vivo experiments. Mol Ther 20:S62–S62

126. Calcedo R, Franco J, Qin Q, Richardson DW, Mason JB, Boyd S, Wilson JM (2015) Preexisting neutralizing antibodies to adeno-associated virus capsids in large animals other than monkeys may confound in vivo gene therapy studies. Hum Gene Ther Methods 26(3):103–105. https://doi.org/10.1089/hgtb.2015.082

127. Cao O, Dobrzynski E, Wang L, Nayak S, Mingle B, Terhorst C, Herzog RW (2007) Induction and role of regulatory CD4+CD25+ T cells in tolerance to the transgene product following hepatic in vivo gene transfer. Blood 110(4):1132–1140. https://doi.org/10.1182/blood-2007-02-073304

128. Hoffman BE, Martino AT, Sack BK, Cao O, Liao G, Terhorst C, Herzog RW (2011) Non-redundant roles of IL-10 and TGF-beta in suppression of immune responses to hepatic AAV-factor IX gene transfer. Mol Ther 19 (7):1263–1272. https://doi.org/10.1038/mt.2011.33

129. Li H, Haurigot V, Doyon Y, Li T, Wong SY, Bhagwat AS, Malani N, Anguela XM, Sharma R, Ivanciu L, Murphy SL, Finn JD, Khazi FR, Zhou S, Paschon DE, Rebar EJ, Bushman FD, Gregory PD, Holmes MC, High KA (2011) In vivo genome editing restores haemostasis in a mouse model of haemophilia. Nature 475(7355):217–221. https://doi.org/10.1038/nature10177

130. Sharma R, Anguela XM, Doyon Y, Wechsler T, DeKelver RC, Sproul S, Paschon DE, Miller JC, Davidson RJ, Shivak D, Zhou S, Rieders J, Gregory PD, Holmes MC, Rebar EJ, High KA (2015) In vivo genome editing of the albumin locus as a platform for protein replacement therapy. Blood 126(15):1777–1784. https://doi.org/10.1182/blood-2014-12-615492

131. Ran FA, Cong L, Yan WX, Scott DA, Gootenberg JS, Kriz AJ, Zetsche B, Shalem O, Wu X, Makarova KS, Koonin EV, Sharp PA, Zhang F (2015) In vivo genome editing using Staphylococcus aureus Cas9. Nature 520 (7546):186–191. https://doi.org/10.1038/nature14299

132. Yang Y, Wang L, Bell P, McMenamin D, He Z, White J, Yu H, Xu C, Morizono H, Musunuru K, Batshaw ML, Wilson JM (2016) A dual AAV system enables the Cas9-mediated correction of a metabolic liver disease in newborn mice. Nat Biotechnol 34 (3):334–338. https://doi.org/10.1038/nbt.3469

133. Barzel A, Paulk NK, Shi Y, Huang Y, Chu K, Zhang F, Valdmanis PN, Spector LP, Porteus MH, Gaensler KM, Kay MA (2015) Promoterless gene targeting without nucleases ameliorates haemophilia B in mice. Nature 517 (7534):360–364. https://doi.org/10.1038/nature13864

134. Markusic DM, Hoffman BE, Perrin GQ, Nayak S, Wang X, LoDuca PA, High KA,

Herzog RW (2013) Effective gene therapy for haemophilic mice with pathogenic factor IX antibodies. EMBO Mol Med 5 (11):1698–1709. https://doi.org/10.1002/emmm.201302859

135. Doerfler PA, Todd AG, Clement N, Falk DJ, Nayak S, Herzog RW, Byrne BJ (2016) Copackaged AAV9 vectors promote simultaneous immune tolerance and phenotypic correction of pompe disease. Hum Gene Ther 27 (1):43–59. https://doi.org/10.1089/hum.2015.103

136. Han SO, Ronzitti G, Arnson B, Leborgne C, Li S, Mingozzi F, Koeberl D (2017) Low-dose liver-targeted gene therapy for pompe disease enhances therapeutic efficacy of ERT via immune tolerance induction. Mol Ther Methods Clin Dev 4:126–136. https://doi.org/10.1016/j.omtm.2016.12.010

137. Crudele JM, Finn JD, Siner JI, Martin NB, Niemeyer GP, Zhou S, Mingozzi F, Lothrop CD Jr, Arruda VR (2015) AAV liver expression of FIX-Padua prevents and eradicates FIX inhibitor without increasing thrombogenicity in hemophilia B dogs and mice. Blood 125(10):1553–1561. https://doi.org/10.1182/blood-2014-07-588194

138. Wasserfall CH, Herzog RW (2009) Gene therapy approaches to induce tolerance in autoimmunity by reshaping the immune system. Curr Opin Investig Drugs 10 (11):1143–1150

139. Keeler GD, Kumar S, Palaschak B, Silverberg EL, Markusic DM, Jones NT, Hoffman BE (2018) Gene therapy-induced antigen-specific tregs inhibit neuro-inflammation and reverse disease in a mouse model of multiple sclerosis. Mol Ther 26(1):173–183. https://doi.org/10.1016/j.ymthe.2017.09.001

140. McIntosh JH, Cochrane M, Cobbold S, Waldmann H, Nathwani SA, Davidoff AM, Nathwani AC (2012) Successful attenuation of humoral immunity to viral capsid and transgenic protein following AAV-mediated gene transfer with a non-depleting CD4 antibody and cyclosporine. Gene Ther 19 (1):78–85. https://doi.org/10.1038/gt.2011.64

141. Corti M, Elder ME, Falk DJ, Lawson L, Smith BK, Nayak S, Conlon TJ, Clement N, Erger K, Lavassani E, Green MM, Doerfler PA, Herzog RW, Byrne BJ (2014) B-cell depletion is protective against anti-AAV capsid immune response: a human subject case study. Mol Ther Methods Clin Dev 1. https://doi.org/10.1038/mtm.2014.33

142. Scallan CD, Jiang H, Liu T, Patarroyo-White-S, Sommer JM, Zhou S, Couto LB, Pierce GF (2006) Human immunoglobulin inhibits liver transduction by AAV vectors at low AAV2 neutralizing titers in SCID mice. Blood 107(5):1810–1817. https://doi.org/10.1182/blood-2005-08-3229

143. Calcedo R, Vandenberghe LH, Gao G, Lin J, Wilson JM (2009) Worldwide epidemiology of neutralizing antibodies to adeno-associated viruses. J Infect Dis 199(3):381–390. https://doi.org/10.1086/595830

144. Wang L, Calcedo R, Wang H, Bell P, Grant R, Vandenberghe LH, Sanmiguel J, Morizono H, Batshaw ML, Wilson JM (2010) The pleiotropic effects of natural AAV infections on liver-directed gene transfer in macaques. Mol Ther 18(1):126–134. https://doi.org/10.1038/mt.2009.245

145. Mingozzi F, Anguela XM, Pavani G, Chen Y, Davidson RJ, Hui DJ, Yazicioglu M, Elkouby L, Hinderer CJ, Faella A, Howard C, Tai A, Podsakoff GM, Zhou S, Basner-Tschakarjan E, Wright JF, High KA (2013) Overcoming preexisting humoral immunity to AAV using capsid decoys. Sci Transl Med 5(194):194ra192. https://doi.org/10.1126/scitranslmed.3005795

146. Meliani A, Boisgerault F, Fitzpatrick Z, Marmier S, Leborgne C, Collaud F, Simon Sola M, Charles S, Ronzitti G, Vignaud A, van Wittenberghe L, Marolleau B, Jouen F, Tan S, Boyer O, Christophe O, Brisson AR, Maguire CA, Mingozzi F (2017) Enhanced liver gene transfer and evasion of preexisting humoral immunity with exosome-enveloped AAV vectors. Blood Adv 1(23):2019–2031. https://doi.org/10.1182/bloodadvances.2017010181

147. Mimuro J, Mizukami H, Hishikawa S, Ikemoto T, Ishiwata A, Sakata A, Ohmori T, Madoiwa S, Ono F, Ozawa K, Sakata Y (2013) Minimizing the inhibitory effect of neutralizing antibody for efficient gene expression in the liver with adeno-associated virus 8 vectors. Mol Ther 21(2):318–323. https://doi.org/10.1038/mt.2012.258

148. Burton M, Nakai H, Colosi P, Cunningham J, Mitchell R, Couto L (1999) Coexpression of factor VIII heavy and light chain adeno-associated viral vectors produces biologically active protein. Proc Natl Acad Sci U S A 96 (22):12725–12730

149. Duan D, Yue Y, Engelhardt JF (2001) Expanding AAV packaging capacity with trans-splicing or overlapping vectors: a quantitative comparison. Mol Ther 4(4):383–391. https://doi.org/10.1006/mthe.2001.0456

150. Nakai H, Iwaki Y, Kay MA, Couto LB (1999) Isolation of recombinant adeno-associated

virus vector-cellular DNA junctions from mouse liver. J Virol 73(7):5438–5447

151. Nakai H, Montini E, Fuess S, Storm TA, Grompe M, Kay MA (2003) AAV serotype 2 vectors preferentially integrate into active genes in mice. Nat Genet 34(3):297–302. https://doi.org/10.1038/ng1179
152. Moscioni D, Morizono H, McCarter RJ, Stern A, Cabrera-Luque J, Hoang A, Sanmiguel J, Wu D, Bell P, Gao GP, Raper SE, Wilson JM, Batshaw ML (2006) Long-term correction of ammonia metabolism and prolonged survival in ornithine transcarbamylase-deficient mice following liver-directed treatment with adeno-associated viral vectors. Mol Ther 14 (1):25–33. https://doi.org/10.1016/j.ymthe.2006.03.009
153. Bell P, Wang L, Lebherz C, Flieder DB, Bove MS, Wu D, Gao GP, Wilson JM, Wivel NA (2005) No evidence for tumorigenesis of AAV vectors in a large-scale study in mice. Mol Ther 12(2):299–306. https://doi.org/10.1016/j.ymthe.2005.03.020
154. Li H, Malani N, Hamilton SR, Schlachterman A, Bussadori G, Edmonson SE, Shah R, Arruda VR, Mingozzi F, Wright JF, Bushman FD, High KA (2011) Assessing the potential for AAV vector genotoxicity in a murine model. Blood 117(12):3311–3319. https://doi.org/10.1182/blood-2010-08-302729
155. Nathwani AC, Rosales C, McIntosh J, Rastegarlari G, Nathwani D, Raj D, Nawathe S, Waddington SN, Bronson R, Jackson S, Donahue RE, High KA, Mingozzi F, Ng CY, Zhou J, Spence Y, McCarville MB, Valentine M, Allay J, Coleman J, Sleep S, Gray JT, Nienhuis AW, Davidoff AM (2011) Long-term safety and efficacy following systemic administration of a self-complementary AAV vector encoding human FIX pseudotyped with serotype 5 and 8 capsid proteins. Mol Ther 19 (5):876–885. https://doi.org/10.1038/mt.2010.274
156. Chandler RJ, LaFave MC, Varshney GK, Trivedi NS, Carrillo-Carrasco N, Senac JS, Wu W, Hoffmann V, Elkahloun AG, Burgess SM, Venditti CP (2015) Vector design influences hepatic genotoxicity after adeno-associated virus gene therapy. J Clin Invest 125(2):870–880. https://doi.org/10.1172/JCI79213
157. Chandler RJ, Sands MS, Venditti CP (2017) Recombinant adeno-associated viral integration and genotoxicity: insights from animal models. Hum Gene Ther 28(4):314–322. https://doi.org/10.1089/hum.2017.009
158. Gil-Farina I, Fronza R, Kaeppel C, Lopez-Franco E, Ferreira V, D'Avola D, Benito A, Prieto J, Petry H, Gonzalez-Aseguinolaza G, Schmidt M (2016) Recombinant AAV integration is not associated with hepatic genotoxicity in nonhuman primates and patients. Mol Ther 24(6):1100–1105. https://doi.org/10.1038/mt.2016.52
159. Nault JC, Datta S, Imbeaud S, Franconi A, Mallet M, Couchy G, Letouze E, Pilati C, Verret B, Blanc JF, Balabaud C, Calderaro J, Laurent A, Letexier M, Bioulac-Sage P, Calvo F, Zucman-Rossi J (2015) Recurrent AAV2-related insertional mutagenesis in human hepatocellular carcinomas. Nat Genet 47(10):1187–1193. https://doi.org/10.1038/ng.3389
160. Nault JC, Mami I, La Bella T, Datta S, Imbeaud S, Franconi A, Mallet M, Couchy G, Letouze E, Pilati C, Verret B, Blanc JF, Balabaud C, Calderaro J, Laurent A, Letexier M, Bioulac-Sage P, Calvo F, Zucman-Rossi J (2016) Wild-type AAV insertions in hepatocellular carcinoma do not inform debate over genotoxicity risk of vectorized AAV. Mol Ther 24(4):660–661. https://doi.org/10.1038/mt.2016.47

Chapter 21

AAV-Mediated Gene Delivery to the Lung

Laura P. van Lieshout, Jakob M. Domm, and Sarah K. Wootton

Abstract

Adeno-associated virus (AAV) vectors are an efficient method of gene delivery to various tissues including the lung. Mouse models are often used as a preliminary preclinical model in order to advance AAV lung gene therapy vectors. In this chapter we describe an AAV purification protocol using heparin affinity chromatography as well as an intranasal and intratracheal method of delivering AAV vectors to the lungs of mice.

Key words Adeno-associated virus, AAV, AAV6.2FF, Lung gene therapy, Lung delivery, Lung gene transfer, Heparin column chromatography, AAV purification, Intranasal instillation

1 Introduction

1.1 General Introduction

Adeno-associated virus (AAV) is a highly efficient gene therapy vector that is utilized to deliver transgenes to a variety of different tissues in vivo. Numerous AAV serotypes have been characterized, each with a unique transduction profile and gene expression kinetics [1]. AAV capsids can be engineered by introducing point mutations or through directed evolution to alter specific characteristics and enhance its utility as a gene therapy vector [2, 3].

1.2 AAV-Mediated Gene Transfer in the Lung

Many AAV serotypes transduce the lung to varying degrees; however, AAV6 is one of the most efficient serotypes used for lung gene transfer [4–6]. AAV6 efficiently transduces both airway and alveolar type II cells [4, 7], which are often the target cell populations for lung gene therapies. Mice are frequently used as preclinical models to validate the efficacy of gene therapy vectors; therefore, efficient delivery of AAV to the mouse lung is critical. Here we describe two reliable methods for delivering AAV vectors to the lung: intranasal and intratracheal administration.

Electronic Supplementary Material: The online version of this chapter (https://doi.org/10.1007/978-1-4939-9139-6_21) contains supplementary material, which is available to authorized users.

Michael J. Castle (ed.), *Adeno-Associated Virus Vectors: Design and Delivery*, Methods in Molecular Biology, vol. 1950, https://doi.org/10.1007/978-1-4939-9139-6_21,

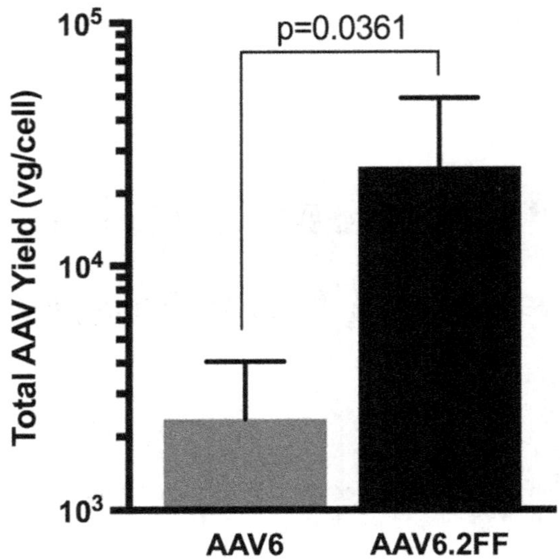

Fig. 1 AAV6.2FF vectors are produced more efficiently than AAV6. Supernatant and cell lysate from AAV6.2FF- and AAV6-luciferase producer cells were harvested 5 days post-transfection ($n = 3$) and analyzed by Taqman qPCR to determine the respective production yields

1.3 AAV6.2FF

AAV6.2FF is a rationally designed triple mutant AAV6 capsid that displays a ten-fold increase in lung transduction compared to the parental AAV6 vector [8]. AAV6.2FF binds to heparin sulfate, allowing heparin affinity chromatography to be used to purify vectors for in vivo use [9]. Additionally, AAV6.2FF vectors are produced more efficiently than their AAV6 counterparts, resulting in greater overall production yields with equal input (Fig. 1). As such, AAV6.2FF is ideally suited for use as a lung gene therapy vector.

2 Materials

2.1 AAV6.2FF Vector Production

1. Plasmids: AAV genome plasmid, RepCap packaging plasmid, and helper plasmid, e.g., pLAdeno5 or pAdDeltaF6.
2. CO_2 incubator.
3. Biological safety cabinet.
4. Pipette aid and sterile serological pipettes.
5. HEK293 cells (ATCC® CRL-1573).
6. Complete Dulbecco's modified eagle medium (DMEM) supplemented with 7% fetal bovine serum (FBS) or cosmic calf serum, 100 μg/mL penicillin-streptomycin, and 2 mM L-glutamine.

Table 1
Recipe for HBSS without NaCl

Chemical	Formula weight (g/mol)	Concentration (g/L)	Molarity
Calcium chloride ($CaCl_2$) (anhyd.)	111	0.14	0.0013
Potassium chloride (KCl)	74.55	0.4	0.0054
Potassium phosphate (monobasic) (KH_2PO_4)	136.1	0.05	0.00034
Magnesium chloride ($MgCl_2{\cdot}6H_2O$)	203	0.1	0.0005
Magnesium sulfate ($MgSO_4{\cdot}7H_2O$)	246	0.1	0.0004
$NaHCO_3$	84.01	0.35	0.0041
Sodium phosphate dibasic (Na_2HPO_4)	142.0	0.048	0.00034
D-glucose (dextrose)	180.2	1.0	0.0055

7. 15 cm dishes.
8. Hemocytometer and inverted microscope.
9. Linear MW 40000 Polyethylenimine (PEI Max): 1 mg/mL, Milli-Q water, pH 6.90 to 7.10, 0.22 μm filtered.
10. Pipettes and filtered tips.
11. Cell scraper.
12. 50 mL polystyrene conical tubes.
13. 37 °C water bath.
14. Dry ice and 95% ethanol.
15. Vortex mixer.
16. High-speed centrifuge.
17. 500 mL 0.22 μm PES filters.

2.2 AAV6.2FF Vector Purification

1. HBSS without NaCl (Table 1).
2. Hank's balanced salt solution (HBSS) **with** Mg^{2+} and Ca^{2}.
3. HBSS **without** Mg^{2+} and Ca^{2}.
4. 1 M NaOH.
5. Basal DMEM.
6. 5% Tween 20 in HBSS **with** Mg^{2+} and Ca^{2+}, 0.22 μm filtered.
7. 200 mM NaCl in HBSS **with** Mg^{2+} and Ca^{2+}, 0.22 μm filtered.
8. 300 mM NaCl in HBSS **with** Mg^{2+} and Ca^{2+}, 0.22 μm filtered.
9. 0.5% N-lauroylsarcosine sodium salt in HBSS **without** Mg^{2+} and Ca^{2+}, 0.22 μm filtered.

10. 2 M NaCl in Milli-Q water, 0.22 μm filtered.
11. 0.1% Triton X-100 in Milli-Q water, 0.22 μm filtered.
12. 20% Ethanol in Milli-Q water, 0.22 μm filtered.
13. 5 mL HiTrap® Heparin High Performance column (GE Healthcare).
14. Peristaltic pump.
15. Cole Parmer MasterFlex Precision Tubing L/S 14, 25 ft (*see* **Note 1**).
16. Amicon Ultra-4 (100,000 NMWL) Centrifugal Filter Units.
17. 15 and 50 mL polystyrene conical tubes.
18. Waste container.
19. Ring stand with clamp.

2.3 Intranasal Delivery of AAV Vector

1. Mice (BALB/c, 6 weeks old).
2. AAV vector (diluted to 80 μL per dose in HBSS **with** Mg^{2+} and Ca^2; to be delivered in 2× 40 μL instillations; *see* **Note 2**).
3. Medication and equipment for anesthesia.
4. Anesthesia chamber (or feline anesthesia mask).
5. Mouse nose cone.
6. P20 micropipette and filtered tips.
7. Recovery cage and heat lamp.

2.4 Intratracheal Delivery of AAV Vector

1. Mice (BALB/c, 6 weeks old).
2. AAV vector (diluted to 40–80 μL per dose in HBSS **with** Mg^{2+} and Ca^2).
3. Medication and equipment for anesthesia.
4. Anesthesia chamber and nose cone for mice.
5. Surgical tape.
6. Surgical board.
7. Electric clippers.
8. Betadine or equivalent disinfectant.
9. Surgical scissors and curved forceps.
10. 29G tuberculin syringe.
11. 5-0 monocryl sutures.
12. Recovery cage and heat lamp.

3 Methods

3.1 AAV6.2FF Vector Production

3.1.1 Cell Preparation

1. **Day 1**: Seed 40 × 15 cm dishes with 1×10^7 HEK293 cells per dish in 18 mL of complete DMEM so that the cell monolayer is ~90% confluent the following day. Incubate in a humidified 5% CO_2 incubator overnight.

3.1.2 Transfection

1. **Day 2**: Prepare the transfection solution in a 50 mL conical tube. For each 15 cm dish, add the Rep-Cap packaging plasmid, the helper plasmid, and the AAV genome plasmid in a 1:1:1 molar ratio (for a total 40 μg/dish) to basal DMEM in a final volume of 880 μL. Add 120 μL PEI Max (based on a 3:1 ratio of PEI to DNA) (*see* **Note 3**) to the DNA-DMEM solution in a dropwise manner and then invert the tube repeatedly to mix (*see* **Note 4**). Incubate the transfection solution at room temperature for 10 min, then add 1 mL per dish dropwise to HEK293 cells. Rotate plates from side to side to disperse transfection reagent before returning to the incubator.
2. 4–8 h post-transfection, remove the transfection medium and replace with 16 mL of basal DMEM (*see* **Notes 5** and **6**).

3.1.3 Harvest

1. **Day 6–8**: Harvest each 15 cm dish by dislodging the cell monolayer into the supernatant using a rubber policeman or cell scraper. Transfer the combined cells and supernatant (SN) into 50 mL conical tubes. Do not exceed 45 mL total per tube.
2. Freeze the 50 mL conical tubes containing the cell and SN suspension in a dry ice ethanol bath. After the suspension has completely frozen and undergoes a color change from red to yellow, remove the 50 mL tubes and place them in a 37 °C water bath to thaw. Once thawed, vortex each tube for 15 s. Repeat this freeze/thaw cycle twice more.
3. Centrifuge crude cell suspension at 8,000 × g for 30 min to pellet cell debris. Immediately dispense the clarified supernatant from each of the 50 mL conical tubes into one clean 500 mL bottle. Filter the pooled clarified supernatant through a 0.22 μm PES filter (*see* **Note 7**) and proceed to purification (*see* **Note 8**) or freeze at −80 °C.

3.2 AAV6.2FF Vector Purification

This protocol is specifically used to purify AAV6 and AAV6.2FF vectors; however, it can also be used for other AAV serotypes that bind heparin sulfate (i.e., AAV2 and AAV-DJ) (*see* **Note 9**).

3.2.1 Amicon Preparation

1. Passivate an Amicon Ultra-4 (100,000 NMWL) centrifugal device with 5 mL of 5% Tween 20 for a minimum of 30 min

Fig. 2 Peristaltic pump and tubing set up for AAV purification

at room temperature (*see* **Note 10**). Start passivation before beginning the column purification procedure.

2. When the column purification procedure has concluded (Subheading 3.2.2), remove the passivation solution and wash the Amicon twice with 5 mL of HBSS **with** Ca^{2+} and Mg^{2+} by centrifuging for 2 min at 900 × *g* in a swinging bucket clinical centrifuge. Remove excess liquid in the holding tube prior to adding AAV vector to the filter device.

3.2.2 Heparin Column Purification

1. Assemble tubing through the peristaltic pump (Fig. 2), and run the following solutions through the tubing at 1–2 mL/min:
   - 20 mL of 1 M NaOH.
   - 100 mL of Milli-Q water.
   - 100 mL of basal DMEM.
2. Remove the heparin column from its package and inspect it for irregularities, such as areas where the column might have dried out. Snap off both the end and inlet stopper and connect sterile tubing to the column at the tubing connectors, holding it in place above the waste container using a ring stand. Run the following solutions through the column at 1–2 mL/min (*see* **Notes 11** and **12**):
   - 50 mL of basal DMEM to equilibrate column.
   - 0.22 μm filtered AAV containing clarified supernatant.
   - 50 mL of HBSS without Mg^{2+} and Ca^{2+}.

- 25 mL of 0.5% N-lauroylsarcosine.
- 50 mL of HBSS without Mg^{2+} and Ca^{2+}.
- 50 mL of HBSS with Mg^{2+} and Ca^{2+}.
- 50 mL of 200 mM NaCl/HBSS with Mg^{2+} and Ca^{2+}.

3. Label 5 × 15 mL tubes with the vector name and elution fraction #1–5. Elute AAV using 25 mL of 300 mM NaCl/HBSS with Mg^{2+} and Ca^{2+}, and collect the eluate in 5 mL fractions in the previously labelled 15 mL tubes (*see* **Notes 13** and **14**).
4. **Optional** To clean the heparin column for reuse, run the following solutions through the column at 1–2 mL/min (*see* **Note 15**):
   - 100 mL of 2 M NaCl.
   - 500 mL of 0.1% Triton X-100 (warmed to 37 °C).
   - 100 mL of Milli-Q water.
   - 25 mL of 20% ethanol (storage buffer).
5. Seal column with stop plug female 1/16″ and finger tighten. Store column at 4 °C.
6. Disassemble tubing and store in 1 M NaOH.

3.2.3 AAV Concentration Using an Amicon

1. Combine elution fractions #2 and #3 and add 5 mL to the passivated Amicon from Subheading 3.2.1 (*see* **Note 16**). Centrifuge in a swinging bucket centrifuge at 900 × *g* for 2 min. Discard the flowthrough.
2. Repeat centrifugation step until all 10 mL of the combined eluate has been applied to the Amicon. Continue centrifugation until the eluate is concentrated to a final volume of approximately 500 μL.
3. Once the desired volume has been achieved, pipette up and down multiple times (~30 s) to dislodge any vector that may have stuck to the membrane and transfer the virus from the Amicon to a sterile 1.5 mL tube.
4. Measure the volume of virus retrieved from the Amicon, then add an equal volume of HBSS without NaCl (Table 1) back to the Amicon. Pipette up and down as before to dislodge and recover any residual virus. Add this solution to the original virus sample and mix well. The virus is now in HBSS containing 150 mM NaCl and is ready for use in vivo.
5. Dispense vector in 100 μL aliquots (*see* **Note 17**), label, and store at −80 °C. Prepare one 10 μL aliquot to be used for AAV vector titration.

6. Titrate the purified AAV vector prior to administration to mice using a universal inverted terminal repeat (ITR) specific Taqman qPCR assay as described (*see* Chapter 4) [10].

3.3 Intranasal Delivery of AAV Vector

1. **Person A:** Anesthetize an individual mouse in an anesthesia chamber, with 2.5% isoflurane and an oxygen flow rate of 1 L per minute.
2. **Person B**: Using a filtered pipette tip, prepare 40 μL (1/2 of the total dose) of vector for intranasal instillation.
3. **Person A:** Remove the mouse from the anesthesia chamber once it is stably unconscious (*see* **Note 18**). Scruff the mouse at the back of the neck using the dominant hand while placing the thumb of the nondominant hand firmly over the mouth (*see* **Note 19**) and the dominant pointer finger on the bridge of the nose to provide gentle pressure from the top of the nose, thereby preventing the mouse from mouth breathing and aspirating the AAV vector (Fig. 3). Rest the mouse's body on the base of the dominant hand at a 45° angle to ensure proper airway access.
4. **Person B:** Hold the pipette so the tip is at a 90° angle to the nose of the mouse. Slowly pipette a drop of AAV vector onto the nose and wait for the liquid to be inhaled prior to placing the next drop. Repeat until all 40 μL has been delivered (Video 1).

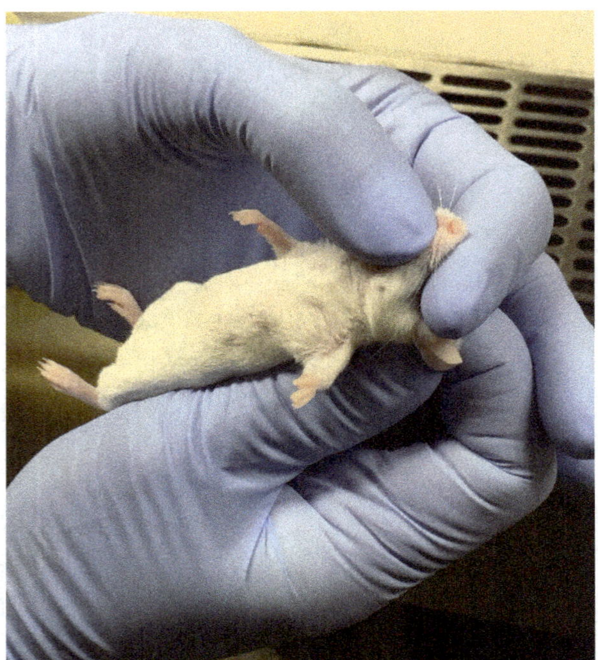

Fig. 3 Hand positioning of person A for intranasal delivery

5. **Person A:** Continue to hold the mouse in this position for 30–60 s after vector delivery to prevent swallowing. The mouse will likely regain consciousness during this time. Subsequently place the animal into an empty recovery cage warmed by a heat lamp for a minimum of 10 min.
6. Repeat **steps 1–5** until all mice have received the first 40 μL half dose.
7. Repeat **steps 1–5** again to deliver the second 40 μL half dose to all mice in the same order they received the first dose. At this point the full 80 μL dose has been administered. Place in heated recovery cage until fully conscious.

3.4 Intratracheal Delivery of AAV Vector

1. In an anesthesia chamber, anesthetize mouse with 2.5% isoflurane and an oxygen flow rate of 1 L per min until the mouse reaches a state of unconsciousness.
2. Prepare the 40–80 μL AAV dose in a 29G tuberculin syringe.
3. Transfer the mouse from the anesthesia chamber to the surgical board on its back, and place a mouse nose cone over the nose to maintain anesthesia.
4. Tape the front legs of the mouse to the surgical board in an extended position (Fig. 4a).
5. Using electric clippers, shave the neck to create a clean surgical field.
6. Disinfect the surgical field with betadine.

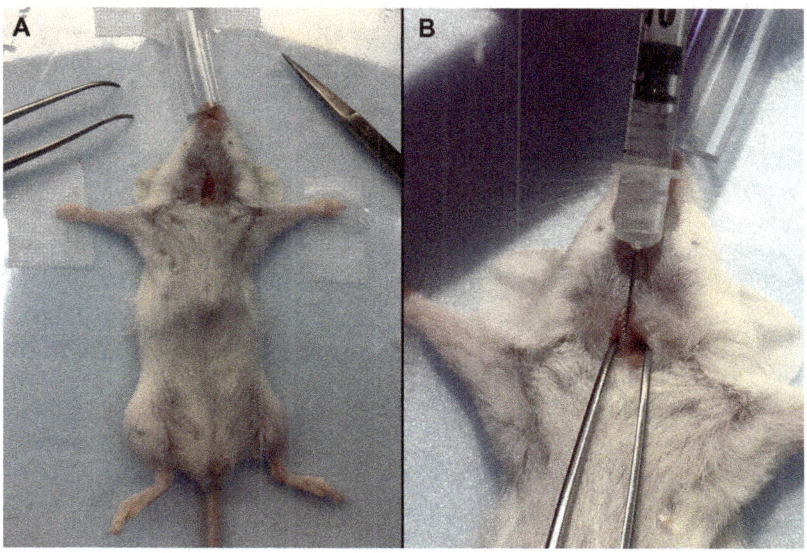

Fig. 4 Intratracheal delivery. (**a**) Incision site to expose the trachea. (**b**) Needle positioning for injection into the trachea

7. Check that the mouse is fully unconscious by pinching the back foot. If the mouse responds to the pinch wait until it is further anesthetized.
8. Grasp the skin on the neck between the front legs using curved forceps and make a small, 5–7 mm, incision using surgical scissors (Fig. 4a).
9. Visualize the tracheal rings and gently immobilize the trachea using curved forceps in the nondominant hand.
10. Using the dominant hand hold the syringe with the bevel facing upwards at a 45° angle to the trachea with the needle facing towards the tail. Insert the needle a few millimeters at a 45° angle (Fig. 4b) (*see* **Note 20**).
11. Slowly inject the AAV into the trachea. Wait 5 s before slowly removing the needle.
12. Place 2–4 simple sutures to close the incision site (*see* **Note 21**).
13. Remove the mouse from the nose cone and place in a recovery cage warmed by a heat lamp until fully conscious.

4 Notes

1. The diameter of this tubing is critical to ensure expected distribution of AAV in the eluate.
2. The dose is split into two instillations to increase vector distribution. Volumes of greater than 40 μL are not advisable to avoid swallowing.
3. PEI Max should be equilibrated to 37 °C prior to use.
4. If DNA crashes out of solution, such that long strands of precipitated DNA are visible, discontinue using the transfection mixture and prepare a fresh master mix transfection solution, but for a smaller number of dishes.
5. Removing the PEI Max helps to maintain the health of the producer cells. Additionally, using basal medium drastically reduces the concentration of fetal bovine serum, which can interfere with AAV heparin affinity interactions.
6. Gently replace the basal DMEM by allowing the medium to drip down the side of the plate in order to ensure the monolayer is not disrupted.
7. Cell pellets are not securely adhered to tube after centrifugation. Be cautious when transferring AAV suspension to the filter to ensure that cell debris is not added. More than one filter may be required.

8. If unable to proceed directly to purification, filtered AAV supernatant can be stored at 4 °C overnight, or −80 °C for longer periods of time.
9. Optimization of elution conditions may be required.
10. Amicon can be passivated at room temperature during the entire purification process or overnight at 4 °C.
11. Do not increase the flow rate beyond 1–2 per minute. This may destroy the column and reduce AAV binding. Do not let the column run dry. If the column runs dry at any point during the purification procedure, you will not be able to retrieve vector from the column.
12. The pump must be stopped while changing input solutions to avoid introducing air bubbles. Air bubbles should be prevented from being pushed through the column as this will disrupt virus binding and decrease viral yield.
13. Secure the top of the input container with Parafilm to ensure the tubing stays secured at the bottom of the container. Do not allow air bubbles to enter the tubing.
14. The hold-up volume of tubing should be less than 5 mL to ensure the expected distribution of AAV in the elution fractions.
15. Heparin column cleanup and reuse is optional. The columns can be reused 3–4 times with proper cleaning and storage.
16. Elution fractions #2 and #3 contain the majority of the AAV vector. Avoid concentrating elution fraction #1 as it contains nontarget proteins.
17. It is very important to vigorously pipette liquid up and down on the Amicon membrane, as AAV can stick to the Amicon. Passivation of the Amicon is critical as it helps prevent the AAV vector from sticking to the membrane.
18. The mice should not be too heavily anesthetized in order to maintain a steady breathing rate.
19. Make sure fingertips are gloved and the mouse's nose is dry and free of liquid, in order to prevent the vector from wicking away from the delivery site. Person A should wear tight fitting gloves at least one size smaller than they typically wear.
20. Once the needle is inserted, gently move the needle outwards away from the mouse's body to visually confirm proper insertion into the trachea.
21. Alternatively, a surgical staple can be used.

Acknowledgments

This work was supported by the Lung Association – Ontario and Cystic Fibrosis Canada.

References

1. Zincarelli C, Soltys S, Rengo G, Rabinowitz J (2008) Analysis of AAV serotypes 1-9 mediated gene expression and tropism in mice after systemic injection. Mol Ther 16:1073–1080
2. Zhong L, Li B, Mah CS, Govindasamy L, Agbandje-McKenna M, Coope M, Herzog RW, Zolotukhin I, Warrington KH, Weigel-Van Aken KA, Hobbs JA, Zolotukhin S, Muzyczka N, Srivastava A (2008) Next generation of adeno-associated virus 2 vectors: point mutations in tyrosines lead to high-efficiency transduction at lower doses. Proc Natl Acad Sci U S A 105:7827–7832
3. Grimm D, Lee JS, Wang L, Desai T, Akache B, Storm TA, Kay MA (2008) In vitro and in vivo gene therapy vector evolution via multispecies interbreeding and retargeting of adeno-associated viruses. J Virol 82:5887–5911
4. Li W, Zhang L, Wu Z, Pickles RJ, Samulski RJ (2011) AAV-6 mediated efficient transduction of mouse lower airways. Virology 417:327–333
5. Halbert C, Madtes D, Vaughan A, Wang Z, Storb R, Tapscott SJ, Miller AD (2010) Expression of human alpha1-antitrypsin in mice and dogs following AAV6 vector-mediated gene transfer to the lungs. Mol Ther 18:1165–1172
6. Halbert CL, Allen JM, Miller AD (2001) Adeno-associated virus type 6 (AAV6) vectors mediate efficient transduction of airway epithelial cells in mouse lungs compared to that of AAV2 vectors. J Virol 75:6615–6624
7. Santry LA, Ingrao JC, Yu DL, de Jong JG, van Lieshout LP, Wood GA, Wootton SK (2017) AAV vector distribution in the mouse respiratory tract following four different methods of administration. BMC Biotechnol 17:43. https://doi.org/10.1186/s12896-017-0365-2
8. van Lieshout LP, Domm JM, Rindler TN, Frost KL, Sorenson DL, Medina SJ, Booth SA, Bridges JP, Wootton SK (2018) A novel triple-mutant AAV6 capsid induces rapid and potent transgene expression in the muscle and respiratory tract of mice. Mol Ther Methods Clin Dev 9:323–329
9. van Lieshout LP, Soule G, Sorensen D, Frost KL, He S, Tierney K, Safronetz D, Booth SA, Kobinger GP, Qiu X, Wootton SK (2018) Intramuscular adeno-associated virus-mediated expression of monoclonal antibodies provides 100% protection against ebola virus infection in mice. J Infect Dis 217:916–925
10. Aurnhammer C, Haase M, Muether N, Hausl M, Rauschhuber C, Huber I, Nitschko H, Busch U, Sing A, Ehrhardt A, Baiker A (2012) Universal real-time PCR for the detection and quantification of adeno-associated virus serotype 2-derived inverted terminal repeat sequences. Hum Gene Ther Methods 23:18–28

Chapter 22

Simplified Purification of AAV and Delivery to the Pancreas by Intraductal Administration

Ping Guo, John Wiersch, Xiangwei Xiao, and George Gittes

Abstract

Genetic manipulation is a very powerful tool for studying diabetes, pancreatitis, and pancreatic cancer. Here we discuss the use of an adeno-associated virus (AAV) vector to modify gene expression, such as to introduce a green fluorescence protein (GFP) in wild-type mice, cre recombinase in loxP mice, or to inactivate a gene with shRNA. The use of viruses for genetic modification allows for time-specific genetic changes which have advantages over time-consuming and often complex cross-breeding strategies. Here we provide a detailed approach for this process from viral production and purification through pancreatic ductal infusion. Our protocol allows efficient delivery of AAV to mediate GFP or cre expression for cell lineage tracing in the mouse pancreas or for the delivery of transgenes under a specific promoter to these cells.

Key words Adeno-associated virus (AAV), PEG aqua partitioning, AAV viral purification, Pancreatic ductal infusion, Pancreatic cell lineage tracing

1 Introduction

Diabetes mellitus has become a major health problem worldwide [1, 2]. Similarly, pancreatic cancer is the fourth leading cause of cancer mortality [3]. As such, research on the pancreas is paramount to reduce the societal burden of these two conditions. Genetic manipulation of pancreatic cells is an essential tool in the researcher's armamentarium. Below, we discuss a method for temporally sensitive genetic modification of pancreatic cells utilizing AAVs.

Gene delivery using AAVs in animals is already well described [4, 5]. In brief, an AAV is a nonpathogenic gene delivery method that has been used in both bench and clinical research. Both in vitro and in vivo experiments require high purity preparations of AAV to achieve desired effects. Current methods for purification of AAVs entail centrifugation over a gradient (either CsCl or iodixanol) or the use of chromatography. These methods are cumbersome and

Michael J. Castle (ed.), *Adeno-Associated Virus Vectors: Design and Delivery*, Methods in Molecular Biology, vol. 1950, https://doi.org/10.1007/978-1-4939-9139-6_22,

expensive, requiring special equipment and technical skills not readily available in many labs. While these methods are time-consuming, they may also fail to yield a highly pure viral product. Currently, no commercial purification kits (without using ultrahigh-speed centrifugation) are available for AAV serotypes other than AAV2 or AAV-DJ. We developed and published a simplified method for highly efficient purification of AAVs and demonstrated the vector's effectiveness in adult and embryonic mice [6–13]. Our method does not require ultrahigh-speed gradient centrifugation nor chromatography. Instead, we purify the AAV vector using polyethylene glycol (PEG) and aqueous two-phase partitioning to remove soluble proteins from the PEG8000 precipitated virus-protein mixture. This procedure rapidly yields up to 95% recovery of highly purified AAV. Our entire purification process, including HEK-293 cell transfection, can be completed readily within one week.

We were able to achieve high pancreatic duct labeling efficiency using an AAV vector with a duct-specific sox9 promoter infused into the mouse common biliary/pancreatic duct [12, 14]. As cre/loxP is a site-specific recombinase system of proven efficacy, it is frequently used in mouse studies requiring time-dependent or lineage-specific genetic modification. However, this system only allows conditional activation or inactivation of genes in a cell type and/or temporal-specific manner in one lineage at a time. Although combination of the cre/loxP system with other site-specific recombinase systems such as flp/FRT or dre/rox could theoretically perform multiple genetic manipulations concurrently, the cost of time and breeding complexity present formidable challenges [15–17].

In this protocol, we present a technique based on previous publications [7, 10, 12, 14], which uses pancreatic intraductal infusion to deliver cell-tagging viruses carrying transgenes or shRNA under a specific promoter directly to pancreatic cells [12–14]. When this approach is used in transgenic mice (floxed mice), it allows for either double lineage tracing or lineage tracing combined with temporally activated genetic manipulation.

2 Materials

Prepare all solutions using ultrapure water (prepared by purifying deionized water to attain a sensitivity of 18 MΩ-cm at 25 °C) and analytical grade reagents. Prepare and store all reagents at room temperature unless otherwise indicated. Diligently follow all waste disposal regulations when disposing waste materials.

In this protocol, all animal procedures were performed with prior approval of the University of Pittsburgh Institutional Animal Care and Use Committee.

2.1 Cell Culture and Triple Plasmid PEI Transfection

1. DMEM cell culture medium.
2. Opti-MEM reduced serum medium.
3. Fetal bovine serum (FBS).
4. T-150 flasks.
5. Sterile cell culture hood.
6. Cell culture incubator.
7. Countertop cell culture centrifuge with cooling system.
8. 1 mg/mL polyethyleneimine (PEI), pH 7.4.
9. Vector plasmid carrying the expression cassette flanked by viral ITRs, such as pAAV-CMV-ZsGreen, pAAV-Sox9-GFP, pAAV-CMV-Cre, or pAAV-Sox9-Cre.
10. Packaging plasmid carrying the serotype 6 AAV rep and cap genes.
11. Helper plasmid carrying adenovirus helper proteins.

2.2 Virus Harvesting and Purification

1. DMEM cell culture medium.
2. Benzonase.
3. 10% deoxycholate
4. Lysis buffer: 50 mM Tris–HCl, 150 mM NaCl, 2 mM $MgCl_2$, pH 8.0.
5. HEPES solution: 150 mM NaCl, 20 mM EDTA, 50 mM HEPES, 1% Sodium lauroyl sarcosinate.
6. Chloroform.
7. PEG8000-NaCl solution: 40% polyethylene glycol (w/vol) in 2.5 M NaCl, average mol wt 8000 (*see* **Note 1**).
8. PEG8000 solution: 40% polyethylene glycol (w/vol) in H20, average mol wt 8000 (Sigma) (*see* **Note 1**).
9. 50% ammonium sulfate.
10. 50 mL conical tubes.
11. 1.5 and 2.0 mL microcentrifuge tubes.
12. Soft-walled 1.5 mL microcentrifuge tubes.
13. Plastic syringes.
14. 18G 25 mm needle.
15. 0.22 μM syringe filter.
16. Desktop centrifuge with cooling system.
17. 37 °C water bath with shaker.
18. Orbital rotator.
19. Centrifugal filter unit with Ultracel-50K membrane.
20. Slide-A-Lyzer Dialysis Cassette, 10,000 MWCO.
21. AAVpro titration kit (Takara).

2.3 Pancreatic Ductal AAV Infusion

1. Infusion catheter.
2. Infusion syringe pump.
3. Plastic syringe.
4. 30G needle.
5. 31G blunt-ended cannula.
6. 4″ × 4″ Gauze.
7. Nair hair removal cream.
8. Surgical tape.
9. 6–0 Silk suture with BV-1 needle.
10. 4–0 Vicryl suture with RB-1 needle.
11. 1.5 and 2.0 mL microcentrifuge tubes.
12. Cotton applicator Q-tips.
13. Dissecting microscope.
14. Dietrich micro bulldog clamp.
15. Olsen-Hegar combination scissor and needle holder.
16. Arruga micro forceps.
17. Hudson forceps.
18. Micro dissecting scissors.
19. Isoflurane.
20. Plexiglass heat pad.
21. Betadine.
22. 70% ethanol.
23. Ketoprofen.
24. Warm saline.
25. 8-Week-old C57BL/6J mice.
26. 8-Week-old ROSA-CAG-LSL-tdTomato reporter mice (Jackson Laboratory Strain 007909).

3 Methods

3.1 Cell Culture and Triple Plasmid PEI Transfection

The following procedures are for twenty T-150 flasks but can be scaled up to forty T-150 flasks or more.

1. Maintain HEK293 cells in DMEM with 10% FBS.
2. When cells reach 80% confluency, change the culture medium to serum-free DMEM (10 mL for one T-150 flask) in the morning before transfection.
3. For transfection, mix the vector plasmid, packaging plasmid, helper plasmid, and PEI in Opti-MEM medium (*see* **Note 2**).

Incubate 20 min at room temperature (RT). Aspirate medium from flasks. Add 10 mL of serum-free DMEM to each flask. Add 1 mL of transfection plasmids-PEI solution to each flask. Return the flasks to incubator.

4. The following day, add 1 mL of FBS to each flask. Return to incubator for 3 days.

3.2 Virus Harvesting and Purification

1. After 3 days, remove flasks from the incubator and, using a large cell scraper, remove cells from surface. Using a 10 mL pipette, wash down the flask to ensure all cells are removed. Collect cells with medium and transfer to a 50 mL conical tube. Cell suspensions from four flasks should be combined in a single 50 mL conical tube.
2. Approximately 250 mL of culture medium from twenty T-150 flasks will be collected (*see* **Note 3**). Centrifuge the culture medium with cells at 3000 × *g* for 10 min to pellet the cells and cellular debris.
3. Pour the supernatant into seven clean 50 mL conical tubes. Distribute the supernatant so that six tubes have 40 mL and one tube has about 10 mL.
4. Add 0.5 mL of lysis buffer to the cell pellets, then combine the pellets and buffer into one of the tubes. Rinse the other six tubes with another 0.5 mL of lysis buffer per tube, then transfer all buffer into the pellet collection tube.
5. Place the cell pellet suspension tube in a −80 °C freezer and store the culture medium supernatant tubes in a 4 °C refrigerator.
6. Once frozen, remove the cell pellet suspension from the freezer and thaw in a 37 °C water bath. Repeat for a total of three freeze/thaw cycles to release virus particles from the cells.
7. After the third thaw cycle, add 50 U/mL Benzonase and incubate at 37 °C for 30 min with shaking.
8. Add deoxycholate to a concentration of 1% and again incubate at 37 °C for 45 min with shaking (*see* **Note 4**).
9. Transfer the contents of the 50 mL cell suspension tube to multiple 2.0 mL microcentrifuge tubes. Spin in a microcentrifuge at 14,000 rpm (17,000 × *g*) for 10 min at RT.
10. Transfer the supernatant from each cell suspension tube into a single clean 50 mL tube and measure volume. Discard the pellets and microcentrifuge tubes.
11. Remove the 7 culture medium supernatant tubes from 4 °C (*see* **step 4**) and place on ice (*see* **Note 4**). Add an equal volume of the cell suspension supernatant (*see* **step 10**) to each culture medium supernatant tube (*see* **Note 5**).

12. Add 10 mL of PEG8000-NaCl solution to the tubes containing 40 mL of supernatant from **step 3**. Add 2.5 mL of PEG8000-NaCl solution to the tube containing only 10 mL of supernatant from **step 3**. Incubate the tubes at 4 °C on an orbital rotator for 45 min.
13. Remove tubes from the cold room and spin in a cell culture countertop centrifuge at 1500 × *g* for 5 min at 4 °C.
14. Aspirate supernatant in a dedicated tissue culture hood appropriate for virus handling. Add 0.5 mL of HEPES solution to the pellets. Mix thoroughly with a pipette to detach pellets, and consolidate all solution and pellets into a single 50 mL tube. Rinse each tube with another 0.5 mL of HEPES solution and transfer each rinse to the collection tube.
15. Vortex the tube for 1 min to break up pellets. Measure the volume with a pipette.
16. Add an equal volume of chloroform to the tube and vortex for 2 min.
17. Transfer the viral solution, which should appear like milk (Fig. 1a), to microcentrifuge tubes and spin in a microcentrifuge at 14,000 rpm (17,000 × *g*) for 10 min at RT.
18. After centrifugation, carefully transfer the supernatant (Fig. 1b) to a 50 mL tube and add the recommended volume of premixed 40% PEG8000 solution (prewarmed at room temperature) and 50% ammonium sulfate as shown in Table 1 (*see* **Note 6**).

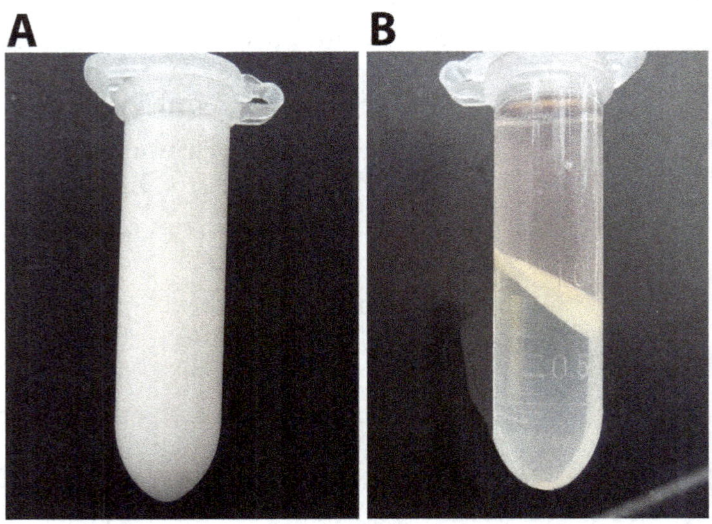

Fig. 1 Hydrophobic protein elimination. After chloroform treatment the solution becomes a milky white color (**a**). After centrifugation, the milky solution separates into an aqueous layer (top, light pink color) and a transparent bottom chloroform layer (**b**)

Table 1
Creation of a 10% PEG, 13.2% ammonium sulfate viral partitioning solution. Ratio of raw AAV solution, PEG8000 (10%), and ammonium sulfate (13.2%)

Vol. of raw virus (mL)	Vol. of 50% ammonium sulfate (mL)	Vol. of 40% PEG8000 solution (mL)	Total vol. (mL)	Final % PEG	Final % ammonium sulfate
1	0.543423913	0.515	2.058423913	10.00766	13.2
1.25	0.679279891	0.64375	2.573029891	10.00766	13.2
1.5	0.81513587	0.7725	3.08763587	10.00766	13.2
1.75	0.950991848	0.90125	3.602241848	10.00766	13.2
2	1.086847826	1.03	4.116847826	10.00766	13.2
2.25	1.222703804	1.15875	4.631453804	10.00766	13.2
2.5	1.358559783	1.2875	5.146059783	10.00766	13.2
2.75	1.494415761	1.41625	5.660665761	10.00766	13.2
3	1.630271739	1.545	6.175271739	10.00766	13.2
3.25	1.766127717	1.67375	6.689877717	10.00766	13.2
3.5	1.901983696	1.8025	7.204483696	10.00766	13.2
3.75	2.037839674	1.93125	7.719089674	10.00766	13.2
4	2.173695652	2.06	8.233695652	10.00766	13.2
4.25	2.30955163	2.18875	8.74830163	10.00766	13.2
4.5	2.445407609	2.3175	9.262907609	10.00766	13.2
4.75	2.581263587	2.44625	9.777513587	10.00766	13.2
5	2.717119565	2.575	10.29211957	10.00766	13.2
5.25	2.852975543	2.70375	10.80672554	10.00766	13.2
5.5	2.988831522	2.8325	11.32133152	10.00766	13.2
5.75	3.1246875	2.96125	11.8359375	10.00766	13.2
6	3.260543478	3.09	12.35054348	10.00766	13.2
6.25	3.396399457	3.21875	12.86514946	10.00766	13.2
6.5	3.532255435	3.3475	13.37975543	10.00766	13.2
6.75	3.668111413	3.47625	13.89436141	10.00766	13.2
7	3.803967391	3.605	14.40896739	10.00766	13.2
7.25	3.93982337	3.73375	14.92357337	10.00766	13.2
7.5	4.075679348	3.8625	15.43817935	10.00766	13.2
7.75	4.211535326	3.99125	15.95278533	10.00766	13.2
8	4.347391304	4.12	16.4673913	10.00766	13.2

(continued)

Table 1 (continued)

Vol. of raw virus (mL)	Vol. of 50% ammonium sulfate (mL)	Vol. of 40% PEG8000 solution (mL)	Total vol. (mL)	Final % PEG	Final % ammonium sulfate
8.25	4.483247283	4.24875	16.98199728	10.00766	13.2
8.5	4.619103261	4.3775	17.49660326	10.00766	13.2
8.75	4.754959239	4.50625	18.01120924	10.00766	13.2
9	4.890815217	4.635	18.52581522	10.00766	13.2
9.25	5.026671196	4.76375	19.0404212	10.00766	13.2
9.5	5.162527174	4.8925	19.55502717	10.00766	13.2
9.75	5.298383152	5.02125	20.06963315	10.00766	13.2
10	5.43423913	5.15	20.58423913	10.00766	13.2

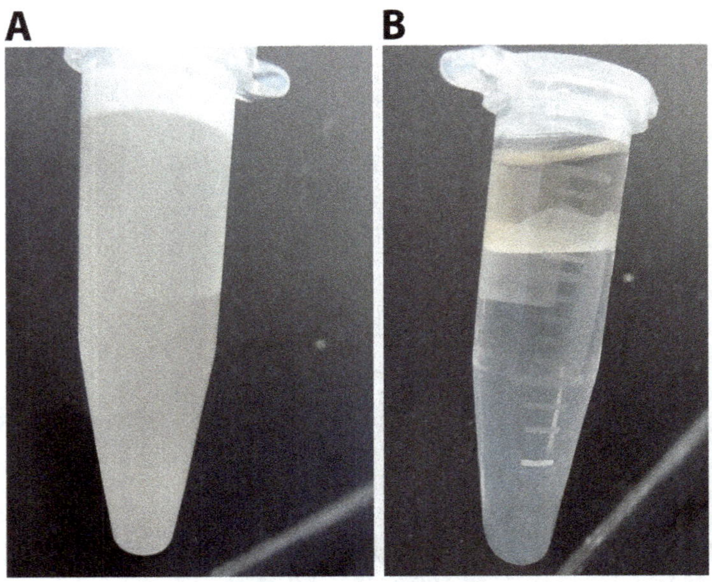

Fig. 2 Nonviral component elimination and partitioning. Before partitioning, the aqueous solution contains raw AAV virus and the PEG8000-ammonium sulfate mixture (**a**). After partitioning, the bottom aqueous solution contains purified AAV virus and the top layer contains condensed PEG8000. Proteins form a thin barrier layer between the solutions and also precipitate at the bottom (**b**)

19. Vortex vigorously for 2 min, then transfer the mixture (Fig. 2a) equally to soft-walled 1.5 mL microcentrifuge tubes and spin in a microcentrifuge at 14,000 rpm (17,000 × *g*) for 10 min at RT.

20. Withdraw the clear bottom aqua phase (Fig. 2b) containing the virus with an 18G needle and syringe, and transfer to a 50 mL conical tube (*see* **Note 7**).
21. Transfer the aqueous phase product to a Slide-a-Lyzer 10,000 MWCO dialysis cassette. Dialyze the aqueous phase product against 1× PBS at RT to remove salt for 4 h. Change PBS buffer and dialyze for another 4 h.
22. Transfer the dialysis cassette to DMEM and place on a rotator for 2 h (*see* **Note 8**).
23. To concentrate the virus, transfer to a centrifugal filter unit with Ultracel-50K membrane. Spin in a cell culture countertop centrifuge at 2500 × *g* for 15 min at RT. Check that product has been concentrated to between 1 and 1.5 mL. If the volume is higher, centrifuge again.
24. When the product volume is below 1.5 mL, move the tube to the cell culture hood. Using a plastic syringe and 0.22 μM syringe filter, pass the product through the filter into a properly labeled 1.5 mL microcentrifuge tube.
25. Titration of AAV can be performed using an AAVpro qPCR titration kit and following the manufacturer's instructions (*see also* Chapter 4). If the virus has a GFP tag, different dilutions of virus can be applied to HEK293 cells to determine transduction efficiency.

3.3 Pancreatic Ductal AAV Infusion

1. Prepare approximately 100 μl per mouse of ~1×10^{12} vector genomes (vg) of AAV in DMEM with phenol red.
2. Set up infusion equipment as shown in Fig. 3. First, load the desired volume of virus into the syringe mounted to the infusion catheter and secure it firmly into the infusion syringe pump. The proper infusion rate should be programmed and then the infusion setup should be tested to ensure that it is free of obstruction.
3. Anesthetize mice with 2–2.5% (w/vol) isoflurane by inhalation.
4. Place a mouse on a 37 °C plexiglass heat pad that has been cleaned three times with alternate changes of Betadine and 70% ethanol. Immobilize the mouse on the plexiglass heat pad using surgical tape, and use Nair to remove hair from the lower chest and abdomen. Clean and disinfect the now hairless area with gauze soaked in 70% ethanol.
5. Laparotomy: Use scissors to incise the skin at the midline of the abdomen. Next, cut the abdominal muscle along the linea alba to create a midline upper abdominal laparotomy incision of 1.5 cm.
6. Exposure of the pancreatic duct using Q-tips to gently bring the stomach into the incision. Rotate and stretch the

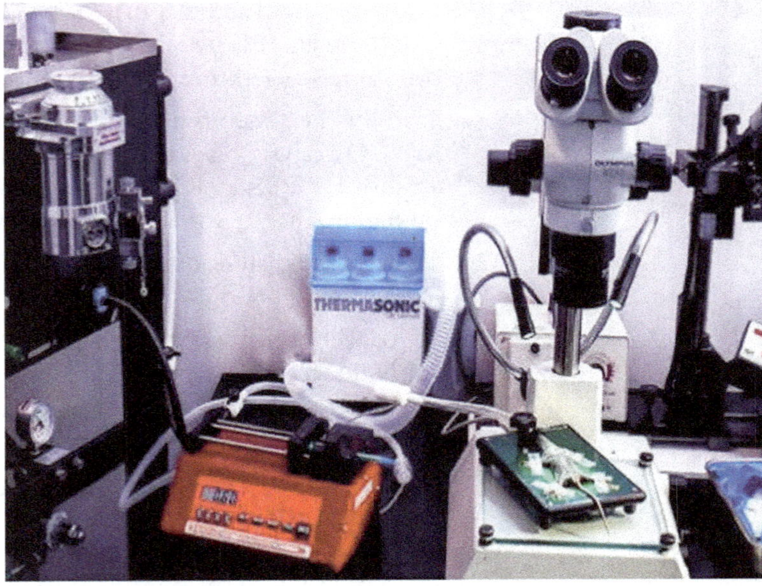

Fig. 3 Organization of the equipment for mouse pancreatic ductal infusion

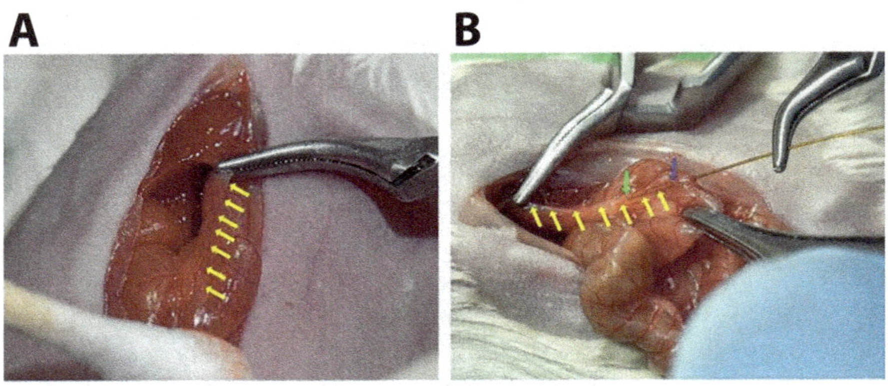

Fig. 4 Setup for pancreatic ductal infusion. A micro bulldog clamp is placed on the bile duct near the liver to prevent perfusion of the liver (**a**). A 30G needle is used to make a small duodenal hole opposite to the sphincter of Oddi. Next, a 31G blunt-ended catheter is carefully placed into the biliary-pancreatic duct through the sphincter of Oddi in the duodenum (**b**)

duodenum to expose the biliary-pancreatic duct and its junction within the duodenum (the sphincter of Oddi), which appears pale (Fig. 4a).

7. Perform a slight backward rotation of the duodenum to expose the distal biliary-pancreatic duct. With a view of the back side of the duodenum and the duodenal portion of the pancreas, the region of the papilla should also be visible (Fig. 4b).
8. To prevent perfusion of the liver, place a micro bulldog clamp on the common bile duct above the branching of the pancreatic

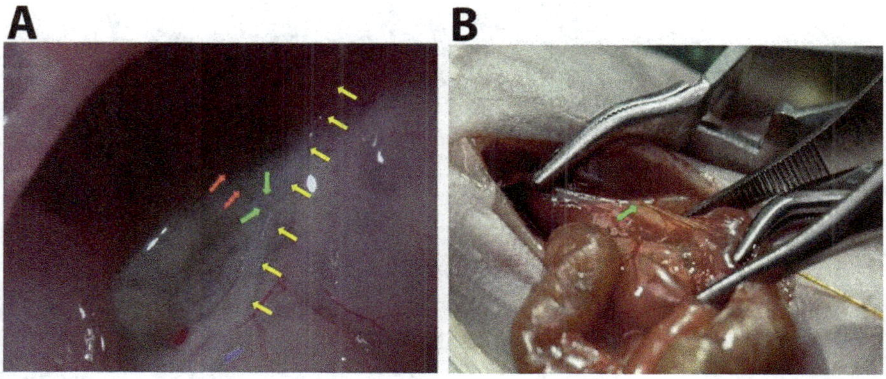

Fig. 5 Detailed positioning of pancreatic ductal infusion. The tip of the catheter should positioned at the origin of the pancreatic branch (red arrows) of the biliary-pancreatic duct (yellow arrows) (**a**). The catheter (tip—green arrows) can then be secured with the second micro bulldog clamp at the sphincter of Oddi (blue arrow) to prevent backflow into the duodenum (**b**)

Table 2
Recommended infusion parameters

AAV vector	Target	AAV concentration (vg/mL)	Vol. for 15–20 g mouse (μL)	Vol. for 20–30 g mouse (μL)	Vol. for 30–45 g mouse (μL)	Vol. for 45–60 g mouse (μL)	Infusion rate (μL/min)
AAV6-Sox9P-X[a]	Duct cells	10^{12}	25	30	35	35	1
AAV8-CMV-X	All pancreatic cells	10^{12}	120	150	165	180	6
AAV8-RIP-X	Beta cells	10^{12}	120	150	165	180	6

[a]Gene of interest

duct. Avoid clamping the portal vein, located behind the common bile duct, while performing this step (Fig. 4b).

9. Use a 30G needle to make a small hole in the duodenum opposite to the sphincter of Oddi, 1–2 mm away from the ampulla of Vater. Next, pass a 31G blunt-ended cannula through this small hole in the duodenum and then the sphincter of Oddi (Fig. 5a). The tip of the catheter should be positioned at the origin of the pancreatic duct branch in the biliary-pancreatic duct (Fig. 5a). If the anatomy is unclear, the tip of the catheter can be placed just past the sphincter of Oddi.
10. Clamp the ampulla around the catheter with another micro bulldog clamp to prevent backflow or leakage into the duodenum during the viral infusion (Fig. 5b). Connect the back end of the catheter to the infusion syringe pump (*see* **Note 9**).
11. Turn on the pump and infuse the predetermined amount of solution at a flow rate determined as shown in Table 2.

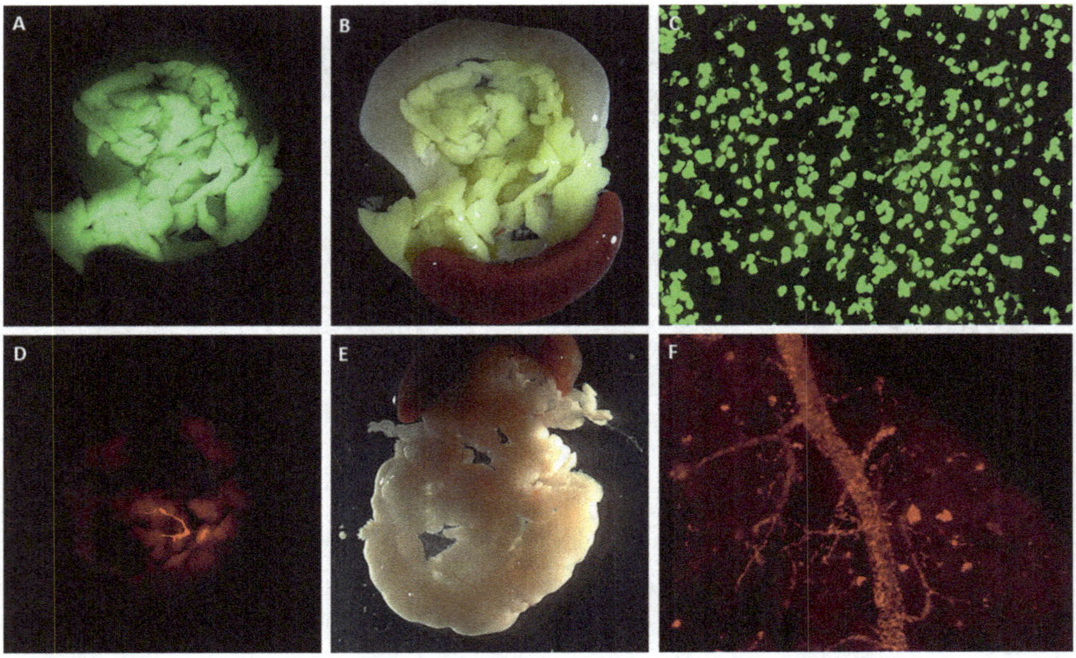

Fig. 6 Visualized mouse pancreatic tissues. AAV6-CMV-GFP infused into a C57BL/6 mouse pancreas was visualized by whole mount green fluorescence (**a**), whole mount bright field (**b**), or tissue section green fluorescence (**c**). AAV6-sox9-Cre infused into a ROSA26-LSL-tdTomato mouse pancreas was visualized by whole mount red fluorescence (**d**), whole mount bright field (**e**), or tissue section red fluorescence (**f**)

12. After the infusion is complete, remove the micro bulldog clamp on the sphincter of Oddi and gently withdraw the catheter from the duct and out of the duodenum. Next, remove the micro bulldog clamp from the biliary-pancreatic duct.
13. Close the hole in the duodenum created by the needle with a 6–0 silk suture.
14. Return the intestines to the abdomen and suture the abdominal wall closed with a 4–0 Vicryl suture using a running stitch.
15. Return the mouse to its cage on a heated pad until it fully recovers. Give ketoprofen at a dose of 5 mg/kg subcutaneously once per day for 3 days after surgery for analgesia.
16. You should see demonstrable results from the procedure approximately one week after viral infusion. Euthanize the animal and harvest the pancreas at the desired experimental endpoint. If using a virus with fluorescent proteins, fluorescence can be visualized under a stereo dissecting microscope (Fig. 6) or with whole mount staining (Fig. 6c, f) (*see* **Notes 10–14**).

4 Notes

1. There are two 40% PEG8000 solutions used in this protocol: "PEG8000-NaCl solution" prepared in-house in large quantities and used for raw virus precipitation (**step 12** in Subheading 3.2), and "PEG8000 solution" purchased from Sigma-Aldrich and used for viral partitioning (**step 18** in Subheading 3.2) (Table 1).
2. The PEI solution must be sterilized using a 0.22 μM filter before transfecting cells. Mix the vector plasmid, helper plasmid, and packaging plasmid in a weight ratio of 1.2:1:1 (12, 10, and 10 μg per T-150 flask), add PEI (120 μl per T-150 flask), mix by vortexing, incubate for at least 15 min at room temperature, and then add dropwise to the flasks. If there is additional DNA-PEI left over, add it to each flask equally.
3. The viral particles are harvested from the cells and media of the T-150 transfection flasks.
4. Deoxycholate stock concentration is 10%. It should be removed from the refrigerator and incubated in a 37 °C water bath during **step 7** (Subheading 3.2) so that it will go into solution more easily.
5. There are six tubes containing 40 mL and one tube containing 10 mL from **step 8**.
6. This will give a final in-tube concentration of 8% supernatant, 10% PEG8000, and 13.2% ammonium sulfate.
7. Use safety precautions to avoid needlestick injury when piercing the tube.
8. This step dialyzes against medium containing phenol red to visualize the viral solution.
9. It is CRITICAL to avoid formation of air bubbles inside the syringe and catheter tubing to achieve satisfactory infusion results.
10. Inhalation should be closely adjusted according to the breathing and heart rate of the mouse during surgery and per local animal anesthesia protocols. A typical infusion takes 30–40 min to complete. Unnecessary delays during the surgery should be avoided.
11. Avoid injury to underlying organs when performing the laparotomy by lifting the abdominal wall away from the intestines while making the incision.
12. The stomach, duodenum, and pancreas should be handled mostly with Q-tips to prevent tissue damage. Avoid handling these tissues with steel instruments to reduce tissue trauma, edema, and bleeding. Careless tissue handling can cause

intraoperative and postoperative bleeding and pancreatitis and prolongs surgical recovery of the mouse. Pancreatitis is very infrequent if the procedure is performed properly. Aside from cotton applicators, gentle use of atraumatic surgical instruments, such as ring forceps, can also be used to manipulate the duodenum with minimal tissue trauma. Care must be taken to avoid overstretching the duodenum which may cause duodenal injury and can distort the biliary-pancreatic duct. Intestines positioned outside the abdominal cavity during the infusion should be covered with gauze moistened with warm saline to avoid substantial heat loss and drying of the tissues.

13. Proper surgical exposure is crucial as a micro bulldog clamp needs to be precisely placed on the common bile duct above the branch point of the pancreatic duct to prevent infusion into the liver (Fig. 4b).
14. Cannulation is the most important and technically difficult part of the protocol. This requires stable hands, careful manipulation, and thorough preoperative planning. The blunt-ended cannula reduces the risk of damage to the duct.

Acknowledgments

We thank Lulin Guo and Aiping Lu for assistance with imaging.

References

1. Weir GC, Bonner-Weir S (2013) Islet beta cell mass in diabetes and how it relates to function, birth, and death. Ann N Y Acad Sci 1281:92–105. https://doi.org/10.1111/nyas.12031
2. Forouhi NG, Wareham NJ (2014) Epidemiology of diabetes. Medicine (Abingdon) 42 (12):698–702. https://doi.org/10.1016/j.mpmed.2014.09.007
3. Werner J, Combs SE, Springfeld C, Hartwig W, Hackert T, Buchler MW (2013) Advanced-stage pancreatic cancer: therapy options. Nat Rev. Clin Oncol 10(6):323–333. https://doi.org/10.1038/nrclinonc.2013.66
4. Balakrishnan B, Jayandharan GR (2014) Basic biology of adeno-associated virus (AAV) vectors used in gene therapy. Curr Gene Ther 14 (2):86–100
5. Zhou SZ, Broxmeyer HE, Cooper S, Harrington MA, Srivastava A (1993) Adeno-associated virus 2-mediated gene transfer in murine hematopoietic progenitor cells. Exp Hematol 21(7):928–933
6. Guo P, El-Gohary Y, Prasadan K, Shiota C, Xiao X, Wiersch J, Paredes J, Tulachan S, Gittes GK (2012) Rapid and simplified purification of recombinant adeno-associated virus. J Virol Methods 183(2):139–146. https://doi.org/10.1016/j.jviromet.2012.04.004
7. Guo P, Xiao X, El-Gohary Y, Paredes J, Prasadan K, Shiota C, Wiersch J, Welsh C, Gittes GK (2013) A simplified purification method for AAV variant by polyethylene glycol aqueous two-phase partitioning. Bioengineered 4(2):103–106. https://doi.org/10.4161/bioe.22293
8. Orabi AI, Sah S, Javed TA, Lemon KL, Good ML, Guo P, Xiao X, Prasadan K, Gittes GK, Jin S, Husain SZ (2015) Dynamic imaging of pancreatic nuclear factor kappaB (NF-kappaB) activation in live mice using adeno-associated virus (AAV) infusion and bioluminescence. J Biol Chem 290(18):11309–11320. https://doi.org/10.1074/jbc.M115.647933
9. Orabi AI, Wen L, Javed TA, Le T, Guo P, Sanker S, Ricks D, Boggs K, Eisses JF, Castro C, Xiao X, Prasadan K, Esni F, Gittes

GK, Husain SZ (2017) Targeted inhibition of pancreatic acinar cell calcineurin is a novel strategy to prevent post-ERCP pancreatitis. Cell Mol Gastroenterol Hepatol 3(1):119–128. https://doi.org/10.1016/j.jcmgh.2016.08.006

10. Xiao X, Chen C, Guo P, Zhang T, Fischbach S, Fusco J, Shiota C, Prasadan K, Dong H, Gittes GK (2017) Forkhead box protein 1 (FoxO1) inhibits accelerated beta cell aging in pancreas-specific SMAD7 mutant mice. J Biol Chem 292(8):3456–3465. https://doi.org/10.1074/jbc.M116.770032
11. Xiao X, Guo P, Chen Z, El-Gohary Y, Wiersch J, Gaffar I, Prasadan K, Shiota C, Gittes GK (2013) Hypoglycemia reduces vascular endothelial growth factor A production by pancreatic beta cells as a regulator of beta cell mass. J Biol Chem 288(12):8636–8646. https://doi.org/10.1074/jbc.M112.422949
12. Xiao X, Guo P, Prasadan K, Shiota C, Peirish L, Fischbach S, Song Z, Gaffar I, Wiersch J, El-Gohary Y, Husain SZ, Gittes GK (2014) Pancreatic cell tracing, lineage tagging and targeted genetic manipulations in multiple cell types using pancreatic ductal infusion of adeno-associated viral vectors and/or cell-tagging dyes. Nat Protoc 9(12):2719–2724. https://doi.org/10.1038/nprot.2014.183
13. Xiao X, Guo P, Shiota C, Zhang T, Coudriet GM, Fischbach S, Prasadan K, Fusco J, Ramachandran S, Witkowski P, Piganelli JD, Gittes GK (2018) Endogenous reprogramming of alpha cells into beta cells, induced by viral gene therapy, reverses autoimmune diabetes. Cell Stem Cell 22(1):78–90. e74. https://doi.org/10.1016/j.stem.2017.11.020
14. Guo P, Xiao X, El-Gohary Y, Criscimanna A, Prasadan K, Rymer C, Shiota C, Wiersch J, Gaffar I, Esni F, Gittes GK (2013) Specific transduction and labeling of pancreatic ducts by targeted recombinant viral infusion into mouse pancreatic ducts. Lab Investig 93(11):1241–1253. https://doi.org/10.1038/labinvest.2013.113
15. Magnuson MA, Osipovich AB (2013) Pancreas-specific Cre driver lines and considerations for their prudent use. Cell Metab 18(1):9–20. https://doi.org/10.1016/j.cmet.2013.06.011
16. Xiao X, Chen Z, Shiota C, Prasadan K, Guo P, El-Gohary Y, Paredes J, Welsh C, Wiersch J, Gittes GK (2013) No evidence for beta cell neogenesis in murine adult pancreas. J Clin Invest 123(5):2207–2217. https://doi.org/10.1172/JCI66323
17. Teta M, Rankin MM, Long SY, Stein GM, Kushner JA (2007) Growth and regeneration of adult beta cells does not involve specialized progenitors. Dev Cell 12(5):817–826. https://doi.org/10.1016/j.devcel.2007.04.011

Chapter 23

rAAV-Mediated Gene Delivery to Adipose Tissue

Wei Huang, Nicholas J. Queen, and Lei Cao

Abstract

Recombinant adeno-associated virus (rAAV) vectors are attractive vehicles for gene therapy. Yet, it is challenging to genetically manipulate adipose tissue in adults due to the low transduction efficiency of naturally occurring AAV serotypes. We recently demonstrated that a novel engineered hybrid serotype Rec2 achieves high transduction of adipose tissue that is superior to naturally occurring serotypes via direct injection to adipose depots. Furthermore, the administration route influences the tropism and efficacy of Rec2 vector: oral administration transduces interscapular brown fat, while intraperitoneal injection preferentially targets visceral fat. Multiple in vivo studies by our lab and others have demonstrated that Rec2 vector provides a powerful tool to genetically manipulate adipose tissue for basic research and potential gene therapies of genetic and acquired diseases. Here we provide detailed protocols for AAV production and delivery to adipose tissue by direct injection, oral administration, and intraperitoneal injection.

Key words rAAV serotype Rec2, Oral administration, Intraperitoneal administration, Direct injection, Visceral fat, Subcutaneous fat, Brown fat

1 Introduction

Adipose tissue is a multifunctional organ that modulates whole body metabolic homeostasis. Approaches to genetically manipulate adipose tissue are essential for mechanistic studies of adipose function. To date, genetic manipulation of adipose tissue has relied primarily on transgenic mouse models. However, transgenic technology can be costly and time consuming, and in transgenic adults it can be difficult to distinguish developmental effects from maintenance effects. In this regard, rAAV vectors are an attractive alternative because they can transduce both dividing and postmitotic tissues with low immunogenicity and long-lasting transgene expression [1]. rAAV has been successfully used to deliver transgenes to liver, heart, skeletal muscle, brain, and eye [1]. Yet the use of rAAV-mediated gene delivery to adipose tissue has been limited due to the low transduction efficiency of naturally occurring AAV serotypes [2–4].

Michael J. Castle (ed.), *Adeno-Associated Virus Vectors: Design and Delivery*, Methods in Molecular Biology, vol. 1950, https://doi.org/10.1007/978-1-4939-9139-6_23,

We have recently developed a novel rAAV vector to achieve highly efficient and specific transduction of adipose tissue. The engineered hybrid serotype Rec2 vector can transduce white adipose tissue and brown adipose tissue far more efficiently than the six natural and engineered serotypes to which it has been compared [5]. Interestingly, the tropism and efficacy of the Rec2 vector are substantially influenced by the administration route [6]. For example, oral administration leads to preferential transduction of brown fat, with absence of transduction in the gastrointestinal track at doses lower than 2×10^{10} viral genomes (vg) per mouse [7]. On the other hand, intraperitoneal injection of Rec2 vector efficiently transduces all visceral white fat depots. One limitation is the transduction of liver when Rec2 vector is injected intravenously [6] or administered to fat at a high dose. To prevent transgene expression in the liver, we generated a new vector that contains two expression cassettes: one uses the chicken β-actin (CBA) promoter to drive the transgene, and the other uses the liver-specific albumin promoter to drive a specific microRNA targeting the woodchuck posttranscriptional regulatory element (WPRE) sequence, which exists only in the transgene expression cassette. This dual-cassette vector achieves highly selective transduction of visceral fat while severely restricting off-target transduction of liver. As proof of concept, intraperitoneal administration of an adipose-targeting Rec2 vector harboring the leptin gene corrects leptin deficiency, obesity, and metabolic syndromes of *ob/ob* mice at a low dose of 2×10^{10} vg per mouse [8].

These results are exciting for several reasons. Firstly, intraperitoneal injection of Rec2 vector can achieve high transduction in all visceral fat depots, which are less accessible than subcutaneous depots. Secondly, the dose is equivalent to that previously used in direct fat pad injection and is at least 1–2 orders of magnitude lower than typical doses used for systemic delivery of AAV vectors. Thirdly, liver transgene expression is minimized. Our collaborators in the United States and France have examined whether other cell types in adipose tissue could be transduced by Rec2 vector using FACS or reporter mice. Rec2 vector does not appear to transduce adipose progenitor cells or adipose macrophages (personal communications). Thus, the combination of Rec2 serotype and the dual-cassette expression plasmid can achieve adipocyte-specific transduction by direct intra-adipose injection or by systemic administration, such as intraperitoneal injection. We have provided Rec2 vectors to several labs, facilitating the *in vivo* study of adipose tissue functions [9–12].

2 Materials

2.1 rAAV Production (see Note 1)

1. AAV production plasmids: Adeno helper plasmid, cis plasmid, trans plasmid.
2. Culture dishes (150 mm × 25 mm, Polystyrene, nonpyrogenic).
3. Culture medium: DMEM medium with 4.5 g/L glucose supplemented with 1 mM sodium pyruvate, 10% fetal bovine serum (FBS), and 10,000 U penicillin/10 mg streptomycin per 100 mL.
4. 2× HEPES Buffered Saline (HBS Buffer): 50 mM HEPES, 0.28 M NaCl, 1.5 mM $Na_2HPO_4{\cdot}7H_2O$. Bring to 500 mL with sterile water, adjust pH to 7.05, then pass through a 0.2 μM filter and aliquot 50 mL for storage at −20 °C (*see* **Note 2**).
5. 2.5 M $CaCl_2$: bring to 100 mL with sterile water, then pass through 0.2 μM filter and store at −20 °C (*see* **Note 2**).
6. Lysis buffer: 20 mM NaCl, 50 mM Tris–HCl, 1 M $MgCl_2$, pH 8.5, 0.2 μM filtered.
7. Viral dilution buffer: 2× PBS with 1 mM $MgCl_2$ and 0.001% Pluronic F-68.
8. PBS (cell culture grade).
9. QIAGEN EndoFree plasmid kits for preparation of all plasmids.
10. 0.2 μM bottle top filters.
11. 0.22 μM syringe filter units.
12. Amicon 100 K NMWL centrifugal filter unit (Millipore).
13. Ultrafree-MC centrifugal filter until, sterile (Millipore).
14. Conical tubes.
15. Screw-cap sample tubes.
16. Benzonase.
17. Vortex mixer.
18. Cell scrapers (39 cm handle, 3 cm blade).
19. Dry ice.
20. Methanol.
21. 37 °C water bath.
22. 10× PBS-MK: 10× PBS with 10 mM $MgCl_2$ and 25 mM KCl (*see* **Note 3**).
23. 1× PBS-MK with 2 M NaCl (*see* **Note 3**).
24. 60% iodixanol: OptiPrep Density Gradient Medium (Sigma).

25. Syringes and 20G syringe needles.
26. Soldering iron with ISOTIP.
27. Clamp stand.
28. Optima L-90K ultracentrifuge (Beckman Coulter).
29. 70Ti rotor (Beckman Coulter).
30. Allegra X-15R Centrifuge (Beckman Coulter).
31. 25 × 89 mm Polyallomer Quick-Seal Centrifuge Tubes (Beckman Coulter).
32. 96 well qPCR plates.
33. qPCR film.
34. qPCR machine.
35. 12.5 mM EDTA.
36. DNase I (amplification grade) with 10× digestion buffer.
37. 20 mg/mL Proteinase K solution with 10× digestion buffer.
38. TE buffer (pH 8.0).
39. Thermocycler, such as StepOne Plus Real Time PCR system (Applied Biosystems).
40. SYBR Green.
41. Biological safety level II cell culture hood.

2.2 Direct Injection

1. Surgical forceps.
2. Surgical scissors.
3. Surgical pad for a sterile plane.
4. Cotton swabs.
5. Ketamine.
6. Xylazine.
7. Sterile saline.
8. Isoflurane.
9. Betadine (Povidone-iodine).
10. Viral dilution buffer: 2× PBS, 1 mM $MgCl_2$, 0.001% Pluronic F-68.
11. 70% ethanol.
12. Non-rebreathing circuit with nose cone.
13. Isoflurane vaporizer.
14. Oxygen.
15. Small animal surgical clippers.
16. Micropipette and micropipette tips.
17. 1.7 mL microcentrifuge tubes.

18. 3/10 mL, 8 mm, 31G syringes.
19. Absorbable 4-0 PDS II FS-2 suture.
20. Absorbable 6-0 polypropylene P-13 suture.
21. Hot bead sterilizer.
22. Buprenorphine and/or ibuprofen.
23. Fast Green FCF or equivalent dye.

2.3 Oral Gavage

1. 22G × 1.5″ metal animal feeding needles, such as Cadence Science #7920 or #9921.
2. 1 mL syringe with Luer-Lok tip.
3. 1.7 mL microcentrifuge tube.
4. Micropipette and micropipette tips.
5. Viral dilution buffer: 2× PBS, 1 mM $MgCl_2$, 0.001% Pluronic F-68.

2.4 Intraperitoneal Injection

1. 1/2 mL, 12.7 mm, 28G syringe.
2. 1.7 mL microcentrifuge tube.
3. Micropipette and micropipette tips.
4. Viral dilution buffer: 2× PBS, 1 mM $MgCl_2$, 0.001% Pluronic F-68.

3 Methods

3.1 rAAV Production

3.1.1 Transfection and Cell Lysis

1. Passage cells at 80–90% confluence and split one plate into five plates (*see* **Note 4**). Change culture medium before transfection.
2. Set up transfection "master mix" using the following template in the order listed:
   - Sterile water: for final volume of 1 mL per plate.
   - Adeno Helper plasmid: 20 μg per plate.
   - AAV trans plasmid: 15 μg per plate.
   - AAV cis plasmid: 10 μg per plate.
   - 2.5 M $CaCl_2$: 0.12 mL per plate.
3. Add $CaCl_2$ dropwise and mix well. Add an equal volume of 2× HBS buffer to the DNA/$CaCl_2$ mix. Add 2× HBS dropwise while gently agitating by vortex mixer. Allow precipitate to sit at room temperature for 20 min while a fine, even, whitish mixture appears. Distribute 2 mL evenly over each plate and gently mix the precipitate by moving the plate side to side. Incubate cells for 12–16 h. Replace the transfection mix with culture medium and continue culturing for at least 48–72 h.

Table 1
Preparation of iodixanol solutions

Solution name	54% Iodixanol (9:1 dilution of 60% stock in 10× PBS-MK) (mL)	1× PBS-MK with 2 M NaCl (mL)	1× PBS-MK (no NaCl) (mL)	Final volume (mL)
15% iodixanol	24	43.2	19.2	86.4
25% iodixanol	24	–	27.84	51.84
40% iodixanol	32	–	11.2	43.2

4. Remove cells from plates using scrapers (39 cm handle, 3 cm blade) without changing medium. Collect cells into conical tubes and spin at 1000 × *g* for 10 min at 4 °C. Discard supernatant and rinse cell pellets with PBS (cell culture grade). Resuspend cells with 2 mL per plate lysis buffer. Continue to next step or store at −20 °C.
5. Perform freeze/thaw cell lysis by alternating three times between dry ice in methanol and a 37 °C water bath.
6. Add benzonase to a final concentration of 50 units/mL, incubate the lysates at 37 °C for 60 min, then spin at 4000 × *g* for 30 min at 4 °C. Collect supernatant and store at 4 °C until the next step.

3.1.2 Iodixanol Purification and Concentration

1. Prepare 15%, 25%, 40%, and 54% iodixanol solutions as shown in Table 1.
2. Transfer 15 mL of cell lysate into a 25 × 89 mm quick-seal tube using a syringe with metal needle.
3. Underlay the lysate with iodixanol solutions in the following order: 9 mL of 15%, 6 mL of 25%, 5 mL of 40%, and 5 mL of 54%. Fill the tube to the top with PBS. Five distinct layers will be present (Fig. 1).
4. Seal the tube with ISOTIP of Soldering Iron. Repeat **steps 1–4** until all lysate has been used.
5. Carefully place the quick-seal tubes into a Beckman 70Ti rotor and ultra-centrifugate using the following parameters: slow acceleration and brake, 350,000 × *g* (69,000 rpm for Beckman Coulter Optima L-90K ultracentrifuge), 60 min at 18 °C.
6. Remove a tube and secure it on a clamp stand.
7. Insert a 20G needle into the top of the tube to release the air in the tube (Fig. 1).
8. Insert a 20G needle attached to a 5 mL syringe into the tube at the bottom edge of the 40% iodixanol layer (Fig. 1) and

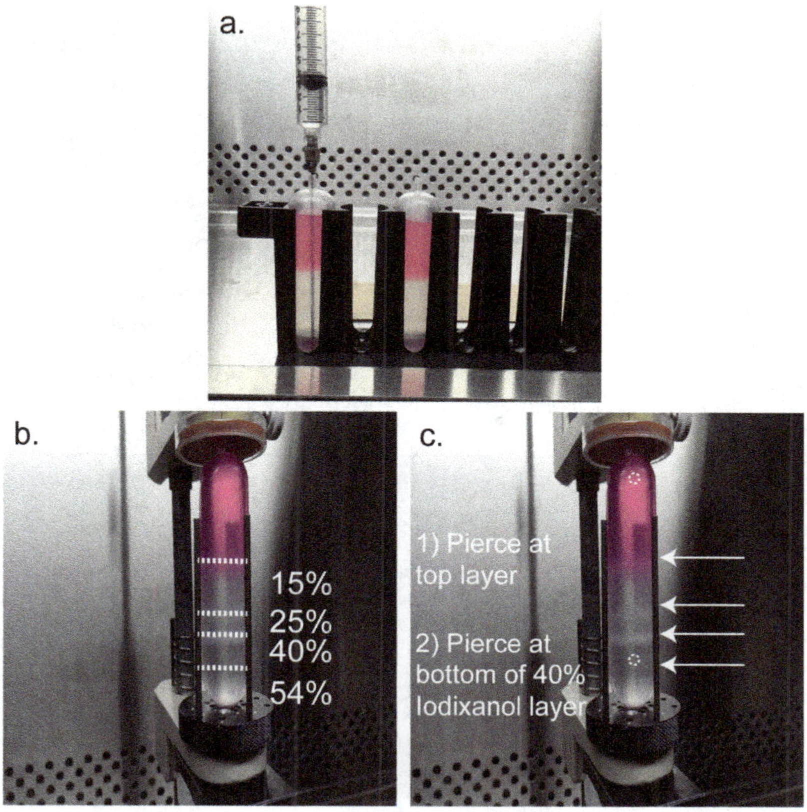

Fig. 1 rAAV Production. (**a**) Underlay the lysates with iodixanol solution to create five total layers (four iodixanol layers and one virus solution layer). (**b**) Iodixanol layers and their interfaces. (**c**) Arrows indicate layer interfaces. Pierce at the top layer with a 20G needle to relieve pressure. Then, pierce at the bottom of the 40% iodixanol layer just above the interface to withdraw virus

withdraw 3–4 mL of the solution from that layer, then inject the solution into an Amicon 100K NMWL centrifugal filter.

9. Add the viral dilution up to the top line of the Amicon filter. Spin at 4000 × *g* for 20–30 min at 18 °C. Remove pass-through buffer. Repeat **step 9** for three total spins.
10. Transfer the sample into a sterile small-volume Ultrafree-MC centrifugal filter unit and spin at 4000 × *g* for 20–30 min at 18 °C. Transfer AAV sample to a screw-cap sample tube and label it with serial number and date. Store the rAAV at 4 °C for up to 6 month or at −20 °C for long-term storage (*see* **Note 5**).

3.1.3 AAV Titration

1. Select a proper sequence in the AAV genome (cis plasmid) for qPCR amplification and create forward and reverse qPCR primers specific for this sequence. For example, we use primers to amplify WPRE, which is present in our AAV vectors.
2. Prepare DNase I reactions (1 unit/μL) in a 96-well plate with each AAV sample in triplicate as follows:

H_2O	89.5 μL
10× digestion buffer	10 μL
DNase	0.5 μL
AAV	2 μL

Place in a thermocycler at 25 °C for 15 min, then cool to 4 °C.

3. Add 2 μL of 12.5 mM EDTA to each well. Place in a thermocycler at 70 °C for 10 min, then cool to 4 °C.
4. Perform Proteinase K treatment (20 mg/mL). To each well on the same plate as above, add:

H_2O	89.5 μL
10× digestion buffer	10 μL
Proteinase K	0.5 μL

Place in a thermocycler at 65 °C for 60 min, 95 °C for 20 min, then cool to 4 °C.

5. Generate a serial dilution of each sample for qPCR. Samples have already been diluted 100× by DNase and Proteinase K treatment.
   - 1000× dilution:

     10 μL of sample from each well, above.

     90 μL of TE buffer (pH 8.0).

     Mix well and briefly spin.
   - 20,000× dilution:

     10 μL of each 1000× dilution

     190 μL of TE buffer (pH 8.0)

     Mix well and briefly spin.
6. Use a known AAV cis plasmid as standard (*see* **Note 6**). Calculate the copy number of the plasmid according to DNA length and concentration using the following formula (assumes average bp weight of 650 Da):

$$\text{Number of Copies} = \left(\text{amount of DNA in ng} \times 6.022 \times 10^{23}\ \text{copies/mol}\right) / \left(\text{length of template in bp} \times 1 \times 10^{9}\text{ng/g} \times 650\ \text{Da/bp}\right).$$

7. Prepare an eight point standard curve by two-fold serial dilution in TE buffer. A highest standard of 1×10^7 vg in 5 μL is recommended.
8. Combine 5 μL of each 20,000× sample or serial standard dilution and 15 μL of master mix (consisting of SYBR Green

and 10 μM primers) per well. Add 5 μL of TE buffer to 15 μL of master mix for No Template Controls.

9. Seal the plate with qPCR film and run on a qPCR machine. Use the "quantification standard curve" setting or equivalent.
10. For each well the qPCR machine will automatically determine the cycle number at which the reaction exceeded the fluorescence threshold. Average the triplicate cycle number readings for each sample, and create a standard curve by plotting cycle number on the *y*-axis against copy number on the *x*-axis using the scatter function of Microsoft Excel (*see* **Note** 7).
11. The copy number in the unknown AAV sample wells may vary. Draw a best-fit curve using 5 points from the standard curve (*see* **Note** 7), so that the unknown AAV sample cycle number falls in the middle of the 5 selected points. Generate a linear formula ($Y = aX + b$) from the standard curve using Excel. A linear regression value of >0.90 is acceptable.
12. Use the best-fit formula to calculate the unknown AAV sample copy number, then multiply by 8×10^6 to convert this number to vg/mL of AAV stock (*see* **Note 8**).

3.2 Direct Injection of Adipose Tissue

1. Create a working solution by diluting the rAAV stock in viral dilution buffer. The optimal concentration is dependent upon the AAV serotype, the promoter and other genomic elements, and the experimental aims; consult the literature for appropriate doses or determine empirically. We recommend an injection volume of 20 μL per adipose depot. Place working solution on ice.
2. Prepare for surgery (*see* **Note 9**). Ensure all surgical tools have been sterilized. Create three sterile fields in the preparation, surgical, and recovery areas. Set up a surgical recovery cage with a heating pad.
3. Inject a mouse intraperitoneally with 80–100 mg/kg ketamine and 5–15 mg/kg xylazine mixture in sterile saline.
4. Shave the mouse exterior near adipose pads with surgical clippers. Take care to avoid cutting the skin.
5. Connect a non-rebreathing circuit to an isoflurane vaporizer and oxygen cylinder. Tape the nose cone to the surgical table (*see* **Note 10**).
6. Place the mouse in the nose cone. Administer 2.5% isoflurane at 0.5 L/min. Confirm anesthesia with a firm toe pinch (*see* **Note 11**).
7. Using cotton swabs, prepare skin above the intended adipose target with three alternating, circular swabs of Betadine and 70% ethanol, then proceed to **steps 8**, **9**, or **10** to make the relevant surgical incision.

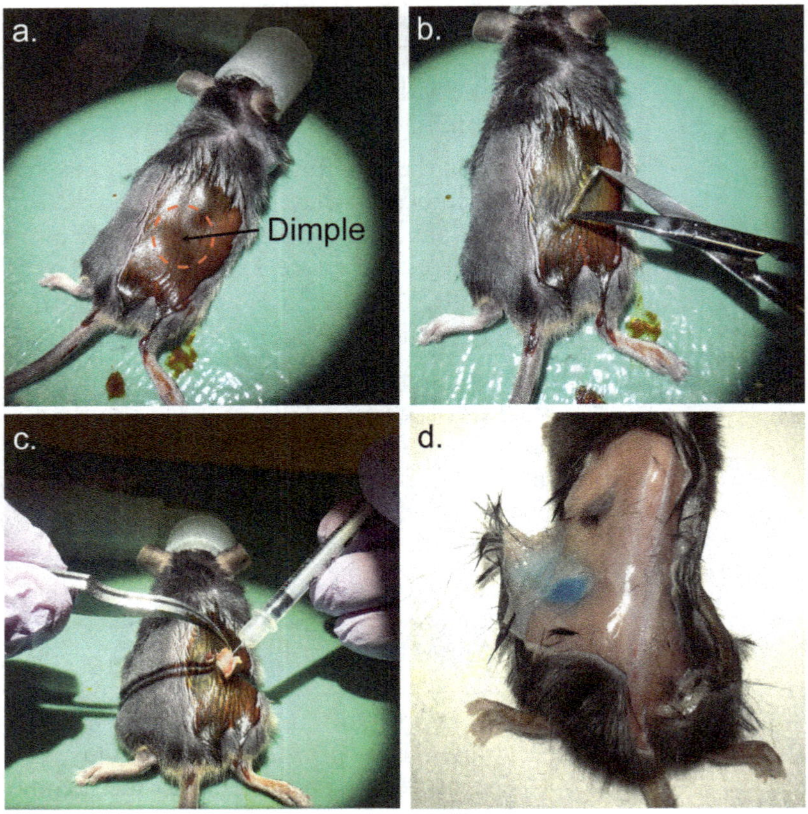

Fig. 2 Direct injection of inguinal adipose tissue depots. (**a**) A dimple on the skin can be used as a visual marker for injection location. (**b**) Use surgical scissors to widen the incision after an initial cut. (**c**) Injection technique. (**d**) Proof of concept for the injection method, using a representative sample and dye

8. *For inguinal adipose depot injections* (Fig. 2), lay the animal flat on its stomach. Make an incision approximately 15 mm lateral to the spinal cord, slightly superior to the hind legs. A small dimple should be visible that can be used for a visual cue. Use forceps and surgical scissors to make a 5 mm incision through the skin and fascia (*see* **Note 12**).
9. *For interscapular brown adipose tissue injections* (Fig. 3), lay the animal flat on its stomach. Make an incision approximately 8 mm inferior to the skull base. The scapulae are a valuable visual guide for injection targeting, as the brown adipose tissue lies between the two bones. Use forceps and surgical scissors to make a 5 mm incision through the skin and fascia (*see* **Notes 12** and **13**).
10. *For retroperitoneal adipose depot injections* (Fig. 4), lay the animal on its side. Make an incision approximately 8 mm lateral to the spinal cord and halfway between the animal's skull and tail base. The inferior, dorsal portion of the rib cage can be used as a tactile/visual guide for injection. Use forceps and surgical

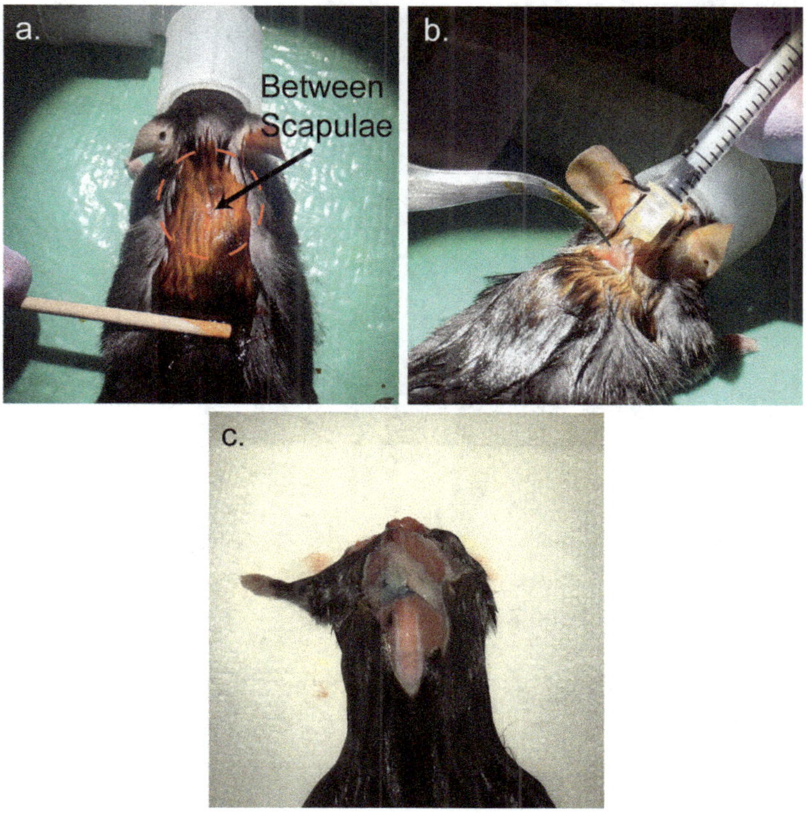

Fig. 3 Direct injection of interscapular brown adipose tissue. (**a**) The scapulae can be used as a visual marker for injection location. (**b**) Injection technique. (**c**) Proof of concept for the injection method, using a representative sample and dye

scissors to make a 5 mm incision through the skin and peritoneum to reach the retroperitoneal space (*see* **Note 12**). Upon opening the incision, a kidney should be visible and can be used as a guide (*see* **Note 14**).

11. The rAAV vector should be kept on ice whenever possible during the procedure. Withdraw the calculated viral dose using a pipette and eject into a 1.7 mL microcentrifuge tube.
12. Collect the rAAV vector from the microcentrifuge tube using a syringe (3/10 mL, 8 mm, 31G) (*see* **Note 15**).
13. Grip the adipose tissue firmly with forceps. Inject the adipose tissue with rAAV. Gently return the adipose tissue to the body cavity.
14. Close the wounds. *For brown and inguinal adipose injections*, close the external wound with an absorbable 4-0 PDS II FS-2 suture using simple interrupted technique. *For retroperitoneal adipose injections*, the experimenter must suture both the internal and external wounds. For the internal sealing of the

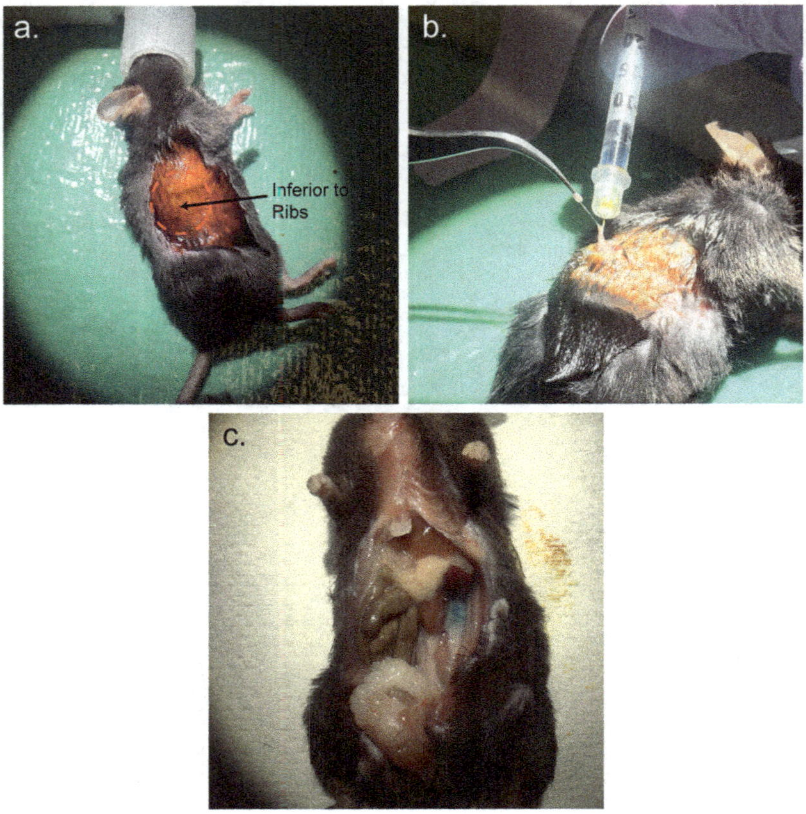

Fig. 4 Direct injection of retroperitoneal adipose tissue depots. (**a**) The ribs can be used as a visual marker for injection location. The injection location is located inferior to the final rib. (**b**) Injection technique. (**c**) Proof of concept for the injection method, using a representative sample and dye

intraperitoneal space, use an absorbable 6-0 polypropylene P-13 suture with simple interrupted technique. Once the intraperitoneal space is closed, close the external wound with an absorbable 4-0 PDS II FS-2 suture using simple interrupted technique.

15. Coat the external surgical wound and closed sutures with Betadine.
16. If required, repeat **steps 6–10** to complete a sham surgery or a second injection on the contralateral adipose tissue. Between mice, sterilize instruments for 30 s with a hot bead sterilizer (*see* **Note 16**).
17. Place the animal into a clean cage on a heating pad until the animal regains normal motor function. Remove the lid from the cage until the animal fully recovers (*see* **Note 17**).
18. Check the surgical wound for closure (*see* **Note 18**) in the days following the surgery. Administer pain relief as needed or required by the IACUC (buprenorphine intraperitoneally or ibuprofen in drinking water).

3.3 Oral Gavage

1. Fast mice for 4–6 h prior to procedure.
2. Create a working solution by diluting the rAAV stock in viral dilution buffer. The optimal concentration is dependent upon the AAV serotype, the promoter and other genomic elements, and the experimental aims; consult the literature for appropriate doses or determine empirically. We recommend an injection volume of 100 μL per mouse for young mice (fewer than 14 weeks old) and 150 μL/mouse for older mice (greater than 14 weeks old). Place working solution on ice.
3. Withdraw the calculated viral dose using a pipette and eject into a 1.7 mL microcentrifuge tube.
4. Attach a gavage needle (also known as an animal feeding needle) to a 1.0 mL Luer-Lok syringe (*see* **Note 19**).
5. Draw 0.25 mL of air into the syringe (*see* **Note 20**).
6. Withdraw the dose from the microcentrifuge tube into the syringe (*see* **Note 15**).
7. Firmly restrain the mouse with the nondominant hand. Insert the ball tip of the gavage needle. Hold the needle parallel to the mouse body. Guide the needle along the roof of the mouth and down the esophagus (*see* **Note 21**).
8. Continue the insertion until the whole shaft of the gavage needle is inside the mouse. The needle should appear to reach a depth near the final rib.
9. Inject the rAAV. Remove the gavage needle slowly, guiding the needle along the esophagus and the roof of the mouth. Replace gavage needle between animals.

3.4 Intraperitoneal Injection

1. Create a working solution by diluting the rAAV stock in viral dilution buffer. The optimal concentration is dependent upon the AAV serotype, the promoter and other genomic elements, and the experimental aims; consult the literature for appropriate doses or determine empirically. We recommend an injection volume of 100 μL per mouse.
2. Withdraw the calculated viral dose using a pipette and eject into a 1.7 mL microcentrifuge tube.
3. Using a syringe (1/2 mL, 12.7 mm, 28G), withdraw the dose from the microcentrifuge tube (*see* **Note 15**).
4. Firmly restrain the mouse with the nondominant hand. Locate the injection site, which should be approximately 2 cm superior to the tail base. In a direction nearly parallel to the skin, completely insert the syringe needle (*see* **Note 22**).
5. Inject the virus slowly. Wait 5 s, and then withdraw the needle (*see* **Note 23**).

4 Notes

1. All reagents and solutions should be molecular biology grade or cell culture grade. All prepared solutions should be filtered through a 0.2 μM filter in a biological safety level II cell culture hood.
2. Accuracy of pH in HSB is very important for DNA/$CaCl_2$ precipitate formation. Temperature is one key factor affecting pH. Warm HSB buffer and 2.5 M $CaCl_2$ to room temperature before transfection. Place the plasmids inside the cell culture hood 1 h before transfection.
3. Use ready-made 10× PBS, 1 M $MgCl_2$, 1 M KCl, and 5 M NaCl to prepare PBS-MK for iodixanol solutions in a cell culture hood.
4. Cells should be passaged at least four times when they are thawed. Number the passage and monitor the transfection efficiency by adding a separate plate of GFP plasmid transfection. The higher transfection efficiency, the higher titer.
5. Seal the screw-cap sample tube with Parafilm to prevent evaporation if the AAV is stored in at 4 °C for more than one month.
6. It is advised to use an AAV virus of known titer as a reference for new AAV sample titration. The reference AAV must receive the same treatment as unknown samples.
7. When using eight standard curve data points, the standard curve will not be linear. We suggest using the five points closest to your sample cycle number to ensure linearity. Using these five points, generate a linear standard curve and equation in Excel.
8. The final sample copy number (per well) is divided by 5 to determine the number of copies in 1 μL diluted sample, multiplied by 20,000 to determine the number of copies in 1 μL undiluted sample, multiplied by 1000 to determine the number of copies in 1 mL undiluted sample, and finally multiplied by 2 because one "copy" of the double-stranded plasmid standard equals two copies of the single-stranded AAV genome.
9. Before attempting an experiment, practice is highly recommended. By using a dye, an experimenter can assess if he/she has properly targeted the adipose depot. In Figs. 2, 3, and 4, dye is used to show a proof of concept for these methods.
10. It may be necessary to tape portions of the non-rebreathing circuit to avoid movement of circuit tubing during surgery. When injecting and suturing an animal, it is easy to accidentally remove the animal's nose from the nose cone. By taping the

tubing to the surgical table, experimenters can ensure animals will not wake due to accidental movement of the circuit.

11. This second method of anesthesia will ensure the animal remains anesthetized for the length of the procedure.
12. It is best to make a small incision, then insert and widen the scissors to enlarge the surgical space to the necessary size (Fig. 2). Take care to balance the need for a clear sightline to the injection site and a small incision size for ease of suturing.
13. Two lobes of the interscapular brown adipose tissue lie beneath a light layer of white adipose tissue. When injecting the brown adipose tissue, make sure to inject both adipose depots before suturing the surgical opening. Making a second incision is not necessary for brown adipose tissue injections.
14. Injecting the incorrect adipose tissue is the most common pitfall for retroperitoneal injections. Thus, when opening the retroperitoneal space, take care to inject the proper adipose tissue. It is common for experimenters to accidentally inject perirenal white adipose tissue. Retroperitoneal adipose tissue is located more dorsal and superior than the perirenal white adipose tissue [13].
15. Do not measure the dose with the syringe alone. Measuring the dose by micropipette is essential to ensure similar injectable doses.
16. Surgical instruments can be shared between several animals if necessary. The local institution's IACUC may suggest other sterilization methods. Please follow institutional requirements.
17. Surgical recovery may take up to one hour, depending on the ketamine/xylazine dose and individual drug metabolism.
18. Open wound frequency depends on the experimenter's suturing technique, injection location, and luck. Some mice will remove sutures due to discomfort. It is very important to monitor for removed (or missing) sutures in the first five days following surgery. If the wound opens naturally or sutures are removed by the animal, the experimenter must perform another surgery to close the wound. The IACUC may require monitoring up to ten days following surgery.
19. A gavage needle with a firm, metal shaft is recommended. Some manufacturers provide models with flexible plastic shafts, but animals tend to chew and distort the feeding tube during the gavage procedure. A firm gavage needle will ensure that this does not occur.
20. This extra air compensates for the dead volume inside the gavage needle and will expel any additional virus that would otherwise be left inside the gavage needle. Alternatively, experimenters can measure the dead volume and make their own

adjustment by loading an equivalent volume of excess rAAV solution.

21. Do not force the needle if resistance is met; instead, remove the needle and try again. If extreme force is used, the risk of tearing the esophagus and injuring the animal increases.
22. Off-target liver or intestine injection is the most common mistake resulting from this method. By inserting the needle at an inferior position on the mouse body, one can ensure the liver will not be injected. By inserting parallel to the mouse body, one can avoid injection into the small or large intestine. After the needle is inserted, pull the needle against the peritoneum; if visible, this suggests that virus is being injected into the intraperitoneal space. The virus will transduce the visceral adipose tissue if it enters the intraperitoneal space as intended.
23. By waiting before withdrawal of the needle, the experimenter can prevent the virus and buffer from leaking out of the intraperitoneal space through the needle insertion hole. Waiting will allow the virus and buffer to disperse into the intraperitoneal space and reduce the likelihood of leaking.

References

1. Mingozzi F, High KA (2011) Therapeutic in vivo gene transfer for genetic disease using AAV: progress and challenges. Nat Rev Genet 12(5):341–355. https://doi.org/10.1038/nrg2988
2. Mizukami H, Mimuro J, Ogura T, Okada T, Urabe M, Kume A, Sakata Y, Ozawa K (2006) Adipose tissue as a novel target for in vivo gene transfer by adeno-associated viral vectors. Hum Gene Ther 17(9):921–928. https://doi.org/10.1089/hum.2006.17.921
3. Zhang FL, Jia SQ, Zheng SP, Ding W (2011) Celastrol enhances AAV1-mediated gene expression in mice adipose tissues. Gene Ther 18(2):128–134. https://doi.org/10.1038/gt.2010.120
4. Jimenez V, Munoz S, Casana E, Mallol C, Elias I, Jambrina C, Ribera A, Ferre T, Franckhauser S, Bosch F (2013) In vivo adeno-associated viral vector-mediated genetic engineering of white and brown adipose tissue in adult mice. Diabetes 62(12):4012–4022. https://doi.org/10.2337/db13-0311
5. Liu X, Magee D, Wang C, McMurphy T, Slater A, During M, Cao L (2014) Adipose tissue insulin receptor knockdown via a new primate-derived hybrid recombinant AAV serotype. Mol Ther Methods Clin Dev 1. https://doi.org/10.1038/mtm.2013.8
6. McMurphy TB, Huang W, Xiao R, Liu X, Dhurandhar NV, Cao L (2017) Hepatic expression of adenovirus 36 E4ORF1 improves glycemic control and promotes glucose metabolism through AKT activation. Diabetes 66 (2):358–371. https://doi.org/10.2337/db16-0876
7. Huang W, McMurphy T, Liu X, Wang C, Cao L (2016) Genetic manipulation of brown fat via oral administration of an engineered recombinant adeno-associated viral serotype vector. Mol Ther 24(6):1062–1069. https://doi.org/10.1038/mt.2016.34
8. Huang W, Liu X, Queen NJ, Cao L (2017) targeting visceral fat by intraperitoneal delivery of novel AAV serotype vector restricting off-target transduction in liver. Mol Ther Methods Clin Dev 6:68–78. https://doi.org/10.1016/j.omtm.2017.06.002
9. Zhu Y, Gao Y, Tao C, Shao M, Zhao S, Huang W, Yao T, Johnson JA, Liu T, Cypess AM, Gupta O, Holland WL, Gupta RK, Spray DC, Tanowitz HB, Cao L, Lynes MD, Tseng YH, Elmquist JK, Williams KW, Lin HV, Scherer PE (2016) Connexin 43 mediates white adipose tissue beiging by facilitating the propagation of sympathetic neuronal signals. Cell Metab 24(3):420–433. https://doi.org/10.1016/j.cmet.2016.08.005

10. Ng R, Hussain NA, Zhang Q, Chang C, Li H, Fu Y, Cao L, Han W, Stunkel W, Xu F (2017) miRNA-32 drives brown fat thermogenesis and trans-activates subcutaneous white fat browning in mice. Cell Rep 19 (6):1229–1246. https://doi.org/10.1016/j.celrep.2017.04.035

11. Zhang Y, Xie L, Gunasekar SK, Tong D, Mishra A, Gibson WJ, Wang C, Fidler T, Marthaler B, Klingelhutz A, Abel ED, Samuel I, Smith JK, Cao L, Sah R (2017) SWELL1 is a regulator of adipocyte size, insulin signalling and glucose homeostasis. Nat Cell Biol 19(5):504–517. https://doi.org/10.1038/ncb3514

12. Xie L, Zhang Y, Gunasekar SK, Mishra A, Cao L, Sah R (2017) Induction of adipose and hepatic SWELL1 expression is required for maintaining systemic insulin-sensitivity in obesity. Channels (Austin) 11(6):673–677. https://doi.org/10.1080/19336950.2017.1373225

13. de Jong JM, Larsson O, Cannon B, Nedergaard J (2015) A stringent validation of mouse adipose tissue identity markers. Am J Physiol Endocrinol Metab 308(12):E1085–E1105. https://doi.org/10.1152/ajpendo.00023.2015

Chapter 24

AAV-Mediated Gene Delivery to the Enteric Nervous System by Intracolonic Injection

Reshma Gore, Maureen S. Riedl, Kelley F. Kitto, Carolyn A. Fairbanks, and Lucy Vulchanova

Abstract

The enteric nervous system of the lower gastrointestinal tract comprises intrinsic neural circuits as well as extrinsic afferent and efferent innervation. The development of strategies for neuronal gene transfer has created new opportunities for functional analysis, circuit mapping, and neuromodulation in the enteric nervous system. Studies of AAV-mediated gene transfer to enteric neurons and dorsal root ganglion neurons (DRG) have provided proofs-of-concept for the utility of AAV vectors for genetic manipulations of the intrinsic and extrinsic components of the enteric nervous system. Here we describe a method for AAV-mediated gene transfer to enteric neurons of the descending colon as well as colon-innervating DRG neurons by injection within the intestinal wall (intracolonic injection).

Key words Enteric, Colon, Dorsal root ganglion, Viral vector

1 Introduction

The enteric nervous system of the lower gastrointestinal tract comprises intrinsic neural circuits as well as extrinsic afferent and efferent innervation [1]. Circuits of intrinsic sensory neurons, interneurons, and motor neurons control gastrointestinal motility and fluid movement between the intestinal lumen and body fluid compartments. The functions of enteric neural networks have been explored extensively for development of therapeutic control of dysregulated motility. Extrinsic efferent innervation in the colon is provided by pre-enteric neurons within the spinal cord, prevertebral ganglia, and pelvic ganglia. Extrinsic primary afferent fibers supplying colon innervation reside in thoracolumbar and lumbosacral dorsal root ganglia (DRG) and transmit sensory information from the colon to the spinal cord under normal conditions and under conditions of visceral hypersensitivity. Modulation of the activity of the spinal afferent innervation of the colon through pharmacotherapy has been pursued for the management of

Michael J. Castle (ed.), *Adeno-Associated Virus Vectors: Design and Delivery*, Methods in Molecular Biology, vol. 1950, https://doi.org/10.1007/978-1-4939-9139-6_24,

abdominal pain and discomfort that accompany chronic gastrointestinal disorders such as irritable bowel syndrome [2, 3].

The development of strategies for neuronal gene transfer has created new opportunities for functional analysis, circuit mapping, and neuromodulation in the enteric nervous system. Adeno-associated virus (AAV) vectors have emerged as a preferred tool for gene transfer to neurons in basic research applications [4] and as promising candidates for therapeutic gene delivery, with a number of ongoing clinical trials [5]. While AAV vectors have been widely used for gene transfer in the central nervous system, applications in the peripheral nervous system have been more limited. AAV-mediated transduction of primary afferent neurons via intrathecal (i.t.) infusion [6–9], sciatic nerve injection [8], or direct intraganglionic injection [10] has been described in a number of studies. AAV vectors have also been used successfully for transduction of enteric neurons by direct injection within the intestinal wall [11] as well as by systemic [12–14], and to a lesser extent intrathecal [15], administration. Therefore, the utility of AAV vectors for genetic manipulations of the intrinsic and extrinsic components of the enteric nervous system has been demonstrated.

Here we describe a method for AAV-mediated gene transfer to enteric neurons of the descending colon as well as colon-innervating DRG neurons by injection within the intestinal wall (intracolonic injection). This route of delivery has been used extensively for injections of neuronal tracers to label colon-innervating DRG neurons [16–20] and has also been employed for transduction of enteric neurons using AAV [11]. In addition to DRG and enteric neurons, limited transduction of nodose ganglion neurons is possible, particularly if the vector is delivered to proximal colon. Gene transfer to extrinsic efferent neurons in prevertebral and pelvic ganglia is also possible, but has not been evaluated. Transduction of pre-enteric neurons in medulla and spinal cord was not observed [11].

Several factors need to be considered when planning intracolonic AAV delivery, including AAV serotype and titer, the promoter used in the viral construct, and the injection volume. AAV serotypes that have been tested in the enteric nervous system include AAV1, 2, 5, 6, 8, and 9 [11, 12]. In adult subjects, neuronal tropism was observed with all serotypes, while AAV6 also transduced enteric glia [11]. Transduction of smooth muscle cells was negligible. The highest transduction efficiency was seen with AAV6 and AAV9 in both the myenteric and submucosal plexus of the enteric nervous system. Preferential targeting of cholinergic enteric neurons by AAV9 has been suggested [12], but differential transduction of subtypes of enteric neurons has not been explored comprehensively. In our experiments, we have also observed transduction of enteric neurons with AAV9 vectors (Fig. 1a), and we consistently observe AAV9-mediated gene transfer to colon-innervating

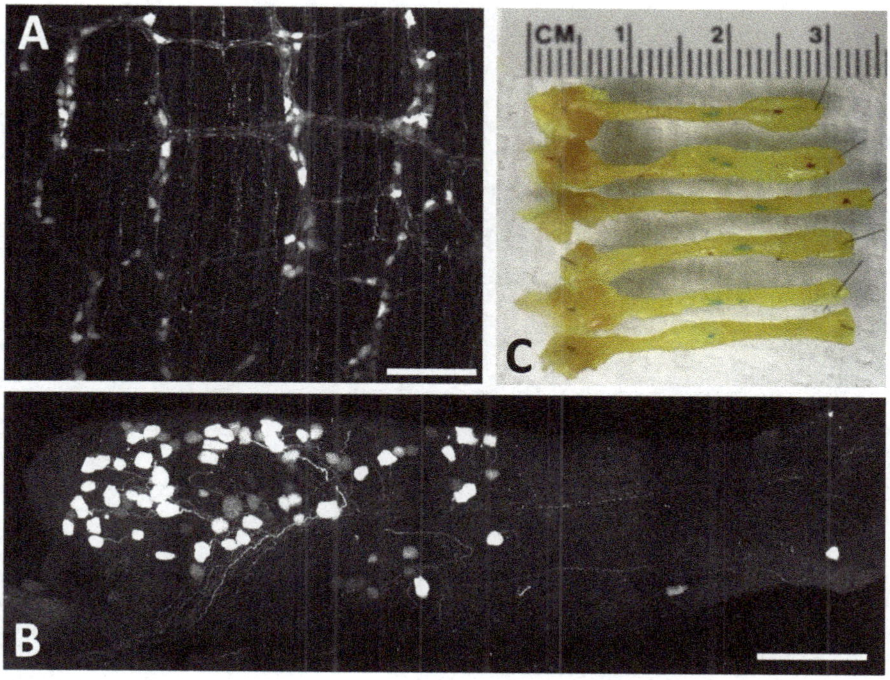

Fig. 1 Representative results of intracolonic AAV delivery. (**a**, **b**) Representative images of tdTomato-expressing neurons in the myenteric plexus of colon (**a**) and in colon-innervating sensory neurons (**b**). Scale bars: 200 μm. (**c**) Colon dissected from AAV9-injected mice. Injection sites are marked in green. In all cases, there was a single injection. The proximal side is on the right

sensory neurons (Fig. 1b). Viral titers of ~10^{12} vector genomes (vg)/mL transduce approximately 30% of enteric neurons, while titers of ~10^{13} vg/mL transduce over 70% of neurons [11]. Although to date only constructs with ubiquitous promoters have been used in the enteric nervous system, non-ubiquitous promoters such as the promoter for the choline acetyltransferase gene may be employed to achieve preferential targeting of subtypes of enteric neurons. Finally, injection volumes of 2–5 μL have been used for intracolonic injections. The injection volume affects the spread of the virus within the enteric nervous system; however direct comparison of different injection volumes has not been performed.

2 Materials

1. Male and female CD-1 or C57/Bl6 mice, 4–8 weeks old, weighing between 17 and 40 g. Mice younger than 3.5 weeks or weighing less than 15 g are challenging to inject due to their small size.
2. AAV vector. A titer of 10^{12} vg/mL or higher is recommended (*see* **Note 1**).

3. Microcentrifuge.
4. Isoflurane.
5. 70% ethanol (v/v).
6. Betadine (povidone-iodine).
7. Analgesic agent, such as Meloxicam.
8. Artificial tears or ocular lubricant.
9. Stereoscope or dissecting microscope.
10. Vaporizer for delivery of anesthetic gas.
11. Scavenger system for anesthetic gas.
12. Heating pads.
13. Oxygen tank.
14. Glass bead sterilizer or chemical sterilizer.
15. Anesthesia nose cone.
16. Sterile cotton-tipped applicator.
17. Sterile cotton gauze, or alternative method to control bleeding.
18. Sterile saline.
19. Nair hair removal cream.
20. 4-0 violet monofilament absorbable suture (Ethicon).
21. 10 μL Hamilton gastight removable needle syringe.
22. 30G, 0.5″ Hamilton needle, point style 4.
23. 3M™ Vetbond™ tissue adhesive.
24. 0.5 mL syringe for dispensing Vetbond.
25. 0.5 mL syringe for administration of analgesics.
26. Scalpel holder and blades.
27. 2 pairs curved blunt forceps.
28. Spring scissors.
29. Self-restraining cross action retractor with 4″ blunt prongs, such as Miltex® 11–16.
30. 2 pairs of forceps for wound closure.
31. Needle holder with suture cutter.
32. Parafilm.
33. Micropipette and micropipette tips.
34. Paper towels or other means to elevate and stabilize the animal in the nose cone.
35. Recovery chamber with heating pad.
36. Tattoo ink to mark injection site.

3 Methods

3.1 Setup and Anesthesia

1. Sterilize surgical tools using an autoclave, glass bead sterilizer, or chemical sterilizer.
2. Autoclave Hamilton syringe before and after vector application to avoid cross contamination if several vectors are injected with the same syringe.
3. Position nose cone over the heating pad.
4. Administer volatile anesthesia (3.5–4% Isoflurane) using an induction chamber. Once fully anesthetized, place the animal in a nose cone secured in place over a heating pad.
5. Confirm depth of anesthesia by checking paw withdrawal reflex with toe pinch.
6. Once desired depth of anesthesia is achieved, apply ocular lubricant.
7. Place the mouse in dorsal recumbency.
8. Apply hair removal cream for 2–3 min.
9. Wipe the hair removal cream with deionized or sterile water.
10. Clean the abdominal surface with 70% ethanol.
11. Alternate betadine and alcohol wipes three times.

3.2 Surgical Procedure

1. Thaw an amount of AAV sufficient for all injections to be performed during the day. Spin the tubes with vector in a microcentrifuge to ensure that fluid is not retained in the tube cap. Store on ice until use.
2. After confirming deep anesthesia, make a midline incision on the lower abdomen under a microscope to expose the skin overlaying the peritoneum. Start the incision below the sternum and continue posteriorly until the preputial gland is exposed.
3. Using spring scissors, open the peritoneal cavity by making a midline incision along the linea alba starting below the xiphoid process and extending ~5 cm caudally until the preputial gland. Do not cut or damage the preputial gland.
4. Using blunt curved forceps, move midline organs laterally to visualize the distal colon.
5. Insert retractor into the abdominal cavity to gently move the intestines and surrounding organs to provide a clear view of and access to the descending colon. Ensure that the teeth of the retractor do not press upon the organs directly as this may cause organ damage (*see* **Note 2**).
6. Pipette 4 μL of AAV onto a piece of parafilm and withdraw into a 10 μL Hamilton syringe. This ensures an accurate volume of vector administration (*see* **Note 3**).

7. With a small paint brush, brush tattoo ink onto the tip of the needle to mark the injection site. Marked injection sites are shown in Fig. 1c.
8. With curved blunt forceps, gently hold the distal portion of the colon and tug cranially to gain access to the caudal portion of the distal colon (2 cm from the anal verge, as close as possible to the pelvic bone).
9. While holding onto a portion of the colon with blunt curved forceps, insert the needle of the Hamilton syringe aborally below the serosa at a 30–45 degree angle, then tilt the syringe to ~20 degrees. Tunnel the needle 2–3 mm subserosally and inject the vector slowly over 30–45 s. If the vector is correctly injected, a slight bulge can be observed at the injection site (*see* **Note 4**).
10. Hold the needle in place for at least 1 min and then slowly retract the needle. Gently wipe the injection site with a cotton swab to remove any excess or leaked vector and flush the abdominal cavity with sterile saline.
11. Close the peritoneum with 4-0 absorbable suture.
12. To close overlying skin, lift the skin on one side of the incision and dispense Vetbond. Quickly bring together the two edges of skin to appose the overlying skin. Grab and hold the apposed skin with blunt forceps for 15–30 s, then pull on the adjacent skin to test for successful adhesion before transferring the mouse to the recovery chamber.

3.3 Recovery and Postoperative Monitoring

1. Subcutaneously administer 2 mg/kg Meloxicam or an alternative analgesic in the back of the neck (scruff).
2. Place the animal in a recovery chamber with heating pad, and monitor until the animal has recovered and is mobile and alert. If the animal looks scruffy or struggles to recover from anesthesia, inject 400–500 μL of sterile saline (s.c.) and continue monitoring.
3. Return the mouse to a clean cage. Administer analgesic for 3 days postoperatively and monitor the incision site, urination, defecation, food intake, and general activity (*see* **Note 5**).

3.4 Representative Results

We have used this protocol to inject AAV9 vectors carrying various constructs. After injections of AAV9-CAG-tdTomato (obtained from the Vector Core at the University of North Carolina), we can observe tdTomato expression in colon enteric neurons as well as colon-innervating dorsal root ganglia (DRG) neurons 3–4 weeks after injections. Representative images of tdTomato fluorescence are shown in Fig. 1a, b (*see* **Note 6**).

4 Notes

1. AAV vectors should be stored in small aliquots at -80 °C to avoid freeze/thaw cycles.
2. Access to the distal colon is similar in males and females. Males are slightly larger than their age matched female littermates; however, in females, the distal colon can be easily visualized by retracting the uterus.
3. It is very difficult to draw small volumes into a Hamilton syringe from an Eppendorf tube. In addition, the tip of the needle can be bumped against the tube, compromising its sharpness. Once the desired volume is pipetted on parafilm, the bevel of the needle can be submerged in the fluid, which is then easily drawn into the syringe. The parafilm should be wiped with 70% ethanol before pipetting the vector onto it to prevent contamination of the injectate.
4. The absence of a bulge after vector injection suggests that the vector leaked out or was injected deep in the colon wall or in the lumen.
5. If the abdominal skin incision opens in the days following the surgery, reclose the overlying skin with Vetbond under anesthesia and in an aseptic environment.
6. This procedure can be incorporated in an intersectional targeting strategy where a vector carrying Cre-recombinase and a vector carrying a Cre-dependent transgene are delivered via distinct but converging routes. For example, intracolonic delivery combined with intrathecal or intraganglionic delivery would result in selective expression of the Cre-dependent transgene in colon-innervating DRG neurons. In this case, it is recommended that the cre-dependent vector be delivered via the route that is expected to drive the most efficient transduction of the target cells. Alternatively, injections of the two viruses at different locations within the colon could potentially transduce selectively ascending or descending enteric neurons. Injection volumes of 1 μL or less are recommended for this approach, which would be facilitated by an infusion system such as the Pump 11 Pico Plus Elite Dual (Harvard Apparatus). If two vectors of the same serotype are to be used, they must be injected on the same day. The initial exposure to AAV may induce a response that prevents transduction with the second AAV vector.

Acknowledgments

This work was supported by NIH-SPARC U18EB021716 and T32DA007234.

References

1. Furness JB, Callaghan BP, Rivera LR, Cho HJ (2014) The enteric nervous system and gastrointestinal innervation: integrated local and central control. Adv Exp Med Biol 817:39–71. https://doi.org/10.1007/978-1-4939-0897-4_3
2. Akbar A, Yiangou Y, Facer P, Walters JR, Anand P, Ghosh S (2008) Increased capsaicin receptor TRPV1-expressing sensory fibres in irritable bowel syndrome and their correlation with abdominal pain. Gut 57(7):923–929. https://doi.org/10.1136/gut.2007.138982
3. Kiyatkin ME, Feng B, Schwartz ES, Gebhart GF (2013) Combined genetic and pharmacological inhibition of TRPV1 and P2X3 attenuates colorectal hypersensitivity and afferent sensitization. Am J Physiol Gastrointest Liver Physiol 305(9):G638–G648. https://doi.org/10.1152/ajpgi.00180.2013
4. McCown TJ (2011) Adeno-Associated Virus (AAV) Vectors in the CNS. Curr Gene Ther 11(3):181–188
5. Kotterman MA, Schaffer DV (2014) Engineering adeno-associated viruses for clinical gene therapy. Nat Rev Genet 15(7):445–451. https://doi.org/10.1038/nrg3742
6. Schuster DJ, Dykstra JA, Riedl MS, Kitto KF, Honda CN, McIvor RS, Fairbanks CA, Vulchanova L (2013) Visualization of spinal afferent innervation in the mouse colon by AAV8-mediated GFP expression. Neurogastroenterol Motil 25(2):e89–e100. https://doi.org/10.1111/nmo.12057
7. Storek B, Reinhardt M, Wang C, Janssen WG, Harder NM, Banck MS, Morrison JH, Beutler AS (2008) Sensory neuron targeting by self-complementary AAV8 via lumbar puncture for chronic pain. Proc Natl Acad Sci U S A 105 (3):1055–1060. https://doi.org/10.1073/pnas.0708003105
8. Towne C, Pertin M, Beggah AT, Aebischer P, Decosterd I (2009) Recombinant adeno-associated virus serotype 6 (rAAV2/6)-mediated gene transfer to nociceptive neurons through different routes of delivery. Mol Pain 5:52
9. Vulchanova L, Schuster DJ, Belur LR, Riedl MS, Podetz-Pedersen KM, Kitto KF, Wilcox GL, McIvor RS, Fairbanks CA (2010) Differential adeno-associated virus mediated gene transfer to sensory neurons following intrathecal delivery by direct lumbar puncture. Mol Pain 6:31. https://doi.org/10.1186/1744-8069-6-31
10. Kollarik M, Ru F, Brozmanova M (2010) Vagal afferent nerves with the properties of nociceptors. Auton Neurosci 153(1–2):12–20. https://doi.org/10.1016/j.autneu.2009.08.001
11. Benskey MJ, Kuhn NC, Galligan JJ, Garcia J, Boye SE, Hauswirth WW, Mueller C, Boye SL, Manfredsson FP (2015) Targeted gene delivery to the enteric nervous system using AAV: a comparison across serotypes and capsid mutants. Mol Ther 23(3):488–500. https://doi.org/10.1038/mt.2015.7
12. Gombash SE, Cowley CJ, Fitzgerald JA, Hall JC, Mueller C, Christofi FL, Foust KD (2014) Intravenous AAV9 efficiently transduces myenteric neurons in neonate and juvenile mice. Front Mol Neurosci 7:81. https://doi.org/10.3389/fnmol.2014.00081
13. Gombash SE, Cowley CJ, Fitzgerald JA, Lepak CA, Neides MG, Hook K, Todd LJ, Wang GD, Mueller C, Kaspar BK, Bielefeld EC, Fischer AJ, Wood JD, Foust KD (2017) Systemic gene delivery transduces the enteric nervous system of guinea pigs and cynomolgus macaques. Gene Ther 24(10):640–648. https://doi.org/10.1038/gt.2017.72
14. Chan KY, Jang MJ, Yoo BB, Greenbaum A, Ravi N, Wu WL, Sanchez-Guardado L, Lois C, Mazmanian SK, Deverman BE, Gradinaru V (2017) Engineered AAVs for efficient noninvasive gene delivery to the central and peripheral nervous systems. Nat Neurosci 20 (8):1172–1179. https://doi.org/10.1038/nn.4593
15. Schuster DJ, Dykstra JA, Riedl MS, Kitto KF, Belur LR, McIvor RS, Elde RP, Fairbanks CA, Vulchanova L (2014) Biodistribution of adeno-associated virus serotype 9 (AAV9) vector after intrathecal and intravenous delivery in mouse. Front Neuroanat 8:42. https://doi.org/10.3389/fnana.2014.00042
16. Christianson JA, Liang R, Ustinova EE, Davis BM, Fraser MO, Pezzone MA (2007) Convergence of bladder and colon sensory innervation

occurs at the primary afferent level. Pain 128 (3):235–243. https://doi.org/10.1016/j.pain.2006.09.023

17. Christianson JA, McIlwrath SL, Koerber HR, Davis BM (2006) Transient receptor potential vanilloid 1-immunopositive neurons in the mouse are more prevalent within colon afferents compared to skin and muscle afferents. Neuroscience 140(1):247–257. https://doi.org/10.1016/j.neuroscience.2006.02.015
18. Christianson JA, Traub RJ, Davis BM (2006) Differences in spinal distribution and neurochemical phenotype of colonic afferents in mouse and rat. J Comp Neurol 494 (2):246–259. https://doi.org/10.1002/cne.20816
19. La JH, Schwartz ES, Gebhart GF (2011) Differences in the expression of transient receptor potential channel V1, transient receptor potential channel A1 and mechanosensitive two pore-domain K+ channels between the lumbar splanchnic and pelvic nerve innervations of mouse urinary bladder and colon. Neuroscience 186:179–187. https://doi.org/10.1016/j.neuroscience.2011.04.049
20. Tan LL, Bornstein JC, Anderson CR (2008) Distinct chemical classes of medium-sized transient receptor potential channel vanilloid 1-immunoreactive dorsal root ganglion neurons innervate the adult mouse jejunum and colon. Neuroscience 156(2):334–343. https://doi.org/10.1016/j.neuroscience.2008.06.071

Index

A

Michael J. Castle (ed.), *Adeno-Associated Virus Vectors: Design and Delivery*, Methods in Molecular Biology, vol. 1950, https://doi.org/10.1007/978-1-4939-9139-6, © Springer Science+Business Media, LLC, part of Springer Nature 2019

B

C

D

E

K

L

M

N

W

CPSIA information can be obtained
at www.ICGtesting.com
Printed in the USA
LVHW061503280219
609071LV00002B/2/P

9 781493 991389